Flavour

Flavour

From Food to Perception

EDITED BY

Elisabeth Guichard

Center for Taste and Feeding Behaviour, CNRS, INRA, University of Bourgogne Franche-Comté, Dijon, France

Christian Salles

Center for Taste and Feeding Behaviour, CNRS, INRA, University of Bourgogne Franche-Comté, Dijon, France

Martine Morzel

Center for Taste and Feeding Behaviour, CNRS, INRA, University of Bourgogne Franche-Comté, Dijon, France

Anne-Marie Le Bon

Center for Taste and Feeding Behaviour, CNRS, INRA, University of Bourgogne Franche-Comté, Dijon, France

WILEY Blackwell

Registered office: John Wiley & Sons, Ltd, The Atrium, Southern Gate, Chichester, West Sussex, PO19 8SQ, UK

Editorial offices: 9600 Garsington Road, Oxford, OX4 2DQ, UK
The Atrium, Southern Gate, Chichester, West Sussex, PO19 8SQ, UK
111 River Street, Hoboken, NJ 07030-5774, USA

For details of our global editorial offices, for customer services and for information about how to apply for permission to reuse the copyright material in this book please see our website at www.wiley.com/wiley-blackwell.

Library of Congress Cataloging-in-Publication Data

Names: Guichard, Elisabeth, 1956- editor. | Salles, Christian, editor. | Morzel, Martine, editor. | Bon, Anne Marie Le, editor. | Le Bon, Anne Marie, editor.
Title: Flavour : from food to perception / edited by Elisabeth Guichard, Christian Salles, Martine Morzel, Anne Marie Le Bon.
Description: Chichester, West Sussex ; Hoboken, NJ : John Wiley & Sons Inc., 2017. | Includes bibliographical references and index.
Identifiers: LCCN 2016032586| ISBN 9781118929414 (cloth) | ISBN 9781118929407 (Adobe PDF) | ISBN 9781118929391 (epub)
Subjects: LCSH: Taste. | Flavor. | Chemical senses.
Classification: LCC QP456 .F53 2017 | DDC 612.8/7–dc23 LC record available at https://lccn.loc.gov/2016032586

A catalogue record for this book is available from the British Library.

Wiley also publishes its books in a variety of electronic formats. Some content that appears in print may not be available in electronic books.

Cover image: Art_of_Sun/Getty Images

Set in 9.5/13pt, MeridienLTStd by SPi Global, Chennai, India

1 2017

Contents

List of Contributors

Carmen Barba
Center for Taste and Feeding Behaviour, CNRS, INRA, University of Bourgogne Franche-Comté, Dijon, France

Maik Behrens
Department Molecular Genetics, German Institute of Human Nutrition Potsdam-Rehbruecke, Nuthetal, Germany

Loïc Briand
Center for Taste and Feeding Behaviour, CNRS, INRA, University of Bourgogne Franche-Comté, Dijon, France

Francis Canon
Center for Taste and Feeding Behaviour, CNRS, INRA, University of Bourgogne Franche-Comté, Dijon, France

Isabelle Cayeux
Firmenich SA Research & Development Department, Geneva, Switzerland

Frédérique Datiche
Center for Taste and Feeding Behaviour, CNRS, INRA, University of Bourgogne Franche-Comté, Dijon, France

Patricia Duchamp-Viret
Centre de Recherche en Neurosciences de Lyon (CRNL), UMR CNRS 5292, INSERM U1020, Université Lyon1, Lyon Cedex, France

Philippe Faure
Center for Taste and Feeding Behaviour, CNRS, INRA, University of Bourgogne Franche-Comté, Dijon, France

Johannes Frasnelli
CÉAMS, Research Center, Sacré-Coeur Hospital, Montréal, Québec, Canada
Department of Anatomy, Université du Québec à Trois-Rivières, Trois-Rivières, Québec, Canada

Jean Gascuel
Center for Taste and Feeding Behaviour, CNRS, INRA, University of Bourgogne Franche-Comté, Dijon, France

Florence Gingras-Lessard
CÉAMS, Research Center, Sacré-Cœur Hospital, Montréal, Québec, Canada
Faculté de Médecine, Université de Montréal, Montréal, Québec, Canada

Anna Gracka
Faculty of Food Science and Nutrition, Poznań University of Life Sciences, Poznań, Poland

Xavier Grosmaitre
Center for Taste and Feeding Behaviour, CNRS, INRA, University of Bourgogne Franche-Comté, Dijon, France

Elisabeth Guichard
Center for Taste and Feeding Behaviour, CNRS, INRA, University of Bourgogne Franche-Comté, Dijon, France

Christian Hannig
Polyclinic of Operative Dentistry and Pediatric Dentistry, Medical Faculty Carl Gustav Carus, TU Dresden, Dresden, Germany

Matthias Hannig
Clinic of Operative Dentistry, Periodontology and Preventive Dentistry, University Hospital, Saarland University, Homburg/Saar, Germany

Hassan-Ismail Hanser
Center for Taste and Feeding Behaviour, CNRS, INRA, University of Bourgogne Franche-Comté, Dijon, France

Jean-Marie Heydel
Center for Taste and Feeding Behaviour, CNRS, INRA, University of Bourgogne Franche-Comté, Dijon, France

Thomas Hummel
Smell & Taste Clinic, Department of Otorhinolaryngology, TU Dresden, Dresden, Germany

Henryk Jeleń
Faculty of Food Science and Nutrition, Poznań University of Life Sciences, Poznań, Poland

Anni Laffitte
Center for Taste and Feeding Behaviour, CNRS, INRA, University of Bourgogne Franche-Comté, Dijon, France

Anne-Marie Le Bon
Center for Taste and Feeding Behaviour, CNRS, INRA, University of Bourgogne Franche-Comté, Dijon, France

Wolfgang Meyerhof
Department Molecular Genetics, German Institute of Human Nutrition Potsdam-Rehbruecke, Nuthetal, Germany

Martine Morzel
Center for Taste and Feeding Behaviour, CNRS, INRA, University of Bourgogne Franche-Comté, Dijon, France

Fabrice Neiers
Center for Taste and Feeding Behaviour, CNRS, INRA, University of Bourgogne Franche-Comté, Dijon, France

Eric Neyraud
Center for Taste and Feeding Behaviour, CNRS, INRA, University of Bourgogne Franche-Comté, Dijon, France

Laurianne Paravisini
Center for Taste and Feeding Behaviour, CNRS, INRA, University of Bourgogne Franche-Comté, Dijon, France
Department of Food Science and Nutrition, University of Minnesota, Saint Paul, Minnesota, United States

Andrea Romano
Department of Food Quality and Nutrition, Research and Innovation Centre, Fondazione Edmund Mach (FEM), San Michele all'Adige, Italy
Faculty of Science and Technology, Freie Universität Bozen – Libera Università di Bolzano, Bozen-Bolzano, Italy

Christian Salles
Center for Taste and Feeding Behaviour, CNRS, INRA, University of Bourgogne Franche-Comté, Dijon, France

Han-Seok Seo
Department of Food Science, University of Arkansas, Fayetteville, Arkansas, U.S.A

Charles Spence
Crossmodal Research Laboratory, Department of Experimental Psychology, University of Oxford, Oxford, UK

Christian Starkenmann
Firmenich SA Research & Development Department, Geneva, Switzerland

Thierry Thomas-Danguin
Center for Taste and Feeding Behaviour, CNRS, INRA, University of Bourgogne Franche-Comté, Dijon, France

Preface

Flavour perception plays a key role in the acceptability of food by consumers. Flavour science has become a broad research subject aiming to provide a comprehensive understanding of flavour from its generation in food to its perception during eating. All the aspects of flavour perception are presented, including aroma, taste and trigeminal sensation, from the general composition of food to the perception at the peri-receptor, receptor and central level taking into account the recent evolution in flavour science. The book responds to a growing need for pluridisciplinary approaches to better understand the mechanisms involved in flavour perception. It provides to scientists and industries working on flavour and food a general view of the mechanisms of flavour perception including multimodal interactions between sensory modalities. This will offer new tools and approaches for the reformulation of foods, for example increasing the organoleptic acceptability of products with low salt, fat and sugar content or elaborated with new sources of fat and proteins.

The first part brings basic knowledge on the chemosensory systems involved in the perception of flavour molecules, complemented with the latest scientific papers. This includes the structural and functional anatomy of olfactory, gustatory and trigeminal systems in vertebrates and the olfactory and gustatory peri-receptor mechanisms.

The second part, focused on flavour compounds in food, provides basic knowledge on the structural, physico-chemical and sensory properties of flavour (aroma, taste and trigeminal) compounds and their interactions with the different macromolecules present in the food matrix.

The third part describes in more details the in-mouth mechanisms leading to the release of aroma and taste compounds from the food matrix during the eating process and their transformation by bodily fluids of the oral cavity before reaching the chemosensory receptors.

The fourth part covers different aspects of flavour perception, from orthonasal and retronasal perception, biological and analytical aspects involved in the perception of mixture of odorants and tastants to multimodal sensory interactions.

List of Abbreviations

3D	3 dimensions
γ_i	activity coefficient
ABT	1-aminobenzotriazole
Aco	anterior cortical nucleus of amygdala
ADH	alcohol dehydrogenase
AEDA	aroma extract dilution analysis
AI	agranular insular cortex
AL	antennal lobe
AMP	adenosine-5′-monophosphate
AOB	accessory olfactory bulb
AON	anterior olfactory nucleus
AOS	allene oxide synthase
APCI-MS	atmospheric pressure chemical ionization-mass spectrometry
aPCx	anterior piriform cortex
*a*PRP	acidic proline-rich protein
ATP	adenosine triphosphate
AUC	area under the curve
BLA	basolateral nucleus
BNPP	bis-p-nitro-phenylphosphate
BOLD	blood-oxygen-level dependent
*b*PRP	basic proline-rich protein
Ca	calcium
cAMP	cyclic adenosine monophosphate
CaSR	calcium sensing receptor
CAVI	carbonic anhydrase VI
CD	cyclodextrin
C_G	concentration of the volatile in the headspace
cGMP	cyclic guanosine monophosphate
CHARM	combined hedonic aroma response measurement
CIT	citral
C_M	concentration of the volatile compound in the matrix
CNGC	cyclic nucleotide gated channels
CNV	cranial nerve 5
CYP	cytochrome P450 enzyme
Da	dalton
DE	dextrose equivalent
DF	detection frequency
DIMS	direct injection-mass spectrometry
d_{liq}	liquid density (kg L^{-1})

DOSY	diffusion ordered spectroscopy investigation
DP	deep pyramidal cell
EEG	electro-encephalogram
EgCG	epigallocatechin gallate
EGF	epidermal growth factor
ENaC	epithelial sodium channel
EOG	electro-olfactogram
EPL	external plexiform layer
ER	endoplasmic reticulum
ERP	event-related potentials
ESP	exocrine-gland-secreting peptide
FB	feedback interneuron
FDA	food and drug administration
FF	feed forward interneuron
FFT	furfurylthiol
FID	flame ionisation detection
fMRI	functional magnetic resonance imaging
FPR	formyl peptide-like receptor
FSCM	finger span cross modality
GBC	globose basal cell
GC-GOOD	gas chromatography-global olfactometry omission detection
GC-MS	gas chromatography-mass spectrometry
GC-O	gas chromatography-olfactometry
GC-PO	gas chromatography-pedestal olfactometry
GC-R	gas chromatography recomposition-olfactometry
GDP	guanosine diphosphate
GG	Grueneberg ganglion
GL	glomerular layer
GMP	guanosine 5′-monophosphate
GPCR	G protein-coupled receptor
*g*PRP	glycosylated proline-rich protein
Gr	granule cells
GRAS	generally regarded as safe
GrL	granular cell layer
GSH	glutathione
GST	glutathione-S-transferase
HEK	human embryonic kidney cell
HPL	hydroperoxide lyase
HPLC	high performance liquid chromatography
Hst	histatin
IAA	isoamyl acetate
IBMX	3-isobutyl-1-methylxanthine
IDP	intrinsically disordered protein
IL	infralimbic cortex
IMAC	in-mouth air cavity
Imax	maximum intensity of the release profiles

IMP	inosine-5′-monophosphate
IPL	internal plexiform layer
ISO	International Standards Organization
K	gas matrix partition coefficient
K_d	dissociation constant
k_i	molar concentration
K_i	molar fraction
k_m	mass fraction
L	liter
LCN1	lipocalin-1
LEC	lateral entorhinal cortex
LH	lateral hypothalamus
LISS	ligand-induced selective signaling
LN	local interneurons
LOT	lateral olfactory tract
M_a	molar mass of air (28.8 g mol^{-1})
MC	mitral cell
MCL	mitral cell layer
MDR	multidrug resistance
mGluR	metabotropic glutamate receptor
M_{liq}	molar mass of liquid (g mol^{-1})
MMP-3	matrix metalloproteinase 3
MOE	main olfactory epithelium
Mp	multipolar cell
MRI	magnetic resonance imaging
MRP	multidrug resistance-associated protein
MSG	monosodium glutamate
MUC	mucin
MUC1	mucin 1
MUC5B	mucin 5B
MUC7	mucin 7
NaCl	sodium chloride
NHDC	neohesperidin dihydrochalcone
NIF	nasal impact frequency
NLOT	nucleus of the lateral olfactory tract
NMP	negative mucosal potential
NMR	nuclear magnetic resonance
NPY	neuropeptide Y
NTS	nucleus tractus solitarius
OASIS	original aroma simultaneously input to the sniffing port method
OAV	odour activity value
OB	olfactory bulb
OBP	odorant-binding protein
OC	olfactory cortex
OCT	octanol/al
OE	olfactory epithelium

OEC	olfactory ensheathing cell
OFC	orbito-frontal cortex
OM	olfactory mucosa
OME	odorant metabolizing enzyme
ONL	olfactory nerve layer
OR	olfactory receptor
ORN	olfactory receptor neuron
OSN	olfactory sensory neuron
OT	odour threshold
OT	olfactory tubercle
P	pressure (Pa)
p_i^s	saturated vapour pressure of compound i (Pa)
PCx	piriform cortex
PEEK	poly-ether-ether-ketone
PET	positron emission tomography
pgC	periglomerular cells
Pgp	P-glycoprotein
PI3	phosphoinositide-3
PI3K	phosphoinositide-3-kinase
PIP3	phosphatidylinositol-(3, 4, 5)-trisphosphate
PLC	phospholipase C
PLco	posterolateralcortical nucleus of amygdala
PN	projection neuron
pPCx	posterior piriform cortex
PPI	polyproline I
PPII	polyproline II
PRH2 protein	proline-rich protein HaeIII subfamily 2
PROP	6-n-propylthiouracil
PRP	proline-rich protein
PRV	phase ratio variation
PTC	phenylthiocarbamide
PTFE	polytetrafluoroethylene
PTR-MS	proton transfer reaction mass spectrometry
QSAR	quantitative structure activity relationships
QSPR	quantitative structure property relationships
R	gas constant ($8.314\,J\,mol^{-1}\,K^{-1}$)
RAS	retronasal aroma simulator
RI	retention index
RMS	rostral migratory stream
RNA	ribonucleic acid
RNAi	RNA interference
SCM	single chain monellin
SCOPe	structural classification of proteins
sIgA	secretory immunoglobulin A
SL	semilunar cell
SNIF	surface of nasal impact frequency

SNP	single nucleotide polymorphism
SO	septal organ
SON	supraoptic nucleus
SP	superficial pyramidal cell
SPLUNC2	short palate, lung and nasal epithelium clone 2
SPME	solid phase micro-extraction
ST	sulfotransferase
SVP	saturated vapour pressure
T	temperature (K)
T1R1	taste receptor type 1, member 1
T1R2	taste receptor type 1, member 2
T1R3	taste receptor type 1, member 3
TAS1R	taste 1 receptor
TAS2R	taste 2 receptor
TDS	temporal dominance of sensations
TGF-α	transforming growth factor-alpha
TI	time-intensity
TM	trans-membrane
Tmax	time to reach maximum intensity
ToF	time-of-flight
TRP	transient receptor potential
Trp	tryptophan
TT	tenia tecta
UGT	UDP-glucuronosyltransferase
V1	first branch of the trigeminal nerve
V2	second branch of the trigeminal nerve
V3	third branch of the trigeminal nerve
VEG	von Ebner's glands
VEGP	von Ebner's gland protein
VNO	vomeronasal organ
VNR	vomeronasal receptor
VSN	vomeronasal sensory neuron
VTD	venus flytrap domain
WL	wiskey lactone
XME	xenobiotic metabolizing enzyme

CHAPTER 1

Olfactory system in mammals: structural and functional anatomy

Anne-Marie Le Bon, Frédérique Datiche, Jean Gascuel & Xavier Grosmaitre
Center for Taste and Feeding Behaviour, CNRS, INRA, University of Bourgogne Franche-Comté Dijon, France

1.1 Introduction

The survival and reproductive success of living organisms, including human beings, depends on the detection of sensory stimuli. Living organisms do not eat or reproduce with whatever is available; instead, they show considerable selectivity by taking advantage of their chemical and physical senses. In this regard, the sense of smell and its capacity to detect myriad of odorant molecules is of critical importance for humans and most animal species. This sense significantly contributes to the identification of food and assessment of its palatability, as well as to the detection of chemical compounds carrying specific information concerning dangers, social interactions and reproductive behaviours. In mammals, these diverse roles are accomplished by a complex olfactory system. The primary tissue responsible for sensing volatile odorants is the olfactory epithelium (OE) which is localized in the nasal cavity. Sensory neurons residing in the OE convey olfactory information to the olfactory bulb (OB) which, in turn, transfers this information towards multiple higher cortical regions collectively referred to as the olfactory cortex. Other olfactory subsystems such as the vomeronasal organ coexist with the main OE in many species. These subsystems are separate entities that are dedicated to distinct functional roles.

The principal aim of this review is to gather the results of very recent as well as major studies on the processing of olfactory information by the olfactory system and to highlight its plasticity. We first describe the physiology of the main OE and the molecular mechanisms of odorant detection. We then show how endogenous and exogenous factors may induce different forms of plasticity of the OE. We also outline the main features of other olfactory subsystems. Next, we examine how the olfactory signal generated at the peripheral level is transformed at the first processing center in the brain, the OB. Finally, we provide an overview of the

Flavour: From food to perception, First Edition.
Edited by Elisabeth Guichard, Christian Salles, Martine Morzel, and Anne-Marie Le Bon.

higher olfactory pathways involved in the processing of olfactory information and we consider the pathways that shape odour perception.

1.2 Organization and function of the peripheral olfactory system

1.2.1 Physiology of the peripheral olfactory system

Stimulation of the olfactory system begins when odorant molecules are detected by the olfactory neuroepithelium located in the upper part of the nasal cavity. The odorant molecules can reach the epithelium by two pathways: *via* the nose (orthonasal olfaction) and *via* the mouth (retronasal olfaction). Odorants perceived by the orthonasal pathway originate from the external world whereas odorants perceived retronasally emanate from food or drink (aroma compounds) (see Chapter 13 for more details on these pathways).

The nose and the nasal cavity are separated into two halves along the midline by a cartilaginous structure called the nasal septum. The lateral wall of each nasal cavity is typically shaped by three bony protuberances termed the inferior, middle and superior turbinates. Animals can have more turbinates, for example, the rat has four. The turbinates and the septum are covered with an epithelium. Depending on its location, this epithelium is either nonsensory (respiratory) or sensory (olfactory). The nonsensory respiratory portion of the nasal cavity warms, cleans and humidifies the inspired air.

There is widespread acknowledgement that the human OE is located in the superior region of the nasal cavity, predominantly on the dorsal side of the nasal vault, the septum, and the superior turbinate. However, recent studies have reported a more extending distribution of OE on the middle turbinate (Escada et al. 2009). Actually, the location of the OE is variable among people. Besides, its organization is thought to change over time: ageing induces conversion to or ingrowth of respiratory epithelium and loss of olfactory neurons (Nakashima et al. 1991). Environmental compounds or pathophysiological processes such as infection or inflammation can also modify the distribution of OE. The OE in the adult has therefore a non-contiguous and patchy distribution. Globally, the human olfactory region covers between 1 and 2 cm^2 in each cavity (Moran et al. 1982). This area is modest relative to those of other vertebrates such as rodents and dogs (Gross et al. 1982, Harkema 1991).

In the superior part of the nasal cavity, a horizontal bone, called the cribriform plate of the ethmoid, separates the OE from the brain. The cribriform plate is a highly perforated bone: the perforations provide access for the olfactory nerve bundles to the OB. This is the only site in the body where the central nervous system is in direct contact with the outer surface. The nerves serving the olfactory region are called the first cranial nerves or the olfactory nerves. They concentrate multiple axons of olfactory neurons located in the lamina propria. These axons convey the nerve impulse generated by the odorant detection into the OB.

1.2.2 Structure of the olfactory epithelium

The human OE has a structure similar to that of other vertebrates (Morrison and Costanzo 1992). It is a pseudo-stratified columnar epithelium that lies on a dense connective tissue, the lamina propria. Together, the OE and the lamina propria form the olfactory mucosa (OM). The human OE is about 60 μm in height and has a slight yellow-brownish colour. It is composed of several distinct cell types, notably olfactory sensory neurons (OSNs), sustentacular cells (a type of nonsensory supporting cells), microvillar cells, two types of stem cells (horizontal basal cell and globose basal cell) as well as Bowman's glands and duct cells (Figure 1.1B).

Vertebrate OSNs are slender and bipolar neurons spread in the epithelium with a density of 10^6-10^7 per cm^2. Their cell bodies are generally located within the lower two thirds of the neuroepithelium. At the apical surface of the epithelium, about 10-25 cilia protrude from the OSN dendrites (Morrison and Costanzo 1992). These olfactory cilia float in the mucus which covers the epithelial surface and their plasma membrane contains the olfactory receptors (ORs). On the opposite side, the axons of OSNs penetrate through the basement membrane into the lamina propria where they are ensheathed by the olfactory ensheathing cells (OECs) (Figure 1.1B). OSNs and OECs together with fibroblasts form the olfactory nerve bundles. Serous glands called olfactory glands or Bowman's glands, bundles of the accessory olfactory nerve (surrounded by accessory OECs), as well as trigeminal nerve bundles (surrounded by Schwann cells) are also located within the lamina propria. The olfactory nerve bundles project through the cribriform plate towards the OB where the OSNs' axons synapse with mitral/tufted cells and interneurons (Figure 1.1A).

Stem cells divide to give rise to sustentacular cells and immature OSNs which mature and migrate apically. In rodent OE, there are two kinds of stem cells: horizontal basal cells and globose basal cells (GBCs). These two types are morphologically and functionally distinct. In humans, however, only one basal cell type has been reported. These human basal cells morphologically resemble the GBCs in the rat (Hahn et al. 2005).

Additional cell types, called microvillar cells, have also been described in the olfactory neuroepithelium of vertebrates. These cells, which are located near the epithelial surface, are flask shaped and have an apical tuft of microvilli extending into the nasal cavity. They provide trophic factors such as neuropeptide Y (NPY) to the OE under the control of odorant or trigeminal nerve stimulation, or both (Montani et al. 2006). Microvillar cells might therefore play a role in the regulation of cellular homeostasis in the OE.

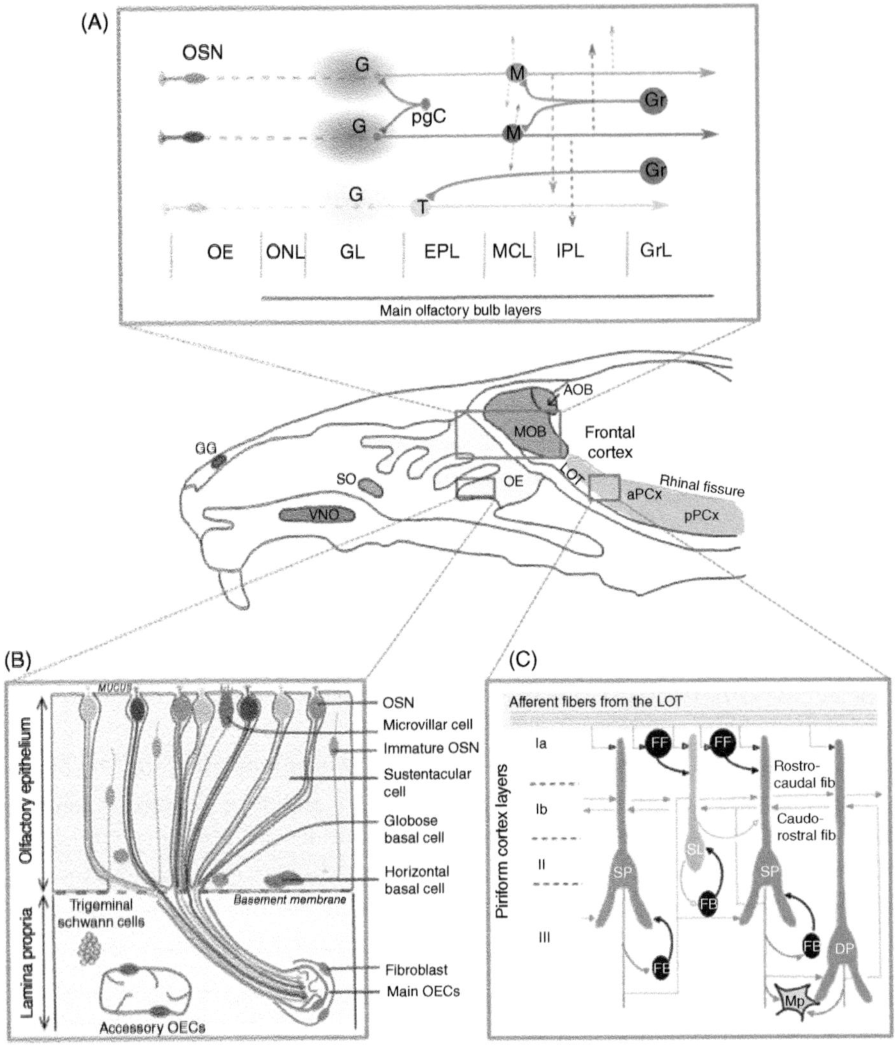

Figure 1.1 Schematic drawing of the rodent olfactory system (sagittal cross section through the nasal region of the head, lower jaw is not shown). Inset **A** shows the different cell layers observed in the olfactory bulb and the neuronal connections. Inset **B** represents the various cell types and structures located in the olfactory mucosa. Inset **C** schematizes the connectivity of glutamatergic neurons in the PCx (Source: Adapted from Ekberg and St John 2014, Haberly 2001). Abbreviations: AOB, accessory olfactory bulb; aPCx, anterior piriform cortex; DP, deep pyramidal cells; EPL, external plexiform layer; FB, feedback interneurons; FF, feed forward interneurons; G, glomeruli; GG, Grueneberg ganglion; GL, glomerular layer; Gr, granule cells; GrL, granular cell layer; IPL, internal plexiform layer; LOT, lateral olfactory tract; M, mitral cells; MCL, mitral cell layer; Mp, multipolar cells; OB, olfactory bulb; OE, olfactory epithelium; ONL, olfactory nerve layer; pgC, periglomerular cells; pPCx, posterior piriform cortex; OSN, olfactory sensory neuron; SL, semilunar cells; SO, septal organ; SP, superficial pyramidal cells; T, tufted cells. A colored version of this figure can be found in the online version of this chapter.

Like other epithelia, the peripheral OE constantly regenerates itself throughout life. OSNs only live for 1-3 months after which they undergo apoptosis and are replaced by new neurons originating from basal cells (Mackay-Sim and Kittel 1991). The continuous turnover of OSNs protects the OE against damage induced by environmental factors that can result in cell death. This replenishment after damage is critical to maintain the functional integrity of the OE.

1.2.3 Molecular mechanisms of odorant detection

The airborne odorants diffuse into the aqueous nasal mucus before reaching olfactory cilia where ORs are localized. In the mucus, proteins called odorant-binding proteins (OBPs) are thought to carry odorants, which are commonly hydrophobic molecules, through the mucus towards the ORs (Heydel et al. 2013). In addition to the solubilisation of odorants, OBPs may have other functional roles. Recent studies have revealed that OBPs directly interact with ORs thus modulating their function (Vidic et al. 2008) or contribute to the clearance of odorants from the microenvironment of the receptor (Strotmann and Breer 2011).

Binding of odorants to specific ORs is a key event that induces olfactory signaling. ORs were first identified from rats in 1991 by Linda Buck and Richard Axel (Buck and Axel 1991) who received the Nobel Prize in 2004 for this discovery. These authors revealed that OR genes belong to a large multigene family that encode G protein coupled receptors (GPCRs). Further studies confirmed that OR genes constitute the largest multigene family in mammals. Comparison of diverse genome sequences showed that the numbers of OR genes vary greatly among species (Niimura 2012). Rats and mice have 1,400–1,700 OR genes in their genomes, cows and horses have higher numbers (2,200-2,600) and recently, it has been reported that the genome of African elephants contains more than 4,200 OR genes (Table 1.1). Compared with other mammals, primates tend to have smaller numbers of OR genes (600-800). A fraction of mammalian OR genes has been shown to be pseudogenes (i.e., genes that are not functional). The fraction of OR pseudogenes varies widely among species (Niimura et al. 2014). In human genome, more than half (52%) of the entire set of OR genes are pseudogenes, leading to 396 intact (potentially functional) OR genes (Matsui et al. 2010).

An important feature of OSNs is the fact that each cell expresses only one allele of a single OR gene: this has been proven through extensive studies in the mouse OSNs (Chess et al. 1994, Malnic et al. 1999, Serizawa et al. 2004). OSNs expressing the same OR would be distributed randomly within one of four circumscribed zones in the OE (Ressler et al. 1993, Vassar et al. 1993). However, some studies suggest that OR gene expression zones broadly overlap rather than bear sharp zonal boundaries (Iwema et al. 2004, Miyamichi et al. 2005). All OSNs expressing the same OR in turn converge upon spatially invariant glomeruli in the OB, the

Table 1.1 Numbers of OR genes in the genome sequence from 13 placental mammalian species (Source: Adapted from Niimura et al. 2014). An intact gene was defined as a sequence starting from an initiation codon and ending with a stop codon that did not contain any disrupting mutation. A pseudogene was defined as a sequence with a nonsense mutation, frameshift, deletion within conserved regions, or some combination thereof. A truncated gene represents a partial sequence of an intact gene.

Species	Total number	Intact genes		Truncated genes		Pseudogenes	
		number	%	number	%	number	%
Human	821	396	*48.2*	0	*0*	425	*51.8*
Chimpanzee	813	380	*46.7*	19	*2.34*	414	*50.9*
Orangutan	821	296	*36.1*	37	*4.51*	488	*59.4*
Macaque	606	309	*51.0*	17	*2.81*	280	*44.7*
Marmoset	624	366	*58.7*	27	*4.33*	231	*36.9*
Mouse	1,366	1,130	*82.7*	0	*0*	236	*17.3*
Rat	1,767	1,207	*68.3*	52	*2.94*	508	*28.7*
Guinea pig	2,162	796	*36.8*	26	*1.20*	1,340	*62*
Rabbit	1,046	768	*73.4*	22	*2.10*	256	*24.5*
Horse	2,658	1,066	*40.1*	23	*0.87*	1,569	*59*
Dog	1,100	811	*73.7*	11	*1.00*	278	*25.3*
Cow	2,284	1,186	*51.9*	41	*1.80*	1,057	*46.3*
Elephant	4,267	1,948	*45.7*	89	*2.09*	2,230	*52.3*

site of the first synaptic relay in olfactory sensory processing (Mombaerts et al. 1996, Ressler et al. 1994, Vassar et al. 1994). Thus, activation of specific ORs by an odorant elicits a characteristic pattern of activity in the OB.

The OR functionality was demonstrated through a number of *in vitro* and *in vivo* studies. Odorants may be recognized by multiple ORs, and one OR may recognize multiple odorants (Kajiya et al. 2001, Malnic et al. 1999). This implies that different odorants are recognized by different combinations of ORs. This scheme of combinatorial coding is now widely admitted to explain how odorants are encoded at the peripheral level. However, some ORs (such as the human receptor OR7D4) have been shown to bind to a limited number of structurally related odorants (Keller et al. 2007). ORs can therefore be classified into two groups: ORs that are broadly tuned and ORs that are narrowly tuned. However, the way a receptor can recognize an odorant still remains poorly understood and further studies are necessary to investigate the physicochemical laws that govern OR-ligand interactions.

ORs belong to the class-A of the GPCR family that includes a number of diverse membrane receptors. Bovine rhodopsin or β2-adrenergic receptors, class-A GPCRs whose structural features have been widely investigated, were used as templates to perform homology modeling. These experiments indicated that ORs fold into quite similar tertiary structures, consisting of seven trans-membrane (7-TM) helices connected by extra-cellular and intra-cellular loops (Baud et al. 2011, Singer 2000). The 7-TM helices form a bundle in which a pocket is dedicated to odorant binding. Studies combining molecular modeling and site-directed mutagenesis helped specifying the nature of the binding sites of some ORs (Baud et al. 2011, Gelis et al. 2012, Katada et al. 2005, Launay et al. 2012). The binding pockets were predicted to be located between TM3, TM5 and TM6 and the main amino acids in contact with the ligands could be identified. For a given OR, the binding mode differs from one odorant to another but some amino acids, all hydrophobic, are involved in binding whatever the ligand (Charlier et al. 2012).

In the cilia of OSNs, ORs are coupled to a specific G-protein called G_{olf}. When a cognate ligand binds to an OR, this interaction activates the $G_{\alpha olf}$ subunit which elevates intracellular cAMP through type III adenylate cyclase enzymatic reaction. Binding of cAMP to the cyclic nucleotide-gated channel allows influx of cations, mainly calcium, into OSNs. Elevation of intracellular calcium induces the opening of the calcium-gated chloride channel that produces an efflux of chloride ions to amplify cellular depolarization (Kleene 2008). The cAMP pathway is thought to be the main signalization pathway involved in peripheral olfactory transduction. However, some studies suggest the involvement of cAMP-independent signaling pathways, including guanylate cyclase and phospholipase C (PLC) signaling, in olfactory transduction (Lin et al. 2004, Meyer et al. 2000). Recently, it has been demonstrated that a subset of mouse OSNs located in the most ventral zone of OE can mediate both the phospholipase C signaling pathway and the cAMP pathway upon binding to structurally similar ligands (Yu et al. 2014). In consequence, some ORs could possess conformational plasticity leading to preferential interactions with different downstream elements, depending on ligand that binds to the OR.

1.2.4 Plasticity of the olfactory epithelium

Several endogenous and exogenous factors induce different forms of plasticity at the OE level.

1.2.4.1 Development and ageing

Evidence has been accumulated that the peripheral olfactory system is functional before birth. Behaviour studies have shown that prenatal olfactory experience provoked by odorants present in the amniotic fluid contributes to postnatal preferences and behaviours such as suckling and feeding (Logan et al. 2012, Schaal et al. 2000). In mouse embryonic development (lasting 19 days from conception), the OE is fully formed at embryonic day 10 (E10) and at around E14, multiple short cilia can be observed on neuron dendrites (Cuschieri and Bannister 1975). Several works reported that ORs and components of the main olfactory signaling pathway (such as protein G_{olf}, adenylate cyclase III and cyclic nucleotide-gated ion channels) are expressed in the OE at the same stage

(Saito et al. 1998, Schwarzenbacher et al. 2005). In line with these observations, electrophysiological studies performed in the OE or individual OSNs of rodents gave evidence for odorant responses at E16 (Gesteland et al. 1982, Lam and Mombaerts 2013). These recordings indicate that late-stage mouse embryos possess functional OSNs and the ability to detect odorants.

A handful of studies have addressed the OR expression profile throughout life. Newborn rats express fewer OR genes than adult and ageing rats, and generally at a lower level (Rimbault et al. 2009). However, a small subset of OR genes are expressed specifically or even overexpressed in newborns. In C57B6L/N mice raised under well-controlled conditions, the majority of OR gene expression (58.4%) remained stable throughout life while 32.8% presented downward profiles and 7.2% upward profiles (Khan et al. 2013). A recent study performed in human OE supports these results. In samples collected from individuals aged from 39 to 81 years, authors showed that the expression of most OR genes is stable with age. However the expression level of a small number of ORs significantly decrease or increase (Verbeurgt et al. 2014). The overall conclusion of these studies is that OR gene expression in mammal OE seems rather stable throughout life.

Nevertheless, decline of olfactory function is common in elderly humans. This decreased sensitivity with ageing has been postulated to be due partly to structural and cellular changes occurring in the OE rather than OR expression level. These changes are probably also associated with alterations occurring in the central components of the olfactory system. Studies on animal models and human biopsies support a gradual degradation of the OE that could account for olfactory loss. A significant age-related loss of OSNs in the affected areas, which results in a thinner epithelium (Rosli et al. 1999), and a strong reduction in the sensitivity of human OSNs (Rawson et al. 2012) have been demonstrated. In aged mice, OSNs expressing a defined OR exhibit a lower density while the functional properties of these neurons did not change (Lee et al. 2009). These peripheral changes might contribute to poor odour discrimination and identification in the elderly.

1.2.4.2 Nutritional and metabolic status

In mammals, odour perception also depends closely on nutritional status. Fasting results in an increased ability to detect odours, some of which are food-related. Meanwhile, satiety with one type of food reduces the ability to detect the odour specially associated with that food type (Mulligan et al. 2002, O'Doherty et al. 2000). Recently published works suggest that the olfactory system is intimately linked with the endocrine systems that regulate energy balance. Hunger and satiety status are signaled by blood-circulating peptide hormones. Receptors for metabolically important hormones such as ghrelin, orexins, NPY, insulin, leptin, and cholecystokinin have been shown to be expressed in the OM (Palouzier-Paulignan et al. 2012). These molecules have access to the OM through the peripheral circulation but the local production of insulin within this tissue has also been reported (Lacroix et al. 2008). Using an *ex vivo* intact epithelium preparation, Savigner and colleagues (Savigner et al. 2009) showed that bath perfusion of insulin or leptin, both anorexigenic factors, decreased the odorant response. These peptides also reduced the odorant-induced activity in the OM in well-fed animals (Lacroix et al. 2008, Savigner et al. 2009). Conversely, NPY, an orexigenic peptide, increases the electrophysiological response of OSNs to odorants in fasted adult rats (Negroni et al. 2012). In addition, the expression of metabolic hormone receptors in mammal OM can be regulated by nutritional status. An overexpression of insulin, leptin and NPY receptors has been observed in OM from fasted rats (Baly et al. 2007, Lacroix et al. 2008, Negroni et al. 2012).

Chronic energy imbalance can also alter the sensitivity of the peripheral olfactory sensory system. A limitation of the duration of daily food intake was found to provoke a modulation of olfactory-mediated behaviours regarding food odours in rats (Badonnel et al. 2012). This restriction was accompanied by a slight decrease in insulin receptor expression in the OM, suggesting that this hormone could be part of this process. Olfactory dysfunctions were also reported in mice fed a high-fat diet for 24 weeks (Thiebaud et al. 2014). Marked loss of OSNs and their axonal projections and a concomitant reduction in electro-olfactogram amplitude were observed in these mice. These structural and functional changes at the OM level could evoke dysfunctions in olfactory driven behaviour. Taken as a whole, these different studies demonstrate that nutritional and metabolic state can modulate olfactory perception by regulating the sensitivity of the peripheral olfactory system.

1.2.4.3 Exogenous compounds

Embedded in the epithelium lining of the nasal cavity, OSNs are continuously exposed to environmental factors such as odorant molecules or non-odorant volatile chemicals. These exogenous compounds can modify OSNs' properties or even induce damage to the OM in case of long-term and high-level exposition.

The effects of odorant enrichment on OSNs have been analyzed in a number of studies. Olfactory stimulation generally promotes the survival of OSNs (Francois et al. 2013, Watt et al. 2004). It also induces higher sensitivity of the epithelium to the odorant used for the exposure, suggesting an increase in the target OR expression (Wang et al. 1993). The effects of odorant exposure on a specific OSN population diverge depending on the population considered: odorant exposure either increases survival of OSNs (Francois et al. 2013, Watt et al. 2004) or decreases the number of OSNs (Cadiou et al. 2014, Cavallin et al. 2010). However, in spite of a decrease in OSN number, exposed target neurons were found to respond to their ligand with higher sensitivity, broader dynamic range, faster rise time, and shorter responses (Cadiou et al. 2014). This suggests that neurons could take part in the compensation of their lower density by sending more information to the OB. Through this form of plasticity, OSNs can adapt to their environment.

Exposure to volatile compounds, including chemicals, solvents and environmental contaminants, may induce various lesions in the OE such as inflammation, necrosis, atrophy and proliferation (for reviews, see Gaskell 1990, Harkema et al. 2006). Global necrosis of the OE has been observed after exposure to irritants such as chlorine and sulfur dioxide. In contrast, cell-specific toxicity may occur in the OE. Notably, it has been reported that inhalation of acetone selectively damages progenitor cells of the OE in mice (Buron et al. 2009). Intranasal administration of satratoxin-G, a mycotoxin produced by the black mold *Stachybotrys chartarum* that grows in water-damaged housings, induces widespread apoptosis in OSNs. This apoptosis was associated with an acute, neutrophilic rhinitis in the nasal airways of Rhesus monkeys (Carey et al. 2012). Several metals, such as cadmium, have also been associated with olfactory function impairment in exposed workers (Gobba 2006). Experimentally, cadmium instillation resulted in an important but recoverable cell loss in mouse OE (Bondier et al. 2008). Accumulation of cadmium in the mice OB has also been observed in the same study, suggesting that cadmium can be transported through the OSN axons to the OB. Like cadmium, a number of metals and other chemicals (Minn et al. 2002), as well as pathogenic microbes (Dando et al. 2014), can enter the brain *via* the olfactory pathway. Among other causes, this phenomenon is suspected to contribute to the development of a number of neurodegenerative diseases, most notably Alzheimer's and Parkinson's diseases (Prediger et al. 2012).

1.2.5 Subsystems in the main olfactory epithelium

We are presenting here three subsystems coexisting with the main OE (MOE) in many mammals: the vomeronasal organ (VNO), the Grueneberg ganglion (GG) and the septal organ (SO) also called Masera organ (Figure 1.1). In humans however, these subsystems are either nonfunctional, for example the VNO, or do not appear to exist (GG and SO).

1.2.5.1 The vomeronasal organ

The VNO was described two hundred years ago by Ludvig von Jacobson (Trotier and Doving 1998) in many mammalian species, especially rodents. In rodents, the VNO is a bilateral tubular structure located ventrally on the nasal septum. A bilayer chemosensory neuroepithelium covers the medial wall of the VNO. The epithelium comprises thousands of microvillar sensory neurons (VSNs for Vomeronasal Sensory Neurons) whose axons project to the accessory OB (AOB). During development, the VNO can be observed in humans, but then shows many signs of regression and even absence of epithelial neurons or nerve fibers that would allow neural information to be transported to the brain (Trotier 2011).

At least three types of vomeronasal receptors (VNRs) have been described, mainly in rodents. All VNRs belong to the GPCR family. The first type, V1r, is associated with G proteins of the Gi type and is only expressed in the apical neuroepithelium (Dulac and Axel 1995). It projects to the anterior portion of the AOB. In humans, only five members of this family have an intact open reading frame (Rodriguez and Mombaerts 2002). They are expressed only in the main OE but their function is unknown (Rodriguez et al. 2000). The second type, V2r, is expressed in the basal neuroepithelium with G proteins of the Go type and VSNs expressing V2rs project to the posterior portion of the AOB (Dulac and Torello 2003, Halpern and Martinez-Marcos 2003). Few mammalian species maintain a functional family of these receptors, which are not present and functional in humans (Shi and Zhang 2007). Signal transduction in V1r and V2r expressing neurons relies on a PLC-mediated pathway involving a G protein (Gi or Go) leading to production of secondary messengers. These messengers eventually open transient receptor potential cation channels (TRPC2) localized in the microvilli of the sensory neurons. The VNRs of the third type belong to the family of formyl peptide-like receptors (FPR) (Liberles et al. 2009, Riviere et al. 2009). These receptors are present in all mammals and are generally expressed in the immune system. Their expression is related to olfaction only in rodents.

Historically, the VNO has been considered as an organ specific for the detection of social cues such as pheromones. However, since the discovery of different types of receptors detecting pheromones in the MOE and others detecting general odorants in the VNO, there is some overlap of ligands of the MOE and the VNO (Ma 2007). The main difference between the MOE and the VNO is that odorants can access to the chemosensory neurons in the VNO only once they are dissolved in the mucus and drawn into the lumen of the organ. This represents an active process involving vasoconstriction of sinuses or blood vessels. The process allows the characterization of very heavy nonvolatile molecules such as peptides.

Relatively few ligands able to activate vomeronasal neurons have been identified. Some of these ligands are volatile compounds, such as 2-heptanone, a ligand of V1rb2 (Boschat et al. 2002). Most ligands however are heavy molecules, including short peptides. The ESP (exocrine-gland-secreting peptides) family is secreted by different glands and is present in tears, nasal mucus and saliva (Kimoto et al. 2005). Some members of this family are involved in sexual behaviours (Haga et al. 2010). Major histocompatibility complex-related peptides are also mentioned (Leinders-Zufall et al. 2004). Several ligands connected to inflammation or pathogens such as the peptide fMLF have been associated with specific FPRs (Riviere et al. 2009). Other potential ligands of VNRs are generated in urine, either volatiles or non-volatiles such as steroids, or peptides involved in aggressive behaviours (Ibarra-Soria et al. 2014). The receptors involved in the detection of these ligands are not identified at the moment.

1.2.5.2 The septal organ and the grueneberg ganglion

These two subsystems were primarily described in rodents, and there is so far no description of them in humans and other primates. Their role in olfaction was described only recently.

The SO (or Masera organ) is a small area of olfactory epithelium isolated from the main OE in the respiratory epithelium. It is present in the nasal cavity of many species of mammals particularly rodents (Ma 2010). There is no evidence of the presence of a Masera organ in humans. The position of the SO in the nasal cavity is unique: located ventrally just behind the vomeronasal organ and near the choana, it is located on the pathway of the air during breathing at rest. This anatomical situation suggested that the SO may have a role to alert the animal of the presence of an odorant.

The SO appears during embryonic development and reaches its maximum development in young adults. Its epithelium has many similarities with that of the MOE: it contains many ciliated olfactory neurons but also some rare microvillar cells (Ma 2007). However, there are some structural differences with the MOE (number of layers, morphology of OSNs). Neurons in the SO express the canonical transduction cascade of the main OE (Grosmaitre et al. 2007, Ma et al. 2003). The expression levels of the 120 receptors identified is very characteristic: a small group of receptors covers approximately 95% of these cells. But the rule of one-receptor gene expressed in one neuron remains valid. The axons of the SO neurons target a small group of glomeruli called "septal glomeruli" and a large group of weakly stained glomeruli receiving axons from both SO and MOE (Levai and Strotmann 2003).

SO neurons respond to a large number of odorants either in electroolfactogram (Marshall and Maruniak 1986) or in single cell recordings (Grosmaitre et al. 2007, Ma et al. 2003): the SO neurons seem to be generalists, due to the expression of broadly tuned odorant receptors (Grosmaitre et al. 2009). These physiological results confirm the alerting role of the SO as an odorant detector leaving the discrimination task to the MOE.

The Grueneberg ganglion (also spelled Grüneberg ganglion, GG) was at first thought to be part of the peripheral nervous system without any connection with olfaction (Grüneberg 1973). This group of cells is located at the dorsal tip of the nasal cavity and close to the opening of the naris. Recently, clues from gene-targeted mice suggested a role in olfaction (Munger et al. 2009). Although they express olfactory marker proteins, the GG cells have a very different morphology from that of classical OSNs: i) their shape is ovoid without a dendrite but with short cilia directly connected to the soma; ii) they are combined and attached to each other in grape-like clusters; iii) they are separated from the nasal cavity by a keratinized epithelial layer made of glial cells. These clusters of cells appear during the embryonic development and reach their maximum at the perinatal stage, suggesting a role in the mother-infant interactions. The one-neuron one-receptor rule also applies to GG neurons. The majority of

GG neurons express a VNR; some express a classical OR and others a trace amine-associated receptor (Fleischer and Breer 2010). The putative transduction pathway proteins depend on the receptor expressed, but the majority seems to use a cGMP pathway (Fleischer and Breer 2010).

GG neurons project their axon to the OB into glomeruli located in the necklace glomeruli area. This suggests that GG neurons are involved in mother-pups interactions. This role was also supported by the detection by these neurons of a decrease in temperature through the isolation of pups from the mother enhancing the response to specific odorants (Fleischer and Breer 2010). In adult, GG neurons were shown to respond to an alarm pheromone (Brechbuhl et al. 2008) with structural similarities to predator chemosensory signals (Brechbuhl et al. 2013).

To date, the question of the role of the GG is not fully resolved: at an early age, GG neurons respond to a drop in temperature, and specific odorant molecules. At adult age, they respond to an alarm pheromone emitted by conspecifics in distress. This does not completely unveil the role of the GG while strongly suggesting its involvement in the mother-pup relationship.

1.3 Anatomical and functional organization of the main olfactory bulb

1.3.1 Background

OSNs send their axons to OB where they synapse onto second order neurons at the level of glomeruli which are neuropilar subunits. The number of glomeruli in the mice OB has been estimated to 1,810 (Royet et al. 1988). As the total number of OSNs in the mice OE is about 5.10^6 unilaterally (Kawagishi et al. 2014), it can be calculated that a single glomerulus receives converging inputs from 2,760 OSN axons. In the human OB, the number of glomeruli varies between individuals; the average number of glomeruli has been estimated to 5,500 (Maresh et al. 2008). At the level of the OB, olfactory signals are processed by interneurons (periglomerular cell, granular cells, external tufted cells, short axon cells, Van Gehuchten cells, Blanes cells) before being exported to higher centers of the brain by output neurons (mitral cells, tufted cells). The main interneurons are the periglomerular cells and the granular cells. The periglomerular cells are small cells which surround glomeruli. They have highly arborizing dendrites in one glomerulus and may extend towards 3-6 glomeruli where they synapse onto apical trunk of projecting neurons and other granular cells. The granular cells are also small cells; they have a large apical dendrite which projects onto tufted and mitral cells but they do not have axons.

The overall organization of rodent OB is laminated (for review see Greer et al. 2008). From the periphery to the center of the bulb, the following layers can be found (Figure 1.1A): the olfactory nerve layer (ONL) which is composed of the mass of axons projecting from the OE, the glomerular layer (GL) where the glomeruli are located, the external plexiform layer (EPL) which contains the tufted neuron cell bodies, the mitral cell layer (MCL) which contains the cell bodies of mitral cells, the internal plexiform layer (IPL) where the axons migrating to the cortex fasciculate, the granular cell layer (GrL) which contains the granular cell bodies, and the rostral migratory stream layer (RMS). A comparable laminar organization is found in the human OB. However, it is less rigorous in the segregation of cell populations and also often lacks the circumferential organization of layers found in rodents and the medial-lateral symmetry of the rodent OB (Maresh et al. 2008).

Mitral cells are the most prominent population of output neurons. They have a single apical dendrite which invades a single glomerulus where it arborizes. About 20–25 mitral cells project into each glomerulus. The axon extends from the basal pole of the neuron and joins other axons forming the lateral olfactory tract (LOT). Mitral cells also have secondary dendrites which extend laterally in the EPL. Tufted cells are the second population of output neurons in number. Their shapes are similar to those of mitral cells but they are thought to mediate parallel circuits for processing olfactory information. The cortical target of mitral and tufted cells differs. Tufted cells project to the most rostral part of the olfactory cortex and the more medial olfactory tubercle while mitral cells distribute broadly throughout the olfactory cortex.

The synaptic organization of microcircuits in the OB is highly complex and far from being fully understood. The OB exhibits a circuitry that supports extensive inhibitory lateral interactions before the information is transmitted to the rest of the brain (Gire et al. 2013). This lateral inhibition is mainly due to the large population of interneurons.

1.3.2 The architecture of the olfactory bulb supports its function

As mentioned above, all OSNs expressing the same OR send their projection axon into the OB in a very limited number of glomeruli (~2 in each OB) (for review see Mombaerts 2004). The spatial position of a given glomerulus is not randomly distributed in an area corresponding in size to the equivalent of 30 glomeruli (Mombaerts 2006). Such invariant organization among different individuals of the same species strongly suggests a highly wired organization of the topography of projection from OE to OB. What are the consequences of this organization in term of odorant coding? Each OSN can detect a range of odorants and a given odorant can activate different OSNs. The activation of OSNs by odorants is therefore a combinatorial event (Kajiya et al. 2001, Malnic et al. 1999). Nevertheless, each OSN is thought to have its own specificity in terms of odorant sensitivity. Then, the organization of OSN projection onto the bulb raises the question of a functional topological organization between OE and OB. At least there is an anatomical topology which could be named "receptoro-topy" (Murthy 2011) since all OSNs expressing the same OR project onto the "same" glomeruli. Does such anatomical topology support functional topology? If odotopy which refers to the activation of subset of glomeruli by a given odorant is generally accepted, the chemotopy which refers to activation of subset of glomeruli according to properties of odorant molecules is more controversial (Murthy 2011).

1.3.3 The overall architecture of the olfactory system is genetically determined

The relatively stereotyped organization of the olfactory system has been further supported by the discovery of the "receptoro-topy" between OE and OB. From the developmental point of view, one can wonder what are the mechanisms underlying such an invariant organization. In other words, what are the mechanisms allowing all the OSNs expressing the same OR to converge onto one or two glomeruli in specified location in the OB? The mechanisms involved appear to be complex and combinatorial. First, it has been known for long now that there is a correspondence between the position of the OSN in the OE and the dorsoventral position of the glomerulus where they converge into at the OB level (Astic et al. 1987). Second, the guidance of OSN along the dorso-ventral axis appears to be dependent of 2 sets of repulsive ligand/receptor pairs, that is, Slits/Robo2 and Sema3F/Nrp2 (Takeuchi et al. 2010). Third, concerning the antero posterior-position of the glomeruli, the pre-targeted axon sorting is due to classic axon guidance molecules such as Np1 and Sema3A (Sakano 2010). These are thought to be under the control of intracellular signaling involving cAMP (Col et al. 2007, Zou et al. 2007). The axon sorting along the antero-posterior axis occurs before the OB, at the level of the axons bundles (Imai et al. 2009) and is independent of stimulus driven activity (Nakashima et al. 2013).

1.3.4 The fine architecture of the olfactory system is environmentally determined

Even in invertebrates, non-programmed activity-dependent factors are involved in the development of the nervous system. In the mammalian olfactory system, this is true both at the level of the OE and the OB. For instance, since the 1980's, it is known that naris closure induce a reduction of 10% in the number of OSNs in the OE (Farbman et al. 1988). At the level of the OB, the convergence of all OSNs expressing the same OR to one or two glomeruli sets up progressively during post-natal development. This convergence establishes gradually and takes place differently depending on the OR. Finally, this convergence is activity-dependent since olfactory deprivation prevents its typical organization (Zou et al. 2004), even if this latest statement has been controversial (Lin et al. 2000). Nevertheless, the involvement of neuronal activity by the way of competition mechanisms between active and inactive neurons has been beautifully demonstrated (Zhao and Reed 2001). More recently, it has been shown that ligand driven activity of OSNs appears to control the glomerular segregation and that this is mediated by G_{olf} transducing cascade (Nakashima et al. 2013).

1.3.5 In the olfactory system, development never ends

The OE has been the first place where continuous neurogenesis has been demonstrated (Hinds et al. 1984). This peripheral neurogenesis is thought to be linked to the fact that OSNs are continuously exposed to external air and particularly to toxic components. This exposure induces an increased neuronal death which should be compensated by neurogenesis in order to keep the system functional. Later, another nest of adult neurogenesis has been demonstrated in the brain in the sub-ventricular zone. From there, 3 migratory pathways send newborn cells in different directions. Among them, the rostral migratory pathway forwards new cells to the OB. While migrating, the newborn cells differentiate into neurons and integrate the OB circuits. They appear to mature mainly as granule cells.

There is a controversy regarding the status of an intact rostral migratory stream in the human OB. However, a wider consensus exists showing that newly differentiated neurons are found in the adult human OB, even among the elderly. This indicates that the human OB is a dynamic structure with a capacity for plasticity throughout life (Maresh et al. 2008). Such neurogenesis is meaningful regarding the plasticity of the olfactory system in response to changes in the olfactory environment. For example, an exposure to an odorant-enriched environment increases significantly both the number of newborn neurons integrated in the circuits and the learning performances (Rochefort et al. 2002). Conversely, a naris closure induces a reduction in the number of new neurons integrated into the OB circuits (Gheusi et al. 2000, Gheusi and Rochefort 2002).

1.4 Central odour processing

1.4.1 The primary olfactory cortex

In rodents, the primary olfactory cortex consists in brain areas that are direct targets of the OB. The bulbar outputs are conveyed by the LOT. The LOT is a myelinated fiber tract reaching diverse brain structures (Figure 1.2): the anterior olfactory nucleus, the tenia tecta, the olfactory tubercle, the anterior and posterior piriform cortex, the nucleus of the lateral olfactory tract, the anterior cortical amygdaloid nucleus, the posterolateral cortical amygdaloid nucleus, and the lateral entorhinal cortex (Haberly 2001). The LOT appears as a non-homogeneous tract. Thus axons from bulbar neurons (mitral/tufted cells) seem to be located in distinct parts of the LOT, suggesting different pathways to send information to higher olfactory regions (Nagayama et al. 2010). Each area of the primary

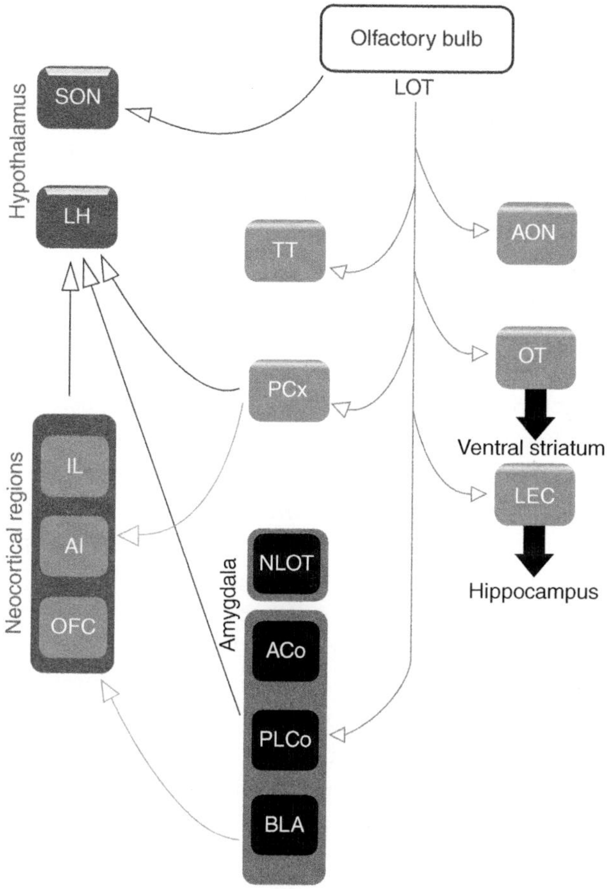

Figure 1.2 Schematic illustration of the main anatomical efferents arising from the main olfactory bulb in rodents. The primary cortex consists in regions receiving bulbar outputs conveyed by the LOT. The piriform cortex relays the olfactory information to several neocortical areas involved in complex processes such as multisensory integration, flavour perception and decision-making. It is also connected with the lateral hypothalamus that plays a role in feeding behaviour. Within the amygdala, bulbar efferents primarily target the superficial cortical nuclei (Aco, PLco) which are connected to deep nuclei such as BLA. The NLOT is located in the rostral part of the amygdala. Unlike other parts of olfactory amygdala, it does not project directly to the hypothalamus. Olfactory information is also transmitted to reward circuit through the olfactory tubercle efferents reaching the ventral striatum. The entorhinal cortex, as a gateway to hippocampus, allows olfactory-related mnesic processes. Abbreviations: Aco, anterior cortical nucleus of amygdala; AI, agranular insular cortex; AON: anterior olfactory nucleus; BLA, basolateral amygdala; IL, infralimbic cortex; LEC, lateral entorhinal cortex; LH, lateral hypothalamus; LOT, lateral olfactory tract; NLOT, nucleus of the lateral olfactory tract; OFC, orbitofrontal cortex; OT, olfactory tubercle; PCx, piriform cortex; PLco, posterolateralcortical nucleus of amygdala; SON, supraoptic nucleus; TT, tenia tecta. A colored version of this figure can be found in the online version of this chapter.

olfactory pathway is believed to process odours and to interpret the activity map originating from the OB in different manners.

In humans, the organization is not completely similar to rodents since the olfactory cortical areas that are direct targets of the OB are less developed. In the human brain, the main region receiving bulbar information through the lateral olfactory tract consists of the piriform cortex which is located at the junction of the inferior frontal and temporal lobes (Chen et al. 2010, Zatorre et al. 1992).

1.4.1.1 The piriform cortex

In rodents, the piriform cortex (PCx) is the largest cortical area of the olfactory cortex. It is located on the ventrolateral surface of the brain close to the LOT (Figure 1.1C). The PCx is divided into anterior (aPCx) and posterior (pPCx) subdivisions which differ in their anatomical features. In contrast to cortical regions in other sensory systems, it does not receive sensory inputs *via* the thalamus but direct synaptic inputs from the OB. Mitral cells have been shown to terminate in rather broad patches in the PCx (Buonviso et al. 1991) and collaterals are distributed to wide areas of the PCx (Nagayama et al. 2010). Axonal projections of mitral/tufted cells from the OB to the PCx appears to be sparsely distributed and overlapping (Stettler and Axel 2009). Individual cortical neurons in highly restricted areas of the PCx receive direct inputs representing glomeruli that are distributed throughout the OB with no apparent topographical organization (Ghosh et al. 2011, Miyamichi et al. 2011, Sosulski et al. 2011). No marked topography in odour-evoked activity has been demonstrated in the PCx and the spatial organization pattern of activity induced by odour in the OB is not conserved (Illig and Haberly 2003).

Cellular types and cytoarchitecture

The PCx is a trilaminar paleocortex (Figure 1.1C). The upper part of the layer I (layer Ia) contains afferent fibers from the LOT. The layer II contains the principal PCx glutamatergic neuronal types which are superficial pyramidal (SP) cells. The SP cells are characterized by well-developed dendritic trees and are the target of more associational excitatory inputs (Bekkers and Suzuki 2013, Suzuki and Bekkers 2011). Besides SP, another glutamatergic cells, the semilunar (SL) cells, are also found in the layer II (Suzuki and Bekkers 2006). The SL cells receive stronger afferent excitatory bulbar inputs and weak associational inputs than SP cells. In the layer III, deep pyramidal (DP) cells and multipolar (Mp) cells are observed (Protopapas and Bower 2000). DP cells have a rather similar connectivity as SP cells.

As in other cortices, synaptic inhibition is observed in the PCx. Indeed the PCx contains GABAergic interneurons that are present across all layers and provide feed-forward and feedback inhibition of the principal cells (Suzuki and Bekkers 2012). Thus, neurogliaform and horizontal interneurons in the layer Ia receive LOT inputs and provide feed-forward inhibition of the distal apical dendrites of SL and SP cells. Feedback inhibitory interneurons are restricted to deeper associational layers and involve a variety of interneurons such as bi-tufted, soma targeted fast-spiking, axons targeting chandelier, dendrite targeting regular-spiking and deep neurogliaform cells (Larriva-Sahd 2010).

A main feature of the PCx is a dense network of associational fibers. Indeed, the SP cells give rise to massive axon collaterals that form synapses on other pyramidal cells across wide areas of the PCx and form an extensive circuitry of recurrent connections (Johnson et al. 2000). Thus, each pyramidal cell makes a small number of synaptic contacts on a large number (more than 1000) of other cells at various locations within the PCx (Johnson et al. 2000). The network of intrinsic connections can enhance or suppress bulbar inputs and subsequently influence the recruitment of PCx principal neurons by afferent bulbar inputs (Franks et al. 2011). This might allow detection of temporally patterned OB inputs, shaping odour-evoked responses and encoding odorant identity.

Both PCx subdivisions are not similar regarding the organization and intra-cortical connectivity. The pPCx receives dense connections from the aPCx and has more recurrent connections, suggesting an associative role. On the contrary, the aPCx receives more afferent inputs from the OB and less associational inputs. In aPCx, the strength of inhibitory connections onto pyramidal cells is different along its rostro-caudal axis. Thus, pyramidal cells located at more caudal level of aPCx receives greater inhibition than cells at rostral location (Luna and Pettit 2010).

Besides intrinsic cortico-cortical fibers, the pyramidal cells give rise to extrinsic associational fibers which are restricted to layers Ib, II and III and connect the PCx with other regions of the primary cortex. In addition, commissural fibers originating from the aPCx layer II can reach the contralateral pPCx.

Role in olfactory processing

In the environment, odours mainly result from the perception of odorant mixtures. The PCx is assumed to build odour representations from sensory fragment allowing perceptual stability and behavioural adaptability even if changes in mixture components occur (Wilson and Sullivan 2011). The PCx receives convergent inputs from random collections of glomeruli and, as a result, it might be expected that odour representation mainly depends on its behavioural significance. The extensive network of excitatory association fibers support the view that PCx might construct unitary odour objects from the chemical components processed at earlier stages of the olfactory system. This auto associative network might allow complex processes such as pattern completion (possibility to fill the gap for partial inputs) and pattern discrimination (possibility to extract information from background) (Bekkers and Suzuki 2013, Chapuis and Wilson 2012, Uchida et al. 2014).

In rodents, PCx subdivisions have been shown to differ in their role regarding coding of odorant identity and odour quality. The process of familiarization differentially involves both subdivisions. As odorant mixtures become more familiar, aPCx neurons show habituation and develop enhanced ability to discriminate those mixtures from their environment (Wilson 2000). In the pPCx, more broadly tuned neurons might encode odorant qualities of the stimuli (Kadohisa and Wilson 2006). In humans, data from functional magnetic resonance imaging suggest that the PCx also shows a functional heterogeneity along its rostro-caudal axis (Gottfried et al. 2002).

Odour representation in the PCx has been shown to be modified through experience. Plasticity occurs at the level of the associated connections (Stripling and Galupo 2008). Long-lasting modifications of neuronal activity and synaptic efficiency have been shown to occur in various learning contexts (Barkai and Saar 2001, Martin et al. 2004). The PCx may mediate different learned behaviour in the absence of sensory inputs (Choi et al. 2011). It may function as an associative area rather that a classical primary sensory cortex (Barkai et al. 1994), synthesizing features from olfactory cues and linking them with other brain functions.

1.4.1.2 Other regions of the primary olfactory cortex

The anterior olfactory nucleus

The anterior olfactory nucleus (AON) is placed in the olfactory peduncle (i.e. the region connecting the OB with the basal forebrain). It consists in several subdivisions. The AON is reciprocally connected to both the ipsi- and contralateral regions of OB and PCx. There is a highly topographic axonal projection of mitral/tufted cells on the AON (*pars externa*) following the glomerular map. As a consequence, the AON maintains a dorso-ventral topography (Miyamichi et al. 2011, Yan et al. 2008). Regarding the PCx, dense functional connections from the AON exist with the aPCx in comparison with the pPCx (Hagiwara et al. 2012). In view of its connections, the AON is in position to broadly influence the central processing of odour information by preprocessing the bulbar inputs before sending to other cortical areas (Kay and Brunjes 2014).

The AON contains two main neuronal populations: excitatory projection neurons and inhibitory interneurons (at least 5 classes of inhibitory cells were observed). It has been proposed that the AON provides feed-forward modification of information from the OB to the PCx. The AON is also assumed to participate to the localization of odour sources by comparing the ipsi-nostril to contra-nostril inputs of the same odorant category (Kikuta et al. 2010).

The olfactory tubercle

Together with the accumbens, the olfactory tubercle (OT) is referred as the ventral striatum. It is assumed to be involved in the induction of appetitive and fearful motivated behaviours (Ikemoto 2007, Wesson and Wilson 2011). Thus, food odour information processed in the OT appears to influence the dopaminergic circuits for reward expectation (Giessel and Datta 2014).

The anterior and posterolateral cortical amygdaloid nuclei
The anterior cortical amygdaloid nucleus projects to melanin-concentrating hormone-containing neurons in the lateral hypothalamus. This suggests a role in the modulation of feeding behaviours (Niu et al. 2012). The posterolateral cortical amygdaloid nucleus has relationships with the ventral striatum which might play a role in processing the reinforcing properties of olfactory stimuli (Ubeda-Banon et al. 2007). Trans-synaptic tracing has shown that neurons from the cortical amygdala mainly receive inputs from the dorsal OB (Miyamichi et al. 2011). Mice lacking OSNs that project to the dorsal OB lose their innate avoidance for odour from predator urine and spoiled food, suggesting that cortical amygdala nuclei may preferentially process olfactory information that directs innate behaviours.

The lateral entorhinal cortex
The lateral entorhinal cortex (LEC) receives direct input from OB and piriform cortex. The LEC has multiple reciprocal connections with hippocampus, amygdala and perirhinal cortex (Kerr et al. 2007). It also projects back to the PCx and OB (Agster and Burwell 2009). As a result, it is thought to play a crucial role in the olfactory memory and modulation of odour processing (Chapuis et al. 2013). The LEC exerts an inhibitory effect on PCx responses to the OB stimulations (Mouly and Di Scala 2006) and this might participate to the modulation of olfactory learning and memory (Wirth et al. 1998). To sum up, the major role of LEC is mnesic, cognitive and multimodal processing of olfactory cues (Martin and Ravel 2014).

1.4.2 Beyond the primary olfactory cortex

Besides regions of the primary olfactory cortex, the olfactory inputs reach various brain areas. These areas are thought to mediate complex functions related to the integration of sensory cues with behaviour, emotional or motivational significance, multisensory association and memory.

1.4.2.1 The basolateral amygdala

The superficial amygdala nuclei (anterior and posterolateral) relay the olfactory inputs coming from the OB to deeper amygdala nuclei such as the basolateral nucleus (BLA) (McDonald 1998). The BLA is a major area for odour-taste associations, i.e. flavour integration. This nucleus further plays a role in emotional learning involving olfaction such as conditioned odour aversion (Sevelinges et al. 2009), taste-potentiated odour aversion (Dardou et al. 2007, Shionoya and Datiche 2009) but also socially transmitted food preference (Wang et al. 2006) or conditioned flavour learning (Lienard et al. 2014). The BLA shares extensive reciprocal connections with the orbitofrontal cortex and these regions likely contribute to both mnesic and affective processes (Cardinal et al. 2002). The orbitofrontal cortex and BLA play partially overlapping roles in the use of incentive information that supports normal discrimination performance.

1.4.2.2 The hippocampus

The LEC is the gateway for olfactory inputs to the hippocampus which is a major region for formation of associative memories. A wide range of evidence indicates that this pathway sustains olfactory mnemonic processing (Gold et al. 2011, Raineki et al. 2010).

1.4.2.3 The orbito-frontal cortex

The PCx has reciprocal connections with the orbito-frontal cortex (OFC) which is a multimodal cortical area with neurons responding to several types of sensory cues including the olfactory ones (Rolls 2012). The OFC receives convergent inputs from olfactory and gustatory cortices; both sensory modalities can be combined to give rise to the sensation of flavour. Neurons from the OFC are also believed to encode reward expectancy and to link sensory representations to behavioural inputs (Mainen and Kepecs 2009, Schoenbaum et al. 2003).

1.4.2.4 The hypothalamus

Several regions from the primary olfactory cortex (at least the anterior olfactory nucleus, the piriform cortex, the olfactory tubercle and the anterior cortical nucleus of the amygdala) have been shown to project onto the lateral hypothalamus (Barone et al. 1981, Price et al. 1991). A direct projection from the main OB to the supraoptic nucleus of the rat has also been reported (Smithson et al. 1989). Olfactory inputs further target vasopressin neurons from the paraventricular hypothalamic nucleus (Bader et al. 2012). Together, these connections with hypothalamic nuclei might allow to affect feeding, reproductive activity, and autonomic reflexes triggered by olfactory signals (Palouzier-Paulignan et al. 2012).

1.5 Conclusion

In the last three decades, our knowledge about the functional architecture of the primary and accessory olfactory systems in mammals, including humans, has grown rapidly. The discovery of OR genes by Buck and Axel (Buck and Axel 1991) has been a major step in deciphering the molecular mechanisms that govern odorant coding at the peripheral level. This led to propose the concept of combinatorial scheme for odour coding. The OB is the first central relay, where olfactory inputs are spatially organized, noise-filtered, and sharpened. A number of studies support the existence of a coarse topographic map from the receptor level to this first stage of processing (named "receptoro-topy") but how such a map translates to a functional olfactory map continues to be difficult to resolve (Murthy 2011). The topographic connectivity between OSNs and OB glomeruli is retained in projections to the amygdala and anterior olfactory nucleus, but lost in the projections to piriform cortex. One step beyond, the network of central connections participate to complex integration processes such as recognition of odours and odour-guided decisions, allowing adaptive behaviours crucial for survival (food intake, maternal bonding, etc ...).

A notable feature of the olfactory system is the continual neurogenesis that occurs during adulthood in the OE (from a locally dividing pool of progenitor cells) and in the OB (from cells born in the sub-ventricular zone). Adult neurogenesis modulated by olfactory inputs has been reported in the olfactory cortex and in brain structures related to emotion (amygdala), reward (striatal system), learning and memory (hippocampus and entorhinal cortex) (Arisi et al. 2012). Recent works demonstrated that several internal and external factors can also modulate olfactory signals throughout the processing pathways. The different levels of the olfactory system are therefore dynamic structures with features reflecting innate and environmental as well as developmental influences. As a consequence, it can be assumed that the structural and functional properties of the olfactory system may slightly differ among individuals and change over time. Future studies are therefore needed to better assess the impact of aging and nutrition on the olfactory function and more specifically how they shape the encoding of odour information.

References

Agster, K. L. and Burwell, R. D. (2009) Cortical efferents of the perirhinal, postrhinal, and entorhinal cortices of the rat. *Hippocampus, 19*(12), pp. 1159–1186.

Arisi, G. M., Foresti, M. L., Mukherjee, S. and Shapiro, L. A. (2012) The role of olfactory stimulus in adult mammalian neurogenesis. *Behavioural Brain Research, 227*(2), pp. 356–362.

Astic, L., Saucier, D. and Holley, A. (1987) Topographical relationships between olfactory receptor-cells and glomerular foci in the rat olfactory-bulb. *Brain Research, 424*(1), pp. 144–152.

Bader, A., Klein, B., Breer, H. and Strotmann, J. (2012) Connectivity from OR37 expressing olfactory sensory neurons to distinct cell types in the hypothalamus. *Frontiers in Neural Circuits, 6*.

Badonnel, K., Lacroix, M.-C., Monnerie, R., Durieux, D., Caillol, M. and Baly, C. (2012) Chronic restricted access to food leading to undernutrition affects rat neuroendocrine status and olfactory-driven behaviors. *Hormones and Behavior, 62*(2), pp. 120–127.

Baly, C., Aioun, J., Badonnel, K., Lacroix, M.-C., Durieux, D., Schlegel, C., Salesse, R. and Caillol, M. (2007) Leptin and its receptors are present in the rat olfactory mucosa and modulated by the nutritional status. *Brain Research, 1129*(1), pp. 130–141.

Barkai, E., Bergman, R. E., Horwitz, G. and Hasselmo, M. E. (1994) Modulation of associative memory function in a biophysical simulation of rat piriform cortex. *Journal of Neurophysiology, 72*(2), pp. 659–677.

Barkai, E. and Saar, D. (2001) Cellular correlates of olfactory learning in the rat piriform cortex. *Reviews in the Neurosciences, 12*(2), pp. 111–120.

Barone, F. C., Wayner, M. J., Scharoun, S. L., Guevaraaguilar, R. and Aguilarbaturoni, H. U. (1981) Afferent connections to the lateral hypothalamus - A horseradish-peroxidase study in the rat. *Brain Research Bulletin, 7*(1), pp. 75–88.

Baud, O., Etter, S., Spreafico, M., Bordoli, L., Schwede, T., Vogel, H. and Pick, H. (2011) The mouse eugenol odorant receptor: structural and functional plasticity of a broadly tuned odorant binding pocket. *Biochemistry, 50*(5), pp. 843–853.

Bekkers, J. M. and Suzuki, N. (2013) Neurons and circuits for odor processing in the piriform cortex. *Trends in Neurosciences, 36*(7), pp. 429–438.

Bondier, J.-R., Michel, G., Propper, A. and Badot, P.-M. (2008) Harmful effects of cadmium on olfactory system in mice. *Inhalation Toxicology, 20*(13), pp. 1169–1177.

Boschat, C., Pelofi, C., Randin, O., Roppolo, D., Luscher, C., Broillet, M. C. and Rodriguez, I. (2002) Pheromone detection mediated by a V1r vomeronasal receptor. *Nature Neuroscience, 5*(12), pp. 1261–1262.

Brechbuhl, J., Klaey, M. and Broillet, M. C. (2008) Grueneberg ganglion cells mediate alarm pheromone detection in mice. *Science, 321*(5892), pp. 1092–1095.

Brechbuhl, J., Moine, F., Klaey, M., Nenniger-Tosato, M., Hurni, N., Sporkert, F., Giroud, C. and Broillet, M. C. (2013) Mouse alarm pheromone shares structural similarity with predator scents. *Proceedings of the National Academy of Sciences of the United States of America, 110*(12), pp. 4762–4767.

Buck, L. and Axel, R. (1991) A novel multigene family may encode odorant receptors: a molecular basis for odor recognition. *Cell, 65*(1), pp. 175–187.

Buonviso, N., Revial, M. F. and Jourdan, F. (1991) The projections of mitral cells from small local regions of the olfactory bulb - An anterograde tracing study using PHA-L (Phaseolus vulgaris leukoagglutinin). *European Journal of Neuroscience, 3*(6), pp. 493–500.

Buron, G., Hacquemand, R., Pourie, G. and Brand, G. (2009) Inhalation exposure to acetone induces selective damage on olfactory neuroepithelium in mice. *Neurotoxicology, 30*(1), pp. 114–120.

Cadiou, H., Aoude, I., Tazir, B., Molinas, A., Fenech, C., Meunier, N. and Grosmaitre, X. (2014) Postnatal odorant exposure induces peripheral olfactory plasticity at the cellular level. *Journal of Neuroscience, 34*(14), pp. 4857–4870.

Cardinal, R. N., Parkinson, J. A., Hall, J. and Everitt, B. J. (2002) Emotion and motivation: the role of the amygdala, ventral striatum, and prefrontal cortex. *Neuroscience and Biobehavioral Reviews, 26*(3), pp. 321–352.

Carey, S. A., Plopper, C. G., Hyde, D. M., Islam, Z., Pestka, J. J. and Harkema, J. R. (2012) Satratoxin-G from the black mold Stachybotrys chartarum induces rhinitis and apoptosis of olfactory sensory neurons in the nasal airways of Rhesus monkeys. *Toxicologic Pathology, 40*(6), pp. 887–898.

Cavallin, M. A., Powell, K., Biju, K. C. and Fadool, D. A. (2010) State-dependent sculpting of olfactory sensory neurons is attributed to sensory enrichment, odor deprivation, and aging. *Neuroscience Letters, 483*(2), pp. 90–95.

Chapuis, J., Cohen, Y., He, X., Zhang, Z., Jin, S., Xu, F. and Wilson, D. A. (2013) Lateral entorhinal modulation of piriform cortical activity and fine odor discrimination. *Journal of Neuroscience, 33*(33), pp. 13449–13459.

Chapuis, J. and Wilson, D. A. (2012) Bidirectional plasticity of cortical pattern recognition and behavioral sensory acuity. *Nature Neuroscience, 15*(1), pp. 155–161.

Charlier, L., Topin, J., Ronin, C., Kim, S.-K., Goddard, W. A., III., Efremov, R. and Golebiowski, J. (2012) How broadly tuned olfactory receptors equally recognize their agonists. Human OR1G1 as a test case. *Cellular and Molecular Life Sciences, 69*(24), pp. 4205–4213.

Chen, C. C., Huang, F., Zheng, J. W., Fu, S. Q., Kong, F. Z., Chen, Z. X., Yang, X. D. and Zang, C. S. (2010) Sectional anatomy of the olfactory pathways. *Journal of Neurosurgical Sciences, 54*(1), pp. 39–44.

Chess, A., Simon, I., Cedar, H. and Axel, R. (1994) Allelic inactivation regulates olfactory receptor gene expression. *Cell, 78*(5), pp. 823–834.

Choi, G. B., Stettler, D. D., Kallman, B. R., Bhaskar, S. T., Fleischmann, A. and Axel, R. (2011) Driving opposing behaviors with ensembles of piriform neurons. *Cell, 146*(6), pp. 1003–1014.

Col, J. A. D., Matsuo, T., Storm, D. R. and Rodriguez, I. (2007) Adenylyl cyclase-dependent axonal targeting in the olfactory system. *Development (Cambridge, England), 134*(13), pp. 2481–2489.

Cuschieri, A. and Bannister, L. H. (1975) Development of olfactory mucosa in mouse - Light-microscopy. *Journal of Anatomy, 119*(APR), pp. 277–286.

Dando, S. J., Mackay-Sim, A., Norton, R., Currie, B. J., St John, J. A., Ekberg, J. A., Batzloff, M., Ulett, G. C. and Beacham, I. R. (2014) Pathogens penetrating the central nervous system: infection pathways and the cellular and molecular mechanisms of invasion. *Clinical Microbiology Reviews, 27*(4), pp. 691–726.

Dardou, D., Datiche, F. and Cattarelli, M. (2007) Does taste or odor activate the same brain networks after retrieval of taste potentiated odor aversion? *Neurobiology of Learning and Memory, 88*(2), pp. 186–197.

Dulac, C. and Axel, R. (1995) A novel family of genes encoding putative pheromone receptors in mammals. *Cell, 83*(2), pp. 195–206.

Dulac, C. and Torello, A. T. (2003) Molecular detection of pheromone signals in mammals: from genes to behaviour. *Nature Reviews Neuroscience, 4*(7), pp. 551–562.

Ekberg, J. A. and St John, J. A. (2014) Crucial roles for olfactory ensheathing cells and olfactory mucosal cells in the repair of damaged neural tracts. *Anatomical Record, 297*(1), pp. 121–128.

Escada, P. A., Lima, C. and da Silva, J. M. (2009) The human olfactory mucosa. *European Archives of Oto-Rhino-Laryngology, 266*(11), pp. 1675–1680.

Farbman, A. I., Brunjes, P. C., Rentfro, L., Michas, J. and Ritz, S. (1988) The effect of unilateral naris occlusion on cell-dynamics in the developing rat olfactory epithelium. *Journal of Neuroscience, 8*(9), pp. 3290–3295.

Fleischer, J. and Breer, H. (2010) The Grueneberg ganglion: a novel sensory system in the nose. *Histology and Histopathology, 25*(7), pp. 909–915.

Francois, A., Laziz, I., Rimbaud, S., Grebert, D., Durieux, D., Pajot-Augy, E. and Meunier, N. (2013) Early survival factor deprivation in the olfactory epithelium enhances activity-survival driven survival. *Frontiers in Cellular Neuroscience, 7*.

Franks, K. M., Russo, M. J., Sosulski, D. L., Mulligan, A. A., Siegelbaum, S. A. and Axel, R. (2011) Recurrent circuitry dynamically shapes the activation of piriform cortex. *Neuron, 72*(1), pp. 49–56.

Gaskell, B. A. (1990) Nonneoplastic changes in the olfactory epithelium - Experimental studies. *Environmental Health Perspectives, 85*, pp. 275–289.

Gelis, L., Wolf, S., Hatt, H., Neuhaus, E. M. and Gerwert, K. (2012) Prediction of a ligand-binding niche within a human olfactory receptor by combining site-directed mutagenesis with dynamic homology modeling. *Angewandte Chemie (International Edition), 51*(5), pp. 1274–1278.

Gesteland, R. C., Yancey, R. A. and Farbman, A. I. (1982) Development of olfactory receptor neuron selectivity in the rat fetus. *Neuroscience, 7*(12), pp. 3127–3136.

Gheusi, G., Cremer, H., McLean, H., Chazal, G., Vincent, J. D. and Lledo, P. M. (2000) Importance of newly generated neurons in the adult olfactory bulb for odor discrimination. *Proceedings of the National Academy of Sciences of the United States of America, 97*(4), pp. 1823–8.

Gheusi, G. and Rochefort, C. (2002) Neurogenesis in the adult brain. Functional consequences. *Journal de la Société de Biologie, 196*(1), pp. 67–76.

Ghosh, S., Larson, S. D., Hefzi, H., Marnoy, Z., Cutforth, T., Dokka, K. and Baldwin, K. K. (2011) Sensory maps in the olfactory cortex defined by long-range viral tracing of single neurons. *Nature, 472*(7342), pp. 217–220.

Giessel, A. J. and Datta, S. R. (2014) Olfactory maps, circuits and computations. *Current Opinion in Neurobiology, 24,* pp. 120–132.

Gire, D. H., Restrepo, D., Sejnowski, T. J., Greer, C., De Carlos, J. A. and Lopez-Mascaraque, L. (2013) Temporal processing in the olfactory system: can we see a smell? *Neuron, 78*(3), pp. 416–432.

Gobba, F. (2006) Olfactory toxicity: long-term effects of occupational exposures. *International Archives of Occupational and Environmental Health, 79*(4), pp. 322–331.

Gold, P. E., Countryman, R. A., Dukala, D. and Chang, Q. (2011) Acetylcholine release in the hippocampus and prelimbic cortex during acquisition of a socially transmitted food preference. *Neurobiology of Learning and Memory, 96*(3), pp. 498–503.

Gottfried, J. A., Deichmann, R., Winston, J. S. and Dolan, R. J. (2002) Functional heterogeneity in human olfactory cortex: An event-related functional magnetic resonance imaging study. *Journal of Neuroscience, 22*(24), pp. 10819–10828.

Greer, C. A., Whitman, M. C., Rela, L., Imamura, I. and Rodriguez Gil, D. (2008) Architecture of the olfactory bulb. in Firestein, S. and Beauchamp, G. K., (eds.) *The senses: a comprehensive reference,* Canada: Elsevier. pp. 623–640.

Grosmaitre, X., Fuss, S. H., Lee, A. C., Adipietro, K. A., Matsunami, H., Mombaerts, P. and Ma, M. (2009) SR1, a mouse odorant receptor with an unusually broad response profile. *Journal of Neuroscience, 29*(46), pp. 14545–52.

Grosmaitre, X., Santarelli, L. C., Tan, J., Luo, M. and Ma, M. (2007) Dual functions of mammalian olfactory sensory neurons as odor detectors and mechanical sensors. *Nature Neuroscience, 10*(3), pp. 348–354.

Gross, E. A., Swenberg, J. A., Fields, S. and Popp, J. A. (1982) Comparative morphometry of the nasal cavity in rats and mice. *Journal of Anatomy, 135*(Pt 1), pp. 83–88.

Grüneberg, H. (1973) A ganglion probably belonging to the N. terminalis system in the nasal mucosa of the mouse. *Zeitschrift fur Anatomie und Entwicklungsgeschichte, 140*(1), pp. 39–52.

Haberly, L. B. (2001) Parallel-distributed processing in olfactory cortex: New insights from morphological and physiological analysis of neuronal circuitry. *Chemical Senses, 26*(5), pp. 551–576.

Haga, S., Hattori, T., Sato, T., Sato, K., Matsuda, S., Kobayakawa, R., Sakano, H., Yoshihara, Y., Kikusui, T. and Touhara, K. (2010) The male mouse pheromone ESP1 enhances female sexual receptive behaviour through a specific vomeronasal receptor. *Nature, 466*(7302), pp. 118–122.

Hagiwara, A., Pal, S. K., Sato, T. F., Wienisch, M. and Murthy, V. N. (2012) Optophysiological analysis of associational circuits in the olfactory cortex. *Frontiers in Neural Circuits, 6.*

Hahn, C. G., Gomez, G., Restrepo, D., Friedman, E., Josiassen, R., Pribitkin, E. A., Lowry, L. D., Gallop, R. J. and Rawson, N. E. (2005) Aberrant intracellular calcium signaling in olfactory neurons from patients with bipolar disorder. *American Journal of Psychiatry, 162*(3), pp. 616–618.

Halpern, M. and Martinez-Marcos, A. (2003) Structure and function of the vomeronasal system: an update. *Progress in Neurobiology, 70*(3), pp. 245–318.

Harkema, J. R. (1991) Comparative aspects of nasal airway anatomy: relevance to inhalation toxicology. *Toxicologic Pathology, 19*(4 Pt 1), pp. 321–36.

Harkema, J. R., Carey, S. A. and Wagner, J. G. (2006) The nose revisited: A brief review of the comparative structure, function, and toxicologic pathology of the nasal epithelium. *Toxicologic Pathology, 34*(3), pp. 252–269.

Heydel, J.-M., Coelho, A., Thiebaud, N., Legendre, A., Le Bon, A.-M., Faure, P., Neiers, F., Artur, Y., Golebiowski, J. and Briand, L. (2013) Odorant-binding proteins and xenobiotic metabolizing enzymes: implications in olfactory perireceptor events. *Anatomical Record, 296*(9), pp. 1333–1345.

Hinds, J. W., Hinds, P. L. and McNelly, N. A. (1984) An autoradiographic study of the mouse olfactory epithelium - Evidence for long-lived receptors. *Anatomical Record, 210*(2), pp. 375–383.

Ibarra-Soria, X., Levitin, M. O. and Logan, D. W. (2014) The genomic basis of vomeronasal-mediated behaviour. *Mammalian Genome, 25*(1-2), pp. 75–86.

Ikemoto, S. (2007) Dopamine reward circuitry: Two projection systems from the ventral midbrain to the nucleus accumbens-olfactory tubercle complex. *Brain Research Reviews, 56*(1), pp. 27–78.

Illig, K. R. and Haberly, L. B. (2003) Odor-evoked activity is spatially distributed in piriform cortex. *Journal of Comparative Neurology, 457*(4), pp. 361–373.

Imai, T., Yamazaki, T., Kobayakawa, R., Kobayakawa, K., Abe, T., Suzuki, M. and Sakano, H. (2009) Pre-target axon sorting establishes the neural map topography. *Science, 325*(5940), pp. 585–590.

Iwema, C. L., Fang, H. S., Kurtz, D. B., Youngentob, S. L. and Schwob, J. E. (2004) Odorant receptor expression patterns are restored in lesion-recovered rat olfactory epithelium. *Journal of Neuroscience, 24*(2), pp. 356–369.

Johnson, D. M. G., Illig, K. R., Behan, M. and Haberly, L. B. (2000) New features of connectivity in piriform cortex visualized by intracellular injection of pyramidal cells suggest that "primary" olfactory cortex functions like "association" cortex in other sensory systems. *Journal of Neuroscience, 20*(18), pp. 6974–6982.

Kadohisa, M. and Wilson, D. A. (2006) Separate encoding of identity and similarity of complex familiar odors in piriform cortex. *Proceedings of the National Academy of Sciences of the United States of America, 103*(41), pp. 15206–15211.

Kajiya, K., Inaki, K., Tanaka, M., Haga, T., Kataoka, H. and Touhara, K. (2001) Molecular bases of odor discrimination: Reconstitution of olfactory receptors that recognize overlapping sets of odorants. *Journal of Neuroscience, 21*(16), pp. 6018–6025.

Katada, S., Hirokawa, T., Oka, Y., Suwa, M. and Touhara, K. (2005) Structural basis for a broad but selective ligand spectrum of a mouse olfactory receptor: mapping the odorant-binding site. *Journal of Neuroscience, 25*(7), pp. 1806–1815.

Kawagishi, K., Ando, M., Yokouchi, K., Sumitomo, N., Karasawa, M., Fukushima, N. and Moriizumi, T. (2014) Stereological quantification of olfactory receptor neurons in mice. *Neuroscience, 272*, pp. 29–33.

Kay, R. B. and Brunjes, P. C. (2014) Diversity among principal and GABAergic neurons of the anterior olfactory nucleus. *Frontiers in Cellular Neuroscience, 8*.

Keller, A., Zhuang, H., Chi, Q., Vosshall, L. B. and Matsunami, H. (2007) Genetic variation in a human odorant receptor alters odour perception. *Nature, 449*(7161), pp. 468–72.

Kerr, K. M., Agster, K. L., Furtak, S. C. and Burwell, R. D. (2007) Functional neuroanatomy of the Parahippocampal region: The lateral and medial Entorhinal areas. *Hippocampus, 17*(9), pp. 697–708.

Khan, M., Vaes, E. and Mombaerts, P. (2013) Temporal patterns of odorant receptor gene expression in adult and aged mice. *Molecular and Cellular Neuroscience, 57*, pp. 120–129.

Kikuta, S., Sato, K., Kashiwadani, H., Tsunoda, K., Yamasoba, T. and Mori, K. (2010) Neurons in the anterior olfactory nucleus pars externa detect right or left localization of odor sources. *Proceedings of the National Academy of Sciences of the United States of America, 107*(27), pp. 12363–12368.

Kimoto, H., Haga, S., Sato, K. and Touhara, K. (2005) Sex-specific peptides from exocrine glands stimulate mouse vomeronasal sensory neurons. *Nature, 437*(7060), pp. 898–901.

Kleene, S. J. (2008) The electrochemical basis of odor transduction in vertebrate olfactory cilia. *Chemical Senses, 33*(9), pp. 839–859.

Lacroix, M. C., Badonnel, K., Meunier, N., Tan, F., Schlegel-Le Poupon, C., Durieux, D., Monnerie, R., Baly, C., Congar, P., Salesse, R. and Caillol, M. (2008) Expression of insulin system in the olfactory epithelium: first approaches to its role and regulation. *Journal of Neuroendocrinology, 20*(10), pp. 1176–1190.

Lam, R. S. and Mombaerts, P. (2013) Odorant responsiveness of embryonic mouse olfactory sensory neurons expressing the odorant receptors S1 or MOR23. *European Journal of Neuroscience, 38*(2), pp. 2210–2217.

Larriva-Sahd, J. A. (2010) Chandelier and interfascicular neurons in the adult mouse piriform cortex. *Frontiers in Neuroanatomy, 4.*

Launay, G., Teletchea, S., Wade, F., Pajot-Augy, E., Gibrat, J.-F. and Sanz, G. (2012) Automatic modeling of mammalian olfactory receptors and docking of odorants. *Protein Engineering Design and Selection, 25*(8), pp. 377–386.

Lee, A. C., Tian, H., Grosmaitre, X. and Ma, M. (2009) Expression patterns of odorant receptors and response properties of olfactory sensory neurons in aged mice. *Chemical Senses, 34*(8), pp. 695–703.

Leinders-Zufall, T., Brennan, P., Widmayer, P., S , P. C. , Maul-Pavicic, A., Jager,M., Li, X. H., Breer, H., Zufall, F. and Boehm, T. (2004) MHC class I peptides as chemosensory signals in the vomeronasal organ. *Science, 306*(5698), pp. 1033–1037.

Levai, O. and Strotmann, J. (2003) Projection pattern of nerve fibers from the septal organ: DiI-tracing studies with transgenic OMP mice. *Histochemistry and Cell Biology, 120*(6), pp. 483–492.

Liberles, S. D., Horowitz, L. F., Kuang, D., Contos, J. J., Wilson, K. L., Siltberg-Liberles, J., Liberles, D. A. and Buck, L. B. (2009) Formyl peptide receptors are candidate chemosensory receptors in the vomeronasal organ. *Proceedings of the National Academy of Sciences of the United States of America, 106*(24), pp. 9842–9847.

Lienard, F., Desmoulins, L., Penicaud, L. and Datiche, F. (2014) Role of the basolateral amygdala in retrieval of conditioned flavors in the awake rat. *Behavioural Brain Research, 268,* pp. 40–47.

Lin, D. M., Wang, F., Lowe, G., Gold, G. H., Axel, R., Ngai, J. and Brunet, L. (2000) Formation of precise connections in the olfactory bulb occurs in the absence of odorant-evoked neuronal activity. *Neuron, 26*(1), pp. 69–80.

Lin, W. H., Arellano, J., Slotnick, B. and Restrepo, D. (2004) Odors detected by mice deficient in cyclic nucleotide-gated channel subunit A2 stimulate the main olfactory system. *Journal of Neuroscience, 24*(14), pp. 3703–3710.

Logan, D. W., Brunet, L. J., Webb, W. R., Cutforth, T., Ngai, J. and Stowers, L. (2012) Learned recognition of maternal signature odors mediates the first suckling episode in mice. *Current Biology, 22*(21), pp. 1998–2007.

Luna, V. M. and Pettit, D. L. (2010) Asymmetric rostro-caudal inhibition in the primary olfactory cortex. *Nature Neuroscience, 13*(5), pp. 533–535.

Ma, M. (2007) Encoding olfactory signals via multiple chemosensory systems. *Critical Reviews in Biochemistry and Molecular Biology, 42*(6), pp. 463–480.

Ma, M. (2010) Multiple olfactory subsystems convey various sensory signals. in Menini, A., (ed.) *The Neurobiology of Olfaction, Boca Raton* (Florida, USA). pp. 225–240.

Ma, M., Grosmaitre, X., Iwema, C. L., Baker, H., Greer, C. A. and Shepherd, G. M. (2003) Olfactory signal transduction in the mouse septal organ. *Journal of Neuroscience, 23*(1), pp. 317–324.

Mackay-Sim, A. and Kittel, P. (1991) Cell-dynamics in the adult-mouse olfactory epithelium - A quantitative autoradiographic study. *Journal of Neuroscience, 11*(4), pp. 979–984.

Mainen, Z. F. and Kepecs, A. (2009) Neural representation of behavioral outcomes in the orbitofrontal cortex. *Current Opinion in Neurobiology, 19*(1), pp. 84–91.

Malnic, B., Hirono, J., Sato, T. and Buck, L. B. (1999) Combinatorial receptor codes for odors. *Cell, 96*(5), pp. 713–723.

Maresh, A., Rodriguez Gil, D., Whitman, M. C. and Greer, C. A. (2008) Principles of glomerular organization in the human olfactory bulb--implications for odor processing. *Plos One, 3*(7), pp. e2640.

Marshall, D. A. and Maruniak, J. A. (1986) Masera's organ responds to odorants. *Brain Research, 366*(1-2), pp. 329–332.

Martin, C., Gervais, R., Chabaud, P., Messaoudi, B. and Ravel, N. (2004) Learning-induced modulation of oscillatory activities in the mammalian olfactory system: The role of the centrifugal fibres. *Journal of Physiology, 98*(4-6), pp. 467–478.

Martin, C. and Ravel, N. (2014) Beta and gamma oscillatory activities associated with olfactory memory tasks: different rhythms for different functional networks? *Frontiers in Behavioral Neuroscience, 8*.

Matsui, A., Go, Y. and Niimura, Y. (2010) Degeneration of olfactory receptor gene repertories in primates: no direct link to full trichromatic vision. *Molecular Biology and Evolution, 27*(5), pp. 1192–1200.

McDonald, A. J. (1998) Cortical pathways to the mammalian amygdala. *Progress in Neurobiology, 55*(3), pp. 257–332.

Meyer, M. R., Angele, A., Kremmer, E., Kaupp, U. B. and Muller, F. (2000) A cGMP-signaling pathway in a subset of olfactory sensory neurons. *Proceedings of the National Academy of Sciences of the United States of America, 97*(19), pp. 10595–10600.

Minn, A., Leclerc, S., Heydel, J. M., Minn, A. L., Denizcot, C., Cattarelli, M., Netter, P. and Gradinaru, D. (2002) Drug transport into the mammalian brain: the nasal pathway and its specific metabolic barrier. *Journal of Drug Targeting, 10*(4), pp. 285–296.

Miyamichi, K., Amat, F., Moussavi, F., Wang, C., Wickersham, I., Wall, N. R., Taniguchi, H., Tasic, B., Huang, Z. J., He, Z., Callaway, E. M., Horowitz, M. A. and Luo, L. (2011) Cortical representations of olfactory input by trans-synaptic tracing. *Nature, 472*(7342), pp. 191–196.

Miyamichi, K., Serizawa, S., Kimura, H. M. and Sakano, H. (2005) Continuous and overlapping expression domains of odorant receptor genes in the olfactory epithelium determine the dorsal/ventral positioning of glomeruli in the olfactory bulb. *Journal of Neuroscience, 25*(14), pp. 3586–3592.

Mombaerts, P. (2004) Odorant receptor gene choice in olfactory sensory neurons: the one receptor-one neuron hypothesis revisited. *Current Opinion in Neurobiology, 14*(1), pp. 31–36.

Mombaerts, P. (2006) Axonal wiring in the mouse olfactory system. *Annual Review of Cell and Developmental Biology, 22*, pp. 713–737.

Mombaerts, P., Wang, F., Dulac, C., Chao, S. K., Nemes, A., Mendelsohn, M., Edmondson, J. and Axel, R. (1996) Visualizing an olfactory sensory map. *Cell, 87*(4), pp. 675–686.

Montani, G., Tonelli, S., Elsaesser, R., Paysan, J. and Tirindelli, R. (2006) Neuropeptide Y in the olfactory microvillar cells. *European Journal of Neuroscience, 24*(1), pp. 20–24.

Moran, D. T., Rowley, J. C., Jafek, B. W. and Lovell, M. A. (1982) The fine structure of the olfactory mucosa in man. *Journal of Neurocytology, 11*(5), pp. 721–46.

Morrison, E. E. and Costanzo, R. M. (1992) Morphology of olfactory epithelium in humans and other vertebrates. *Microscopy Research and Technique, 23*(1), pp. 49–61.

Mouly, A. M. and Di Scala, G. (2006) Entorhinal cortex stimulation modulates amygdala and piriform cortex responses to olfactory bulb inputs in the rat. *Neuroscience, 137*(4), pp. 1131–1141.

Mulligan, C., Moreau, K., Brandolini, M., Livingstone, B., Beaufrere, B. and Boirie, Y. (2002) Alterations of sensory perceptions in healthy elderly subjects during fasting and refeeding - A pilot study. *Gerontology, 48*(1), pp. 39–43.

Munger, S. D., Leinders-Zufall, T. and Zufall, F. (2009) Subsystem organization of the mammalian sense of smell. *Annual Review of Physiology, 71*, pp. 115–140.

Murthy, V. N. (2011) Olfactory maps in the brain. *Annual Review of Neuroscience, 34*, pp. 233–58.

Nagayama, S., Enerva, A., Fletcher, M. L., Masurkar, A. V., Igarashi, K. M., Mori, K. and Chen, W. R. (2010) Differential axonal projection of mitral and tufted cells in the mouse main olfactory system. *Frontiers in Neural Circuits, 4.*

Nakashima, A., Takeuchi, H., Imai, T., Saito, H., Kiyonari, H., Abe, T., Chen, M., Weinstein, L. S., Yu, C. R., Storm, D. R., Nishizumi, H. and Sakano, H. (2013) Agonist-independent GPCR activity regulates anterior-posterior targeting of olfactory sensory neurons. *Cell, 154*(6), pp. 1314–1325.

Nakashima, T., Tanaka, M., Inamitsu, M. and Uemura, T. (1991) Immunohistopathology of variations of human olfactory mucosa. *European Archives of Oto-Rhino-Laryngology, 248*(6), pp. 370–375.

Negroni, J., Meunier, N., Monnerie, R., Salesse, R., Baly, C., Caillol, M. and Congar, P. (2012) Neuropeptide Y enhances olfactory mucosa responses to odorant in hungry rats. *Plos One, 7*(9).

Niimura, Y. (2012) Olfactory receptor multigene family in vertebrates: from the viewpoint of evolutionary genomics. *Current Genomics, 13*(2), pp. 103–114.

Niimura, Y., Matsui, A. and Touhara, K. (2014) Extreme expansion of the olfactory receptor gene repertoire in African elephants and evolutionary dynamics of orthologous gene groups in 13 placental mammals. *Genome Research, 24*(9), pp. 1485–1496.

Niu, J.-G., Yokota, S., Tsumori, T., Oka, T. and Yasui, Y. (2012) Projections from the anterior basomedial and anterior cortical amygdaloid nuclei to melanin-concentrating hormone-containing neurons in the lateral hypothalamus of the rat. *Brain Research, 1479,* pp. 31–43.

O'Doherty, J., Rolls, E. T., Francis, S., Bowtell, R., McGlone, F., Kobal, G., Renner, B. and Ahne, G. (2000) Sensory-specific satiety-related olfactory activation of the human orbitofrontal cortex. *Neuroreport, 11*(4), pp. 893–897.

Palouzier-Paulignan, B., Lacroix, M.-C., Aime, P., Baly, C., Caillol, M., Congar, P., Julliard, A. K., Tucker, K. and Fadool, D. A. (2012) Olfaction under metabolic influences. *Chemical Senses, 37*(9), pp. 769–797.

Prediger, R. D. S., Aguiar, A. S., Jr., Matheus, F. C., Walz, R., Antoury, L., Raisman-Vozari, R. and Doty, R. L. (2012) Intranasal administration of neurotoxicants in animals: support for the olfactory vector hypothesis of Parkinson's disease. *Neurotoxicity Research, 21*(1), pp. 90–116.

Price, J. L., Slotnick, B. M. and Revial, M. F. (1991) Olfactory projections to the hypothalamus. *Journal of Comparative Neurology, 306*(3), pp. 447–461.

Protopapas, A. D. and Bower, J. M. (2000) Physiological characterization of layer III non-pyramidal neurons in piriform (olfactory) cortex of rat. *Brain Research, 865*(1), pp. 1–11.

Raineki, C., Holman, P. J., Debiec, J., Bugg, M., Beasley, A. and Sullivan, R. M. (2010) Functional emergence of the hippocampus in context fear learning in infant rats. *Hippocampus, 20*(9), pp. 1037–1046.

Rawson, N. E., Gomez, G., Cowart, B. J., Kriete, A., Pribitkin, E. and Restrepo, D. (2012) Age-associated loss of selectivity in human olfactory sensory neurons. *Neurobiology of Aging, 33*(9), pp. 1913–1919.

Ressler, K. J., Sullivan, S. L. and Buck, L. B. (1993) A zonal organization of odorant receptor gene expression in the olfactory epithelium. *Cell, 73*(3), pp. 597–609.

Ressler, K. J., Sullivan, S. L. and Buck, L. B. (1994) Information coding in the olfactory system - Evidence for a stereotyped and highly organized epitope map in the olfactory bulb. *Cell, 79*(7), pp. 1245–1255.

Rimbault, M., Robin, S., Vaysse, A. and Galibert, F. (2009) RNA profiles of rat olfactory epithelia: individual and age related variations. *BMC Genomics, 10.*

Riviere, S., Challet, L., Fluegge, D., Spehr, M. and Rodriguez, I. (2009) Formyl peptide receptor-like proteins are a novel family of vomeronasal chemosensors. *Nature, 459*(7246), pp. 574–7.

Rochefort, C., Gheusi, G., Vincent, J. D. and Lledo, P. M. (2002) Enriched odor exposure increases the number of newborn neurons in the adult olfactory bulb and improves odor memory. *Journal of Neuroscience, 22*(7), pp. 2679–2689.

Rodriguez, I., Greer, C. A., Mok, M. Y. and Mombaerts, P. (2000) A putative pheromone receptor gene expressed in human olfactory mucosa. *Nature Genetics, 26*(1), pp. 18–9.

Rodriguez, I. and Mombaerts, P. (2002) Novel human vomeronasal receptor-like genes reveal species-specific families. *Current Biology, 12*(12), pp. R409–R411.

Rolls, E. T. (2012) Taste, olfactory and food texture reward processing in the brain and the control of appetite. *Proceedings of the Nutrition Society, 71*(4), pp. 488–501.

Rosli, Y., Breckenridge, L. J. and Smith, R. A. (1999) An ultrastructural study of age-related changes in mouse olfactory epithelium. *Journal of Electron Microscopy, 48*(1), pp. 77–84.

Royet, J. P., Souchier, C., Jourdan, F. and Ploye, H. (1988) Morphometric study of the glomerular population in the mouse olfactory bulb: numerical density and size distribution along the rostrocaudal axis. *The Journal of Comparative Neurology, 270*(4), pp. 559–568.

Saito, H., Mimmack, M., Kishimoto, J., Keverne, E. B. and Emson, P. C. (1998) Expression of olfactory receptors, G-proteins and AxCAMs during the development and maturation of olfactory sensory neurons in the mouse. *Developmental Brain Research, 110*(1), pp. 69–81.

Sakano, H. (2010) Neural map formation in the mouse olfactory system. *Neuron, 67*(4), pp. 530–542.

Savigner, A., Duchamp-Viret, P., Grosmaitre, X., Chaput, M., Garcia, S., Ma, M. and Palouzier-Paulignan, B. (2009) Modulation of spontaneous and odorant-evoked activity of rat olfactory sensory neurons by two anorectic peptides, insulin and leptin. *Journal of Neurophysiology, 101*(6), pp. 2898–906.

Schaal, B., Marlier, L. and Soussignan, R. (2000) Human foetuses learn odours from their pregnant mother's diet. *Chemical Senses, 25*(6), pp. 729–737.

Schoenbaum, G., Setlow, B., Nugent, S. L., Saddoris, M. P. and Gallagher, M. (2003) Lesions of orbitofrontal cortex and basolateral amygdala complex disrupt acquisition of odor-guided discriminations and reversals. *Learning and Memory, 10*(2), pp. 129–140.

Schwarzenbacher, K., Fleischer, J. and Breer, H. (2005) Formation and maturation of olfactory cilia monitored by odorant receptor-specific antibodies. *Histochemistry and Cell Biology, 123*(4-5), pp. 419–428.

Serizawa, S. U., Miyamichi, K. and Sakano, H. (2004) One neuron-one receptor rule in the mouse olfactory system. *Trends in Genetics, 20*(12), pp. 648–653.

Sevelinges, Y., Desgranges, B. and Ferreira, G. (2009) The basolateral amygdala is necessary for the encoding and the expression of odor memory. *Learning and Memory, 16*(4), pp. 235–242.

Shi, P. and Zhang, J. (2007) Comparative genomic analysis identifies an evolutionary shift of vomeronasal receptor gene repertoires in the vertebrate transition from water to land. *Genome Research, 17*(2), pp. 166–74.

Shionoya, K. and Datiche, F. (2009) Inactivation of the basolateral amygdala impairs the retrieval of recent and remote taste-potentiated odor aversion memory. *Neurobiology of Learning and Memory, 92*(4), pp. 590–596.

Singer, M. S. (2000) Analysis of the molecular basis for octanal interactions in the expressed rat I7 olfactory receptor. *Chemical Senses, 25*(2), pp. 155–165.

Smithson, K. G., Weiss, M. L. and Hatton, G. I. (1989) Supraoptic nucleus afferents from the main olfactory bulb .1. Anatomical evidence from anterograde and retrograde tracers in rat. *Neuroscience, 31*(2), pp. 277–287.

Sosulski, D. L., Bloom, M. L., Cutforth, T., Axel, R. and Datta, S. R. (2011) Distinct representations of olfactory information in different cortical centres. *Nature, 472*(7342), pp. 213–216.

Stettler, D. D. and Axel, R. (2009) Representations of odor in the piriform cortex. *Neuron, 63*(6), pp. 854–864.

Stripling, J. S. and Galupo, M. P. (2008) Differential potentiation of early and late components evoked in olfactory cortex by stimulation of cortical association fibers. *Brain Research, 1246*, pp. 70–79.

Strotmann, J. and Breer, H. (2011) Internalization of odorant-binding proteins into the mouse olfactory epithelium. *Histochemistry and Cell Biology, 136*(3), pp. 357–369.

Suzuki, N. and Bekkers, J. M. (2006) Neural coding by two classes of principal cells in the mouse piriform cortex. *Journal of Neuroscience, 26*(46), pp. 11938–11947.

Suzuki, N. and Bekkers, J. M. (2011) Two layers of synaptic processing by principal neurons in piriform cortex. *Journal of Neuroscience, 31*(6), pp. 2156–2166.

Suzuki, N. and Bekkers, J. M. (2012) Microcircuits mediating feedforward and feedback synaptic inhibition in the piriform cortex. *Journal of Neuroscience, 32*(3), pp. 919–931.

Takeuchi, H., Inokuchi, K., Aoki, M., Suto, F., Tsuboi, A., Matsuda, I., Suzuki, M., Aiba, A., Serizawa, S., Yoshihara, Y., Fujisawa, H. and Sakano, H. (2010) Sequential Arrival and Graded Secretion of Sema3F by Olfactory Neuron Axons Specify Map Topography at the Bulb. *Cell, 141*(6), pp. 1056–1067.

Thiebaud, N., Johnson, M. C., Butler, J. L., Bell, G. A., Ferguson, K. L., Fadool, A. R., Fadool, J. C., Gale, A. M., Gale, D. S. and Fadool, D. A. (2014) Hyperlipidemic diet causes loss of olfactory sensory neurons, reduces olfactory discrimination, and disrupts odor-reversal learning. *Journal of Neuroscience, 34*(20), pp. 6970–6984.

Trotier, D. (2011) Vomeronasal organ and human pheromones. *European Annals of Otorhinolaryngology, Head and Neck Diseases, 128*(4), pp. 184–190.

Trotier, D. and Doving, K. B. (1998) Anatomical description of a new organ in the nose of domesticated animals by Ludvig Jacobson (1813). *Chemical Senses, 23*(6), pp. 743–54.

Ubeda-Banon, I., Novejarque, A., Mohedano-Moriano, A., Pro-Sistiaga, P., de la Rosa-Prieto, C., Insausti, R., Martinez-Garcia, F., Lanuza, E. and Martinez-Marcos, A. (2007) Projections from the posterolateral olfactory amygdala to the ventral striatum: neural basis for reinforcing properties of chemical stimuli. *BMC Neuroscience, 8*.

Uchida, N., Poo, C. and Haddad, R. (2014) Coding and transformations in the olfactory system. *Annual Review of Neuroscience, 37*, pp. 363–385.

Vassar, R., Chao, S. K., Sitcheran, R., Nunez, J. M., Vosshall, L. B. and Axel, R. (1994) Topographic organization of sensory projections to the olfactory bulb. *Cell, 79*(6), pp. 981–991.

Vassar, R., Ngai, J. and Axel, R. (1993) Spatial segregation of odorant receptor expression in the mammalian olfactory epithelium. *Cell, 74*(2), pp. 309–318.

Verbeurgt, C., Wilkin, F., Tarabichi, M., Gregoire, F., Dumont, J. E. and Chatelain, P. (2014) Profiling of olfactory receptor gene expression in whole human olfactory mucosa. *Plos One, 9*(5).

Vidic, J., Grosclaude, J., Monnerie, R., Persuy, M. A., Badonnel, K., Baly, C., Caillol, M., Briand, L., Salesse, R. and Pajot-Augy, E. (2008) On a chip demonstration of a functional role for Odorant Binding Protein in the preservation of olfactory receptor activity at high odorant concentration. *Lab on a Chip, 8*(5), pp. 678–688.

Wang, H. W., Wysocki, C. J. and Gold, G. H. (1993) Induction of olfactory receptor sensitivity in mice. *Science, 260*(5110), pp. 998–1000.

Wang, Y., Fontanini, A. and Katz, D. B. (2006) Temporary basolateral amygdala lesions disrupt acquisition of socially transmitted food preferences in rats. *Learning and Memory, 13*(6), pp. 794–800.

Watt, W. C., Sakano, H., Lee, Z. Y., Reusch, J. E., Trinh, K. and Storm, D. R. (2004) Odorant stimulation enhances survival of olfactory sensory neurons via MAPK and CREB. *Neuron, 41*(6), pp. 955–967.

Wesson, D. W. and Wilson, D. A. (2011) Sniffing out the contributions of the olfactory tubercle to the sense of smell: Hedonics, sensory integration, and more? *Neuroscience and Biobehavioral Reviews, 35*(3), pp. 655–668.

Wilson, D. A. (2000) Odor specificity of habituation in the rat anterior piriform cortex. *Journal of Neurophysiology, 83*(1), pp. 139–145.

Wilson, D. A. and Sullivan, R. M. (2011) Cortical processing of odor objects. *Neuron, 72*(4), pp. 506–519.

Wirth, S., Ferry, B. and Di Scala, G. (1998) Facilitation of olfactory recognition by lateral entorhinal cortex lesion in rats. *Behavioural Brain Research, 91*(1-2), pp. 49–59.

Yan, Z., Tan, J., Qin, C., Lu, Y., Ding, C. and Luo, M. (2008) Precise circuitry links bilaterally symmetric olfactory maps. *Neuron, 58*(4), pp. 613–624.

Yu, Y., Boyer, N. P. and Zhang, C. (2014) Three structurally similar odorants trigger distinct signaling pathways in a mouse olfactory neuron. *Neuroscience, 275*, pp. 194–210.

Zatorre, R. J., Jonesgotman, M., Evans, A. C. and Meyer, E. (1992) Functional localization and lateralization of human olfactory cortex. *Nature, 360*(6402), pp. 339–340.

Zhao, H. Q. and Reed, R. R. (2001) X inactivation of the OCNC1 channel gene reveals a role for activity-dependent competition in the olfactory system. *Cell, 104*(5), pp. 651–660.

Zou, D.-J., Chesler, A. T., Le Pichon, C. E., Kuznetsov, A., Pei, X., Hwang, E. L. and Firestein, S. (2007) Absence of adenylyl cyclase 3 perturbs peripheral olfactory projections in mice. *Journal of Neuroscience, 27*(25), pp. 6675–6683.

Zou, D. J., Feinstein, P., Rivers, A. L., Mathews, G. A., Kim, A., Greer, C. A., Mombaerts, P. and Firestein, S. (2004) Postnatal refinement of peripheral olfactory projections. *Science, 304*(5679), pp. 1976–1979.

CHAPTER 2

Odorant metabolizing enzymes in the peripheral olfactory process

Jean-Marie Heydel, Hassan-Ismail Hanser, Philippe Faure & Fabrice Neiers
Center for Taste and Feeding Behaviour, CNRS, INRA, University of Bourgogne Franche-Comté Dijon, France

2.1 Introduction

It is generally accepted that the most peripheral step in the olfactory process is the binding of an odorant molecule to olfactory receptors associated with the olfactory sensory neurons of the olfactory epithelium (OE) (Buck and Axel 1991). Thereafter, the process includes the transduction of the signal and its integration first in the olfactory bulb and then in the primary olfactory cortex (piriform cortex, anterior olfactory nucleus, etc.) (Buck 2005, Haberly 2001). However, we know now that many other events which could impact the olfactory process occur in the OE. Starting from the observation that the peripheral modulation of smell has been widely underestimated, a recent review of the literature suggested numerous peripheral modulators (Lucero 2013). Those modulators include the extrinsic innervation of the OE, airflow modulation of odorant delivery, cellular signaling (auto-, para-, endocrine) and perireceptor events. Although peripheral modulation on the olfactory process most likely results from a combination of all the different mechanisms, it has often been observed that each of the mechanisms alone can significantly affect the odorant response.

In the whole organism, xenobiotic metabolizing enzymes (XMEs) are primarily involved in the detoxification of foreign chemicals (xenobiotics). Their activity also regulates and/or terminates the biological effects of endobiotics, including hormones or other signaling molecules. Thus, it is logical to hypothesize that similar enzymes in the olfactory tissue, which is constantly exposed to chemicals coming from the environment, could be involved both in the detoxification and modulation of odorant bioavailability and consequently in the olfactory process. Odorant metabolizing enzymes (OMEs) may modulate the bioavailability of odorants by catalyzing their biotransformation, thus allowing their elimination. In this manuscript, we focus on odorant metabolizing enzymes (OMEs), corresponding to the XMEs present in the nasal tissues having odorants as substrates and participating in the perireceptor olfactory process. However,

Flavour: From food to perception, First Edition.
Edited by Elisabeth Guichard, Christian Salles, Martine Morzel, and Anne-Marie Le Bon.

other proteins present in the receptor environment, such as odorant-binding proteins (OBPs), are involved in odorant bioavailability modulation (Heydel et al. 2013). Although the odorant binding properties of OBPs in mucus are now quite well understood, additional potential roles have been proposed regarding odorant delivery to olfactory receptors, odorant cellular internalization and transduction. Only a few OBPs are present in the vertebrate olfactory system, which would normally make them easier to study than the numerous OMEs. However, knowledge regarding OMEs has recently progressed more rapidly.

2.2 Odorant metabolizing enzymes

2.2.1 Xenobiotic metabolism

In the organism, an essential process is the detoxification of foreign chemicals that are potentially toxic due to their bioactivity and their lipophilicity. This process involves a network of proteins and enzymes acting together to biotransform xenobiotics, making them inactive, hydrophilic and easily eliminated from the cells and finally from the body. A diagram of this metabolic process in the context of odorants is presented in Figure 2.1. Classically, the xenobiotic metabolism process is described in three phases. Phase I corresponds to the functionalization step of a lipophilic xenobiotic. For example, cytochrome P450 enzymes (CYP) introduce hydroxyl functional groups (Anzenbacher and Anzenbacherova 2001), but other forms of functionalization may consist in oxidation, reduction or hydrolysis (Testa and Kramer 2007). In phase II, the conjugation step, the phase I metabolite is conjugated to hydrophilic groups by transferase enzymes, such as sulfotransferases (ST), UDP-glucuronosyltransferases (UGT), and glutathione-S-transferases (GST) as well as methyltransferases and acetyltransferases (Green et al. 2005, Heydel et al. 2010). During phase III, the resulting hydrophilic phase II metabolite is then excreted from the cell *via* efflux transporters, such as P-glycoprotein also termed multidrug resistance protein (MDR) and multidrug resistance-associated protein (MRP) (Jones and George 2004, Thiebaud et al. 2011). In some instances, phase I (mainly involving CYP) or phase II biotransformation can lead to the synthesis of bioactive metabolites presenting high toxicity (Rooney et al. 2004).

Xenobiotics can reach the cells of an organism *via* many routes, including inhaled air, food, skin, etc. Therefore, xenobiotic metabolizing enzymes can be found almost everywhere through the body.

2.2.2 XMEs participate in odorant biosynthesis

In this section, we propose that XMEs may contribute to the biosynthesis of odorants. All organisms from prokaryotes to plants and mammals emit volatile compounds, some of which are key components of their flavour and some

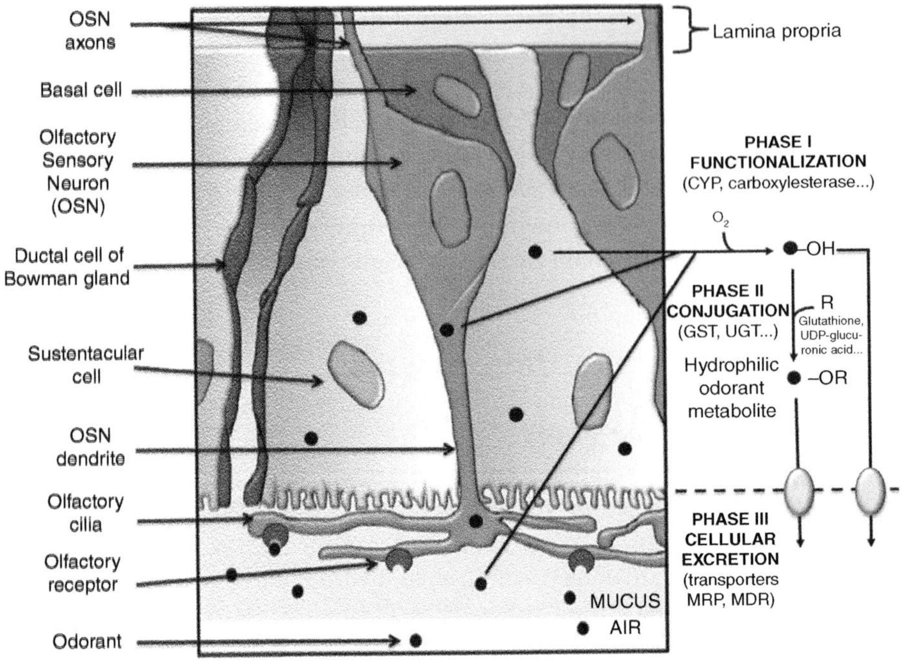

Figure 2.1 Schematic representation of the OE structure. The right part of the figure presents the different phases of the metabolism towards an odorant xenobiotic. Phase III presents the hypothetical excretion of phase I and phase II metabolites in the olfactory mucus. CYP, cytochrome P450; GST, glutathione-S-transferase, UGT, UDP-glucuronosyltransferase; MRP, multidrug resistance-associated protein; MDR, multidrug resistance protein.

of which have been demonstrated to act as chemical cues for intra- and/or inter-species communication signals.

In many plants, the oxygenation of fatty acids in the lipoxygenase pathway leads to the formation of volatile aldehydes through the activity of hydroperoxide lyases (HPL), members of the CYP family (for a review see Blee 1998). C6-aldehydes such as hexanal and 3-hexenal and C9-aldehydes, such as (2E,6Z)-nonadienol or 2E-nonenal, are HPL-derived products that are components of the aroma of several plant species and are implicated in plant defenses against insects such as aphids in tobacco and potato (Hildebrand et al. 1993, Vancanneyt et al. 2001) and beetles in wheat grain (Hubert et al. 2008). Some of these aldehydes can then be reduced to the corresponding alcohols by alcohol dehydrogenases (ADH), another class of phase I XMEs. The volatile alcohols are often implicated in fruit flavour during ripening, as exemplified by hexanol and Z-3-hexenol in tomatoes (Speirs et al. 1998).

A different branch of the lipoxygenase pathway in plants generates volatile compounds derived from jasmonic acids through the activity of another CYP enzyme family: allene oxide synthases (AOS, for review see Blee 1998). Methyl jasmonate, an AOS derivative and a component of the jasmine and black tea scent, is implicated in interplant communication signals (Farmer and Ryan 1990, Yao et al. 2012).

Humans also emit volatiles primarily originating in the skin. Skin odour can be a source of information and can be associated with age, sex, diet, anthropologic group and disease (for reviews see Kippenberger et al. 2012, Shirasu and Touhara 2011). Volatiles emitted from the body are largely the result of the bacterial flora present on the skin. Some of these bacteria use XMEs to synthesize volatiles from odourless precursor molecules produced primarily by the apocrine and eccrine sweat glands and sebaceous glands. For example, odorous steroids such as 5α-androstenol and 5α-androstenone are products of hydroxysteroid dehydrogenases that are phase I enzymes (Austin and Ellis 2003). Different chemical classes of volatiles are emitted by bacteria (Audrain et al. 2015, Schulz and Dickschat 2007). Volatiles result of the catabolism of nutrients present in their environment. Thus, for instance, some bacteria from buccal and gut flora contribute to metabolize alcohol and produce volatiles such as acetaldehyde and acetate, through alcohol dehydrogenase (Vakevainen et al. 2001) and aldehyde dehydrogenase (Nosova et al. 1996) phase I enzymes' activity, respectively.

Many insects and mammals have been demonstrated to emit various semiochemical volatiles perceived by conspecifics (pheromones) or other animals, as indicated by the more than 7500 species and 6500 semiochemical compounds recorded in the Pherobase, a database of pheromones and semiochemicals (El-Sayed 2014). However, only a few studies on the biosynthetic pathway of semiochemicals in insects are found in literature. Among these studies, some involve XME and particularly CYP. This pathway is the source of a poorly volatile hydrocarbon, (Z)-9-tricosene, in the housefly *Musca domestica* (Ahmad et al. 1987) and some volatile spiroacetals in certain *Bactrocera sp* flies (Booth et al. 2009). Drug metabolism may also be a source of volatiles. For example, the administration of cyclophosphamide, an antineoplastic pro-drug, generates volatile metabolites (e.g., N-2-chloroethylaziridine, found in urine) after activation by CYP enzymes (Lu and Chan 2006). Another example is isoflurane, a drug used as a volatile anesthetic that leads to the formation of a volatile metabolite (trifluoroacetate) excreted in urine as the result of CYP metabolism (Arici et al. 2013). Besides, volatile organic compounds are markers of lung cancer, present in human breath and coming from lipid peroxidation of polyunsaturated fatty acids of cell membrane. It has been shown that the alkanes and methylalkanes generated by this oxidative stress mechanism are metabolized by CYP in alkyl alcohols also detected in human breath (Phillips et al. 2010). The corresponding CYP are induced in breast cancer.

Mammals such as the mouse *Mus musculus* (Zhang et al. 2007), deer *Odocoileus virginianus* (Gassett et al. 1996, Jemiolo et al. 1995), bobcat *Lynx rufus* (Mattina et al. 1991), and wolf *Canis lupus* (Raymer et al. 1984) release a large profile of volatiles of various chemical forms including hydrocarbons, terpenes, esters, aromatics, aldehydes, ketones, alcohols, acids, amines and thiols in urine and through secretory glands. Some of these chemicals are well-characterized in terms of behavioural responses, whereas little is yet known about their biosynthetic pathways. However, regarding the catalytic properties of XMEs and the broad chemicals families of volatiles emitted, it is reasonable to suggest that XMEs participate in the synthetic pathway of these volatiles.

2.2.3 Anatomy of the olfactory epithelium and expression of OMEs

Figure 2.1 presents a schematic representation of the OE, showing the different extracellular and cellular locations where OMEs are expressed.

Nasal mucus is an aqueous interface between the odorants entering the nasal cavity in the air and the olfactory receptors located in the cilia of the olfactory neurons. Human mucus has indispensable functions, including cleaning and humidifying inhaled air and protecting the epithelium. Mucus is mostly composed of water (95%), mucopolysaccharides (2%), salts and proteins. The proteins in human mucus are very diverse and include protease inhibitors and antibodies as well as enzymes. Among the different proteins detected in the human mucus, some, for example lysozyme, are responsible for antibacterial protection, while others are involved in protein folding as chaperone proteins (Hsp70) or in resistance to oxidative stress, such as peroxiredoxin. Carboxylesterase activity detected in rat olfactory mucus suggests that some OMEs are secreted (Bogdanffy 1990). This type of activity has been confirmed in mouse mucus; a recent study showed that an ester-containing odorant can be hydrolyzed *in vivo* into the corresponding alcohol and acid, which are found in mucus (Nagashima and Touhara 2010). According to a proteomic study performed in human mucus, an aldehyde dehydrogenase and two GSTs are found in this fluid (Debat et al. 2007). The potential for some OMEs to be secreted into the mucus is discussed in paragraph 4.2.

In the OE, the basal cells differentiate in sustentacular cells and olfactory sensory neurons (OSNs). Sustentacular cells, also called supporting cells, contribute to the structure of the epithelium and participate in the overall homeostasis of the OE. OME expression was localized to the apical part, as shown by immunochemistry and *in situ* hybridization methods, as well as to the acinar and ductal cells of the Bowman glands, which are responsible for mucus production (Bogdanffy 1990, Getchell et al. 1993, Zupko et al. 1991). A few studies have demonstrated the expression of OMEs in cell bodies of OSNs (Heydel et al. 2001, Lazard et al. 1991) and in their cilia using a proteomic approach (Mayer et al. 2009, Mayer et al. 2008). In the OE, immunolocalization demonstrated a distinct localization for phase III transporters. MRP1 was found in the supranuclear portions of sustentacular cells, while MRP3 and MRP5 were found in the OSN. MDR1 was localized to the capillary endothelium of the lymphatic vessels in the subepithelial region (Thiebaud et al. 2011).

Microvillar cells (not represented in Figure 2.1) are now well-identified structures of the OE. These cells demonstrate non-neuronal properties, but their function remains unknown (Asan and Drenckhahn 2005, Kwon et al. 2005). To our knowledge, no OME activity was positively identified in these cells.

2.3 Functional role of OMEs in the olfactory process

This section focuses on the mechanism of odorant enzymatic biotransformation associated with perireceptor events. This mechanism has been well studied in insects (Chertemps et al. 2012, Durand et al. 2012, Maibeche-Coisne et al. 2004) by reason of the great importance of olfaction in the survival of these species, but remains poorly understood in vertebrates. Even if there are similarities between olfactory cells and enzyme structures between these species, the structure of the tissue is very different and does not allow unambiguous extrapolation. We will consider here only the knowledge acquired in vertebrates.

For each of the studies discussed below, summarized information is shown in Table 2.1.

2.3.1 The initial hypothesis for olfactory signal termination: the UDP-glucuronosyltransferases

When UGT2A1, a UDP-glucuronosyltransferase cloned from rat OE, was found to be specifically localized in this tissue, its role in olfactory signal termination was hypothesized (Lazard et al. 1991). This phase II enzyme may catalyze the conjugation of UDP-glucuronic acid to odorants, inactivating them and preventing further binding to the olfactory receptors.

2.3.1.1 UGT2A1 preferential localization in OE

UGT2A1 was localized to the apical cytoplasmic region of the sustentacular cells and to the Bowman's glands (Lazard et al. 1991). Subsequent studies showed that UGT2A1 expression was also detectable in OSNs, including olfactory cilia (Heydel et al. 2001, Mayer et al. 2009, Mayer et al. 2008), and also in the rat olfactory bulb (Heydel et al. 2001) and brain (Heydel et al. 2010). It is finally ubiquitously expressed through the human respiratory and aerodigestive tracts (Bushey et al. 2011, Thum et al. 2006). Its expression and regulation was also demonstrated in rat liver (Thiebaud et al. 2010). Nevertheless, UGT2A1 is preferentially expressed in the OE, where it appears to be weakly inducible by classical UGT inducers (Thiebaud et al. 2010).

2.3.1.2 A dominant UGT in OE

Lazard et al. (1991) heterologously expressed UGT2A1 in COS-7 cells to measure its conjugation activity toward different odorant molecules (Table 2.1).

Table 2.1 Summary of the different studies investigating OME function in the olfactory process.

Odorants	Animal models	Odorant metabolizing proteins	Localization	Methods	References and main conclusions
α-naphthol 4-methyl umbelliferone 4-nitrophenol 4-hydroxybiphenyl Eugenol Guaiacol 2-ethyl hexanol Borneol	Rat	UDP-glucuronosyltransferase (UGT2A1)	Olfactory epithelium Sustentacular cells and Bowman glands Liver	– Gene cDNA cloning – Northern Blot – Immunohistology – Olfactory transduction test: activation of adenylyl cyclase – Heterologous expression of UGT2A1 in COS-7 cells: glucuronidation activity on thin layer chromatography with ^{14}C-labelled UDP-glucuronic acid	Lazard et al. (1991) UGT2A1 is predominant in OE and may be involved in signal termination
24 chemically unrelated odorants	Rat	UDP-glucuronosyltransferase (UGT2A1 and UGT1A6)	Olfactory epithelium	– Age-related changes – RT-PCR – Glucuronidation activity – Recording of multiunit activity of mitral cells in OB	Leclerc et al. (2002) UGT2A1 is a major odorant degrading enzymes and may participate in signal termination
2-methoxyaceto-phenone "Ketone 55"	Human	Cytochrome P450 (CYP2A13)	Olfactory epithelium	– Detection of metabolites in exhaled air: APCI-MS – Gas Chromatogra-phy – Olfactometry (GCO) – Enzyme inhibitor: (E)-3-(cyclopropylm-ethylene) octan-2-one – Sensory evaluation	Schilling et al. (2010) CYP enzymes participate in odour quality evaluation in human

Benzaldehyde Acetyl Isoeugenol	Mouse	Aldehyde oxidase Aldehyde dehydrogenase Aldehyde reductase Carboxylesterase	Olfactory mucus	– Nasal mucus preparation – GC/MS analysis of odorant metabolites – Ca^{2+} imaging in OB – Enzyme inhibitors: bis (p-nitrophenyl) phosphate and acetazolamide for acetate odorants / SKF525A, benzamidine, 3,5-dichlorosalicylic acid, cyanamide, citral for aldehyde odorants – Olfactory discrimination test	Nagashima and Touhara. (2010) Carboxylesterase activity degrades odorants in olfactory mucus. Disturbance of this activity changes glomerular activation and olfactory perception. Odorant metabolites may be involved in olfactory coding
Coumarin 3-hydroxycoumarin 4-hydroxycoumarin 6-hydroxycoumarin 7-hydroxycoumarin 4-methycoumarin 4-methylumbeliferone Quinoline Quinoline-1-oxide 3-hydroxyquinoline 8-hydroxyquinoline	Rat	Cytochrome P450 UDP-glucuronosyltransferase Carboxylesterase	Olfactory epithelium Liver	– Metabolism activity measurement, LC-MS – Enzyme inhibitors: 1aminobenz-o-triazole (ABT) for CYP activity, bis-p-nitro-phenyl-phosphate (BNPP) for carboxylesterase activity – Submerged Electroolfactogram recording	Thiebaud et al. (2013) XMEs significantly participate in (i) the sensitivity and the quality of the EOG response (ii) maintaining olfactory sensitivity by avoiding saturation of the peripheral olfactory system

Similar levels of activity towards the same odorants were observed in OE microsomes, suggesting that UGT2A1 accounts for the glucuronidation of a large panel of odorants. Accordingly, UGT2A1 mRNA expression was compared with glucuronidation activity toward 24 odorant molecules in the rat OE during development (1 day, 1 week, 2 weeks, 3 months, 12 months, 24 months). It was observed that the regulation profile matched perfectly for most of the molecules, suggesting that UGT2A1 is the main UGT involved in odorant glucuronidation. These results emphasize the dominant role of UGT2A1 in OE xenobiotic glucuronidation and suggest that it is likely to be a critical enzyme in odorant metabolism.

2.3.1.3 Higher efficiency of odorant glucuronidation in OE

In addition, Lazard's study showed that odorants are more efficiently glucuronidated in the OE than in the liver. This preference of OE enzymes for odorant substrates was previously observed (Ben-Arie et al. 1993, Nef et al. 1989) and later confirmed in comparison with olfactory bulb or brain (Leclerc et al. 2002).

2.3.1.4 Functional termination of olfactory signal

The biotransformation of an odorant by glucuronidation leads to a completely different, highly polar molecule likely lacking odorant properties. Lazard et al. (1991) tested this hypothesis using an in vitro neuro-olfactory cilia system and showed that unlike the parent odorants, the glucuronidated odorants did not elicit the production of cyclic AMP by the signal transduction pathway. Accordingly, we have shown that when presented in vapour phase to rats, an odorant known to be efficiently glucuronidated in the OE elicited significantly lower electrophysiological responses recorded from olfactory bulb mitral cells than an odorant with a lower glucuronidation rate, which provoked higher olfactory responses (Leclerc et al. 2002). Finally, we recently showed that the glucurono-conjugated derivatives of the odorants coumarin and quinoline were unable to elicit electro-olfactogram responses in rats (Thiebaud et al. 2013). Collectively, these results demonstrate that glucuronidation is a mechanism of olfactory signal termination.

2.3.2 Impact of OMEs on peripheral olfactory responses

One recent study addressed several questions regarding the role of XMEs in olfaction, including the ability of metabolites to trigger olfactory signals (Thiebaud et al. 2013). The results and conclusions mainly rely on electro-olfactogram (EOG) recordings performed using rat OE because the objective was to assess the impact of odorant metabolism, in particular the activity of phase I enzymes, on olfactory reception. This method measures the summated odorant-induced generator potentials of the OSNs on the OE surface (Scott and Scott-Johnson

2002). We performed submerged EOG because it allowed us to quantify responses elicited by non-volatile and low-volatility compounds and to deliver XME inhibitors.

2.3.2.1 High *in vitro* metabolizing activity towards odorant molecules in OE

Cytochrome P450-dependent metabolism

In vitro experiments were performed using the microsomal fraction prepared from OE. We demonstrated the ability of rat OE to metabolize the odorants coumarin and quinoline by identifying their oxygenated metabolites using LC-mass spectrometry. The rate of biotransformation of both coumarin and quinoline was much higher in the OE than in the liver, confirming the high efficiency of the OE in odorant metabolism. Moreover, CYP involvement in this metabolism was established using 1-aminobenzotriazole (ABT), a general CYP inhibitor. The formation of coumarin and quinoline metabolites was efficiently (80–90%) inhibited under our experimental conditions. According to previous studies and by extrapolation from experiments performed in liver, CYP2A3 isoform may be a good candidate for both coumarin and quinoline metabolism under our experimental conditions (Hirano and Mizutani 2003, von Weymarn et al. 1999, Zhuo et al. 1999).

Carboxylesterase-dependent metabolism

The phase I enzymes, carboxylesterases, hydrolyze volatile esters to their corresponding acids (Dahl et al. 1987). Analysis with the odorant isoamyl acetate showed acetic acid release in the reaction medium following incubation of the odorant with olfactory or hepatic microsomal fractions. Interestingly, in contrast to the results for coumarin and quinoline, the esterase activity for this odorant was equivalent in both tissues. Isoamyl acetate hydrolysis was inhibited by bis-p-nitro-phenylphosphate (BNPP), a specific carboxylesterase inhibitor. In contrast, ABT had no effect on isoamyl acetate hydrolysis, confirming that this reaction was not dependent on CYP enzymes.

2.3.2.2 Odorants and their metabolites elicit different olfactory responses

As described above, in 1991, Lazard et al. had hypothesized the involvement of UGTs in olfactory signal termination, logically thinking that conjugation of odorants with glucuronic acid could hinder interactions with ORs. They indirectly confirmed their hypothesis *in vitro* by showing that the glucuronidated odorants were unable to trigger the production of cAMP, a second messenger involved in signal transduction. Accordingly, glucurono-conjugated metabolites of coumarin and quinoline were unable to elicit EOG responses in our study. However, although the EOG responses of hydroxylated derivatives of coumarin, quinoline and isoamyl acetate were approximately 50% lower than those elicited by the parent compounds, they still induced a significant response. Accordingly, various studies have shown that hydroxylated odorants are efficient OR agonists, and in some cases, the hydroxyl group is even necessary for adequate OR activation (Katada et al. 2005, Saito et al. 2009, Touhara et al. 1999). This observation suggests that depending on the odorant, phase I metabolism could be involved in both signal termination and in olfactory coding *via* the metabolites produced within the receptor environments.

2.3.2.3 Inhibition of odorant-metabolizing enzymes modulates peripheral olfactory responses

Immediately after treatment with the inhibitors of the corresponding enzymes, we observed a significant increase of EOG responses for coumarin, quinoline and isoamyl acetate. This increase may result from an increased number of recruited receptors or from modification of the odorant affinity for ORs. Following repeated stimulation with odorants, we observed a significant decrease in the EOG responses for the three odorants. This effect could be a consequence of neuronal adaptation that occurs when OSNs are continuously exposed to odorants, which prevents the saturation of cellular transduction machinery (Zufall and Leinders-Zufall 2000).

In our opinion, in both cases, treatment with inhibitors likely leads to a prolongation of the presence of the odorants in the perireceptor environment, which initially induces the EOG response and then leads to OSN adaptation because of odorant accumulation. Thus, under physiological conditions, XMEs may significantly participate in maintaining olfactory sensitivity by preventing the saturation of the peripheral olfactory system.

2.3.3 A unique human approach

In humans, a very encouraging study (unfortunately with few details regarding materials and methods) was published quite recently (Schilling et al. 2010). The authors originally selected technical approaches never used before in this field of research. The targeted enzyme of the investigation was the odorant hydroxylation enzyme CYP2A13, which is predominantly expressed in OE and metabolizes numerous odorants (Table 2.1).

The authors ambitiously planned to detect 2-hydroxyacetophenone, the CYP2A13 metabolite of 2-methoxyacetophenone, in exhaled breath immediately after inhalation. Interestingly, atmospheric pressure chemical ionization mass spectrometry (APCI-MS) allowed the detection of this metabolite and demonstrated very rapid metabolite synthesis kinetics. The authors also observed that among the odorant substrates of CYP2A13, a particular ketone and its metabolite exhibited distinct but overlapping odours. They therefore tested whether the biotransformation by CYP2A13 was a component of the ketone perception. The challenge was to synthesize a volatile CYP2A13 inhibitor with low odour properties (Chougnet et al. 2009, Schilling et al. 2010) to combine it with the ketone for sensory evaluation tests. The tests revealed that two-thirds of the panelists reported that the note corresponding to the metabolite was not perceived in the presence of the inhibitor (E)-3-(cyclopropylmethylene), although inter-individual variability was observed. This last point could be explained by different CYP2A13 phenotypes among the panelist, supporting our hypothesis regarding the potential impact of XME polymorphisms in olfactory perception discussed in paragraph 4.1. Phenotypic exploration would have been interesting to correlate with the results of the sensory tests. It was unfortunately not reported whether the metabolite of the ketone was detected in exhaled breath.

Compared with approaches in animals (Nagashima and Touhara 2010, Thiebaud et al. 2013), which focused on OE or the associated mucus, it is not clear whether the observations of this human study were the result of nasal metabolism or involved the whole respiratory tract. However, this novel and convincing approach is very promising and demonstrates that techniques such as APCI-MS and sensory evaluation are useful tools to explore human olfactory perception in this research field. Moreover, the study demonstrated that the efficiency of the OME activity is consistent with the olfactory kinetics.

2.3.4 Olfactory mucus odorant-degrading enzymes: an integrated approach in mouse

An important recent study (Nagashima and Touhara 2010) was based on previous observations showing differences in the specificity between odorant-activated ORs and their corresponding glomeruli (Oka et al. 2006). Because these differences seemed to involve events occurring in the olfactory mucus, the authors hypothesized a role for OMEs. The two odorants benzaldehyde (aldehyde group) and acetyl isoeugenol (acetate group) were chosen for their large differences in glomerular activation and discrepancies between *in vivo* and *in vitro* OR activation. These authors showed efficient enzymatic metabolism of both odorants in the nasal mucus of mice. Moreover, they demonstrated that the enzymatic conversion of the odorants occurred *in vitro* (in collected mucus) as well as *in vivo* following the nasal administration of odorants or exposure to the vapour phase. Nasal administration of inhibitors of enzymes involved in the metabolism of the studied odorants was used. Interestingly, inhibitors of aldehyde oxidase, aldehyde dehydrogenase or aldehyde reductase metabolism failed to efficiently inhibit benzaldehyde conversion. Although benzaldehyde is not primarily metabolized by glutathione transferases, one may speculate that if these phase II enzymes can be secreted in mucus, as discussed in paragraph 2.2 and 4.2, they could be involved in its metabolism. Nevertheless, the authors demonstrated the nasal efficiency of a carboxylesterase inhibitor towards acetyl isoeugenol metabolism. Specifically, using Ca^{2+} imaging of the odorant responses in the olfactory bulb, they showed that the glomerular activation patterns induced by acetyl isoeugenol were modulated after inhibitor treatment. Interestingly, depending on the mouse, more glomeruli were activated by acetyl isoeugenol after inhibitor treatment. This experiment emphasized a fast and functional role of odorant enzymatic metabolism in odorant detection.

Initially, the authors tested the hypothesis regarding olfactory integration using a behavioural test. In the olfactory discrimination test, after training to associate odorants with a sugar reward, the time spent by a mouse to investigate odorants was proportional to the efficiency of the discrimination between the odorants paired to the sugar and those not paired to the sugar. Mice treated with the carboxylesterase inhibitor were able to discriminate between the odorants but spent significantly less time investigating acetyl isoeugenol, suggesting changes in the olfactory note or its intensity. As concluded by the authors, both imaging and behavioural approaches strongly suggested that olfactory perception could rely both on the detection of the odorant and its metabolites.

2.4 OME characteristics potentially influencing their function in olfaction

2.4.1 OME genetic polymorphisms

The ability to detect some odorants as well as the detection threshold varies among individuals due to genetic differences (Ache and Young 2005). Many authors have shown that odour perception is particularly influenced by the genotypes of olfactory receptors (Eriksson et al. 2010, Keydar et al. 2013, Menashe et al. 2007, Pelchat et al. 2011). The best known example is the relationship between the alleles of the olfactory receptor OR7D4 and the perception of androstenone (Keller et al. 2007, Knaapila et al. 2012).

Several factors have been identified as contributing to variability in XME activity. The factors contributing to the inter-individual variations include age, sex, enzyme expression inducers and genetic polymorphisms. Because these modulations of XME activity differ among individuals, for instance depending on chemicals to which they are exposed, drug treatments or genotype, we postulated that the function of OMEs in olfaction may also be differentially affected and could lead to inter-individual differences in olfactory perception. In this section, we review polymorphisms that affect known OMEs.

The mammalian OE is unique among extrahepatic tissues, as it has high levels and tissue-selective forms of CYP. The presence of these enzymes in this chemosensory organ may be important for the maintenance of homeostasis, the metabolic transformation of odorant molecules, and the protection of the brain against inhaled chemicals. However, metabolic activation of foreign chemicals by CYP in this portal-of-entry organ may also lead to toxic consequences in the nasal cavity and neighbouring organs (Baker et al. 2003). CYP2A13 is preferentially expressed in OE. It has been shown that an induction of the *CYP2A13* gene for which a genetic polymorphism is related to the risk of smoking-related lung adenocarcinoma (Wang et al. 2003), may lead to an increased risk of smoking-related nasal and lung tumors. Accordingly, a C→T nucleotide transition in the *CYP2A13* gene causes an Arg257Cys amino acid substitution and thus results in a significantly reduced activity towards the activation of the tobacco carcinogen 4-(methylnitrosamino)-1-(3-pyridyl)-1-butanone and other substrates. The variant *CYP2A13* genotype (*CT* + *TT*) is associated with a substantially reduced risk for lung adenocarcinoma (Wang et al. 2003). CYP2G1 is another abundant, OE-specific CYP enzyme active in the metabolism of sex steroids and xenobiotic substrates in mammals. Two different human CYP2G genes, CYP2GP1 and CYP2GP2, have been characterized, and polymorphisms in these genes are well established. The CYP2GP1 gene contains a single nucleotide deletion in exon 2 (deltaC) and a 2.4-kb deletion between exons 3 and 7 (deltaE4-6), whereas the CYP2GP2 gene contains a nonsense mutation in exon 1 and another in exon 3. These genetic polymorphisms and loss-of-function mutations are widely observed and indicate that a functional CYP2G allele is rare in humans. An analysis of the allelic distribution in different ethnic groups suggested that a functional CYP2G allele, if present, is more likely to be found in Black and Hispanic subjects (Sheng et al. 2000).

UGTs have been well studied in pharmacogenetics and clinical research because they have central roles in the metabolism and detoxification of foreign chemicals including hydrophobic drugs, chemotherapeutic agents and carcinogens. UGT2A1 was originally characterized from rat and human olfactory tissue and was shown to exhibit activity against a range of steroids, drugs and phenol odorants (Jedlitschky et al. 1999, Lazard et al. 1991). In a recent publication, reverse-phase HPLC was used to show that UGT2A1 was involved in the glucuronidation of a variety of polycyclic aromatic hydrocarbons associated with tobacco-induced cancers. The study also assessed the potential functional role of single nucleotide missense polymorphisms on UGT2A1 enzyme activity (Bushey et al. 2011). Cell homogenates prepared from wild-type UGT2A1 75Lys308Gly-over-expressing HEK293 cells showed significant glucuronidation activity toward polycyclic aromatic hydrocarbons. Conversely, a significant decrease in glucuronidation activity was observed for the UGT2A1 75Arg308Gly variant. These results demonstrate that UGT2A1 is an important detoxification enzyme in the metabolism of polycyclic aromatic hydrocarbons, including tobacco carcinogens, within the target tissues and that functional polymorphisms in UGT2A1 may play a role in tobacco-related cancer risk. A polymorphism in UGT2A1 was also reported in a Japanese population (Iida et al. 2002). A high-resolution single nucleotide polymorphism (SNP) map of three UGT gene loci, including UGT2A1, was provided, and a total of 86 SNPs and 14 variations of other types among 96 chromosomes were detected. This collection of UGT2A1 SNPs should serve as an additional resource for studies of complex genetic diseases or odour-perception variations.

In conclusion, despite the high rates of polymorphisms affecting XME, only a limited number of studies have investigated OME polymorphisms. This is certainly a field of research to better characterize olfactory dysfunctions either caused by nasal toxicity or defective OME activity.

2.4.2 OME subcellular localization

We described in paragraphs 2.2 and 3.4 that phase I enzymes, including carboxylesterases, are present and functional in the olfactory mucus (Nagashima and Touhara 2010). Moreover, other enzymes, including GST, are found in the olfactory mucus (Debat et al. 2007). This localization is coherent because certain classes of these enzymes are soluble and because high levels of glutathione were measured in the olfactory mucus (Krishna et al. 1992). However, the secretion of membrane-bound OMEs in the olfactory mucus is more questionable. CYP and other UGTs are indeed known to be localized to the membrane of the endoplasmic reticulum. Interestingly, in rats, the secretion of CYP4B1 was strongly suggested after immunoreactivity for this enzyme was observed in the lumen of the ducts of Bowman's glands (Genter et al. 2006).

To further evaluate this issue, we used the bioinformatics tool SignalP to analyze the localization of all OMEs detected in humans as well as all the human homologues of the rat and mouse OMEs. The results are compiled in Table 2.2. The prediction of the localization of human CYP4B1 presented in Table 2.2 supports the possibility of secretion. This observation is not consistent with a previous *in silico* analysis performed by Genter et al. (2006), which paradoxically did not predict CYP4B1 as secreted. Surprisingly, all of the GSTs are predicted to have an intracellular localization. Once more, this result can be interpreted as contradictory with the previous hypothesis that GSTs are secreted (Debat et al. 2007). Thus, the predictions shown in Table 2.2 must be interpreted carefully; high confidence is justified for all of the proteins predicted to be secreted, based on the signal peptide homology detection compared to a demonstrated signal peptide. However, the proteins predicted not to be secreted can only be considered not to be secreted based on the actual prediction. One should note that GST expression was also observed in human saliva (Fabrini et al. 2014), another interface fluid, supporting a possible extra-cellular localization. Concerning UGTs, most of these enzymes are also predicted to be secreted.

In conclusion, most of the phase I and phase II OMEs are predicted to be secreted. However, the non-secreted prediction for the others does not exclude secretion via a non-classical mechanism. Moreover, OMEs may play a role in odorant biotransformation even if they are not secreted into the mucus. The hydrophobic nature of the odorants would be expected to permit a rapid passage

Table 2.2 Prediction of the human OME subcellular localization. All the phase I and II OMEs detected in the human OE were analyzed. Human orthologues of the rat and mice OMEs also detected in the OE were added. The sequences were analyzed with SIGNALP 4.0. For each protein, the UniProt code/accession number, the predicted localization (S for secreted, NS for non-secreted or SA for anchor secreted), the secretion, anchoring and cleavage probabilities and the cleavage site are indicated.

	OME	Accession number	Prediction	Secretion probability	Anchoring probability	Cleavage probability and site
	CYP1A1	P04798	S	0.956	0.044	0.838 between 28 and 29
	CYP1B1	Q16678	S	0.554	0.374	0.234 between 31 and 32
	CYP2A13	Q16696	S	0.939	0.061	0.725 between 22 and 23
	CYP2A6	P11509	S	0.948	0.052	0.538 between 25 and 26
	CYP2B6	P20813	S	0.916	0.083	0.280 between 24 and 25
	CYP2C9	P11712	S	0.999	0.001	0.674 between 25 and 26
	CYP2E1	P05181	S	0.997	0.003	0.934 between 28 and 29
Phase I	CYP2F1	P24903	S	0.969	0.031	0.493 between 23 and 24
	CYP2J2	P51589	S	0.989	0.01	0.244 between 31 and 32
	CYP3A4	P08684	S	0.943	0.056	0.748 between 29 and 30
	CYP4B1	P13584	S	0.786	0.209	0.212 between 19 and 20
	FMO1	Q01740	S	0.815	0.005	0.474 between 17 and 18
	EPHX1	P07099	SA	0.194	0.546	0.131 between 14 and 15
	ALDH1A1	P00352	NS	0	0	–
	ALDH6	P47895	NS	0	0	–
	ALDH7	P43353	NS	0	0	–
	GSTA1	P08263	NS	0	0	–
	GSTM1	P09488	NS	0.015	0	–
	GSTP1	P09211	NS	0.084	0	–
	NQO1	P15559	NS	0.028	0	–
	UGT2A1	Q9Y4X1	S	0.990	0.001	0.955 between 20 and 21
	UGT2A2	Q9Y4X1-2	NS	0.296	0.002	0.259 between 29 and 30
Phase II	UGT2B17	O75795	S	0.994	0.002	0.560 between 24 and 25
	UGT1A10	Q9HAW8	S	0.999	0.001	0.776 between 25 and 26
	UGT1A8	Q9HAW9	S	0.998	0.001	0.773 between 25 and 26
	UGT1A6	P19224	S	0.828	0.169	0.794 between 26 and 27
	UGT1A4	P22310	S	0.998	0.001	0.738 between 28 and 29
	Phenol-ST	O00338	NS	0	0	

through the membrane, allowing a facilitated interaction with the cytoplasmic and reticulum-bound XMEs. The most probable scenario involves a complementary action of cellular and extracellular XMEs.

2.5 Conclusion and perspectives

This review of the literature showed that considerable progress has been made lately supporting the hypothesis of a significant role for OMEs in the olfactory process. We have observed that similar approaches were used to assess this function in various animal models. Specifically, the authors used inhibitors to modify the activity of the enzymes and measured the consequences on odorant response (by electro-olfactogram, olfactory bulb Ca^{2+} imaging or olfactory behaviour). This approach is efficient but raises the question of indirect effects of inhibitors, including the regulation of other OE proteins or toxicological effects. Thus far, RNA interference (RNAi) has not been used in vertebrates to study the function of these enzymes; the challenge would be to shut down the expression of a specific OME *in vivo* using RNAi and then measure olfactory perception. Olfactory behaviour and sensory evaluation are appropriate methods for evaluating integrated olfactory perception, but they often need heavy experiments to overcome inter-individual variations. Thus, we recently focused on a peculiar odorant triggering a specific behaviour: the mammary pheromone. We showed that the mammary pheromone was efficiently metabolized by glutathione-S-transferases in newborn rabbit OE (Legendre et al. 2014). This pheromone triggers an easily measurable behavioural reflex (Schaal et al. 2003). We are studying the function of GST in olfactory perception using this robust and reliable behaviour. In addition, some studies have shown that OMEs have functions in the termination of odorant signaling, but more interestingly, the odorant metabolites could also be involved in the overall perception. This opens a wide field of research, specifically the odorant properties of metabolites and their interaction with other odorants and their receptors. With the aim to measure the metabolism of odorants and the potential synthesis of volatile metabolites, we developed an *ex vivo* headspace gas-chromatography method (Faure et al. 2016). Taking into account the gaseous state of odorants and the structure of OE (explants of olfactory mucosa), such an analytical method may permit the screening of odorants metabolism alone or in mixture or studying the impact of ageing, pathology, polymorphism or inhibitors on odorant metabolism. Besides, we have observed that some OMEs are expressed in intact OE while others are detected in the mucus, and we emphasized the hypothetical secretion of other OMEs in olfactory mucus with our *in silico* analysis. These OME functions exist in humans and they could impact olfactory perception, as demonstrated using sensory evaluation. Thus, the natural odour perception variability observed in humans could be partly explained by OME polymorphisms. Finally, it is clear that with the improvement of our understanding of these enzymatic mechanisms, new strategies should arise either regarding therapeutic interventions in cases of olfactory dysfunction or improvements beneficial to the food or perfume industries.

Acknowledgements

This work was supported by grants from FEDER (fonds européen de développement régional) and the Regional Council of Burgundy.

References

Ache, B. W. and Young, J. M. (2005) Olfaction: diverse species, conserved principles. *Neuron, 48*(3), pp. 417–430.

Ahmad, S., Kirkland, K. E. and Blomquist, G. J. (1987) Evidence for a sex pheromone metabolizing cytochrome P-450 mono-oxygenase in the housefly. *Archives of Insect Biochemistry and Physiology,6*, pp. 121–140.

Anzenbacher, P. and Anzenbacherova, E. (2001) Cytochromes P450 and metabolism of xenobiotics. *Cellular and Molecular Life Sciences, 58*(5-6), pp. 737–747.

Arici, S., Karaman, S., Dogru, S., Arici, A., Karaman, T., Tapar, H., Suren, M. and Kaya, Z. (2013) Effects of isoflurane in an intoxication model: experimental study. *European Review for Medical and Pharmacological Sciences, 17*(13), pp. 1738–1743.

Asan, E. and Drenckhahn, D. (2005) Immunocytochemical characterization of two types of microvillar cells in rodent olfactory epithelium. *Histochemistry and Cell Biology, 123*(2), pp. 157–168.

Audrain, B., Farag, M. A., Ryu, C. M. and Ghigo, J. M. (2015) Role of bacterial volatile compounds in bacterial biology. *FEMS Microbiology Reviews, 39*(2), pp. 222–233.

Austin, C. and Ellis, J. (2003) Microbial pathways leading to steroidal malodour in the axilla. *The Journal of Steroid Biochemistry and Molecular Biology, 87*(1), pp. 105–110.

Baker, J. R., Satarug, S., Edwards, R. J., Moore, M. R., Williams, D. J. and Reilly, P. E. (2003) Potential for early involvement of CYP isoforms in aspects of human cadmium toxicity. *Toxicology Letters, 137*(1–2), pp. 85–93.

Ben-Arie, N., Khen, M. and Lancet, D. (1993) Glutathione S-transferases in rat olfactory epithelium: purification, molecular properties and odorant biotransformation. *Biochemical Journal, 292* (2), pp. 379–384.

Blee, E. (1998) Phytooxylipins and plant defense reactions. *Progress in Lipid Research, 37*(1), pp. 33–72.

Bogdanffy, M. S. (1990) Biotransformation enzymes in the rodent nasal mucosa: the value of a histochemical approach. *Environmental Health Perspectives, 85*, pp. 177–186.

Booth, Y. K., Kitching, W. and De Voss, J. J. (2009) Biosynthesis of insect spiroacetals. *Natural Product Reports, 26*(4), pp. 490–525.

Buck, L. and Axel, R. (1991) A novel multigene family may encode odorant receptors: a molecular basis for odor recognition. *Cell, 65*(1), pp. 175–187.

Buck, L. B. (2005) Unraveling the sense of smell (Nobel lecture). *Angewandte Chemie International Edition in English, 44*(38), pp. 6128–6140.

Bushey, R. T., Chen, G., Blevins-Primeau, A. S., Krzeminski, J., Amin, S. and Lazarus, P. (2011) Characterization of UDP-glucuronosyltransferase 2A1 (UGT2A1) variants and their potential role in tobacco carcinogenesis. *Pharmacogenetics and Genomics, 21*(2), pp. 55–65.

Chertemps, T., Francois, A., Durand, N., Rosell, G., Dekker, T., Lucas, P. and Maibeche-Coisne, M. (2012) A carboxylesterase, Esterase-6, modulates sensory physiological and behavioral response dynamics to pheromone in Drosophila. *BMC Biology, 10*, pp. 56.

Chougnet, A., Woggon, W. D., Locher, E. and Schilling, B. (2009) Synthesis and in vitro activity of heterocyclic inhibitors of CYP2A6 and CYP2A13, two cytochrome P450 enzymes present in the respiratory tract. *European Journal of Chemical Biology, 10*(9), pp. 1562–1567.

Dahl, A. R., Miller, S. C. and Petridou-Fischer, J. (1987) Carboxylesterases in the respiratory tracts of rabbits, rats and Syrian hamsters. *Toxicology Letters, 36*(2), pp. 129–136.

Debat, H., Eloit, C., Blon, F., Sarazin, B., Henry, C., Huet, J. C., Trotier, D. and Pernollet, J. C. (2007) Identification of human olfactory cleft mucus proteins using proteomic analysis. *Journal of Proteome Research, 6*(5), pp. 1985–1996.

Durand, N., Chertemps, T. and Maibeche-Coisne, M. (2012) Antennal carboxylesterases in a moth, structural and functional diversity. *Communicative and Integrative Biology, 5*(3), pp. 284–286.

El-Sayed, A. M. (2014) The pherobase: database of pheromones and semiochemicals. http://www.pherobase.com.

Eriksson, N., Macpherson, J. M., Tung, J. Y., Hon, L. S., Naughton, B., Saxonov, S., Avey, L., Wojcicki, A., Pe'er, I. and Mountain, J. (2010) Web-based, participant-driven studies yield novel genetic associations for common traits. *PLoS Genetics, 6*(6), p. e1000993.

Fabrini, R., Bocedi, A., Camerini, S., Fusetti, M., Ottaviani, F., Passali, F. M., Topazio, D., Iavarone, F., Francia, I., Castagnola, M. and Ricci, G. (2014) Inactivation of human salivary glutathione transferase P1-1 by hypothiocyanite: a post-translational control system in search of a role. *PLoS One, 9*(11), p. e112797.

Farmer, E. E. and Ryan, C. A. (1990) Interplant communication: airborne methyl jasmonate induces synthesis of proteinase inhibitors in plant leaves. *Proceedings of the National Academy of Sciences U S A, 87*(19), pp. 7713–7716.

Faure, P., Legendre, A., Hanser, H. I., Andriot, I., Artur, Y., Guichard, E., Coureaud, G. and Heydel, J. M. (2016) Odorant metabolism analysis by an automated ex vivo headspace gas-chromatography method. *Chemical Senses, 41*(1), pp. 15–23.

Gassett, J. W., Wiesler, D. P., Baker, A. G., Osborn, D. A., Miller, K. V., Marchinton, R. L. and Novotny, M. (1996) Volatile compounds from interdigital gland of male white-tailed deer (Odocoileus virginianus). *Journal of Chemical Ecology, 22*(9), pp. 1689–1696.

Genter, M. B., Yost, G. S. and Rettie, A. E. (2006) Localization of CYP4B1 in the rat nasal cavity and analysis of CYPs as secreted proteins. *Journal of Biochemal and Molecular Toxicology, 20*(3), pp. 139–141.

Getchell, M. L., Chen, Y., Ding, X., Sparks, D. L. and Getchell, T. V. (1993) Immunohistochemical localization of a cytochrome P-450 isozyme in human nasal mucosa: age-related trends. *Annals of Otology, Rhinology and Laryngology, 102*(5), pp. 368–374.

Green, N., Weech, M. and Walters, E. (2005) Localization and characterization of glutathione-s-transferase isozymes alpha, mu, and pi within the mouse vomeronasal organ. *Neuroscience Letters, 375*(3), pp. 198–202.

Haberly, L. B. (2001) Parallel-distributed processing in olfactory cortex: new insights from morphological and physiological analysis of neuronal circuitry. *Chemical Senses, 26*(5), pp. 551–576.

Heydel, J., Leclerc, S., Bernard, P., Pelczar, H., Gradinaru, D., Magdalou, J., Minn, A., Artur, Y. and Goudonnet, H. (2001) Rat olfactory bulb and epithelium UDP-glucuronosyltransferase 2A1 (UGT2A1) expression: in situ mRNA localization and quantitative analysis. *Molecular Brain Research, 90*(1), pp. 83–92.

Heydel, J. M., Coelho, A., Thiebaud, N., Legendre, A., Le Bon, A. M., Faure, P., Neiers, F., Artur, Y., Golebiowski, J. and Briand, L. (2013) Odorant-binding proteins and xenobiotic metabolizing enzymes: implications in olfactory perireceptor events. *Anatomical Record (Hoboken), 296*(9), pp. 1333–1345.

Heydel, J. M., Holsztynska, E. J., Legendre, A., Thiebaud, N., Artur, Y. and Le Bon, A. M. (2010) UDP-glucuronosyltransferases (UGTs) in neuro-olfactory tissues: expression, regulation, and function. *Drug Metabolism Reviews, 42*(1), pp. 71–94.

Hildebrand, D. F., Brown, G. C., Jackson, D. M. and Hamiltonkemp, T. R. (1993) Effects of some leaf-emitted volatile compounds on Aphid population increase. *Journal of Chemical Ecology, 19*(9), pp. 1875–1887.

Hirano, Y. and Mizutani, T. (2003) Study of inhibition of CYP2A6 by some drugs derived from quinoline. *Journal of Pharmacy and Pharmacology, 55*(12), pp. 1667–1672.

Hubert, J., Munzbergova, Z. and Santino, A. (2008) Plant volatile aldehydes as natural insecticides against stored-product beetles. *Pest Management Science, 64*(1), pp. 57–64.

Iida, A., Saito, S., Sekine, A., Mishima, C., Kitamura, Y., Kondo, K., Harigae, S., Osawa, S. and Nakamura, Y. (2002) Catalog of 86 single-nucleotide polymorphisms (SNPs) in three uridine diphosphate glycosyltransferase genes: UGT2A1, UGT2B15, and UGT8. *Journal of Human Genetics, 47*(10), pp. 505–510.

Jedlitschky, G., Cassidy, A. J., Sales, M., Pratt, N. and Burchell, B. (1999) Cloning and characterization of a novel human olfactory UDP-glucuronosyltransferase. *Biochemical Journal, 340*(3), pp. 837–843.

Jemiolo, B., Miller, K. V., Wiesler, D., Jelinek, I., Novotny, M. and Marchinton, R. L. (1995) Putative chemical signals from white-tailed deer (Odocoileus virginianus). Urinary and vaginal mucus volatiles excreted by females during breeding season. *Journal of Chemical Ecology, 21*(6), pp. 869–879.

Jones, P. M. and George, A. M. (2004) The ABC transporter structure and mechanism: perspectives on recent research. *Cellular and Molecular Life Sciences, 61*(6), pp. 682–699.

Katada, S., Hirokawa, T., Oka, Y., Suwa, M. and Touhara, K. (2005) Structural basis for a broad but selective ligand spectrum of a mouse olfactory receptor: mapping the odorant-binding site. *Journal of Neuroscience, 25*(7), pp. 1806–1815.

Keller, A., Zhuang, H., Chi, Q., Vosshall, L. B. and Matsunami, H. (2007) Genetic variation in a human odorant receptor alters odour perception. *Nature, 449*(7161), pp. 468–472.

Keydar, I., Ben-Asher, E., Feldmesser, E., Nativ, N., Oshimoto, A., Restrepo, D., Matsunami, H., Chien, M. S., Pinto, J. M., Gilad, Y., Olender, T. and Lancet, D. (2013) General olfactory sensitivity database (GOSdb): candidate genes and their genomic variations. *Human Mutation, 34*(1), pp. 32–41.

Kippenberger, S., Havlicek, J., Bernd, A., Thaci, D., Kaufmann, R. and Meissner, M. (2012) 'Nosing Around' the human skin: what information is concealed in skin odour? *Experimental Dermatology, 21*(9), pp. 655–659.

Knaapila, A., Hwang, L. D., Lysenko, A., Duke, F. F., Fesi, B., Khoshnevisan, A., James, R. S., Wysocki, C. J., Rhyu, M., Tordoff, M. G., Bachmanov, A. A., Mura, E., Nagai, H. and Reed, D. R. (2012) Genetic analysis of chemosensory traits in human twins. *Chemical Senses, 37*(9), pp. 869–881.

Krishna, N. S., Getchell, M. L., Tate, S. S., Margolis, F. L. and Getchell, T. V. (1992) Glutathione and gamma-glutamyl transpeptidase are differentially distributed in the olfactory mucosa of rats. *Cell Tissue Res, 270*(3), pp. 475–484.

Kwon, B. S., Kim, M. K., Kim, W. H., Pyo, J. S., Cheon, Y. H., Cha, C. I., Nam, S. Y., Baik, T. K. and Lee, B. L. (2005) Age-related changes in microvillar cells of rat olfactory epithelium. *Neuroscience Letters, 378*(2), pp. 65–69.

Lazard, D., Zupko, K., Poria, Y., Nef, P., Lazarovits, J., Horn, S., Khen, M. and Lancet, D. (1991) Odorant signal termination by olfactory UDP glucuronosyl transferase. *Nature, 349*(6312), pp. 790–793.

Leclerc, S., Heydel, J. M., Amosse, V., Gradinaru, D., Cattarelli, M., Artur, Y., Goudonnet, H., Magdalou, J., Netter, P., Pelczar, H. and Minn, A. (2002) Glucuronidation of odorant molecules in the rat olfactory system: activity, expression and age-linked modifications of UDP-glucuronosyltransferase isoforms, UGT1A6 and UGT2A1, and relation to mitral cell activity. *Molecular Brain Research, 107*(2), pp. 201–213.

Legendre, A., Faure, P., Tiesset, H., Potin, C., Jakob, I., Sicard, G., Schaal, B., Artur, Y., Coureaud, G. and Heydel, J. M. (2014) When the nose must remain responsive: glutathione conjugation of the mammary pheromone in the newborn rabbit. *Chemical Senses, 39*(5), pp. 425–437.

Lu, H. and Chan, K. K. (2006) Pharmacokinetics of N-2-chloroethylaziridine, a volatile cytotoxic metabolite of cyclophosphamide, in the rat. *Cancer Chemotherapy and Pharmacology, 58*(4), pp. 532–539.

Lucero, M. T. (2013) Peripheral modulation of smell: fact or fiction? *Seminars in Cell and Developmental Biology, 24*(1), pp. 58–70.

Maibeche-Coisne, M., Nikonov, A. A., Ishida, Y., Jacquin-Joly, E. and Leal, W. S. (2004) Pheromone anosmia in a scarab beetle induced by in vivo inhibition of a pheromone-degrading enzyme. *Proceedings of the National Academy of Sciences U S A, 101*(31), pp. 11459–11464.

Mattina, M. J. I., Pignatello, J. J. and Swihart, R. K. (1991) Identification of Volatile Components of bobcat (Lynx-Rufus) urine. *Journal of Chemical Ecology, 17*(2), pp. 451–462.

Mayer, U., Kuller, A., Daiber, P. C., Neudorf, I., Warnken, U., Schnolzer, M., Frings, S. and Mohrlen, F. (2009) The proteome of rat olfactory sensory cilia. *Proteomics, 9*(2), pp. 322–334.

Mayer, U., Ungerer, N., Klimmeck, D., Warnken, U., Schnolzer, M., Frings, S. and Mohrlen, F. (2008) Proteomic analysis of a membrane preparation from rat olfactory sensory cilia. *Chemical Senses, 33*(2), pp. 145–162.

Menashe, I., Abaffy, T., Hasin, Y., Goshen, S., Yahalom, V., Luetje, C. W. and Lancet, D. (2007) Genetic elucidation of human hyperosmia to isovaleric acid. *PLoS Biology, 5*(11), p. e284.

Nagashima, A. and Touhara, K. (2010) Enzymatic conversion of odorants in nasal mucus affects olfactory glomerular activation patterns and odor perception. *Journal of Neuroscience, 30*(48), pp. 16391–16398.

Nef, P., Heldman, J., Lazard, D., Margalit, T., Jaye, M., Hanukoglu, I. and Lancet, D. (1989) Olfactory-specific cytochrome P-450. cDNA cloning of a novel neuroepithelial enzyme possibly involved in chemoreception. *Journal of Biological Chemistry, 264*(12), pp. 6780–6785.

Nosova, T., Jokelainen, K., Kaihovaara, P., Jousimies-Somer, H., Siitonen, A., Heine, R. and Salaspuro, M. (1996) Aldehyde dehydrogenase activity and acetate production by aerobic bacteria representing the normal flora of human large intestine. *Alcohol and Alcoholism, 31*(6), pp. 555–564.

Oka, Y., Katada, S., Omura, M., Suwa, M., Yoshihara, Y. and Touhara, K. (2006) Odorant receptor map in the mouse olfactory bulb: in vivo sensitivity and specificity of receptor-defined glomeruli. *Neuron, 52*(5), pp. 857–869.

Pelchat, M. L., Bykowski, C., Duke, F. F. and Reed, D. R. (2011) Excretion and perception of a characteristic odor in urine after asparagus ingestion: a psychophysical and genetic study. *Chemical Senses, 36*(1), pp. 9–17.

Phillips, M., Cataneo, R. N., Saunders, C., Hope, P., Schmitt, P. and Wai, J. (2010) Volatile biomarkers in the breath of women with breast cancer. *Journal of Breath Research, 4*(2), p. 026003.

Raymer, J., Wiesler, D., Novotny, M., Asa, C., Seal, U. S. and Mech, L. D. (1984) Volatile constituents of wolf (Canis lupus) urine as related to gender and season. *Experientia, 40*(7), pp. 707–709.

Rooney, P. H., Telfer, C., McFadyen, M. C., Melvin, W. T. and Murray, G. I. (2004) The role of cytochrome P450 in cytotoxic bioactivation: future therapeutic directions. *Current Cancer Drug Targets, 4*(3), pp. 257–265.

Saito, H., Chi, Q., Zhuang, H., Matsunami, H. and Mainland, J. D. (2009) Odor coding by a Mammalian receptor repertoire. *Science Signaling, 2*(60), p. ra9.

Schaal, B., Coureaud, G., Langlois, D., Ginies, C., Semon, E. and Perrier, G. (2003) Chemical and behavioural characterization of the rabbit mammary pheromone. *Nature, 424*(6944), pp. 68–72.

Schilling, B., Kaiser , R., Natsch, A. and Gautschi, M. (2010) Investigation of odors in the fragrance industry. *Chemoecology, 20*(2), pp. 135–147.

Schulz, S. and Dickschat, J. S. (2007) Bacterial volatiles: the smell of small organisms. *Natural Product Report, 24*(4), pp. 814–842.

Scott, J. W. and Scott-Johnson, P. E. (2002) The electroolfactogram: a review of its history and uses. *Microscopy Research and Technique, 58*(3), pp. 152–160.

Sheng, J., Guo, J., Hua, Z., Caggana, M. and Ding, X. (2000) Characterization of human CYP2G genes: widespread loss-of-function mutations and genetic polymorphism. *Pharmacogenetics, 10*(8), pp. 667–678.

Shirasu, M. and Touhara, K. (2011) The scent of disease: volatile organic compounds of the human body related to disease and disorder. *Journal of Biochemistry, 150*(3), pp. 257–266.

Speirs, J., Lee, E., Holt, K., Yong-Duk, K., Scott, N. S., Loveys, B. and Schuch, W. (1998) Genetic manipulation of alcohol dehydrogenase levels in ripening tomato fruit affects the balance of some flavor aldehydes and alcohols. *Plant Physiology, 117*(3), pp. 1047–1058.

Testa, B. and Krämer, S. D. (2007) The biochemistry of drug metabolism--an introduction: Part 2. Redox reactions and their enzymes. *Chemistry and Biodiversity, 4*(3), pp. 257–405.

Thiebaud, N., Menetrier, F., Belloir, C., Minn, A. L., Neiers, F., Artur, Y., Le Bon, A. M. and Heydel, J. M. (2011) Expression and differential localization of xenobiotic transporters in the rat olfactory neuro-epithelium. *Neuroscience Letters, 505*(2), pp. 180–185.

Thiebaud, N., Sigoillot, M., Chevalier, J., Artur, Y., Heydel, J. M. and Le Bon, A. M. (2010) Effects of typical inducers on olfactory xenobiotic-metabolizing enzyme, transporter, and transcription factor expression in rats. *Drug Metabolism and Disposition, 38*(10), pp. 1865–1875.

Thiebaud, N., Veloso Da Silva, S., Jakob, I., Sicard, G., Chevalier, J., Menetrier, F., Berdeaux, O., Artur, Y., Heydel, J. M. and Le Bon, A. M. (2013) Odorant metabolism catalyzed by olfactory mucosal enzymes influences peripheral olfactory responses in rats. *PLoS One, 8*(3), p. e59547.

Thum, T., Erpenbeck, V. J., Moeller, J., Hohlfeld, J. M., Krug, N. and Borlak, J. (2006) Expression of xenobiotic metabolizing enzymes in different lung compartments of smokers and nonsmokers. *Environmental Health Perspectives, 114*(11), pp. 1655–1661.

Touhara, K., Sengoku, S., Inaki, K., Tsuboi, A., Hirono, J., Sato, T., Sakano, H. and Haga, T. (1999) Functional identification and reconstitution of an odorant receptor in single olfactory neurons. *Proceedings of the National Academy of Sciences U S A, 96*(7), pp. 4040–4045.

Vakevainen, S., Tillonen, J., Blom, M., Jousimies-Somer, H. and Salaspuro, M. (2001) Acetaldehyde production and other ADH-related characteristics of aerobic bacteria isolated from hypochlorhydric human stomach. *Alcoholism: Clinical and Experimental Research, 25*(3), pp. 421–426.

Vancanneyt, G., Sanz, C., Farmaki, T., Paneque, M., Ortego, F., Castanera, P. and Sanchez-Serrano, J. J. (2001) Hydroperoxide lyase depletion in transgenic potato plants leads to an increase in aphid performance. *Proceedings of the National Academy of Sciences U S A, 98*(14), pp. 8139–8144.

von Weymarn, L. B., Felicia, N. D., Ding, X. and Murphy, S. E. (1999) N-Nitrosobenzylmethylamine hydroxylation and coumarin 7-hydroxylation: catalysis by rat esophageal microsomes and cytochrome P450 2A3 and 2A6 enzymes. *Chemical Research in Toxicology, 12*(12), pp. 1254–1261.

Wang, H., Tan, W., Hao, B., Miao, X., Zhou, G., He, F. and Lin, D. (2003) Substantial reduction in risk of lung adenocarcinoma associated with genetic polymorphism in CYP2A13, the most active cytochrome P450 for the metabolic activation of tobacco-specific carcinogen NNK. *Cancer Research, 63*(22), pp. 8057–8061.

Yao, Y., Danna, C. H., Ausubel, F. M. and Kovalchuk, I. (2012) Perception of volatiles produced by UVC-irradiated plants alters the response to viral infection in naive neighboring plants. *Plant Signal Behavior, 7*(7), pp. 741–745.

Zhang, J. X., Rao, X. P., Sun, L. X., Zhao, C. H. and Qin, X. W. (2007) Putative chemical signals about sex, individuality, and genetic background in the preputial gland and urine of the house mouse (Mus musculus). *Chemical Senses, 32*(3), pp. 293–303.

Zhuo, X., Gu, J., Zhang, Q. Y., Spink, D. C., Kaminsky, L. S. and Ding, X. (1999) Biotransformation of coumarin by rodent and human cytochromes P-450: metabolic basis of tissue-selective toxicity in olfactory mucosa of rats and mice. *Journal of Pharmacology and Experimental Therapeutics, 288*(2), pp. 463–471.

Zufall, F. and Leinders-Zufall, T. (2000) The cellular and molecular basis of odor adaptation. *Chemical Senses, 25*(4), pp. 473–481.

Zupko, K., Poria, Y. and Lancet, D. (1991) Immunolocalization of cytochromes P-450olf1 and P-450olf2 in rat olfactory mucosa. *European Journal of Biochemistry, 196*(1), pp. 51–58.

CHAPTER 3

The vertebrate gustatory system

Maik Behrens & Wolfgang Meyerhof
Department Molecular Genetics, German Institute of Human Nutrition Potsdam-Rehbruecke, Nuthetal, Germany

3.1 Introduction

Vertebrates use the sense of taste to detect foods rich in macronutrients and electrolytes and poor in potentially harmful compounds. Taste reception is limited to the five basic taste qualities sweet, bitter, and umami (the taste of L-glutamate in humans and several L-amino acids in rodents) which are mediated by G protein-coupled receptors of the taste 1 (TAS1R) and taste 2 receptor (TAS2R) families, and sour and salty transmitted presumably by ion channels (for a recent review see Liman et al. 2014). Recent studies hint at a gustatory component involved in the perception of fat constituents (for a recent review, see Mattes 2011). The existence of several fat taste receptor candidates for example, potassium channels (Gilbertson et al. 1997), the fatty acid transporter CD36 (Fukuwatari et al. 1997, Laugerette et al. 2005), and G protein-coupled fatty acid receptors (Cartoni et al. 2010, Galindo et al. 2012b, Matsumura et al. 2007) as well as suggested mechanisms for appropriate stimulus presentation (Kawai and Fushiki 2003, Kulkarni and Mattes 2013, Pepino et al. 2012, Voigt et al. 2014) have also been reported. However, further research will be required to evaluate whether stimulus recognition, information transmission, and central processing of fat constituents indeed qualifies the orosensory perception of fat as an additional taste modality or not. It is therefore not considered in this chapter.

Whereas sweet and umami taste indicates the presence of energy-rich nutrients in the form of carbohydrates and amino acids/proteins in the oral cavity, sour and bitter taste signal potentially unripe or bacterially spoiled food or even the presence of toxic substances (Lindemann 1996). The detection of salts in the ingested food or consumed water is important for the maintenance of the electrolyte balance of the body fluids. Because the different taste qualities signal the presence of beneficial and harmful food items, they evoke attraction to sweet and umami stimuli or aversion to sour and bitter tastants. Thus, our taste system, in concert with our other senses such as vision, touch, and, especially smell, is

Flavour: From food to perception, First Edition.
Edited by Elisabeth Guichard, Christian Salles, Martine Morzel, and Anne-Marie Le Bon.

exquisitely prepared to guide the consumption of foods towards the ingestion of healthy, nutritionally relevant food items (Galindo et al. 2012a). Tragically, it appears that this sophisticated innate control system of food choice which evolved during the evolution of species is obviously not well adapted to the situation in well-developed industrialized countries. The easy access to food here results in the enforcement of unhealthy over-eating of high-caloric diets associated with high incidence of obesity and metabolic syndrome in man and pets (Breslin 2013). Therefore, understanding taste perception provides an important access to potential therapeutic options of nutrition-associated diseases.

This chapter will provide a basic overview of the taste system of vertebrates. Firstly, the anatomical organization of peripheral taste receptive structures and the information flow into the central nervous system is described. Then, the various taste receptor molecules starting with the G protein-coupled receptors devoted to the detection of sweet, umami, and bitter substances as well as the ion channels involved in sour and salt perception are described in detail. Finally, the intracellular signal transduction processes linking taste receptor activation to gustatory nerve responses is briefly outlined before the current state and future challenges of taste research are summarized.

3.2 Taste anatomy

The oral cavity of vertebrates is equipped with taste sensitive cells, which are organized in groups of cells, called taste buds (Miller 1995). Each taste bud consists of ~100 cells which can be categorized based on morphological and functional features. The type I cells are believed to exert glia-like functions, type II cells harbor the components of the G protein-coupled signaling cascade, and the type III cells represent the only cell type expressing the constituents of the synaptic vesicle release machinery. Finally, type IV, or basal cells populate taste buds and it is believed that this cell type represents stem cells replenishing mature cell types that turnover rapidly (for a recent review, see Chaudhari and Roper 2010).

It is meanwhile well established that type II cells (Figure 3.1) form discrete subpopulations devoted to detect either sweet, or umami, or bitter stimuli (for a recent review, see Yarmolinsky et al. 2009).

Type III cells do not only act as specific sour sensors (Huang et al. 2006, Huang et al. 2008), they seem to fulfill additional roles in orchestrating intercellular communication within the taste bud (DeFazio et al. 2006, Tomchik et al. 2007). Also for the type I cells an additional role in the detection of salty stimuli is discussed (Vandenbeuch et al. 2008). With the exception of the basal cells (type IV cells) all intragemmal cell types exhibit microvillar extensions at their apical side that reach into the single pore region of the taste bud where direct contacts of tastants with receptors are possible. At the basolateral side afferent nerve fibers enter the taste buds, thereby facilitating the transmission of peripheral taste information to the central nervous system (Miller 1995).

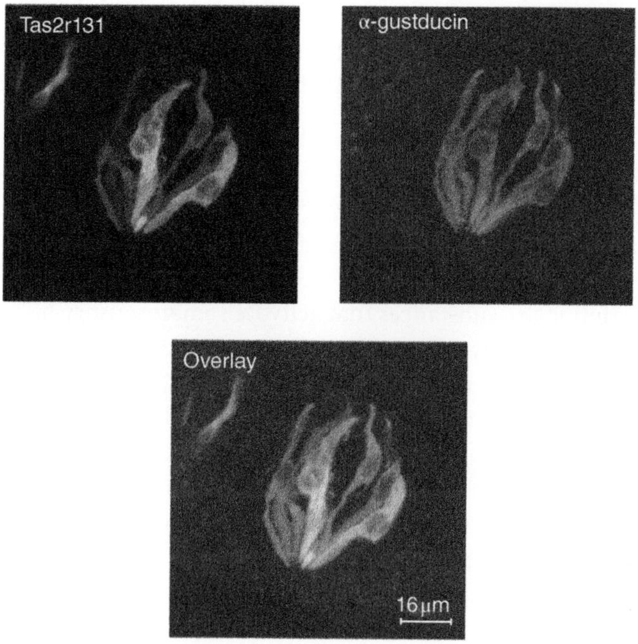

Figure 3.1 Taste buds of the mouse vallate papillae express bitter taste receptor genes in a subset of type II taste receptor cells. Cross-section of a taste bud from the vallate papillae of a knock-in mouse expressing green fluorescent protein (top, left) under the control of the Tas2r131 gene promoter stained with an α-gustducin antiserum (top, right). The overlay image (bottom) shows that Tas2r131 signals are confined to a subset of α-gustducin cells. A colored version of this figure can be found in the online version of this chapter.

Typically, lingual taste buds are housed in epithelial specializations called taste papillae. These can be either rather simple epithelial protrusions named fungiform papillae or complex invaginations forming a trench-like structure flushed by saliva from minor salivary glands such as vallate and foliate papillae. In mammals, numerous fungiform papillae decorate the apical two-thirds of the tongue, whereas the more complex taste papillae are restricted to the back of the tongue with the foliate papillae located on both sides and the vallate papillae close to the terminal sulcus. Taste buds are also embedded in the epithelia of the soft palate, epiglottis, pharynx and larynx (Miller 1995).

The peripheral gustatory information is collected by the peripheral branches of three cranial nerves targeting different regions of the oral cavity. The fungiform papillae of the apical tongue and the soft palate are innervated by the Chorda tympani and greater superficial petrosal branches of the facial nerve, whereas the glossopharyngeal nerve contacts the foliate and vallate papillae of the posterior tongue. Finally, laryngeal and pharyngeal taste buds are supplied by the vagal nerve. The central branches of all three of these cranial nerves convey gustatory information to the first relay station within the central nervous system, the medial-rostral part of the Nucleus tractus solitarius (Miller 1995) in the brainstem. From there, taste information is passed on to local brain stem circuits, the pontine parabrachial nucleus, thalamic and cortical areas as well as to ventral forebrain structures (Miller 1995).

3.3 Tastants

For a thorough description of the chemical structures and sensory properties of taste compounds, the reader is directed to chapter 7 of this book. Here, excluded from this description are substances that activate somatosensory fibers leading to sensations not mediated by taste cells (chemesthesis, e.g., hot sensations caused by chili pepper; for a recent review see (Green 2012)) as well as compounds whose detection is not proven to occur *via* taste cells (e.g., fatty acids, for a recent review see (Mattes 2011); water (Watson et al. 2007)). Even though mammals possess only a single sweet taste receptor, the number of sweet tasting compounds is astonishingly high. In addition to the commonly known mono- and disaccharides, many natural and artificial high-potency sweeteners and sweet proteins bind and activate the sweet taste receptor (for a recent review see (Behrens et al. 2011)). From an evolutionary point of view this variety of sweet tasting substances is very interesting. Whereas the sweetness of mono- and disaccharides perceived by the predator indeed signals the presence of nutritionally relevant carbohydrates, the high-potency natural sweeteners and sweet proteins cause an overestimation of the caloric content of the consumed plant material. In case of the sweet protein thaumatin that is found in the arils covering the seeds of the plant Thaumatococcus daniellii (Benth) (van der Wel and Loeve 1972) one could imagine that it represents a low-energy cost alternative to attract seed dispersers. However, many small molecule sweeteners not present in fruits seem to play other, less obvious roles. This includes the sweet terpenoid glycyrrhizin from the roots of licorice (Glycyrrhiza glabara L.) (Mizutani et al. 1994), and the sweet phenolic substance phyllodulcin from the leaves of the Hortensia plant (Hydrangea macrophylla (Thunb.) (Yamato et al. 1975).

Activators of the umami receptor are fewer and less diverse than those described for the sweet taste receptor (for a recent review see (Behrens et al. 2011)). Depending on the species natural umami stimuli are limited to L-amino acids or a subset of them (Li et al. 2002, Nelson et al. 2002, Oike et al. 2007). Among the umami receptors subjected to functional expression assays, the human TAS1R1/TAS1R3 heteromer displayed the most restricted agonist profile being activated only by L-glutamate and, to a lesser degree, L-aspartate (Li et al. 2002). The rodent and fish counterparts respond to numerous additional L-amino acids (Oike et al. 2007). Interestingly, whereas the Tas1r mediated responses in fish did not shown synergistic effects of L-amino acids with IMP (Oike et al. 2007), the responses of rodent and human Tas1r1/Tas1r3 heteromers are substantially enhanced by IMP (Li et al. 2002, Nelson et al. 2002).

Compared to all other taste stimuli the group of bitter compounds is most complex. Although the majority of natural bitter substances represent organic compounds of different chemical classes, simple salts such as magnesium sulfate or calcium sulfate taste bitter as well (for a review see (Meyerhof 2005)). The question of how many bitter compounds may actually exist cannot be answered with good precision, however, a large database collection of bitter compounds with currently more than 680 entries (Wiener et al. 2012) suggests that a number in the 4-digit range is not too farfetched. With respect to structure bitter substances are frequently found in the groups of sesquiterpene lactones, amino acids, peptides, amines, amides, carbamides, phenolic compounds, and, of course, alkaloids to name just a few. Given the large number of bitter compounds and their structural diversity, it is not surprising that most species possess several or even numerous bitter taste receptors for their recognition (Behrens et al. 2014, Dong et al. 2009, Li and Zhang 2014).

Salt taste is not uniform. Depending on the nature and concentration of the ions constituting the presented stimulus as well as the body status salts are either perceived as attractive or aversive (Beauchamp et al. 1990). The attractive salt taste is limited to low to moderate concentrations of sodium chloride. However, higher concentrations of sodium chloride evoke, similar to potassium chloride distinct aversive sensations (Oka et al. 2013). Indeed, there is evidence accumulating that different sensory cells and different receptor molecules facilitate the recognition of these stimuli explaining the observed perceptual differences (Chandrashekar et al. 2010, Oka et al. 2013).

The sour stimulus is often believed to be solely determined by the concentration of protons (pH-value). However, already many years ago it was observed that weak organic acids represent more effective sour stimuli than inorganic acids such as hydrochloric acid at the same pH (Harvey 1920). Recent data suggest that sour taste is evoked by two components, that is, by protons and by numerous weak organic acids (Chang et al. 2010, Huang et al. 2008).

3.4 Taste receptors

3.4.1 Sweet and umami taste receptors

The first taste receptors which were identified at the molecular level belonged to the *Tas1r* gene family (Hoon et al. 1999). Two genes were found to be expressed in spatially segregated taste receptor cell populations on the tongue of rodents suggesting that they function in different taste transduction pathways. The *Tas1r1* (originally called TR1) gene showed dominant expression in fungiform papillae of the anterior tongue, whereas *Tas1r2* (originally called TR2) gene expressing taste cells occurred mostly in gustatory papillae of the posterior tongue (Hoon et al. 1999). In double-label fluorescent in situ hybridization experiments it was demonstrated that within a taste bud, taste receptor cells practically never co-expressed both genes (Hoon et al. 1999). Further research showed that for the formation of functional taste receptors, however, a third *Tas1r* gene is required, the *Tas1r3*, which was identified after intense research by several groups independently (Bachmanov et al. 2001, Kitagawa et al. 2001, Max et al. 2001, Montmayeur et al. 2001, Nelson et al. 2001, Sainz et al. 2001). Functional experiments confirmed that Tas1r3 is the common subunit of both the sweet taste receptor heteromer formed by the Tas1r2 and Tas1r3 subunits (Nelson et al. 2001), and the umami taste receptor constituted by Tas1r1 and Tas1r3 (Li et al. 2002, Nelson et al. 2002).

The three genes forming the Tas1r family belong to the G protein-coupled receptor family class C and exhibit extensive amino acid sequence and structural similarity with the calcium sensing receptor (CaSR, (Brown et al. 1993)), the V1R pheromone receptor family (Herrada and Dulac 1997, Matsunami and Buck 1997, Ryba and Tirindelli 1997) and the metabotropic glutamate receptors (Houamed et al. 1991, Masu et al. 1991, Tanabe et al. 1992). The Tas1rs possess long extracellular amino termini which form agonist binding domains resembling in shape the insect trapping apparatus of the venus flytrap plant (venus flytrap domain (VTD)), followed by a cysteine-rich domain (CRD) and a 7-transmembrane domain (7-TMD) (for a recent review see (Behrens et al. 2011)).

The functional characterization of the sweet taste receptor heteromer Tas1r2/Tas1r3 in heterologous expression assays confirmed its activation by mono- and disaccharides, natural and artificial sweeteners as well as sweet proteins (Ide et al. 2009, Jiang et al. 2004, Li et al. 2002, Nakajima et al. 2008, Nelson et al. 2001, Winnig et al. 2007, Winnig et al. 2005). In fact, all tested substances perceived as sweet by humans or representing tentatively sweet-like attractive stimuli for rodents activated this receptor in calcium-imaging experiments. Moreover, the concentration response relationship determined *in vitro* agreed reasonably well with human psychophysical and rodent behavioral experiments (Li et al. 2002). Importantly, the comparison of human and rodent sweet taste receptor properties confirmed species-specific differences in particular for the detection of several artificial sweeteners that activate the human but not the rodent sweet taste receptor (Li et al. 2002, Nelson et al. 2001). This discovery enabled further experiments utilizing interspecies receptor-chimeras and site-directed mutagenesis to map the binding sites for various sweet receptor ligands (Jiang et al. 2005a, Jiang et al. 2005b, Jiang et al. 2004, Winnig et al. 2007, Xu et al. 2004). It turned out that the different sweet agonists bind and activate the receptor at different sites. These binding sites are not only located in both subunits of the sweet taste receptor heteromer, but also reside in the extracellular as well as in the transmembrane domains.

In contrast to the sweet taste receptor the umami taste receptor is activated by a rather restricted panel of agonists. In fact, the human TAS1R1/TAS1R3 heteromer responds remarkably selectively to L-glutamate (Li et al. 2002).

Again, species-specific differences in the activation pattern have been observed between human and rodent umami receptors. Compared to the human receptor which responds with strong preference to L-glutamate enhanced by IMP, the rodent Tas1r1/Tas1r3 shows less pronounced preference for L-glutamate compared to other L-amino acids. A well-known feature of umami taste is its enhancement by 5′-ribonucleotides such as IMP, or GMP (Sako and Yamamoto 1999). Indeed, the phenomenon of umami taste-enhancement by 5′-ribonucleotides, well-established in sensory experiments, was also proven by heterologous receptor expression assays confirming an approximately 30-fold shift of the human umami receptor sensitivity towards lower L-glutamate concentrations (Li et al. 2002). Using a combination of functional experiments, site-directed mutagenesis and molecular modeling it was shown that IMP binds close to the agonist L-glutamate in the VTD of the TAS1R1 subunit and seems to stabilize the agonist-bound closed conformation thereby leading to elevated receptor activity (Zhang et al. 2008). Although the TAS1R1/TAS1R3 heteromer reacts specifically to L-glutamate and shows significant enhancement by 5′-ribonucleotides, which is considered a hallmark of umami taste, alternative umami receptors have been proposed. Indeed, additional putative umami receptors have been identified in gustatory tissue of rodents. Both candidate receptors represent truncated variants of metabotropic glutamate receptors, namely mGluR1 and mGluR4, and exhibit lower sensitivities for L-glutamate compared to the full-length brain variants. Accordingly, these receptors have been named taste-mGluR1 (San Gabriel et al. 2009) and taste-mGluR4 (Chaudhari et al. 2000). It was shown that the sensitivity of taste-mGluR4 compared to brain-mGluR4 is much lower with an L-glutamate EC_{50}-concentration in the high micromolar range (Chaudhari et al. 2000), consistent with a role as gustatory low-affinity glutamate receptor. The observation of residual umami responses in a Tas1r1-knockout mouse model (Kusuhara et al. 2013) as well as in Tas1r3 knockout mice (Damak et al. 2003, Delay et al. 2006, Maruyama et al. 2006) confirmed the assumption of more than one receptor for this taste quality, although contrasting evidence showing the complete absence of responses in Tas1r1- and Tas1r3-knockout mice was obtained as well (Zhao et al. 2003).

3.4.2 Bitter taste receptors

The enormous number and structural diversity of bitter substances encountered in nature is recognized by the *Tas2r* gene family (Adler et al. 2000, Chandrashekar et al. 2000, Matsunami et al. 2000). The ever growing number of genome projects across a wide range of chordates revealed that the number of putatively functional *Tas2r* genes fluctuates largely across species (Dong et al. 2009, Li and Zhang 2014). While some animals such as turkey and chicken possess only 2-3 functional bitter taste receptors (Li and Zhang 2014, Shi and Zhang 2006), the genomes of other species can encode more than 50 and up to

70 of these receptors as seen for the Western clawed frog (X. tropicalis) (Behrens et al. 2014, Li and Zhang 2014) and the coelacanth (L. chalumnae) (Syed and Korsching 2014), respectively. An important factor influencing the *Tas2r* gene repertoire sizes seems to be diet with herbivores possessing on average more putatively functional *Tas2r* genes compared to carnivores (Li and Zhang 2014). However, the number of putatively functional *Tas2r* genes does not allow predictions about the possible importance of the bitter tasting abilities since the average tuning breadth of the corresponding receptors (Table 3.1) may influence the number of recognizable bitter substances considerably (Behrens et al. 2014).

The human *TAS2R* gene repertoire is medium sized consisting of ~25 functional bitter taste receptors of which 21 have been associated with activating bitter substances over the last decade (Meyerhof et al. 2010, Thalmann et al. 2013). It turned out that human TAS2Rs can deviate largely in the number of cognate bitter agonists (Meyerhof et al. 2010). In fact, based on the observed agonist profiles the receptors can be grouped into subgroups with numerous, intermediate, and few recognized compounds. The three receptors with the broadest agonist profiles are TAS2R10 (Bufe et al. 2002), TAS2R14 (Behrens et al. 2004), and TAS2R46 (Brockhoff et al. 2007), with each of the receptors recognizing about one-third of all 104 recently tested bitter compounds. Together these three TAS2Rs recognize about 50% of all bitter substances (Meyerhof et al. 2010). It appears therefore justified to envision these receptors as "generalists" contributing strongly to the overall ability of human to perceive bitterness. An example of the enormous versatility of broadly tuned human TAS2Rs is provided in Table 3.2, which shows the currently known 121 agonists of the receptor TAS2R14 (Table 3.2a) together with 111 substances that were experimentally proven to not activate this receptor (Table 3.2b) (Behrens et al. 2004, Intelmann et al. 2009, Kohl et al. 2013, Le Neve et al. 2010, Levit et al. 2014, Meyerhof 2005, Roland et al. 2013, Roland et al. 2011, Soares et al. 2013).

In contrast to these three receptors, the receptors, TAS2R3, TAS2R5 (Meyerhof et al. 2010), TAS2R8 (Pronin et al. 2007), TAS2R13 (Meyerhof et al. 2010), TAS2R41 (Thalmann et al. 2013), TAS2R49 (Meyerhof et al. 2010), and TAS2R50 (Behrens et al. 2009) recognize only very few bitter compounds and therefore

Table 3.1 Categorization of bitter taste receptors of different species based on their tuning width.

Species	Putatively functional TAS2R	Generalists	Specialists
Human	25	3	7
Frog	54	3	3
Chicken	3	3	0
Turkey	2	2	0
Zebra finch	7	0	1

Table 3.2a Known agonists of TAS2R14.

Absinthin	Divinylsulfoxid	Niflumic acid
o-(p-Anisoyl)benzoic acid sodium salt	(−)-Epicatechin gallate	1-Nitronaphthalene
Apigenin	(−)-Epigallocatechin gallate	Noscapine
Arborescin	(±)-Equol	Pantoprazole
Arglabin	Eriodictyol	Papaverine
Aristolochic acid	Eriodictyolchalcone	Parthenolide
Artemorin	Ethylhydrocupreine	Pelorgoninidin chloride
Azathioprine	Falcarindiol	Pemirolast
Benzamide	Flavanone	Phloretin
Benzoate, sodium	Flavone	Picrotin
Benzoin	Flufenamic acid	Picrotoxinin
Biochanin A	Formononetin	Pinocembrin
Butein	Fustin	Piperonylic acid
Camphor	β-1,2,3,4,6-penta-O-galloyl-D-glucopyranose	8-prenylnaringenin
Caffeine	Genistein	Prunetin
Carisoprodol	Glimepiride	Quassin
Cascarillin	Glycitein	Quercetin
(+)-Catechin	Haloperidol	Quinine
Chalcone	Herbacetin	Resveratrol
Chlorhexidine	Hesperetin	Salsalate
Chlorpheniramine	Homoeriodictyol	Scutellarein
Chrysin	Humulone isomeres	Secobarbital sodium
Chryseriol	4-Hydroxyanisol	Silibinin
Clonixin	7-Hydroxyisoflavone	Sulfuretin
Coumarin	4′-Hydroxy-6-methoxyflavone	(+)-Taxifolin
Coumestrol	4′-Hydroxy-7-methoxyflavone	α-(−)-Thujon
Cucurbitacin B	Isoflavone	12β-tigloyloxy-14β-hydroxy-pregn-5-en-20-one
Cyanidin chloride	Isorhamnetin	Tributyrin
Daidzein	Kaempherol	Triethyl citrate
Datiscetin	Liquiritigenin	2,2′,4′-Trihydroxychalcone
Diclofenac	Luteolin	4,2′,5′-Trihydroxychalcone
5,4′-Dihydroxyflavone	6-Methoxyluteolin	3,7,4′-Trihydroxyflavone
6,4′-Dihydroxyflavone	Malathion	5,7,2′-Trihydroxyflavone
7,4′-Dihydroxyflavone	Mefenamic acid	7,3′,4′-Trihydroxyflavone
3,5-Diiodosalicilic acid	Miconazole	6,7,4′-Trihydroxyisoflavone
5,7-Dimethoxyflavone	Morin	7,3′,4′-Trihydroxyisoflavone
6,7-Dimethoxyflavone	Myricetin	7,8,4′-Trihydroxyisoflavone
7,4′-Dimethoxy-isoflavone	1,8-Naphthaldehydic acid	5,7,4′-Trimethoxyflavone
Diphenhydramine	1-naphthoic acid	Trp-Trp-Trp
Diphenidol	Naringenin	Xanthone
Diphenylthiourea		

Table 3.2b Compounds that do not activate TAS2R14.

Acacetin	Dicyclomine hydrochloride	4(6)-Methyl-2-thiouracil
Acesulfame K	Digitonin	Methylthiourea
Acetaminophen	Dihydrophenanthrene	Metronidazole
Acetylgenistin	5,2′-Dihydroxyflavone	Naphazoline
2-Acetylpyrazine	7,4′-Dimethoxy-5-hydroxyisoflavone	Naringin
Acetylthiourea	Dimethyl thioformamide	Neohesperidine
Alfuzosin	Doxepin	Nicotin
Aloin	Emetine	Nimbin
Allylisothiocyanate	(–)-Epicatechin	Orphenadrine
Alverine citrate salt	Erythromycin	Ouabain
Amarogentin	Ethylpyrazine	Oxolinic acid
Amlodipine	N,N′-ethylene thiourea	1,10-Phenanthroline
Amygdalin, D	N-ethylthiourea	D-Phenylalanine
Antipyrin	Famotidine	L-Phenylalanine
Andrographolide	Felodipine	Phenylethyl isothiocyanate
Apigeninidin chloride	Fenspiride	Phenylthiocarbamide
Arbutin	Fluticasone propionate	Picric acid
Atropine	Gardenin A	Procyanidin trimer C2
Benzilic acid	Genistin	Propylthiouracil
Bezafibrate	Genkwanin	Purvalanol A
Brucine	Ginkgolide A	Quinidine
Caffeic acid	Glipizide	Riboflavin
Caprolactam	Glycitin	Saccharin
Chloramphenicol	Gossypetin	D-salicin
Chloroquine	Grossheimin	Sildenafil
Cnicin	Helicin	Sinigrin
Colchicine	Hydrocortisone	α-Solanine
Crispolide	4′-Hydroxyflavanone	Solanidine
Cromolyn	5-Hydroxy-3′-methoxyflavone	(–)-Sparteine
Cucurbitacin E	Hyperforin	Strychnine
Curcumin	Ibuprofen	D-(+)-Sucrose octaacetate
Cyclamate, sodium	Limonin	L-Sulforaphane
Cycloheximide	Lovastatin	Taurine
Daidzin	Malonylgenistin	Tatridin B
Dapsone	Malvidin-3-glucoside	Thiamine
Denatonium benzoate	Methimazole	Thiocyanate, sodium
Dextromethorphan	6-Methoxyflavonol	Yohimbine

behave as "specialists". It is speculated that also the 4 receptors which have resisted deorphanization so far, may belong to this group of receptors. All of the other human TAS2Rs with identified bitter activators respond to an intermediate number of bitter substances and hence, represent the largest group. The two receptors, TAS2R16 (Bufe et al. 2002) and TAS2R38 (Bufe et al. 2005, Kim et al. 2003), were shown to recognize specific chemical features within their agonist molecules. Whereas the TAS2R16 is activated by β-D-glucopyranosides (Bufe et al. 2002), TAS2R38 recognizes the NCS-moiety of isothiocyanates/thiourea compounds (Bufe et al. 2005), both of which occur plentiful in nature. Hence, although from a chemical viewpoint these receptors behave as specialists, the number of bitter agonists that exhibit these structures is considerable.

Tas2r genes are affected by numerous genetic polymorphisms. Among the bitter taste receptors exhibiting functional genetic polymorphisms the TAS2R38 is most famous. Firstly, it has been observed that people deviate in the perceptual sensitivity toward the synthetic bitter substances phenylthiocarbamide (PTC) and 6-n-propylthiouracil (PROP) almost a century ago (Fox 1932). Secondly, the difference in sensitivities between carriers of sensitive gene variants (also known as taster variant) and insensitive variants (also named non-taster variant) are most pronounced since the non-taster receptor variant (TAS2R38-AVI) is indeed non-functional, whereas the taster variant (TAS2R38-PAV) is exquisitely specific and sensitive to PROP and PTC (Bufe et al. 2005). Thirdly, both variants occur with high frequency in the human population and therefore large fractions of PROP/PTC-tasters and non-tasters exist (Wooding et al. 2004). Until now several additional functional polymorphisms in bitter taste receptor genes have been reported affecting the receptors TAS2R9 (Dotson et al. 2008), TAS2R16 (Soranzo et al. 2005), TAS2R31, and TAS2R43 (Pronin et al. 2007, Roudnitzky et al. 2011).

In order to investigate if the non-taster variant of the TAS2R38 is indeed non-functional or if rather the ligand binding site may be modified and hence may lose its ability to respond to PTC, PROP and related compounds whereas other yet unknown agonists could still interact with this receptor, *in silico* modeling and mutagenesis experiments were performed (Fig. 3.2).

Whereas a pure in silico study found indeed evidence for an impaired agonist interaction in the non-taster TAS2R38-AVI variant (Tan et al. 2012), another

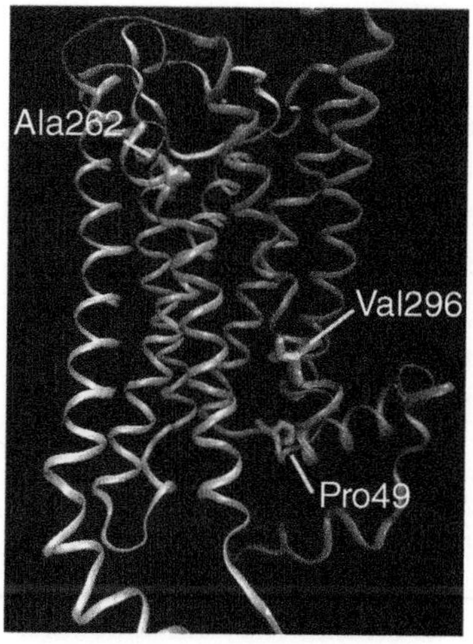

Figure 3.2 Homology model of the human bitter taste receptor TAS2R38. Ribbon model of the TAS2R38 taster variant shown from the side. The extracellular side points to the top of the image, the intracellular side to the bottom. The three amino acid positions that differ between the main taster and non-taster variants are labeled. The non-taster variant would show alanine in position 49, valine in position 262, and isoleucine in position 296 instead. The image is based on a previously published in silico model (Source: Adapted from Biarnes et al. 2010).

study combining computer modeling with *in vitro* mutagenesis and functional heterologous expression assays concluded that those amino acid positions in TAS2R38 affected by the genetic polymorphisms are not involved in agonist interaction. Rather the data proposed that receptor activation is affected in the non-taster TAS2R38-AVI variant (Biarnes et al. 2010, Marchiori et al. 2013). Other recent studies on the structure-function relationship of bitter taste receptors (for a recent review see (Behrens and Meyerhof 2013)) revealed that even broadly tuned TAS2R46 possess only a single agonist binding pocket (Brockhoff et al. 2010) and that the binding site of bitter taste receptors are tailored to accommodate multiple agonists at the expense of potentially higher affinities for individual bitter substances (Born et al. 2013).

3.4.3 Salt taste receptor

The taste of low concentrations of NaCl in rodents which originates from the anterior tongue is transmitted *via* the C. tympani, and can be blocked specifically by the diuretic drug amiloride suggesting that the amiloride-sensitive epithelial sodium channel ENaC is involved in mediating attractive salt taste transduction (for a review see (Lindemann 1996)). Although numerous publications confirmed the potential role of ENaC as salt taste sensor (Kretz et al. 1999, Lin et al. 1999, Stahler et al. 2008), only recently the identification of ENaC-α as an obligatory component for the attractive low salt taste of rodents was confirmed by the generation of taste-specific ENaC-α knockout mice (Chandrashekar et al. 2010). The same group of authors also identified taste bud cells that respond to high concentrations of sodium ions and other cations that are clearly distinct from those responding to low sodium stimuli. Intriguingly, the cells mediating the aversive taste of high salt concentrations seem to represent subpopulations of bitter and sour taste receptor cells as shown by pharmacological blocking experiments (Oka et al. 2013).

3.4.4 Sour taste receptor

Compared to the other taste qualities the mechanism of sour taste transduction is not well understood. This appears surprising as the nature of the stimulus, namely acids, as well as the identity of the sour sensing taste bud cells, the type III or presynaptic cells are known. Moreover, the search for the corresponding sour taste receptor(s) resulted in the identification of multiple candidates such as acid-sensing ion channels (Ugawa et al. 1998), hyperpolarization-activated cyclic nucleotide-gated potassium channels (Stevens et al. 2001), two-pore-domain potassium channels (Lin et al. 2004, Richter et al. 2004) as well as TRP channels of the polycystic kidney disease family (Huang et al. 2006, Ishimaru et al. 2006, LopezJimenez et al. 2006). However, unambiguous molecular identification of one of these candidates as sour taste receptor has not been achieved so far. The situation is further complicated by the fact that sour stimuli may have at least two components. On the one hand there is a channel-mediated influx of extracellular protons (pH) into the sour sensing cells (Chang et al. 2010). On the other hand protonated weak organic acids diffuse through the plasma membrane and cause intracellular acidification by the dissociation of acids in the cytosol (Huang et al. 2008). The ultimate identification of the missing sour taste receptor and the clarification of the sour transduction mechanism represent currently one of the most urgent questions in taste research.

3.5 Taste signal transduction

The signal transduction in type II taste receptor cells expressing the G protein-coupled taste receptors is comparatively well investigated. In fact, several signaling components involved in the transmission of signals originating from taste G protein-coupled receptors (GPCRs) have been discovered prior to the receptor molecules themselves. To date it is well established that several constituents of heteromeric G proteins are expressed in taste receptor cells of vertebrates likely forming different types of these important signaling molecules. Whereas a number of different Gα-subunits such as α-transducin, $G\alpha_{i2}$, $G\alpha_{i3}$, $G\alpha_{15}$, $G\alpha_{q}$, $G\alpha_{11}$, $G\alpha_{14}$ (Asano-Miyoshi et al. 2000, Kusakabe et al. 1998, Kusakabe et al. 2000, Tizzano et al. 2008, Wong et al. 1996), in addition to Gα-gustducin (McLaughlin et al. 1992), were found in type II cells, only two isoforms of Gβ, Gβ1 and Gβ3 (Rossler et al. 2000), and a single Gγ-subunit, Gγ13 were observed (Huang et al. 1999). It remains to be seen whether future research will identify additional G protein subunits and whether all possible combinations are realized *in vivo*. After the activation of heteromeric G proteins, the dissociated βγ-complex in turn activates the membrane associated phospholipase C-β2 (Dotson et al. 2005, Rossler et al. 1998, Zhang et al. 2003) leading to the generation of diacylglycerol and inositol-1,4,5-triphosphate (IP_3). The next step in the cascade is the binding of IP_3 to its receptor, the IP_3R III, located in the membrane of the taste cell endoplasmic reticulum (ER) (Clapp et al. 2001, Hisatsune et al. 2007, Miyoshi et al. 2001). This ligand-activated calcium channel opens and leads to calcium ion efflux from the ER lumen into the cytosol. The elevated cytosolic calcium ions finally result in the opening of a plasma membrane associated cation channel of the transient receptor potential channel family, the TRPM5 (Liu and Liman 2003, Perez et al. 2002, Zhang et al. 2003, Zhang et al. 2007). The corresponding cell depolarization is associated with the release of ATP (Finger et al. 2005), the neurotransmitter of type II taste receptor cells, through hemichannels or CALHM1 at their cell surface (Huang et al. 2007, Romanov et al. 2007). A role of the Gα-subunit α-gustducin in type II taste receptor cell signaling is also discussed. The activity of α-gustducin keeps the level of cyclic AMP in taste receptor cells low. This reduces the activity of protein kinase A, which would otherwise negatively regulate the IP_3R. Hence, it is believed that tonic α-gustducin activity maintains the sensitivity of the calcium signaling pathway (Clapp et al. 2008).

The ATP that is released from taste receptor cells acts in multiple ways (Huang and Roper 2010). On the one hand it is directly exciting afferent nerve fibers that terminate in close proximity of the taste receptor cells. The afferent nerve fibers express heteromeric purinergic $P2X_2/P2X_3$ receptors thereby mediating the signal transmission towards the central nervous system (Finger et al. 2005). On the other hand the released ATP is also capable to act on type III cells in the taste bud, which are likely involved in intragemmal communication, apart from their role as sour taste receptors. ATP also induces a positive feedback loop onto the type II cells themselves modulating its own release (Huang et al. 2009).

3.6 Conclusions

Taste research is a very active rapidly evolving integrative area. Perhaps due to the still relatively recent discovery of receptor molecules, some aspects of the taste perception mechanism seem to be quite well understood, while other aspects contain astonishingly large gaps in knowledge. A good example is the status of research on the taste receptor molecules themselves. Whereas the G protein-coupled receptors have been identified, cloned, deorphanized and now even targeted by taste modifiers, the nature and molecular identity of the sour taste receptor is still unknown. Similarly, whereas we learned that ENaC-α is an essential component of the salt taste transduction machinery in mice (Chandrashekar et al. 2010) the precise channel composition remains unknown. Whether carnivores, herbivores, and omnivores, including man, use the same or similar salt transduction mechanisms as mice also awaits future research. Clearly, additional research efforts are necessary to identify all receptor molecules. Another important open question is how taste information is transmitted into the central nervous system and how it is processed. Here, it appears currently as if the long standing debate whether taste information is transmitted by hard-wired segregated information channels (labeled-line theory) or *via* shared pathways and a centrally decoding (across-fiber pattern theory) mechanism rather points into the direction of a labeled-line type mechanism. Nevertheless, numerous steps along the transmission pathway remain to be elucidated. The finding that taste receptor molecules and the corresponding signaling elements are not restricted to the gustatory system, but are present in an ever growing number of tissues clearly indicates that additional research is required to clarify the role of the taste related molecules in the extra-gustatory tissues and throughout the evolution of vertebrates. It will be important to find out what primary role taste receptors fulfilled during evolution.

Acknowledgements

The authors thank Dr. Anja Voigt for the images assembled in Figure 3.1 and Dr. Alessandro Marchiori for preparation of Figure 3.2.

References

Adler, E., Hoon, M. A., Mueller, K. L., Chandrashekar, J., Ryba, N. J. and Zuker, C. S. (2000) A novel family of mammalian taste receptors. *Cell, 100*, pp. 693-702.

Asano-Miyoshi, M., Abe, K. and Emori, Y. (2000) Co-expression of calcium signaling components in vertebrate taste bud cells. *Neuroscience Letters, 283*, pp. 61-64.

Bachmanov, A. A., Li, X., Reed, D. R., Ohmen, J. D., Li, S., Chen, Z., Tordoff, M. G., de Jong, P. J., Wu, C., West, D. B., Chatterjee, A., Ross, D. A. and Beauchamp, G. K. (2001) Positional cloning of the mouse saccharin preference (Sac) locus. *Chemical Senses, 26*, pp. 925-933.

Beauchamp, G. K., Bertino, M., Burke, D. and Engelman, K. (1990) Experimental sodium depletion and salt taste in normal human volunteers. *American Journal of Clinical Nutrition, 51*, pp. 881-889.

Behrens, M., Brockhoff, A., Batram, C., Kuhn, C., Appendino, G. and Meyerhof, W. (2009) The human bitter taste receptor hTAS2R50 is activated by the two natural bitter terpenoids andrographolide and amarogentin. *Journal of Agricultural and Food Chemistry, 57*, pp. 9860-9866.

Behrens, M., Brockhoff, A., Kuhn, C., Bufe, B., Winnig, M. and Meyerhof, W. (2004) The human taste receptor hTAS2R14 responds to a variety of different bitter compounds. *Biochemical and Biophysical Research Communications, 319*, pp. 479-485.

Behrens, M., Korsching, S. I. and Meyerhof, W. (2014) Tuning properties of avian and frog bitter taste receptors dynamically fit gene repertoire sizes. *Molecular Biology and Evolution, 31*, pp. 3216-3227.

Behrens, M. and Meyerhof, W. (2013) Bitter taste receptor research comes of age: from characterization to modulation of TAS2Rs. *Seminars in Cell & Developmental Biology, 24*, pp. 215-221.

Behrens, M., Meyerhof, W., Hellfritsch, C. and Hofmann, T. (2011) Sweet and umami taste: natural products, their chemosensory targets, and beyond. *Angewandte Chemie (International Edition), 50*, pp. 2220-2242.

Biarnes, X., Marchiori, A., Giorgetti, A., Lanzara, C., Gasparini, P., Carloni, P., Born, S., Brockhoff, A., Behrens, M. and Meyerhof, W. (2010) Insights into the binding of Phenyltiocarbamide (PTC) agonist to its target human TAS2R38 bitter receptor. *PLoS One, 5*, p. e12394.

Born, S., Levit, A., Niv, M. Y., Meyerhof, W. and Behrens, M. (2013) The human bitter taste receptor TAS2R10 is tailored to accommodate numerous diverse ligands. *Journal of Neuroscience, 33*, pp. 201-213.

Breslin, P. A. (2013) An evolutionary perspective on food and human taste. *Current Biology, 23*, pp. R409-418.

Brockhoff, A., Behrens, M., Massarotti, A., Appendino, G. and Meyerhof, W. (2007) Broad tuning of the human bitter taste receptor hTAS2R46 to various sesquiterpene lactones, clerodane and labdane diterpenoids, strychnine, and denatonium. *Journal of Agricultural and Food Chemistry, 55*, pp. 6236-6243.

Brockhoff, A., Behrens, M., Niv, M. Y. and Meyerhof, W. (2010) Structural requirements of bitter taste receptor activation. *Proceedings of the National Academy of Sciences, USA, 107*, pp. 11110-11115.

Brown, E. M., Gamba, G., Riccardi, D., Lombardi, M., Butters, R., Kifor, O., Sun, A., Hediger, M. A., Lytton, J. and Hebert, S. C. (1993) Cloning and characterization of an extracellular Ca(2+)-sensing receptor from bovine parathyroid. *Nature, 366,* pp. 575-580.

Bufe, B., Breslin, P. A., Kuhn, C., Reed, D. R., Tharp, C. D., Slack, J. P., Kim, U. K., Drayna, D. and Meyerhof, W. (2005) The molecular basis of individual differences in phenylthiocarbamide and propylthiouracil bitterness perception. *Current Biology, 15,* pp. 322-327.

Bufe, B., Hofmann, T., Krautwurst, D., Raguse, J. D. and Meyerhof, W. (2002) The human TAS2R16 receptor mediates bitter taste in response to beta-glucopyranosides. *Nature Genetics, 32,* pp. 397-401.

Cartoni, C., Yasumatsu, K., Ohkuri, T., Shigemura, N., Yoshida, R., Godinot, N., le Coutre, J., Ninomiya, Y. and Damak, S. (2010) Taste preference for fatty acids is mediated by GPR40 and GPR120. *Journal of Neuroscience, 30,* pp. 8376-8382.

Chandrashekar, J., Kuhn, C., Oka, Y., Yarmolinsky, D. A., Hummler, E., Ryba, N. J. and Zuker, C. S. (2010) The cells and peripheral representation of sodium taste in mice. *Nature, 464,* pp. 297-301.

Chandrashekar, J., Mueller, K. L., Hoon, M. A., Adler, E., Feng, L., Guo, W., Zuker, C. S. and Ryba, N. J. (2000) T2Rs function as bitter taste receptors. *Cell, 100,* pp. 703-711.

Chang, R. B., Waters, H. and Liman, E. R. (2010) A proton current drives action potentials in genetically identified sour taste cells. *Proceedings of the National Academy of Sciences, USA, 107,* pp. 22320-22325.

Chaudhari, N., Landin, A. M. and Roper, S. D. (2000) A metabotropic glutamate receptor variant functions as a taste receptor. *Nature Neuroscience, 3,* pp. 113-119.

Chaudhari, N. and Roper, S. D. (2010) The cell biology of taste. *Journal of Cell Biology, 190,* pp. 285-296.

Clapp, T. R., Stone, L. M., Margolskee, R. F. and Kinnamon, S. C. (2001) Immunocytochemical evidence for co-expression of Type III IP3 receptor with signaling components of bitter taste transduction. *BMC Neuroscience, 2,* 6.

Clapp, T. R., Trubey, K. R., Vandenbeuch, A., Stone, L. M., Margolskee, R. F., Chaudhari, N. and Kinnamon, S. C. (2008) Tonic activity of Galpha-gustducin regulates taste cell responsivity. *FEBS Letters, 582,* pp. 3783-3787.

Damak, S., Rong, M., Yasumatsu, K., Kokrashvili, Z., Varadarajan, V., Zou, S., Jiang, P., Ninomiya, Y. and Margolskee, R. F. (2003) Detection of sweet and umami taste in the absence of taste receptor T1r3. *Science, 301,* pp. 850-853.

DeFazio, R. A., Dvoryanchikov, G., Maruyama, Y., Kim, J. W., Pereira, E., Roper, S. D. and Chaudhari, N. (2006) Separate populations of receptor cells and presynaptic cells in mouse taste buds. *Journal of Neuroscience, 26,* pp. 3971-3980.

Delay, E. R., Hernandez, N. P., Bromley, K. and Margolskee, R. F. (2006) Sucrose and monosodium glutamate taste thresholds and discrimination ability of T1R3 knockout mice. *Chemical Senses, 31,* pp. 351-357.

Dong, D., Jones, G. and Zhang, S. (2009) Dynamic evolution of bitter taste receptor genes in vertebrates. *BMC Evolutionary Biology, 9,* 12.

Dotson, C. D., Roper, S. D. and Spector, A. C. (2005) PLCβ2-independent behavioral avoidance of prototypical bitter-tasting ligands. *Chemical Senses, 30,* pp. 593-600.

Dotson, C. D., Zhang, L., Xu, H., Shin, Y. K., Vigues, S., Ott, S. H., Elson, A. E., Choi, H. J., Shaw, H., Egan, J. M., Mitchell, B. D., Li, X., Steinle, N. I. and Munger, S. D. (2008) Bitter taste receptors influence glucose homeostasis. *PLoS ONE, 3,* p. e3974.

Finger, T. E., Danilova, V., Barrows, J., Bartel, D. L., Vigers, A. J., Stone, L., Hellekant, G. and Kinnamon, S. C. (2005) ATP signaling is crucial for communication from taste buds to gustatory nerves. *Science, 310,* pp. 1495-1499.

Fox, A. L. (1932) The relationship between chemical constitution and taste. *Proceedings of the National Academy of Sciences, USA, 18,* pp. 115-120.

Fukuwatari, T., Kawada, T., Tsuruta, M., Hiraoka, T., Iwanaga, T., Sugimoto, E. and Fushiki, T. (1997) Expression of the putative membrane fatty acid transporter (FAT) in taste buds of the circumvallate papillae in rats. *FEBS Letters, 414*, pp. 461-464.

Galindo, M. M., Schneider, N. Y., Stahler, F., Tole, J. and Meyerhof, W. (2012a) Taste preferences. *Progress in Molecular Biology and Translational Science, 108*, pp. 383-426.

Galindo, M. M., Voigt, N., Stein, J., van Lengerich, J., Raguse, J. D., Hofmann, T., Meyerhof, W. and Behrens, M. (2012b) G protein-coupled receptors in human fat taste perception. *Chemical Senses, 37*, pp. 123-139.

Gilbertson, T. A., Fontenot, D. T., Liu, L., Zhang, H. and Monroe, W. T. (1997) Fatty acid modulation of K+ channels in taste receptor cells: gustatory cues for dietary fat. *American Journal of Physiology: Cell Physiology, 272*, pp. C1203-1210.

Green, B. G. (2012) Chemesthesis and the chemical senses as components of a "chemofensor complex". *Chemical Senses, 37*, pp. 201-206.

Harvey, R. B. (1920) The relation between the total acidity, the concentration of the hydrogen ion, and the taste of acid solutions. *Journal of the American Chemical Society, 42*, pp. 712-714.

Herrada, G. and Dulac, C. (1997) A novel family of putative pheromone receptors in mammals with a topographically organized and sexually dimorphic distribution. *Cell, 90*, pp. 763-773.

Hisatsune, C., Yasumatsu, K., Takahashi-Iwanaga, H., Ogawa, N., Kuroda, Y., Yoshida, R., Ninomiya, Y. and Mikoshiba, K. (2007) Abnormal taste perception in mice lacking the type 3 inositol 1, 4, 5-trisphosphate receptor. *Journal of Biological Chemistry, 282*, pp. 37225-37231.

Hoon, M. A., Adler, E., Lindemeier, J., Battey, J. F., Ryba, N. J. and Zuker, C. S. (1999) Putative mammalian taste receptors: a class of taste-specific GPCRs with distinct topographic selectivity. *Cell, 96*, pp. 541-551.

Houamed, K. M., Kuijper, J. L., Gilbert, T. L., Haldeman, B. A., O'Hara, P. J., Mulvihill, E. R., Almers, W. and Hagen, F. S. (1991) Cloning, expression, and gene structure of a G protein-coupled glutamate receptor from rat brain. *Science, 252*, pp. 1318-1321.

Huang, A. L., Chen, X., Hoon, M. A., Chandrashekar, J., Guo, W., Trankner, D., Ryba, N. J. and Zuker, C. S. (2006) The cells and logic for mammalian sour taste detection. *Nature, 442*, pp. 934-938.

Huang, L., Shanker, Y. G., Dubauskaite, J., Zheng, J. Z., Yan, W., Rosenzweig, S., Spielman, A. I., Max, M. and Margolskee, R. F. (1999) Ggamma13 colocalizes with gustducin in taste receptor cells and mediates IP3 responses to bitter denatonium. *Nature Neuroscience, 2*, pp. 1055-1062.

Huang, Y. A., Dando, R. and Roper, S. D. (2009) Autocrine and paracrine roles for ATP and serotonin in mouse taste buds. *Journal of Neuroscience, 29*, pp. 13909-13918.

Huang, Y. A., Maruyama, Y., Stimac, R. and Roper, S. D. (2008) Presynaptic (Type III) cells in mouse taste buds sense sour (acid) taste. *Journal of Physiology, 586*, pp. 2903-2912.

Huang, Y. A. and Roper, S. D. (2010) Intracellular Ca(2+) and TRPM5-mediated membrane depolarization produce ATP secretion from taste receptor cells. *Journal of Physiology, 588*, pp. 2343-2350.

Huang, Y. J., Maruyama, Y., Dvoryanchikov, G., Pereira, E., Chaudhari, N. and Roper, S. D. (2007) The role of pannexin 1 hemichannels in ATP release and cell-cell communication in mouse taste buds. *Proceedings of the National Academy of Sciences, USA, 104*, pp. 6436-6441.

Ide, N., Sato, E., Ohta, K., Masuda, T. and Kitabatake, N. (2009) Interactions of the sweet-tasting proteins thaumatin and lysozyme with the human sweet-taste receptor. *Journal of Agricultural and Food Chemistry, 57*, pp. 5884-5890.

Intelmann, D., Batram, C., Kuhn, C., Haseleu, G., Meyerhof, W. and Hofmann, T. (2009) Three TAS2R Bitter Taste Receptors Mediate the Psychophysical Responses to Bitter Compounds of Hops (*Humulus lupulus* L.) and Beer. *Chemosensory Perception, 2*, pp. 118-132.

Ishimaru, Y., Inada, H., Kubota, M., Zhuang, H., Tominaga, M. and Matsunami, H. (2006) Transient receptor potential family members PKD1L3 and PKD2L1 form a candidate sour taste receptor. *Proceedings of the National Academy of Sciences, USA, 103*, pp. 12569-12574.

Jiang, P., Cui, M., Zhao, B., Liu, Z., Snyder, L. A., Benard, L. M., Osman, R., Margolskee, R. F. and Max, M. (2005a) Lactisole interacts with the transmembrane domains of human T1R3 to inhibit sweet taste. *Journal of Biological Chemistry, 280,* pp. 15238-15246.

Jiang, P., Cui, M., Zhao, B., Snyder, L. A., Benard, L. M., Osman, R., Max, M. and Margolskee, R. F. (2005b) Identification of the cyclamate interaction site within the transmembrane domain of the human sweet taste receptor subunit T1R3. *Journal of Biological Chemistry, 280,* pp. 34296-34305.

Jiang, P., Ji, Q., Liu, Z., Snyder, L. A., Benard, L. M., Margolskee, R. F. and Max, M. (2004) The cysteine-rich region of T1R3 determines responses to intensely sweet proteins. *Journal of Biological Chemistry, 279,* pp. 45068-45075.

Kawai, T. and Fushiki, T. (2003) Importance of lipolysis in oral cavity for orosensory detection of fat. *American Journal of Physiology: Regulatory, Integrative and Comparative Physiology, 285,* pp. R447-454.

Kim, U. K., Jorgenson, E., Coon, H., Leppert, M., Risch, N. and Drayna, D. (2003) Positional cloning of the human quantitative trait locus underlying taste sensitivity to phenylthiocarbamide. *Science, 299,* pp. 1221-1225.

Kitagawa, M., Kusakabe, Y., Miura, H., Ninomiya, Y. and Hino, A. (2001) Molecular genetic identification of a candidate receptor gene for sweet taste. *Biochemical and Biophysical Research Communications, 283,* pp. 236-242.

Kohl, S., Behrens, M., Dunkel, A., Hofmann, T. and Meyerhof, W. (2013) Amino acids and peptides activate at least five members of the human bitter taste receptor family. *Journal of Agricultural and Food Chemistry, 61,* pp. 53-60.

Kretz, O., Barbry, P., Bock, R. and Lindemann, B. (1999) Differential expression of RNA and protein of the three pore-forming subunits of the amiloride-sensitive epithelial sodium channel in taste buds of the rat. *Journal of Histochemistry & Cytochemistry, 47,* pp. 51-64.

Kulkarni, B. and Mattes, R. (2013) Evidence for presence of nonesterified fatty acids as potential gustatory signaling molecules in humans. *Chemical Senses, 38,* pp. 119-127.

Kusakabe, Y., Yamaguchi, E., Tanemura, K., Kameyama, K., Chiba, N., Arai, S., Emori, Y. and Abe, K. (1998) Identification of two alpha-subunit species of GTP-binding proteins, Galpha15 and Galphaq, expressed in rat taste buds. *Biochimica et Biophysica Acta, 1403,* pp. 265-272.

Kusakabe, Y., Yasuoka, A., Asano-Miyoshi, M., Iwabuchi, K., Matsumoto, I., Arai, S., Emori, Y. and Abe, K. (2000) Comprehensive study on G protein alpha-subunits in taste bud cells, with special reference to the occurrence of Galphai2 as a major Galpha species. *Chemical Senses, 25,* pp. 525-531.

Kusuhara, Y., Yoshida, R., Ohkuri, T., Yasumatsu, K., Voigt, A., Hubner, S., Maeda, K., Boehm, U., Meyerhof, W. and Ninomiya, Y. (2013) Taste responses in mice lacking taste receptor subunit T1R1. *Journal of Physiology, 591,* pp. 1967-1985.

Laugerette, F., Passilly-Degrace, P., Patris, B., Niot, I., Febbraio, M., Montmayeur, J. P. and Besnard, P. (2005) CD36 involvement in orosensory detection of dietary lipids, spontaneous fat preference, and digestive secretions. *Journal of Clinical Investigation, 115,* pp. 3177-3184.

Le Neve, B., Foltz, M., Daniel, H. and Gouka, R. (2010) The steroid glycoside H.g.-12 from Hoodia gordonii activates the human bitter receptor TAS2R14 and induces CCK release from HuTu-80 cells. *American Journal of Physiology: Gastrointestinal and Liver Physiology, 299,* pp. G1368-G1375.

Levit, A., Nowak, S., Peters, M., Wiener, A., Meyerhof, W., Behrens, M. and Niv, M. Y. (2014) The bitter pill: clinical drugs that activate the human bitter taste receptor TAS2R14. *FASEB Journal, 28,* pp. 1181-1197.

Li, D. and Zhang, J. (2014) Diet shapes the evolution of the vertebrate bitter taste receptor gene repertoire. *Molecular Biology and Evolution, 31,* pp. 303-309.

Li, X., Staszewski, L., Xu, H., Durick, K., Zoller, M. and Adler, E. (2002) Human receptors for sweet and umami taste. *Proceedings of the National Academy of Sciences, USA, 99,* pp. 4692-4696.

Liman, E. R., Zhang, Y. V. and Montell, C. (2014) Peripheral coding of taste. *Neuron, 81*, pp. 984-1000.

Lin, W., Burks, C. A., Hansen, D. R., Kinnamon, S. C. and Gilbertson, T. A. (2004) Taste receptor cells express pH-sensitive leak K+ channels. *Journal of Neurophysiology, 92*, pp. 2909-2919.

Lin, W., Finger, T. E., Rossier, B. C. and Kinnamon, S. C. (1999) Epithelial Na+ channel subunits in rat taste cells: localization and regulation by aldosterone. *Journal of Comparative Neurology, 405*, pp. 406-420.

Lindemann, B. (1996) Taste reception. *Physiological Reviews, 76*, pp. 718-766.

Liu, D. and Liman, E. R. (2003) Intracellular Ca2+ and the phospholipid PIP2 regulate the taste transduction ion channel TRPM5. *Proceedings of the National Academy of Sciences, USA, 100*, pp. 15160-15165.

LopezJimenez, N. D., Cavenagh, M. M., Sainz, E., Cruz-Ithier, M. A., Battey, J. F. and Sullivan, S. L. (2006) Two members of the TRPP family of ion channels, Pkd1l3 and Pkd2l1, are co-expressed in a subset of taste receptor cells. *Journal of Neurochemistry, 98*, pp. 68-77.

Marchiori, A., Capece, L., Giorgetti, A., Gasparini, P., Behrens, M., Carloni, P. and Meyerhof, W. (2013) Coarse-grained/molecular mechanics of the TAS2R38 bitter taste receptor: experimentally-validated detailed structural prediction of agonist binding. *PLoS One, 8*, e64675.

Maruyama, Y., Pereira, E., Margolskee, R. F., Chaudhari, N. and Roper, S. D. (2006) Umami responses in mouse taste cells indicate more than one receptor. *Journal of Neuroscience, 26*, pp. 2227-2234.

Masu, M., Tanabe, Y., Tsuchida, K., Shigemoto, R. and Nakanishi, S. (1991) Sequence and expression of a metabotropic glutamate receptor. *Nature, 349*, pp. 760-765.

Matsumura, S., Mizushige, T., Yoneda, T., Iwanaga, T., Tsuzuki, S., Inoue, K. and Fushiki, T. (2007) GPR expression in the rat taste bud relating to fatty acid sensing. *Biomedical Research, 28*, pp. 49-55.

Matsunami, H. and Buck, L. B. (1997) A multigene family encoding a diverse array of putative pheromone receptors in mammals. *Cell, 90*, pp. 775-784.

Matsunami, H., Montmayeur, J. P. and Buck, L. B. (2000) A family of candidate taste receptors in human and mouse. *Nature, 404*, pp. 601-604.

Mattes, R. D. (2011) Accumulating evidence supports a taste component for free fatty acids in humans. *Physiology & Behavior, 104*, pp. 624-631.

Max, M., Shanker, Y. G., Huang, L., Rong, M., Liu, Z., Campagne, F., Weinstein, H., Damak, S. and Margolskee, R. F. (2001) Tas1r3, encoding a new candidate taste receptor, is allelic to the sweet responsiveness locus Sac. *Nature Genetics, 28*, pp. 58-63.

McLaughlin, S. K., McKinnon, P. J. and Margolskee, R. F. (1992) Gustducin is a taste-cell-specific G protein closely related to the transducins. *Nature, 357*, pp. 563-569.

Meyerhof, W. (2005) Elucidation of mammalian bitter taste. *Reviews of Physiology, Biochemistry and Pharmacology, 154*, pp. 37-72.

Meyerhof, W., Batram, C., Kuhn, C., Brockhoff, A., Chudoba, E., Bufe, B., Appendino, G. and Behrens, M. (2010) The molecular receptive ranges of human TAS2R bitter taste receptors. *Chemical Senses, 35*, pp. 157-170.

Miller, I. J., Jr., (ed.) (1995) *Anatomy of the peripheral taste system, In: Handbook of Olfaction and Gustation*, New York: Dekker.

Miyoshi, M. A., Abe, K. and Emori, Y. (2001) IP(3) receptor type 3 and PLCbeta2 are co-expressed with taste receptors T1R and T2R in rat taste bud cells. *Chemical Senses, 26*, pp. 259-265.

Mizutani, K., Kuramoto, T., Tamura, Y., Ohtake, N., Doi, S., Nakaura, M. and Tanaka, O. (1994) Sweetness of glycyrrhetic acid 3-O-beta-D-monoglucuronide and the related glycosides. *Bioscience, Biotechnology, and Biochemistry, 58*, pp. 554-555.

Montmayeur, J. P., Liberles, S. D., Matsunami, H. and Buck, L. B. (2001) A candidate taste receptor gene near a sweet taste locus. *Nature Neuroscience, 4,* pp. 492-498.

Nakajima, K., Morita, Y., Koizumi, A., Asakura, T., Terada, T., Ito, K., Shimizu-Ibuka, A., Maruyama, J., Kitamoto, K., Misaka, T. and Abe, K. (2008) Acid-induced sweetness of neoculin is ascribed to its pH-dependent agonistic-antagonistic interaction with human sweet taste receptor. *FASEB Journal, 22,* pp. 2323-2330.

Nelson, G., Chandrashekar, J., Hoon, M. A., Feng, L., Zhao, G., Ryba, N. J. and Zuker, C. S. (2002) An amino-acid taste receptor. *Nature, 416,* pp. 199-202.

Nelson, G., Hoon, M. A., Chandrashekar, J., Zhang, Y., Ryba, N. J. and Zuker, C. S. (2001) Mammalian sweet taste receptors. *Cell, 106,* pp. 381-390.

Oike, H., Nagai, T., Furuyama, A., Okada, S., Aihara, Y., Ishimaru, Y., Marui, T., Matsumoto, I., Misaka, T. and Abe, K. (2007) Characterization of ligands for fish taste receptors. *Journal of Neuroscience, 27,* pp. 5584-5592.

Oka, Y., Butnaru, M., von Buchholtz, L., Ryba, N. J. and Zuker, C. S. (2013) High salt recruits aversive taste pathways. *Nature, 494,* pp. 472-475.

Pepino, M. Y., Love-Gregory, L., Klein, S. and Abumrad, N. A. (2012) The fatty acid translocase gene CD36 and lingual lipase influence oral sensitivity to fat in obese subjects. *Journal of Lipid Research, 53,* pp. 561-566.

Perez, C. A., Huang, L., Rong, M., Kozak, J. A., Preuss, A. K., Zhang, H., Max, M. and Margolskee, R. F. (2002) A transient receptor potential channel expressed in taste receptor cells. *Nature Neuroscience, 5,* pp. 1169-1176.

Pronin, A. N., Xu, H., Tang, H., Zhang, L., Li, Q. and Li, X. (2007) Specific alleles of bitter receptor genes influence human sensitivity to the bitterness of aloin and saccharin. *Current Biology, 17,* pp. 1403-1408.

Richter, T. A., Dvoryanchikov, G. A., Chaudhari, N. and Roper, S. D. (2004) Acid-sensitive two-pore domain potassium (K2P) channels in mouse taste buds. *Journal of Neurophysiology, 92,* pp. 1928-1936.

Roland, W. S., van Buren, L., Gruppen, H., Driesse, M., Gouka, R. J., Smit, G. and Vincken, J. P. (2013) Bitter taste receptor activation by flavonoids and isoflavonoids: modeled structural requirements for activation of hTAS2R14 and hTAS2R39. *Journal of Agricultural and Food Chemistry, 61,* pp. 10454-10466.

Roland, W. S., Vincken, J. P., Gouka, R. J., van Buren, L., Gruppen, H. and Smit, G. (2011) Soy isoflavones and other isoflavonoids activate the human bitter taste receptors hTAS2R14 and hTAS2R39. *Journal of Agricultural and Food Chemistry, 59,* pp. 11764-11771.

Romanov, R. A., Rogachevskaja, O. A., Bystrova, M. F., Jiang, P., Margolskee, R. F. and Kolesnikov, S. S. (2007) Afferent neurotransmission mediated by hemichannels in mammalian taste cells. *EMBO Journal, 26,* pp. 657-667.

Rossler, P., Boekhoff, I., Tareilus, E., Beck, S., Breer, H. and Freitag, J. (2000) G protein betagamma complexes in circumvallate taste cells involved in bitter transduction. *Chemical Senses, 25,* pp. 413-421.

Rossler, P., Kroner, C., Freitag, J., Noe, J. and Breer, H. (1998) Identification of a phospholipase C beta subtype in rat taste cells. *European Journal of Cell Biology, 77,* pp. 253-261.

Roudnitzky, N., Bufe, B., Thalmann, S., Kuhn, C., Gunn, H. C., Xing, C., Crider, B. P., Behrens, M., Meyerhof, W. and Wooding, S. P. (2011) Genomic, genetic and functional dissection of bitter taste responses to artificial sweeteners. *Human Molecular Genetics, 20,* pp. 3437-3449.

Ryba, N. J. and Tirindelli, R. (1997) A new multigene family of putative pheromone receptors. *Neuron, 19,* pp. 371-379.

Sainz, E., Korley, J. N., Battey, J. F. and Sullivan, S. L. (2001) Identification of a novel member of the T1R family of putative taste receptors. *Journal of Neurochemistry, 77,* pp. 896-903.

Sako, N. and Yamamoto, T. (1999) Analyses of taste nerve responses with special reference to possible receptor mechanisms of umami taste in the rat. *Neuroscience Letters, 261,* pp. 109-112.

San Gabriel, A., Maekawa, T., Uneyama, H. and Torii, K. (2009) Metabotropic glutamate receptor type 1 in taste tissue. *American Journal of Clinical Nutrition, 90,* pp. 743S-746S.

Shi, P. and Zhang, J. (2006) Contrasting modes of evolution between vertebrate sweet/umami receptor genes and bitter receptor genes. *Molecular Biology and Evolution, 23,* pp. 292-300.

Soares, S., Kohl, S., Thalmann, S., Mateus, N., Meyerhof, W. and De Freitas, V. (2013) Different phenolic compounds activate distinct human bitter taste receptors. *Journal of Agricultural and Food Chemistry, 61,* pp. 1525-1533.

Soranzo, N., Bufe, B., Sabeti, P. C., Wilson, J. F., Weale, M. E., Marguerie, R., Meyerhof, W. and Goldstein, D. B. (2005) Positive Selection on a High-Sensitivity Allele of the Human Bitter-Taste Receptor TAS2R16. *Current Biology, 15,* pp. 1257-1265.

Stahler, F., Riedel, K., Demgensky, S., Neumann, K., Dunkel, A., Taubert, A., Raab, B., Behrens, M., Raguse, J. D., Hofmann, T. and Meyerhof, W. (2008) A Role of the Epithelial Sodium Channel in Human Salt Taste Transduction? *Chemosensory Perception, 1,* pp. 78-90.

Stevens, D. R., Seifert, R., Bufe, B., Muller, F., Kremmer, E., Gauss, R., Meyerhof, W., Kaupp, U. B. and Lindemann, B. (2001) Hyperpolarization-activated channels HCN1 and HCN4 mediate responses to sour stimuli. *Nature, 413,* pp. 631-635.

Syed, A. S. and Korsching, S. I. (2014) Positive Darwinian selection in the singularly large taste receptor gene family of an 'ancient' fish, Latimeria chalumnae. *BMC Genomics, 15,* 650.

Tan, J., Abrol, R., Trzaskowski, B. and Goddard, W. A. (2012) The 3D structure prediction of TAS2R38 bitter receptors bound to agonists phenylthiocarbamide (PTC) and 6-n-Propylthiouracil (PROP). *Journal of Chemical Information and Modeling, 52,* pp. 1875-1885.

Tanabe, Y., Masu, M., Ishii, T., Shigemoto, R. and Nakanishi, S. (1992) A family of metabotropic glutamate receptors. *Neuron, 8,* pp. 169-179.

Thalmann, S., Behrens, M. and Meyerhof, W. (2013) Major haplotypes of the human bitter taste receptor TAS2R41 encode functional receptors for chloramphenicol. *Biochemical and Biophysical Research Communications, 435,* pp. 267-273.

Tizzano, M., Dvoryanchikov, G., Barrows, J. K., Kim, S., Chaudhari, N. and Finger, T. E. (2008) Expression of Galpha14 in sweet-transducing taste cells of the posterior tongue. *BMC Neuroscience, 9,* 110.

Tomchik, S. M., Berg, S., Kim, J. W., Chaudhari, N. and Roper, S. D. (2007) Breadth of tuning and taste coding in mammalian taste buds. *Journal of Neuroscience, 27,* pp. 10840-10848.

Ugawa, S., Minami, Y., Guo, W., Saishin, Y., Takatsuji, K., Yamamoto, T., Tohyama, M. and Shimada, S. (1998) Receptor that leaves a sour taste in the mouth. *Nature, 395,* pp. 555-556.

van der Wel, H. and Loeve, K. (1972) Isolation and characterization of thaumatin I and II, the sweet-tasting proteins from Thaumatococcus daniellii Benth. *European Journal of Biochemistry, 31,* pp. 221-225.

Vandenbeuch, A., Clapp, T. R. and Kinnamon, S. C. (2008) Amiloride-sensitive channels in type I fungiform taste cells in mouse. *BMC Neuroscience, 9,* 1.

Voigt, N., Stein, J., Galindo, M. M., Dunkel, A., Raguse, J. D., Meyerhof, W., Hofmann, T. and Behrens, M. (2014) The role of lipolysis in human orosensory fat perception. *Journal of Lipid Research, 55,* pp. 870-882.

Watson, K. J., Kim, I., Baquero, A. F., Burks, C. A., Liu, L. and Gilbertson, T. A. (2007) Expression of aquaporin water channels in rat taste buds. *Chemical Senses, 32,* pp. 411-421.

Wiener, A., Shudler, M., Levit, A. and Niv, M. Y. (2012) BitterDB: a database of bitter compounds. *Nucleic Acids Research, 40,* pp. D413-419.

Winnig, M., Bufe, B., Kratochwil, N. A., Slack, J. P. and Meyerhof, W. (2007) The binding site for neohesperidin dihydrochalcone at the human sweet taste receptor. *BMC Structural Biology, 7,* 66.

Winnig, M., Bufe, B. and Meyerhof, W. (2005) Valine 738 and lysine 735 in the fifth transmembrane domain of rTas1r3 mediate insensitivity towards lactisole of the rat sweet taste receptor. *BMC Neuroscience, 6,* 22.

Wong, G. T., Gannon, K. S. and Margolskee, R. F. (1996) Transduction of bitter and sweet taste by gustducin. *Nature, 381*, pp. 796-800.

Wooding, S., Kim, U. K., Bamshad, M. J., Larsen, J., Jorde, L. B. and Drayna, D. (2004) Natural selection and molecular evolution in PTC, a bitter-taste receptor gene. *American Journal of Human Genetics, 74*, pp. 637-646.

Xu, H., Staszewski, L., Tang, H., Adler, E., Zoller, M. and Li, X. (2004) Different functional roles of T1R subunits in the heteromeric taste receptors. *Proceedings of the National Academy of Sciences, USA, 101*, pp. 14258-14263.

Yamato, M., Hashigaki, K., Uenishi, J., Yamakawa, I. and Sato, N. (1975) Chemical structure and sweet taste of isocoumarin and related compounds. *VI. Chemical & Pharmaceutical Bulletin (Tokyo), 23*, pp. 3101-3105.

Yarmolinsky, D. A., Zuker, C. S. and Ryba, N. J. (2009) Common sense about taste: from mammals to insects. *Cell, 139*, pp. 234-244.

Zhang, F., Klebansky, B., Fine, R. M., Xu, H., Pronin, A., Liu, H., Tachdjian, C. and Li, X. (2008) Molecular mechanism for the umami taste synergism. *Proceedings of the National Academy of Sciences, USA, 105*, pp. 20930-20934.

Zhang, Y., Hoon, M. A., Chandrashekar, J., Mueller, K. L., Cook, B., Wu, D., Zuker, C. S. and Ryba, N. J. (2003) Coding of sweet, bitter, and umami tastes: different receptor cells sharing similar signaling pathways. *Cell, 112*, pp. 293-301.

Zhang, Z., Zhao, Z., Margolskee, R. and Liman, E. (2007) The transduction channel TRPM5 is gated by intracellular calcium in taste cells. *Journal of Neuroscience, 27*, pp. 5777-5786.

Zhao, G. Q., Zhang, Y., Hoon, M. A., Chandrashekar, J., Erlenbach, I., Ryba, N. J. and Zuker, C. S. (2003) The receptors for mammalian sweet and umami taste. *Cell, 115*, pp. 255-266.

CHAPTER 4

Bioadhesion and oral fluids—perireceptor modulators of taste perception?

Christian Hannig[1], Matthias Hannig[2] & Martine Morzel[3]

[1] *Polyclinic of Operative Dentistry and Pediatric Dentistry, Medical Faculty Carl Gustav Carus, TU Dresden, Dresden, Germany*

[2] *Clinic of Operative Dentistry, Periodontology and Preventive Dentistry, University Hospital, Saarland University, Homburg/Saar, Germany*

[3] *Center for Taste and Feeding Behaviour, CNRS, INRA, University of Bourgogne Franche-Comté, Dijon, France*

4.1 Introduction

As extensively described in the previous chapter, a taste sensation is elicited when tastants interact with taste receptors, primarily located on the tongue. Lingual taste receptors, concentrated in taste buds, bathe in the oral fluid constituted of saliva, desquamating cells, microorganisms etc. The oral fluid can also form a structured biological layer on the tongue surface through bioadhesion mechanisms. It is therefore likely that this specific environment may have an impact, positive or negative, on the integrity, functionality or accessibility of the anatomical taste structures. Previous reviews gave a good overview on the known and proposed interplay between the sense of taste and oral fluids (Bradley and Beidler 2003, Carpenter 2013b, Liman et al. 2014, Matsuo 2000, Salles et al. 2011, Spielman 1990). However, in spite of the coexistence of saliva and tastants, an understanding of their interactions was considered as still incomplete in 1990 (Spielman 1990). This nearly 25 year old statement is still true and applies much more for the mechanisms of bioadhesion occurring in the oral cavity and their effects on taste perception.

Due to differences in taste and salivary physiology between animals and humans (Mese and Matsuo 2007, Schmale et al. 1993, Spielman et al. 1993, Takeda and Hoshino 1975), this chapter focuses on human subjects unless stated, and reviews the peri-receptor events occurring within the oral cavity, considering in particular the impact of saliva, salivary pellicles, bioadhesion phenomena and microflora on the gustatory function.

Flavour: From food to perception, First Edition.
Edited by Elisabeth Guichard, Christian Salles, Martine Morzel, and Anne-Marie Le Bon.

4.2 Saliva: a complex, dynamic and multifunctional fluid

4.2.1 Saliva composition

The oral fluid present in the mouth is referred to as "whole saliva". Yet, it is a mixed fluid composed of secretions from the major and minor salivary glands (for example the lingual von Ebner's glands), of crevicular fluid, and of constituents originating from exfoliating cells, mucosal secretions, from food debris or from oral microorganisms, respectively (Amerongen and Veerman 2002). All in all, the secretion of the major salivary glands dominates the composition of the oral fluids. From a compositional point of view, saliva is an aqueous solution composed of approximately 99% water. It contains inorganic constituents, mainly Na^+, Cl^-, HCO_3^-, K^+, Ca^{2+}, Mg^{2+} and phosphate which are present in the milli molar range of concentration (Amerongen and Veerman 2002, Dawes 1969, Dawes 1974). It also contains nucleic acids of bacterial or human origin (Park et al. 2006) and metabolites. Modern methods allow the evaluation of the human salivary metabolome which is of considerable interest in the context of taste. Typical metabolites in the oral fluids are organic acids (acetate, lactate, succinate and propionate), free amino acids, sugars or fatty acids - the main ones being oleic, stearic and palmitic acids (Neyraud et al. 2013). However, the bulk of the organic fraction of saliva is made up of proteins and peptides. The protein concentration of saliva can range from approximately 1 to 5 g/l depending on the secreting gland and sampling conditions (Dawes 1969, Dawes 1984, Dodds et al. 2005). The protein fraction is characterized by a very high diversity of constituents. Thus, a recent review states that, using different proteomics analytical strategies, more than 3000 different protein and peptide species have been identified in saliva (Amado et al. 2013, Oppenheim et al. 2007). The most abundant proteins of glandular origin are amylase, carbonic anhydrase VI (CAVI), cystatins, histatins, mucins, proline rich proteins (PRPs), and statherin, while other abundant proteins are for example lysozyme, peroxidase, immunoglobulins and serum albumin (Amerongen and Veerman 2002).

4.2.2 Sources of variations in saliva composition

In general, salivary parameters show a high interindividual and intraindividual variability. This complex fluid is subjected to flow and composition variations according to many factors. For example, acute physical or psychosocial stress may result in alteration of saliva composition, in particular in elevated levels of cortisol and amylase (Hellhammer et al. 2009, Nater et al. 2006) but also of neuropeptides such as vasoactive intestinal peptide (Ventre et al. 2013). The smoking status has an impact on saliva composition, as demonstrated for example on its metabolome composition (Takeda et al. 2009). Another factor which has an impact on the composition of saliva is the presence of oral or even systemic pathologies, and the use of medication or medical treatment such as irradiation (Hannig et al. 2006, Sreebny 2000, Sreebny and Valdini 1988, Sreebny et al. 1989).

Even in a healthy situation, saliva is a dynamic fluid which varies for one given subject according to different timescales (Jehmlich et al. 2013). Thus, saliva flow and composition is subjected to circadian (Dallmann et al. 2012, Dawes 1972, Hardt et al. 2005) and circannual (Foglio-Bonda et al. 2013, Jehmlich et al. 2013, Kariyawasam and Dawes 2005) variations, and it also evolves along the different stages of life. This has been documented first in longitudinal studies during infancy (Morzel et al. 2012, Morzel et al. 2011, Ruhl et al. 2005, Sevenhuysen et al. 1984), where the most notable findings were an increase of amylase with age, serum albumin and of S-type cystatins associated to a decrease of cystatins A and B and of MUC7. Comparing children and adults of different ages (Cabras et al. 2009) allowed describing some changes in several salivary compounds (e.g., acidic PRPs, histatins), including a consistent rise in basic PRPs from childhood to adulthood. Comparing elderly vs. young subjects also brought some information, in particular a decline in flow rate with age (Smith et al. 2013). Only limited modification of composition was observed by some authors (Morales-Bozo et al. 2006, Nassar et al. 2014) while others reported for example a significant decline of histatins and mucins with age (Dodds et al. 2005).

In addition to this time-dependent variation, the flow and composition of saliva can further be dependent on the subjects' individual diet. Acidity of the food is the strongest stimulus for salivary secretion. A higher fiber intake is also associated to an increased stimulated flow rate (Laine et al. 2014), while transition from a mixed to a lactovegetarian diet modifies the salivary ionic composition (Johansson and Birkhed 1994), and the transition from a milk-only to a diversified diet in infants increases the abundance of proteins and peptides with antiproteolytic properties (Morzel et al. 2012, Morzel et al. 2011).

Finally, salivary flow and composition are modulated by taste stimulation, as described in Chapter 12 of this book.

4.2.3 Main functions of saliva

Saliva is considered as the guardian of the oral cavity. Saliva, this complex and dynamic fluid, has many functions, reviewed for example by Humphrey and Williamson (2001), Amerongen and Veerman (2002) or Huq et al. (2007). Briefly, saliva ensures protection of the oral cavity against microorganisms through its antibacterial, antifungal and antiviral activities provided mainly by proteins such as histatins or lysozyme – with direct lytic actions on the microorganisms – or mucins – which can aggregate microorganisms and thus facilitate their clearance from the oral cavity by swallowing. Other unspecific antimicrobial components are lactoferrin or defensins whereas sIgA is the main specific antibacterial biomolecule in the saliva and the only secretory immunoglobulin. Salivary peroxidase and myeloperoxidase detoxify free radicals and serve as agonists to lysozyme by formation of antibacterial hypothiocyanate (C. Hannig et al. 2005, Hannig et al. 2008c). Saliva also protects surfaces of the oral cavity. Integrity and lubrication of the hard and soft tissues is for example ensured by formation of the dental and mucosal pellicles (see part 3 of this chapter). Saliva is also involved in maintaining the integrity of the teeth, through control of enamel mineralization and pH buffering. The latter is a generic protective function of saliva and it impacts the perception of acidic foodstuff (Mese and Matsuo 2007). The buffer capacity is modulated by the different buffer systems present in the oral fluids, namely the carbonate, the phosphate and the protein buffer systems.

Another function of saliva lies in its role in digestion, in particular through its amylolytic and lipolytic activity. Concerning amylolysis, that is, the hydrolysis of α-1-4 glycosidic bonds of starch yielding malto-oligosaccharides, there is no clear consensus on the respective contributions of salivary *vs.* pancreatic amylase in starch digestion. Yet, the most abundant enzyme in the oral fluids is α-amylase; it is known that the AMY1 (salivary amylase) gene copy-number and the corresponding protein level in saliva are lower in populations with traditionally low-starch diets (Perry et al. 2007). Subjects with high amylase activity also report a faster and greater decrease in the perceived viscosity of a starch solution during oral processing (Mandel et al. 2010), suggesting that although food remains in the mouth for only a few seconds, salivary amylase may initiate the digestion of starch. Also the texture perception of food rich in starch is modulated by amylase (de Wijk et al. 2004). Furthermore, subjects with high salivary amylase activity have reduced postprandial blood glucose levels following starch ingestion (Mandel and Breslin 2012), which the authors putatively attributed to a better anticipatory metabolic response to signaling by in-mouth glucose release.

Concerning oral lipolysis, that is, the release of free fatty acids from triglycerides, it has been suggested to play a role in fat digestion specifically in newborns and young infants (Smith et al. 1986), where intestinal levels of pancreatic lipase are lower than in adults (Lindquist and Hernell 2010). There is limited evidence for the sources of lipase activity in human saliva. Hamosh and Burns (1997) have detected lipolytic activity in human von Ebner's glands but they investigated autopsy material and *postmortem* changes cannot be excluded. Spielman et al. (1993) detected low levels of lipase activity in von Ebner's glands saliva. Based on these results and on different studies in rats, the von Ebner's glands would be the only source of lipase among salivary glands (Hamosh and Burns 1977, Riva et al. 1985). To date, however, lingual lipase has not been isolated in humans and the origin of lipolysis in the oral cavity, whether host-derived or bacterial, remains unknown.

Finally, it might be speculated that certain endo- and exo-peptidases and glutamate releasing proteases (Schlueter et al. 2012) in the oral fluids may contribute to protein pre-digestion and could be relevant for the perception of umami.

Another direct effect of saliva on the digestive processes is the binding of food components such as tannins or other polyphenolic compounds by salivary proteins, in particular PRPs (Lu and Bennick 1998) or histatins (Yan and Bennick 1995). Interestingly, consumption of food rich in tannins leads to an increased synthesis and release in saliva of PRPs in many animals, as reviewed by Bradley and Beidler (2003). In this context it is noteworthy that polyphenols can inhibit digestive enzymes such as amylase (C. Hannig et al. 2005, Xiao et al. 2013) or proteolytic digestive enzymes (Oh and Hoff 1986).

Focusing now on food perception, the role of saliva has been pointed out in sensory tactile attributes such as creaminess, roughness or astringency (de Wijk and Prinz 2006), mainly through its lubricating properties. These properties are also crucial in the formation of a food bolus that is cohesive and that can pass the esophagus without damage to the mucosa (Pedersen et al. 2002). Mucins (MUC5B, MUC7) and also other glycoproteins and proteins such as statherin and PRPs are of essential relevance for this lubricating process (Amerongen and Veerman 2002, Bradley and Beidler 2003, Tabak et al. 1982). Mucins are of high hydrophilicity and retain much water contributing to insalivation of the bolus (Pedersen et al. 2002). In this sense oral processing of food is a mixture of mechanical crushing and mashing combined with hydration and mixing with biomolecules from the oral fluids (Amerongen and Veerman 2002, Carpenter 2013b, Chen 2009, Matsuo 2000, Salles et al. 2011). Saliva can also bind and/or release food flavour compounds (tastants, aromas, trigeminal compounds), as described in Chapter 12 of this book.

Finally, saliva is also involved in the maintenance of the gustatory function. However it has to be kept in mind that the physiological function of many salivary peptides and proteins has not been investigated until now. Many of them, as described for example by Siqueira and Dawes (2011), might contribute to food and taste perception.

4.3 Interaction of the oral fluids with taste receptors

In 1986, a controversial article has been published reporting the case of 8 patients with no salivary flow from any of the major salivary glands, and exhibiting normal suprathreshold taste performance (Weiffenbach et al. 1986). The authors therefore concluded that the functionality of the taste system is not dependent on the presence of saliva in the oral cavity. However, many other studies have pointed at a role of saliva in the taste function through three main mechanisms (Matsuo 2000, Mese and Matsuo 2007). First, saliva contributes to the maintenance of taste buds or taste receptor cells. Second, it contains elements that can bind and/or activate taste receptors and thus it can modulate the response of the taste receptors to specific tastants. Finally, a hypothesis proposes that saliva or salivary films may modulate accessibility of tastants to taste receptors by creating a sort of "soft shield" at the surface of the taste buds. Table 4.1 presents the saliva components with a documented or proposed role in taste perception, which will be described in detail below.

Table 4.1 Functional components of oral fluids potentially relevant for taste perception in man (findings based only on investigations in animals are specified in brackets).

Constituent	Documented origin	Suggested impact on the taste function
Lipocalin 1	von Ebners' gland	Concentration and delivery of hydrophobic tastants to taste receptors, protection of taste structures against detergent effect of fatty acids
Ebnerin	von Ebners' gland (rat)	Carrier for growth factor produced in von Ebner's gland
Epidermal Growth Factor (EGF)	All salivary glands	Growth factor of taste buds
Transforming Growth Factor-α (TGF-α)	All salivary glands	Growth factor of taste buds
Matrix Metalloproteinase-3 (MMP-3)	All salivary glands Gingival fibroblasts	Regeneration of taste buds
Carbonic anhydrase VI (CAVI)	Major salivary glands mainly von Ebners' gland	Growth factor of taste buds Protection of taste buds against apoptosis through regulation of pH Buffering action: regulation of $[H^+]$ / acid perception
Zinc	Plasma	Functionality of CAVI and possibly of MMP-3
Sodium	Plasma / re-absorption along salivary ducts	Constant stimulation of receptors (self-adaptation to saltiness)
Glutamate	Plasma Proteolysis of salivary proteins?	Constant stimulation of receptors (self-adaptation to umami)
Bitter peptides (e.g., RP, GRP)	Proteolysis of Proline-Rich Proteins?	Constant stimulation of receptors (self-adaptation to bitterness)?
Cystatins	Submandibular glands mainly	Control of proteolysis: modulation of mucosal pellicle structure?
Amylase	Major salivary glands mainly	Release of short starch hydrolysates
Lipase	Unknown: human or bacterial? von Ebners' gland (rat)	Release of free fatty acids from triglycerides
Basic Proline-Rich Proteins	Parotid glands mainly	Binding of polyphenols and other bitter tastants (PROP) Protection of amylase against denaturation by polyphenols
Histatin 5	Major salivary glands	Binding of quinine

4.3.1 The particular status of the von Ebner's glands secretions

Among minor salivary glands, the von Ebner's glands yield exceptional characteristics as they are purely serous in contrast to the other minor salivary glands (Riva and Tandler 2000). Their morphology differs from other small salivary glands (Spielman et al. 1993) and the secretory granules are different from those of the major salivary glands (Riva et al. 1985). This might result in a distinct composition. The secretion rate of the von Ebner's glands is very low (resting saliva: 2.3 μl/ min, stimulated saliva: 4.5 μl/min) (Eliasson and Carlen 2010, Spielman et al. 1993), a fact that is detrimental to studies on the composition of this specific saliva. The glands are located at the extremities of the clefts of circumvallate and foliate papillae. Therefore, their secretions are in immediate contact with the taste buds of these papillae, and there have been many speculations on whether von Ebner's glands' saliva may play a specific role in the gustatory function. Previously, rinsing of the taste buds was regarded as their main function (Riva et al. 1985). For example, it was demonstrated that the neurophysiological response to various tastants measured in circumvallate papillae of rats was reduced when salivary secretion from von Ebner's glands was induced (Gurkan and Bradley 1988b) and the authors proposed that such saliva aids in rinsing the papilla. The composition of the von Ebner's glands secretion has not been fully explored until now: proteome analysis has not been performed for this type of minor salivary glands' secretion (Siqueira et al. 2008). However, it was analyzed directly in oral fluids collected from the duct openings, and indirectly by immunological and histochemical methods adopted on gland parenchyma. Typical salivary components such as lactoferrin, lysozyme, peroxidase, secretory IgA, histatins, CAVI, amylase, and statherin have been detected in studies in humans (Hand et al. 1999, Leinonen et al. 2001, Moro et al. 1984, Piludu et al. 2006, Spielman et al. 1993). Many of them have antiviral, antibacterial and other protective properties. PRPs were detected in VEG saliva of different animals (Ahmad et al. 2004, Azen et al. 1990, Hand et al. 1999).

Compositional specificity of VEG saliva was also sought for, first by studying protein expression in VEG glands from rats (Schmale et al. 1990) and human subjects (Bläker et al. 1993, Li and Snyder 1995). Two proteins specific to (or at least greatly overexpressed in) this saliva were identified. These are the VEG protein, also termed lipocalin 1, and Ebnerin (Sbarbati et al. 1999). By sequence homology with other transporters of hydrophobic molecules, it was suggested that lipocalin 1 protein may be necessary for the concentration and delivery of hydrophobic sapid molecules to the taste system (Schmale et al. 1990), and that lipocalin 1 may protect the membranes of the taste structures against the detergent effect of fatty acids produced locally by the epithelium or brought about by food (Bläker et al. 1993). Interestingly, VEG or lipocalin can act as inhibitor of cysteine proteinases and serves as an oxidative stress induced scavenger of radicals (Amerongen and Veerman 2002, Redl et al. 1999, Veerman et al. 1992).

Concerning Ebnerin, released along the apical region of taste buds in rat circumvallate papillae, its suggested function would be to serve as carrier for putative growth factors produced by the VEGs (Li and Snyder 1995). However, Ebnerin was to our knowledge never identified in human saliva, while the presence of lipocalin 1 was confirmed in VEG saliva (Spielman et al. 1993). Besides VEG saliva that had received special attention in the 90's, constituents of whole saliva have also been proposed as important in the taste function (Matsuo 2000, Mese and Matsuo 2007).

4.3.2 Factors involved in growth and regeneration of taste buds

Well-known growth factors in saliva are the epidermal growth factor (EGF) and the transforming growth factor-alpha (TGF-α). EGF has received special attention since it is present in human saliva at relatively high concentration, in the range of 0.3 to 5 ng/ml (Valdez and Fox 1991). EGF is secreted by the parotid and submandibular glands. Salivary EGF is involved in wound healing within the oral cavity and is induced for example by periodontal surgery (Oxford et al. 1998), but it has also been suggested to act on taste buds maintenance. This was demonstrated in rats, where removal of salivary glands resulted in a great loss and morphological alteration of fungiform taste buds, and where the effects were reversed by administration of EGF in the drinking water (Morris-Wiman et al. 2000). Interestingly, circumvallate taste buds were not affected by EGF deprivation. In another study, exogenous EGF enhanced proliferation of taste cells obtained by primary cultures of rat circumvallate and foliate papillae (Ozdener and Rawson 2012). However, to our knowledge, salivary EGF has never been investigated in relation to the taste function in human subjects.

Several studies, particularly focusing on saliva of patients with taste dysfunctions, suggested that saliva contains other factors of growth and regeneration of taste buds. This is the case for example of carbonic anhydrase VI (CAVI previously also termed gustin) (Bradley and Beidler 2003, Leinonen et al. 2001, Thatcher et al. 1998), which was originally isolated and characterized from parotid saliva (Henkin et al. 1975b). Salivary levels of CAVI are reduced in hypogeusic patients (Shatzman and Henkin 1981) and this decrease is accompanied by development of taste bud anatomical abnormalities (Henkin et al. 1999). CAVI was therefore considered as a trophic factor involved in taste bud growth and development. It was later shown that CAVI is expressed by major salivary glands but also by the human von Ebner's glands (Leinonen et al. 2001). The authors proposed that CAVI may be either a growth factor, or that it may protect the taste buds from apoptosis through its main enzymatic activity resulting in pH regulation. In the same line of ideas, the metalloproteinase MMP-3 was demonstrated to be reduced in whole saliva of patients with taste disorders, and was suggested to serve in continuous regeneration of taste buds (Watanabe et al. 2005). No direct evidence for this assumption was provided, although it is quite consistent with the positive impact of MMP-3 on odontoblast-like cells' proliferation (Hiyama et al. 2013). To be fully functional, carbonic anhydrase requires zinc to be bound at its active site *via* at least one histidine residue (Shatzman and Henkin 1980). MMP-3 is also a zinc-dependent protease. This partly explains why deficiency of salivary zinc is associated to impaired taste function (Shatzman and Henkin 1981, Watanabe et al. 2005). An apparently opposite observation was reported, since salivary zinc levels were higher in subjects unable to perceive the bitterness of the compound 6-n-propyl thiouracil (PROP). In that case, however, the specific allelic form of the CA6 gene in nontasters affected the protein conformation and, thereby, its ability to bind zinc, which resulted in higher free zinc in saliva. Finally, cAMP and cGMP were also found to be greatly reduced in parotid saliva of patients affected by sensory dysfunctions, in particular in patients with hypogeusia (Henkin and Velicu 2012). Intracellularly, cAMP takes part in the signaling cascade of transduction of bitter and sweet stimuli (Herness and Gilbertson 1999). However, it appears rather unlikely that cAMP secreted in the oral cavity can reach the intracellular medium and thus, by homology with the effect of cGMP on proliferation of sensory neuron precursors, Henkin and Velicu (2012) proposed that salivary cyclic nucleotides may also act as putative growth factors of taste structures.

4.3.3 Constant stimulation and conditioning of taste receptors by salivary constituents

It is conceivable that human taste perception is affected by changes in the peri-receptor milieu. Indeed, taste buds are constantly bathed by the oral fluids containing molecules (e.g., NaCl) that can elicit a taste response. However, in normal conditions, human subjects do not perceive the taste of their own saliva, due to the so-called "adaptation" or "self-adaptation" phenomenon (Mese and Matsuo 2007). Adaptation to salivary levels of NaCl was initially evidenced and thoroughly described by comparing detection and recognition threshold of salty taste after adapting the tongue to either water or a 0.01M NaCl solution (Bartoshuk 1978). Based on the lower thresholds obtained after adaptation to water, the author proposed that saltiness can be perceived only at concentrations above average basal salivary levels. Later, this concept was refined and went down to the individual level, where salivary sodium concentrations were measured for five subjects, both in non-chewing and chewing conditions. The study concluded that small changes in sodium levels in secreted saliva were sufficient to alter sensitivity to saltiness (Delwiche and Omahony 1996). This general concept of self-adaptation may in theory apply to any other sapid constituent of saliva. It was formally reported for glutamate, responsible for the umami taste. Subjects with high *vs.* low glutamate endogenous levels in saliva did not differ in terms of electrogustometric thresholds or perceived intensity of monosodium glutamate at suprathreshold concentrations, although their hedonic responses were different (Scinska-Bienkowska et al. 2006).

There are other means by which saliva could act directly on taste receptors although these are mainly hypotheses. Thus, some proteins that can bind sweet taste receptors present some homology with salivary cystatins. This is the case in particular for the sweet-tasting monellin, a member of the cystatin superfamily, which can stabilize the active form of the receptor through the "wedge" model (Tancredi et al. 2004, Temussi 2011). Although the protein conformation rather than sequence homology seems to be important in this model, it could be interesting to investigate further the interaction between salivary cystatins and taste receptors, especially in the context where S-type cystatins abundance has been linked to taste sensitivity in adults (Dsamou et al. 2012) and taste acceptance and sensitivity in infants (Morzel et al. 2014a). In the same line of idea, it may be worthwhile investigating constant activation of bitter taste receptors by short peptides such as those reported by Otagiri et al. (1985), who noted strong bitter taste when arginine is adjacent to a proline residue as in the peptides Arg-Pro (RP) or Gly-Arg-Pro (GRP) for example. Such short sequences are particularly frequent in salivary PRPs, and given the susceptibility of such proteins to proteolytic cleavage (Helmerhorst 2007), it is possible that short peptides may generate a background noise in bitterness perception by the self-adaptation phenomenon again.

Finally, an experiment in rats demonstrated that taste response to sucrose was enhanced in presence of saliva and that this effect was related to the high pH of saliva. The authors suggested that salivary pH was optimal for appropriate binding of a sweet stimulus to its receptor (Matsuo and Yamamoto 1992). However, no comparable study is available for human subjects.

4.3.4 Other salivary markers of taste function

In the past few years, there have been some attempts to correlate at-rest salivary composition to the taste function. Such studies have pointed at several constituents being "markers" of taste performance, although the mechanisms of action are not fully elucidated. These markers are for example proteins, such as CAVI or salivary cystatins which are discussed above. Zinc-α-2glycoprotein is also of special interest, since it was found to be under-expressed in patients complaining from taste disorders (Igarashi et al. 2008), in adult subjects hyposensitive to the taste of oleic acid or in 3-month-old infants readily accepting and therefore presumably less sensitive to the bitter taste of urea (Morzel et al. 2014a). Whether this protein impacts on the taste receptor functionality, or rather interacts with tastants, is unknown.

Another study also included metabolome analysis, and it was found that hyposensitive subjects had higher levels of organic acids (acetate, butyrate) which the authors attributed to a higher bacterial load in the oral cavity (Mounayar et al. 2014). This topic deserves to be investigated further.

4.4 Bioadhesion—A ubiquitous phenomenon in the oral cavity

Bioadhesion is an ubiquitous phenomenon in the aquatic milieu of the oral cavity and occurs on all hard and soft tissues (Bradway et al. 1989, Dawes et al. 1963, Gibbins et al. 2014a, Hannig and Hannig 2009). The first step of bioadhesion is the adsorption of proteins and glycoproteins from the oral fluids onto the substrate. This leads to the formation of a proteinaceous layer free of bacteria, termed pellicle; bacterial adherence and microbial biofilm formation represent the second step (Dawes et al. 1963, Marsh 2004). It is to be expected that pellicle formation on cellular or shedding surfaces will differ from non-cellular and therewith non-shedding surfaces, such as teeth or dental restorative materials (Gibbins et al. 2014b, Hannig and Hannig 2009). From the perspective of taste perireceptor mechanisms, the taste pore material is most relevant followed by mucosal pellicle, whereas teeth and dental materials seem to be of subordinate relevance for taste and taste perception on the first view (Azzali 1997, Matsuo 2000). Nevertheless, teeth and their surfaces are of considerable importance as tactile organs for the perception of attributes such as crispiness, stickiness, firmness or astringency (Gibbins and Carpenter 2013).

In addition, any flavoring substance might adsorb to all oral surfaces limiting the availability for the receptor cells. This applies to taste perception by the taste buds but also to the transfer of the volatile fraction to the *nervus olfactorius*. This adsorption ensures that tastants are available in the oral cavity after swallowing the bolus (Salles et al. 2011). The relevance of studying the dental pellicle in relation to taste also lies in methodological issues. Thus, in contrast to oral soft tissues, defined evaluation and investigation of the pellicle layer on solid substrates can be performed in a standardized manner by adopting an *in situ* model (Hannig and Hannig 2009). Enamel samples or specimens of different dental materials are exposed to the oral fluids with trays or splints worn by the subjects. Afterwards pellicle coated samples can be investigated with many different techniques extraorally (Hannig and Hannig 2009, M. Hannig et al. 2005, Kensche et al. 2013). Accordingly, it is not surprising that there is limited information on the pellicle which is formed on the mucosa due to difficulties in harvesting mucosal pellicles in a standardized manner form the vital cellular surfaces.

4.4.1 Dental pellicle

The dental pellicle serves as a lubricating layer, barrier and buffer on the non-shedding dental surfaces (Hannig and Joiner 2006) and yields a characteristic ultrastructure: an electron dense basal layer is covered by granular and globular structures (Hannig 1999, Hannig and Joiner 2006). The appearance and especially the thickness of the pellicles' outer layers depend on the formation time and on the site in the oral cavity (Hannig 1999). Palatal surfaces yield a rather thin pellicle with up to 50 nm in thickness after 120 min, whereas buccal pellicles are much thicker reaching up to 250 nm in thickness after 120 min. The basal layer is of considerable tenacity (Hannig et al. 2012, M. Hannig et al. 2005). The tenacious adsorption of proteins is driven by different polar and apolar physico-chemical interactions (Hannig and Hannig 2009). Pellicle formation occurs almost instantaneously and fast coating of the dental surfaces is ensured by adsorption of micelle-like globular structures as most salivary proteins are released as aggregates rather than single proteins (Soares et al. 2004, Vitkov et al. 2004). These protein complexes contain characteristic structural and functional components of the dental pellicle such as mucins, sIgA, lactoferrin, amylase, PRPs and lysozyme (Soares et al. 2004). This aspect might also be of relevance for the formation of pellicles on soft tissues. Recent studies investigated the proteome of the enamel pellicle, revealing that the pellicle on permanent enamel contains more than 100 different proteins and that the proteome on deciduous teeth differs considerably (Zimmerman et al. 2013). Structural components of the enamel pellicle are mucins (MUC 5, MUC 7), PRPs, histatins, lactoferrin, statherin, amylase, lysozyme, sIgA and numerous other (glyco)proteins and peptides (Hannig and Joiner 2006, Lendenmann et al. 2000, Li et al. 2004). Many functional components of the saliva are also present in the dental pellicle (C. Hannig et al. 2005). On the one hand, there are antibacterial and protective enzymes such as lysozyme and peroxidase (Hannig et al. 2009b, Hannig et al. 2008c). Also carbonic anhydrases are present in the pellicle with potential impact on buffering mechanisms (Leinonen et al. 1999, Li et al. 2004, Vacca Smith and Bowen 2000). On the other hand, enzymes of carbohydrate metabolism are integral components, high activities of immobilized amylase activity have been detected in the enamel pellicle (Hannig et al. 2004). This ensures optimal accessibility of low molecular starch hydrolysates for glycolytic bacteria. Furthermore, bacterial glycosyltransferases are integrated in the pellicle (Bowen and Koo 2011, Hannig et al. 2008a).

Small amounts of proteolytic and trypsin like enzyme activity are detectable as well as transglutaminase. Both seem to be relevant for maturation of the pellicle (Hannig et al. 2008d). Interestingly, lipase is not present in the *in situ* pellicle (own findings). Lipids are present in the pellicle but less information is available (Hannig and Joiner 2006, Reich et al. 2013, Slomiany et al. 1984). Physiologically, palmitic- (16:0), stearic- (18:0), oleic- (18:1n9c), and erucic acid (22:1n9c) are the most abundant fatty acids in the pellicle layer (Reich et al. 2013).

The pellicle serves as a semipermeable barrier and buffer and protects the tooth surface against acidic noxae to a certain extent (Hannig and Hannig 2014). At pH 3.0 the dental pellicle reduces the mineral loss at the enamel surface about 60%, but demineralized calcium and phosphate ions still pass the protein network as do protons or H_3O^+-ions, respectively (Hannig et al. 2012). Furthermore, longer interaction with acid leads to degeneration of the pellicle layer in a dose- and time- dependent manner (Hannig et al. 2009a, Hannig et al. 2012). This has to be kept in mind when discussing the effect of acids on oral soft tissues and the access of ions and protons to the taste receptor cells.

Application of polyphenolic compounds leads to some denaturation and maturation of the dental pellicle, the ultrastructure is of higher density (Joiner et al. 2006, Joiner et al. 2004, Joiner et al. 2003). Interestingly, peroxidase activity of the pellicle layer is inactivated for up to 45 min after rinsing with a tea rich in polyphenols indicating sustainable denaturation (Hannig et al. 2008b).

4.4.2 Mucosal pellicle

Processing of the food components in the oral cavity is certainly modulated by interactions with the mucous and proteinaceous structures on the soft tissues (Bradway et al. 1989, Matsuo 2000, Tabak et al. 1982). However, the investigation of the mucosal pellicle is much more complicated than that of the dental pellicle, and even a clear definition of this relevant structure is not totally consensual. It is for example difficult to differentiate the mucosal pellicle and the residual fluid or salivary film which is made of moving saliva or salivary components loosely associated to the epithelial cells and/or mucosal pellicle (Pramanik et al. 2010).

Mucosal cells have a dynamic fluid cell membrane mainly composed of phospholipids with numerous receptor proteins, channel proteins and transporter systems (fluid mosaic). Ions and biomolecules released by the cells as well as membrane enzymes have considerable impact on the process of bioadhesion. Accordingly, the underlying substratum of the mucosal pellicle is much more dynamic than the non-shedding dental surfaces. Furthermore, the cells are coated by the glycocalix (Bradway et al. 1989). This polysaccharide matrix is excreted by the cells and composed of glycolipids, glycoproteins and proteoglycans. Last but not least, the mucosa and its level of keratinization are of high variability (Watanabe et al. 2013). There are soft and smooth, non-keratinized surfaces like the buccal sites as well as rugged and keratinized structures such as the tongue or the palatal mucosa, respectively (Bradway et al. 1989, Gibbins et al. 2014b). All these characteristics suggest a great heterogeneity in mucosal pellicles within the oral cavity, which is further exacerbated by the generally high turnover rate of the oral soft tissues with a big impact on the process of bio-adsorption.

Hydrophobic interactions are considered as most relevant driving forces for formation of the mucosal pellicle (Gibbins and Carpenter 2013, Gibbins et al.

2014a, Gibbins et al. 2014b), and epithelial transglutaminase is regarded as relevant enzyme for the crosslinking of membrane proteins and salivary components (Bradway et al. 1989, Bradway et al. 1992, Gibbins et al. 2014b). In contrast to dental enamel where the pellicle is a prerequisite for bacterial adherence, direct attachment of bacteria to epithelial cell membranes is possible in the oral cavity (Chagnot et al. 2013, Hannig and Hannig 2009, Vitkov et al. 2002a, Vitkov et al. 2001, Vitkov et al. 2002b). Filaments and fimbriae of bacteria were shown to interact with the cell membrane. Watanabe et al. (2013) investigated the ultrastructure of the rat tongue mucosa cells and adherent bacteria. In most cases bacteria associated with the epithelial cell membrane were surrounded by glycoproteins. However it is not clear if these glycoproteins are of bacterial, cellular or salivary origin (Watanabe et al. 2013). This is the general dilemma of research on the mucosal pellicle. High resolution electron microscopic techniques are necessary to evaluate the formation and ultrastructure of non-cellular structures on oral mucosal cells. This should be combined with gold-immuno-labelling of typical cellular, bacterial and salivary proteins in order to define that what is considered as mucosal pellicle. Gold-immuno-labelling of a typical pellicle component (MUC5B) allowed defined visualization of the mucosal pellicle (Morzel et al. 2014b). The ultrastructure of the mucosal pellicle covering the characteristic epithelial *microplicae* (Asikainen et al. 2012, Kullaa et al. 2014) was of low density and yielded discontinuity; the thickness showed considerable variability and reached up to 100 nm (Morzel et al. 2014b). In contrast to the enamel pellicle, no electron dense basal layer was observed or it was not distinguishable from the plasma membrane, respectively (Asikainen et al. 2012, Kullaa et al. 2014, Morzel et al. 2014b, Vitkov et al. 2001) These rather scattered inhomogeneous pellicle-like structures of the so called mucosal pellicle were also described in other electron microscopic studies (Vitkov et al. 2001). The ultrastructure was partially filamentous and partially of a fancy, fine granular appearance (Watanabe et al. 2013). It is noteworthy that a recent paper established the term oral mucosal barrier complex, rather than mucosal pellicle (Asikainen et al. 2012).

Analysis of the mucosal pellicle's composition is often based on mucosal structures scrapped of the oral epithelium and washed with different buffers to remove loosely attached fractions (Gibbins et al. 2014a). Abundant and characteristic components of the mucosal pellicle are especially secreted soluble mucins (MUC5B, MUC7) and membrane associated epithelial mucins (MUC1, MUC4, MUC16) but also CAVI, sIgA, and cystatin (Asikainen et al. 2012, Gibbins et al. 2014a, Kullaa et al. 2014, Morzel et al. 2014b). MUC1 is also released by minor salivary glands contributing to the formation of the oral glycocalyx (Sengupta et al. 2001). This underlines that the mucosal pellicle is a mixed coating of salivary and epithelial macromolecules providing a special interface (Kullaa et al. 2014). Other studies also identified amylase and different kinds of PRPs (Bradway et al. 1992) in the mucosal pellicle while proteins that interact mainly with inorganic components of the enamel surface, such as statherin did not appear as abundant (Gibbins et al. 2014a). In the latter study, amylase showed only minimal levels of binding to epithelial cells. Some authors have also tried to simulate the specific process of bioadhesion on the mucosal structures with hydrophobic and hydrophilic particles (Gibbins et al. 2014b). However, this neglects the specific characteristics of a vital tissue. In this setup amylase and proline rich proteins were absent on the hydrophobic surface that served as a model for the mucosal structures. In contrast, considerable amounts of mucins, cystatin and statherin were detected.

Loosely attached mucins will contribute to lubrication of the food during bolus formation before the crushed food is swallowed (Gibbins et al. 2014a). However, based on the limited findings, only speculations are possible concerning the effects of the mucosal pellicle on taste and taste perception. For example, Dsamou et al. (2012) suggested that the enhanced in-mouth proteolysis that they observed in subjects hypersensitive to the bitter taste of caffeine may result in a thinner or looser structure of the mucosal pellicle on the tongue with an impact on accessibility of tastants to the taste buds.

4.4.3 Taste pore material—A specific pellicle structure?

The taste receptor cells are coated by the so called taste pore material (Azzali 1997, Azzali et al. 1996, Matsuo 2000). Little is known about the composition and the ultrastructure. The composition of the taste pore material might differ between the different types of papillae and is determined by the oral fluids in general. As many taste buds of foliate and circumvallate papillae are localized in the trenches of the taste papillae, the von Ebner's gland secretion seems to have a considerable impact on taste pore material, whereas mixed saliva has limited direct contact with the taste buds (Azzali 1997, Azzali et al. 1996, Mese and Matsuo 2007). In contrast, taste buds and, therewith, taste pore material localized at the surface of the anterior part of the tongue and soft palate are mainly influenced by mixed saliva as worked out by Matsuo (Gurkan and Bradley 1987, Gurkan and Bradley 1988a, Matsuo 2000). There are three different types of taste receptor cells: dark type cells, and light type II and III cells (Azzali 1997, Liman et al. 2014). The glycocalix of the different gustatory cells as well as exocytosis could contribute to the composition of the taste pore material. However, Azzali and colleagues (Azzali 1997, Azzali et al. 1996) expressed doubts on whether dark type I cells are relevant for formation of the taste pore material in man as suggested by other authors based on results observed in animals.

Regardless of its exact composition, the particularity of taste pore material is supported by the fact that its carbohydrate composition and the mucous coating on neighboring mucosal cells differ considerably in rabbits (Matsuo 2000, Witt and Miller 1992). The group around Azzali has investigated the ultrastructure of human papillae gustatory cells thoroughly (Azzali 1997, Azzali et al. 1989a, Azzali et al. 1989b). The extension of the microvillous protrusions into the taste canal differs among foliate, vallate and fungiform papillae. In foliate and vallate papillae the microvilli fill the taste canal whereas in fungiform papillae the apex of the microvilli never reaches above 60% of this canal. This certainly has an impact on the process of bioadhesion and on the taste pore material. The authors also described the acellular structures coating the plasma membrane of the microvillous structures. They assumed a high amount of polysaccharides (Azzali et al. 1989a, Azzali et al. 1989b). It is of considerable interest that the proteinaceous taste pore material has no electron dense basal layer and no electron dense structure in contrast to the pellicle on non-shedding surfaces (Arvidson et al. 1981, Azzali 1997, Ovalle and Nahirney 2013, Paran et al. 1975). Accordingly, the barrier-effect of the taste pore material might be less pronounced than that of the electron dense basal pellicle on dental hard tissues. Therefore, it will allow the diffusion of tastants to the receptor cells. On the other hand, the taste pore material may also protect the taste receptor cells against external noxae, and therefore, it can be speculated that some taste stimuli, pre-processed by the oral fluids, may not invade the taste pore unhamperedly (Matsuo 2000).

Based on the findings for the dental pellicle and for the mucosal pellicle (Gibbins et al. 2014a, Gibbins et al. 2014b, Hannig and Hannig 2009, Hannig and Hannig 2014, Hannig and Joiner 2006), it is also possible to hypothesize some functional characteristics of the taste pore material relevant for transmission of tastants (Figure 4.1), though its composition will certainly differ from other pellicles. Many typical pellicle proteins with an impact on taste and taste perception could be present in taste pore material and interact with different metabolites. Molecules with bitter taste such as some polyphenols would be aggregated by PRPs (Bennick 2002, Ferruzzi et al. 2012, Nayak and Carpenter 2008). Similarly to what occurs in whole saliva, quinine may also be bound by histatin (Wada et al. 2010), whereas taste perception of sweetness might be enhanced by amylase which is an abundant component of the enamel pellicle (Hannig et al. 2004). Even proteolytic components of taste pore material or the "taste pore pellicle" might facilitate the perception of umami through release of free amino-acids including glutamate, although proteolytic activity of the enamel pellicle is very limited (C. Hannig et al. 2005, Hannig et al. 2008d). CAVI present in the von Ebner's gland secretion could catalyze buffering reactions diminishing the perception of acidic noxae (Leinonen et al. 2001). Finally, the taste of fat is sometimes regarded as sixth taste modality (Mounayar et al. 2014, Passilly-Degrace et al. 2014). Lipase is present in the taste pores of rodents and potentially releases long chain fatty acids interacting with lipid sensors such as CD36 or GPR120 (Passilly-Degrace et al. 2014). Due to the fact that the von Ebner's glands are regarded as the main, yet hypothetical in human, source for lipase in the oral cavity, this effect might be pronounced in the taste pores and in taste pore material. In line with these considerations, Azzali (1997) postulated that the taste pore material is of more relevance in the transmission of the gustatory stimulus than the intrinsic membrane proteins.

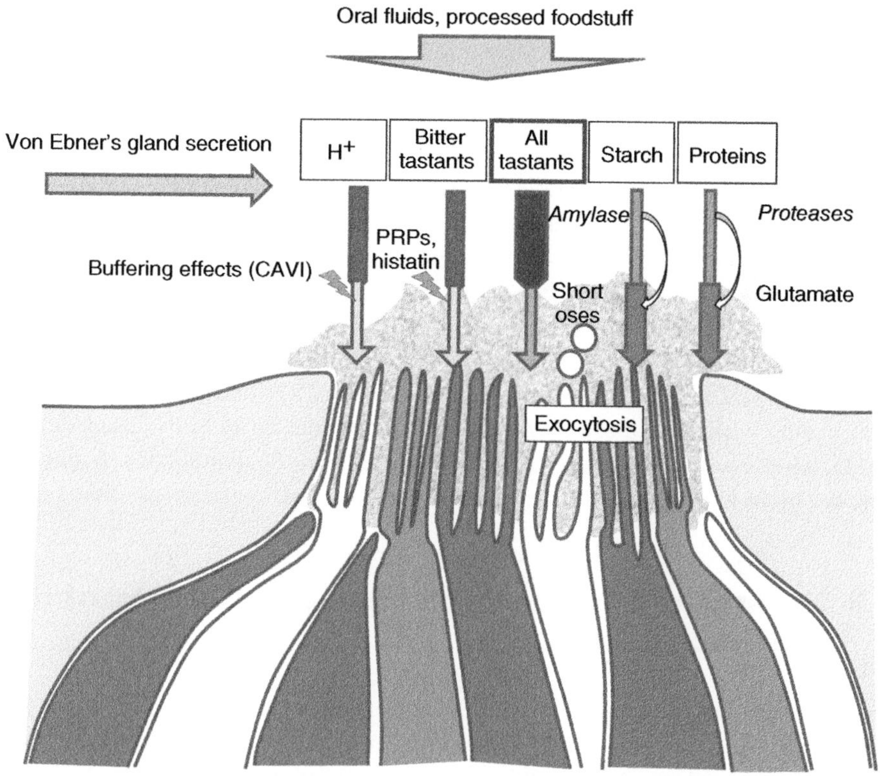

Figure 4.1 Scheme on interactions of tastants and taste pore material (the illustration was modified according to Matsuo 2000). There are three different types of taste receptor cells: dark type cells, and light type II and III cells (Source: Adapted from Azzali 1997, Liman et al. 2014). In general, the taste pore material serves as a barrier and buffer, but due to certain enzymes and proteins tastants might be processed and modified; PRPs: proline rich proteins; CAVI: carbonic anhydrase VI.

4.4.4 Astringency—A perception governed by modification of pellicle structures?

In the context of perireceptor events, the phenomenon of astringency is still rather enigmatic (Carpenter 2013a, Gibbins and Carpenter 2013, Nayak and Carpenter 2008). Based on the current knowledge on the mucosal and the dental pellicle, possible mechanism might be postulated. Thereby, it has to be kept in mind that astringency is perceived by contact of the oral soft tissues with the teeth as well as by mucosal – mucosal interactions. Interaction of tanning agents with the acellular layer on the gingival and mucosal cells might differ. A basal pellicle as observed with the enamel pellicle is not detectable on epithelium cells and the pellicle layer on the soft tissues is assumed to be of lower tenacity. Furthermore, the amount of PRPs available for the interactions with polyphenolic compounds differs between the teeth and the oral soft tissues (Gibbins et al.

2014a, Hannig and Joiner 2006). Accordingly, the astringent compounds might denature, depolarize and in part remove the mucosal pellicle and parts of the cell membrane. In contrast, the pellicle on the dental hard tissues is of higher tenacity, thus astringent foodstuffs have a tanning and denaturing effect on the dental pellicle, but, it is not removed (Joiner et al. 2006, Joiner et al. 2004, Joiner et al. 2003). Vulnerable enzymes are inactivated for a longer time by astringents (Hannig et al. 2008c). The denatured and potentially depolarized surfaces - either teeth or soft tissues – interact with each other, thereby roughness is perceived. The duration depends on the intensity of denaturation. Yet, it is not clear if a denaturation of the pellicle components or temporary depolarization induced by interaction with the tanning agents is responsible for astringency. Condensed tannins interact first and preferentially with acidic PRPs and statherin and in a second step with histatins, as well as glycosylated and basic PRPs (Soares et al. 2011). After a while, the altered pellicle structures on the dental hard tissues are covered by new layers of adsorbed proteins and the astringent effect disappears.

4.5 The impact of bacteria on taste and function of the taste receptors

The previous parts of this chapter focused on physiological interactions and on interplay with anthropogenic biomolecules. Yet, an enormous amount of bacteria and bacterial species as well as other microorganisms such as yeasts are present in the oral cavity as physiological commensals also in the healthy individual. Up to 700 different bacterial species are detectable in the oral cavity of one subject (Beikler and Flemmig 2011, Marsh 2004, Rautemaa and Ramage 2011, Seneviratne et al. 2011). These bacteria are organized in biofilms and occupy the ecological niche at the beginning of the digestive tract to avoid colonization by pathogenic bacteria. In healthy subjects the rugged tongue surface is considered as a specific reservoir of bacteria also for re-colonization of the teeth after oral hygiene (Matsui et al. 2014, Ohno et al. 2003, Quirynen et al. 2004). It is noteworthy that van Leuwenhook, the discoverer of bacteria, gained his epoch-making finding also from dental plaque and that he made a couple of experiments and scientific considerations on taste and taste perception (Bardell 1983). Coating of the tongue during transient infection and impact on taste were investigated by this scientific pioneer. Interestingly, tongue cleaning does not reduce the general bacterial load but it could modify the bacterial species and the density of tongue coating which might act as a barrier for tastants (Quirynen et al. 2004). Typical species in the specific microflora of the tongue are *Lysobacter-type species*, *Streptococcus salivarius*, *Veillonella dispar*, *Actinomyces odontolyticus*, *Atopobium parvulum* and *Veillonella atypica* as well as *Prevotella species* (Donaldson et al. 2005, Riggio et al. 2008). In patients suffering from halitosis and sometimes bad taste, there are especially *Lysobacter-type species*, *S. salivarius*, *Prevotella melaninogenica*,

Prevotella veroralis and *Prevotella pallens* (Riggio et al. 2008) but also *Veillonella alcalescens, Klebsiella pneumoniae,* and *Enterobacter cloacae* detectable (Rosing and Loesche 2011). The release of volatile sulfur compounds, indole, skatole, cadaverine, and putrescine leads to malodor (Rosing and Loesche 2011). Tongue cleaning removes the substrates for this process of putrefaction (Quirynen et al. 2004). Acute and chronic bacterial infections impact taste and taste perception, patients suffering from suppurating infections or dentitio difficilis often complain about perception of bad taste (Henkin et al. 1975a, Smith et al. 1976).

In contrast to common opinion, it is not clear if age *per se* has considerable impact on taste perception, morphology of taste buds, and on salivary flow rate and composition of the saliva (Azzali 1997, Landis et al. 2009, Moore et al. 1982). However, there is an interesting study on the impact of oral health and taste ability in acutely hospitalized older people offering information on the effects of the oral biofilm on taste perception (Solemdal et al. 2012). Many of these patients suffer from poor oral hygiene and heavy plaque accumulation. The study demonstrated clearly that decayed teeth, poor oral hygiene and high load of oral bacteria impaired total taste score. Remarkably, the threshold for sour taste was increased in patients with high growth of bacteria. This was especially associated with high levels of *lactobacilli* and *streptococci.* These species are highly acidogenic and acidophilic inducing an adaptation in sour taste perception (Solemdal et al. 2012). In the cited study, the tongue coating had no impact on taste perception in contrast to other studies (Ohno et al. 2003, Quirynen et al. 2004). Yet, the other studies investigated the effect of tooth cleaning procedures, whereas the study of Solemdal et al. (2012) compared older people with and without pronounced tongue coatings. This underlines the relevance of adaptive mechanisms in the perception of taste. Nevertheless, also *candida albicans* colonization, which is quite common for patients with dentures or after chemotherapy, is a cause for taste disorders (Jensen et al. 2008, Sakashita et al. 2004), and it has been hypothesized that *helicobacter pylori* infection might be associated with distortion on taste perception (Cecchini et al. 2013). In conclusion, alterations of the oral microbiome and bacterial load are of relevance for taste and taste perception. This is in line with the recent suggestion that, in a healthy group of subjects, abundance of metabolites of presumably bacterial origin differed between subjects highly or weakly sensitive to the taste of oleic acid (Mounayar et al. 2014).

In the context of the oral biofilm, pellicle formation and bioadhesion in the oral cavity it is well known that adoption of chlorhexidine induces a transient alteration of taste perception (Helms et al. 1995). The threshold for NaCl is increased as well as the taste intensity of bitter molecules (quinine-HCl). On the one hand chlorhexidine as a strong cation could have an impact on certain ion channels of taste buds, on the other hand its high substantivity and its high affinity to mucosal surfaces have an impact on all pellicle structures described in the present chapter. This surely affects taste perception far beyond ionic interactions (Helms et al. 1995).

4.6 Discussion and ideas for future research

The review of the literature indicates that interactions of food components with the oral fluids, the oral microorganisms and with the organic layers on oral tissues are of great relevance for taste and taste perception. However, all relevant structures have not been investigated completely until now. In fact, there is limited knowledge on ultrastructure, composition and functionality of many key components such as the mucosal pellicle or the taste pore material and the secretion of the minor salivary glands. For example, proteome analysis has already been performed for saliva from the minor labial glands (Eliasson and Carlen 2010), but it is still necessary to evaluate the proteome of the von Ebner's glands.

Thereby, research should focus on *in situ* and *in vivo* studies in humans. Animal model as well as *in vitro* studies do not faithfully mirror the situation in the human oral cavity, since there is a considerable inter-species variability in numerous salivary parameters (Azzali 1997, Hannig and Hannig 2009, Matsuo 2000, Riva et al. 1999). Modern methods such as proteomics, lipidomics, and metabolomics as well as fluorescence microscopy and recent electron microscopic approaches allow verifying the different hypotheses and evaluating the named structures in depth (Mounayar et al. 2014, Oppenheim et al. 2007, Takahashi et al. 2010). These techniques enable the comparison of physiologic and pathophysiologic conditions in the oral cavity. All this will contibute not only to the knowledge on processes in the oral cavity but has also great implications for the whole gastro-intestinal tract as the mouth is the main entry port into the human organism. Also the long and short time impact of diet or certain foods on the oral and intestinal fluids (Neyraud et al. 2006, Perry et al. 2007, Salles et al. 2011) as well as on the human microbiome should be kept in mind. Taste perception is highly variable between subjects and seems to result from an intrincate combination of physiology of the basic taste structures, but also of events in the peri-receptor milieu including the oral fluids, the biological layers on oral tissues and the oral microbiome. With respect to understanding the interplay between these events, we are only at the beginning.

4.7 Conclusion

Oral fluids as well as bioadhesion and biofilm formation in the oral cavity seem to have considerable impact on taste and taste perception. This applies for the general interactions as well as for topical effects at the taste pores. Despite this fact, there are still many open questions. Research on the basic parameters such as physiological composition of the von Ebner's glands secretion or taste pore material is still necessary.

Acknowledgement

The authors would like to thank Prof. Th. Hummel, Dresden, for the dialogue on taste and taste perception.

References

Ahmad, M., Piludu, M., Oppenheim, F. G., Helmerhorst, E. J. and Hand, A. R. (2004) Immunocytochemical localization of histatins in human salivary glands. *Journal of Histochemistry and Cytochemistry, 52*(3), pp. 361–370.

Amado, F. M. L., Ferreira, R. P. and Vitorino, R. (2013) One decade of salivary proteomics: Current approaches and outstanding challenges. *Clinical Biochemistry, 46*(6), pp. 506–517.

Amerongen, A. V. and Veerman, E. C. (2002) Saliva-the defender of the oral cavity. *Oral Diseases, 8*(1), pp. 12–22.

Arvidson, K., Cottler-Fox, M. and Friberg, U. (1981) Fine structure of taste buds in the human fungiform papilla. *Scandinavian Journal of Dental Research, 89*(4), pp. 297–306.

Asikainen, P., Ruotsalainen, T. J., Mikkonen, J. J., Koistinen, A., Ten Bruggenkate, C. and Kullaa, A. M. (2012) The defence architecture of the superficial cells of the oral mucosa. *Medical Hypotheses, 78*(6), pp. 790–792.

Azen, E. A., Hellekant, G., Sabatini, L. M. and Warner, T. F. (1990) mRNAs for PRPs, statherin, and histatins in von Ebner's gland tissues. *Journal of Dental Research, 69*(11), pp. 1724–1730.

Azzali, G. (1997) Ultrastructure and immunocytochemistry of gustatory cells in man. *Annals of Anatomy-Anatomischer Anzeiger, 179*(1), pp. 37–44.

Azzali, G., Bucci, G., Gatti, R., Orlandini, G. and Ferrari, G. (1989a) Fine-Structure of the Excretory System of the Deep Posterior (Ebner) Salivary-Glands of the Human Tongue. *Acta Anatomica, 136*(4), pp. 257–268.

Azzali, G., Gatti, R., Bucci, G. and Orlandini, G. (1989b) Fine-Structure of Bat Deep Posterior Lingual Glands (Von Ebner's). *Journal of Submicroscopic Cytology and Pathology, 21*(4), pp. 669–684.

Azzali, G., Gennari, P. U., Maffei, G. and Ferri, T. (1996) Vallate, foliate and fungiform human papillae gustatory cells. *Minerva Stomatology, 45*, pp. 363–379.

Bardell, D. (1983) The roles of the sense of taste and clean teeth in the discovery of bacteria by Antoni van Leeuwenhoek. *Microbiology Reviews, 47*(1), pp. 121–126.

Bartoshuk, L. M. (1978) The psychophysics of taste. *American Journal of Clinical Nutrition, 31*(6), pp. 1068–1077.

Beikler, T. and Flemmig, T. F. (2011) Oral biofilm-associated diseases: trends and implications for quality of life, systemic health and expenditures. *Periodontology 2000, 55*(1), pp. 87–103.

Bennick, A. (2002) Interaction of plant polyphenols with salivary proteins. *Critical Reviews in Oral Biology & Medicine, 13*(2), pp. 184–196.

Bläker, M., Kock, K., Ahlers, C., Buck, F. and Schmale, H. (1993) Molecular cloning of human von Ebner's gland protein, a member of the lipocalin superfamily highly expressed in lingual salivary glands. *Biochimica et Biophysica Acta, 1172*, pp. 131–137.

Bowen, W. H. and Koo, H. (2011) Biology of Streptococcus mutans-derived glucosyltransferases: role in extracellular matrix formation of cariogenic biofilms. *Caries Research, 45*(1), pp. 69–86.

Bradley, R. M. and Beidler, L. M. (2003) Saliva: its role in taste function. in Doty, R. L., (ed.) *Handbook of olfaction and gustation,*New York: Marcel Dekker. pp. 639–650.

Bradway, S. D., Bergey, E. J., Jones, P. C. and Levine, M. J. (1989) Oral mucosal pellicle. Adsorption and transpeptidation of salivary components to buccal epithelial cells. *The Biochemical Journal, 261*(3), pp. 887–896.

Bradway, S. D., Bergey, E. J., Scannapieco, F. A., Ramasubbu, N., Zawacki, S. and Levine, M. J. (1992) Formation of salivary mucosal pellicle: the role of transglutaminase. *The Biochemal Journal, 284* (Pt 2), pp. 557–564.

Cabras, T., Pisano, E., Boi, R., Olianas, A., Manconi, B., Inzitari, R., Fanali, C., Giardina, B., Castagnola, M. and Messana, I. (2009) Age-Dependent Modifications of the Human Salivary Secretory Protein Complex. *Journal of Proteome Research, 8*(8), pp. 4126–4134.

Carpenter, G. H. (2013a) Do Transient Receptor Protein (Trp) Channels Play a Role in Oral Astringency? *Journal of Texture Studies, 44*(5), pp. 334–337.

Carpenter, G. H. (2013b) The Secretion, Components, and Properties of Saliva. *Annual Review of Food Science and Technology, Vol 4*, 4, pp. 267–276.

Cecchini, M. P., Pellegrini, C., Bassetto, M. A., Osculati, F., Sbarbati, A., Marcolini, L., Pegoraro, M., Fontana, R. and De Franceschi, L. (2013) Might Helicobacter pylori infection be associated with distortion on taste perception? *Medical Hypotheses, 81*(3), pp. 496–499.

Chagnot, C., Zorgani, M. A., Astruc, T. and Desvaux, M. (2013) Proteinaceous determinants of surface colonization in bacteria: bacterial adhesion and biofilm formation from a protein secretion perspective. *Frontiers in Microbiology, 4*, p. 303.

Chen, J. S. (2009) Food oral processing - A review. *Food Hydrocolloids, 23*(1), pp. 1–25.

Dallmann, R., Viola, A. U., Tarokh, L., Cajochen, C. and Brown, S. A. (2012) The human circadian metabolome. *Proceedings of the National Academy of Sciences of the United States of America, 109*(7), pp. 2625–2629.

Dawes, C. (1969) The effects of flow rate and duration of stimulation on the concentrations of protein and the main electrolytes in human parotid saliva. *Archives of Oral Biology, 14*(3), pp. 277–294.

Dawes, C. (1972) Circadian rhythms in human salivary flow rate and composition. *The Journal of Physiology, 220*, pp. 529–545.

Dawes, C. (1974) The effects of flow rate and duration of stimulation on the concentrations of protein and the main electrolytes in human submandibular saliva. *Archives of Oral Biology, 19*(10), pp. 887–895.

Dawes, C. (1984) Stimulus effects on protein and electrolyte concentrations in parotid saliva. *The Journal of Physiology, 346*, pp. 579–588.

Dawes, C., Jenkins, G. N. and Tonge, C. H. (1963) The nomenclature of the integuments of the enamel surface of teeth. *British Dental Journal, 16*, pp. 66–68.

de Wijk, R. A. and Prinz, J. F. (2006) Mechanisms underlying the role of friction in oral texture. *Journal of Texture Studies, 37*(4), pp. 413–427.

de Wijk, R. A., Prinz, J. F., Engelen, L. and Weenen, H. (2004) The role of alpha-amylase in the perception of oral texture and flavour in custards. *Physiology & Behavior, 83*(1), pp. 81–91.

Delwiche, J. and Omahony, M. (1996) Changes in secreted salivary sodium are sufficient to alter salt taste sensitivity: Use of signal detection measures with continuous monitoring of the oral environment. *Physiology & Behavior, 59*(4-5), pp. 605–611.

Dodds, M. W. J., Johnson, D. A. and Yeh, C. K. (2005) Health benefits of saliva: a review. *Journal of Dentistry, 33*(3), pp. 223–233.

Donaldson, A. C., McKenzie, D., Riggio, M. P., Hodge, P. J., Rolph, H., Flanagan, A. and Bagg, J. (2005) Microbiological culture analysis of the tongue anaerobic microflora in subjects with and without halitosis. *Oral Diseases, 11* Suppl 1, pp. 61–63.

Dsamou, M., Palicki, O., Septier, C., Chabanet, C., Lucchi, G., Ducoroy, P., Chagnon, M. C. and Morzel, M. (2012) Salivary protein profiles and sensitivity to the bitter taste of caffeine. *Chemical Senses, 37*(1), pp. 87–95.

Eliasson, L. and Carlen, A. (2010) An update on minor salivary gland secretions. *European Journal of Oral Sciences, 118*(5), pp. 435–442.

Ferruzzi, M. G., Bordenave, N. and Hamaker, B. R. (2012) Does flavor impact function? Potential consequences of polyphenol-protein interactions in delivery and bioactivity of flavan-3-ols from foods. *Physiology & Behavior, 107*(4), pp. 591–597.

Foglio-Bonda, P. L., Migliario, M., Rocchetti, V., Pattarino, F. and Foglio-Bonda, A. (2013) Daily and annually variation of unstimulated whole saliva flow rate and pH and their relation with body profile in healthy young adults. *European Review for Medical and Pharmacological Sciences, 17*(18), pp. 2538–2545.

Gibbins, H. L. and Carpenter, G. H. (2013) Alternative Mechanisms of Astringency - What Is the Role of Saliva? *Journal of Texture Studies, 44*(5), pp. 364–375.

Gibbins, H., Proctor, G., Yakubov, G., Wilson, S. and Carpenter, G. (2014a) Concentration of salivary protective proteins within the bound oral mucosal pellicle. *Oral Diseases, 20*(7):pp. 707–713.

Gibbins, H. L., Yakubov, G. E., Proctor, G. B., Wilson, S. and Carpenter, G. H. (2014b) What interactions drive the salivary mucosal pellicle formation? *Colloids Surfaces B Biointerfaces, 120*, pp. 184–192.

Gurkan, S. and Bradley, R. M. (1987) Autonomic control of von Ebner's lingual salivary glands and implications for taste sensation. *Brain Research, 419*(1-2), pp. 287–293.

Gurkan, S. and Bradley, R. M. (1988a) Effects of electrical stimulation of autonomic nervous system on degranulation of von Ebner's gland acini. *Brain Research, 473*(1), pp. 127–133.

Gurkan, S. and Bradley, R. M. (1988b) Secretion of VonEbners glands influence responses from taste buds in rat circumvallate papilla. *Chemical Senses, 13*(4), pp. 655–661.

Hamosh, M. and Burns, W. A. (1977) Lipolytic activity of human lingual glands (Ebner). *Laboratory Investigation, 37*(6), pp. 603–608.

Hand, A. R., Pathmanathan, D. and Field, R. B. (1999) Morphological features of the minor salivary glands. *Archives of Oral Biology, 44* Suppl 1, pp. S3–S10.

Hannig, C., Attin, T., Hannig, M., Henze, E., Brinkmann, K. and Zech, R. (2004) Immobilisation and activity of human alpha-amylase in the acquired enamel pellicle. *Archives of Oral Biology, 49*(6), pp. 469–475.

Hannig, C., Berndt, D., Hoth-Hannig, W. and Hannig, M. (2009a) The effect of acidic beverages on the ultrastructure of the acquired pellicle-an in situ study. *Archives of Oral Biology, 54*(6), pp. 518–526.

Hannig, C. and Hannig, M. (2009) The oral cavity-a key system to understand substratum-dependent bioadhesion on solid surfaces in man. *Clinical Oral Investigations, 13*(2), pp. 123–139.

Hannig, C., Hannig, M. and Attin, T. (2005) Enzymes in the acquired enamel pellicle. *European Journal of Oral Sciences, 113*(1), pp. 2–13.

Hannig, C., Ruggeri, A., Al-Khayer, B., Schmitz, P., Spitzmüller, B., Deimling, D., Huber, K., Hoth-Hannig, W., Bowen, W. H. and Hannig, M. (2008a) Electron microscopic detection and activity of glucosyltransferase B, C, and D in the in situ formed pellicle. *Archives of Oral Biology, 53*(11), pp. 1003–1010.

Hannig, C., Spitzmüller, B., Al-Ahmad, A. and Hannig, M. (2008b) Effects of Cistus-tea on bacterial colonization and enzyme activities of the in situ pellicle. *Journal of Dentistry, 36*(7), pp. 540–545.

Hannig, C., Spitzmüller, B. and Hannig, M. (2009b) Characterisation of lysozyme activity in the in situ pellicle using a fluorimetric assay. *Clinical Oral Investigations, 13*(1), pp. 15–21.

Hannig, C., Spitzmüller, B., Knausenberger, S., Hoth-Hannig, W., Hellwig, E. and Hannig, M. (2008c) Detection and activity of peroxidase in the in situ formed enamel pellicle. *Archives of Oral Bioogyl, 53*(9), pp. 849–858.

Hannig, C., Spitzmüller, B., Miller, M., Hellwig, E. and Hannig, M. (2008d) Intrinsic enzymatic crosslinking and maturation of the in situ pellicle. *Archives of Oral Biology, 53*(5), pp. 416–422.

Hannig, C., Wagenschwanz, C., Pötschke, S., Kümmerer, K., Kensche, A., Hoth-Hannig, W. and Hannig, M. (2012) Effect of safflower oil on the protective properties of the in situ formed salivary pellicle. *Caries Research, 46*(5), pp. 496–506.

Hannig, M. (1999) Ultrastructural investigation of pellicle morphogenesis at two different intraoral sites during a 24-h period. *Clinical Oral Investigations, 3*(2), pp. 88–95.

Hannig, M., Dounis, E., Henning, T., Apitz, N. and Stosser, L. (2006) Does irradiation affect the protein composition of saliva? *Clinical Oral Investigations, 10*(1), pp. 61–65.

Hannig, M. and Hannig, C. (2014) The pellicle and erosion. *Monographs in Oral Science, 25*, pp. 206–214.

Hannig, M. and Joiner, A. (2006) The structure, function and properties of the acquired pellicle. *Monographs in Oral Science, 19*, pp. 29–64.

Hannig, M., Khanafer, A. K., Hoth-Hannig, W., Al-Marrawi, F. and Acil, Y. (2005) Transmission electron microscopy comparison of methods for collecting in situ formed enamel pellicle. *Clinical Oral Investigations, 9*(1), pp. 30–37.

Hardt, M., Witkowska, H. E., Webb, S., Thomas, L. R., Dixon, S. E., Hall, S. C. and Fisher, S. J. (2005) Assessing the effects of diurnal variation on the composition of human parotid saliva: Quantitative analysis of native peptides using iTRAQ reagents. *Analytical Chemistry, 77*(15), pp. 4947–4954.

Hellhammer, D. H., Wüst, S. and Kudielka, B. M. (2009) Salivary cortisol as a biomarker in stress research. *Psychoneuroendocrinology, 34*(2), pp. 163–171.

Helmerhorst, E. J. (2007) Whole saliva proteolysis. Wealth of information for diagnostic exploitation. *Annals of the New York Academy of Sciences, 1098*(1), pp. 454–460.

Helms, J. A., Della-Fera, M. A., Mott, A. E. and Frank, M. E. (1995) Effects of chlorhexidine on human taste perception. *Archives of Oral Biology, 40*(10), pp. 913–920.

Henkin, R. I., Larson, A. L. and Powell, R. D. (1975a) Hypogeusia, dysgeusia, hyposmia, and dysosmia following influenza-like infection. *The Annals of Otology, Rhinology and Laryngology, 84*(5 Pt 1), pp. 672–682.

Henkin, R. I., Lippoldt, R. E., Bilstad, J. and Edelhoch, H. (1975b) A Zinc Protein Isolated from Human Parotid Saliva. *Proceedings of the National Academy of Sciences of the United States of America, 72*(2), pp. 488–492.

Henkin, R. I., Martin, B. M. and Agarwal, R. P. (1999) Decreased parotid saliva gustin/carbonic anhydrase VI secretion: An enzyme disorder manifested by gustatory and olfactory dysfunction. *The American Journal of the Medical Sciences, 318*(6), pp. 380–391.

Henkin, R. I. and Velicu, I. (2012) Etiological relationships of parotid saliva cyclic nucleotides in patients with taste and smell dysfunction. *Archives of Oral Biology, 57*(6), pp. 670–677.

Herness, M. S. and Gilbertson, T. A. (1999) Cellular mechanisms of taste transduction. *Annual Review of Physiology, 61*, pp. 873–900.

Hiyama, T., Ozeki, N., Mogi, M., Yamaguchi, H., Kawai, R., Nakata, K., Kondo, A. and Nakamura, H. (2013) Matrix Metalloproteinase-3 in Odontoblastic Cells Derived from Ips Cells: Unique Proliferation Response as Odontoblastic Cells Derived from ES Cells. *PLoS ONE, 8*(12).

Humphrey, S. P. and Williamson, R. T. (2001) A review of saliva: Normal composition, flow, and function. *Journal of Prosthetic Dentistry, 85*(2), pp. 162–169.

Huq, N. L., Cross, K. J., Ung, M., Myroforidis, H., Veith, P. D., Chen, D., Stanton, D., He, H., Ward, B. R. and Reynolds, E. C. (2007) A review of the salivary proteome and peptidome and saliva-derived peptide therapeutics. *International Journal of Peptide Research and Therapeutics, 13*(4), pp. 547–564.

Igarashi, A., Ito, K., Funayama, S., Hitomi, Y., Ikui, A., Ikeda, M. and Nomura, S. (2008) The salivary protein profiles in the patients with taste disorders: The comparison of salivary protein profiles by two-dimensional gel electrophoresis between the patients with taste disorders and healthy subjects. *Clinica Chimica Acta, 388*(1-2), pp. 204–206.

Jehmlich, N., Dinh, K. H., Gesell-Salazar, M., Hammer, E., Steil, L., Dhople, V. M., Schurmann, C., Holtfreter, B., Kocher, T. and Volker, U. (2013) Quantitative analysis of the intra- and inter-subject variability of the whole salivary proteome. *Journal of Periodontal Research, 48*(3), pp. 392–403.

Jensen, S. B., Mouridsen, H. T., Bergmann, O. J., Reibel, J., Brunner, N. and Nauntofte, B. (2008) Oral mucosal lesions, microbial changes, and taste disturbances induced by adjuvant chemotherapy in breast cancer patients. *Oral Surgery Oral Medicine Oral Pathology Oral Radiology, 106*(2), pp. 217–226.

Johansson, G. and Birkhed, D. (1994) Effect of a long-term change from a mixed to a lactovegetarian diet on human saliva. *Archives of Oral Biology, 39*(4), pp. 283–288.

Joiner, A., Elofsson, U. M. and Arnebrant, T. (2006) Adsorption of chlorhexidine and black tea onto in vitro salivary pellicles, as studied by ellipsometry. *European Journal of Oral Sciences, 114*(4), pp. 337–342.

Joiner, A., Muller, D., Elofsson, U. M. and Arnebrant, T. (2004) Ellipsometry analysis of the in vitro adsorption of tea polyphenols onto salivary pellicles. *European Journal of Oral Sciences 112*(6), pp. 510–515.

Joiner, A., Muller, D., Elofsson, U. M., Malmsten, M. and Arnebrant, T. (2003) Adsorption from black tea and red wine onto in vitro salivary pellicles studied by ellipsometry. *European Journal of Oral Sciences, 111*(5), pp. 417–422.

Kariyawasam, A. P. and Dawes, C. (2005) A circannual rhythm in unstimulated salivary flow rate when the ambient temperature varies by only about 2 °C. *Archives of Oral Biology, 50*(10), pp. 919–922.

Kensche, A., Basche, S., Bowen, W. H., Hoth-Hannig, W., Hannig, M. and Hannig, C. (2013) Fluorescence microscopic visualization of non-cellular components during initial bioadhesion in situ. *Archives of Oral Bioogyl, pp.* 1271–1281.

Kullaa, A. M., Asikainen, P., Herrala, M., Ukkonen, H. and Mikkonen, J. J. (2014) Microstructure of Oral Epithelial Cells as an Underlying Basis for Salivary Mucosal Pellicle. *Ultrastructural Pathology*, pp. 1–5.

Laine, M. A., Tolvanen, M., Pienihakkinen, K., Soderling, E., Niinikoski, H., Simell, O. and Karjalainen, S. (2014) The effect of dietary intervention on paraffin-stimulated saliva and dental health of children participating in a randomized controlled trial. *Archives of Oral Biology, 59*(2), pp. 217–225.

Landis, B. N., Welge-Luessen, A., Bramerson, A., Bende, M., Mueller, C. A., Nordin, S. and Hummel, T. (2009) "Taste Strips" - a rapid, lateralized, gustatory bedside identification test based on impregnated filter papers. *Journal of Neurology, 256*(2), pp. 242–248.

Leinonen, J., Kivela, J., Parkkila, S., Parkkila, A. K. and Rajaniemi, H. (1999) Salivary carbonic anhydrase isoenzyme VI is located in the human enamel pellicle. *Caries Research, 33*(3), pp. 185–190.

Leinonen, J., Parkkila, S., Kaunisto, K., Koivunen, P. and Rajaniemi, H. (2001) Secretion of carbonic anhydrase isoenzyme VI (CA VI) from human and rat lingual serous von Ebner's glands. *Journal of Histochemistry and Cytochemistry, 49*(5), pp. 657–662.

Lendenmann, U., Grogan, J. and Oppenheim, F. G. (2000) Saliva and dental pellicle--a review. *Advances in Dental Research, 14*, pp. 22–28.

Li, J., Helmerhorst, E. J., Troxler, R. F. and Oppenheim, F. G. (2004) Identification of in vivo pellicle constituents by analysis of serum immune responses. *Journal of Dental Research, 83*(1), pp. 60–64.

Li, X. J. and Snyder, S. H. (1995) Molecular-Cloning of Ebnerin, a Von Ebners Gland Protein Associated with Taste-Buds. *Journal of Biological Chemistry, 270*(30), pp. 17674–17679.

Liman, E. R., Zhang, Y. V. and Montell, C. (2014) Peripheral coding of taste. *Neuron, 81*(5), pp. 984–1000.

Lindquist, S. and Hernell, O. (2010) Lipid digestion and absorption in early life: an update. *Current Opinion in Clinical Nutrition and Metabolic Care, 13*(3), pp. 314–320.

Lu, Y. and Bennick, A. (1998) Interaction of tannin with human salivary proline-rich proteins. *Archives of Oral Biology, 43*(9), pp. 717–728.

Mandel, A. L. and Breslin, P. A. S. (2012) High Endogenous Salivary Amylase Activity Is Associated with Improved Glycemic Homeostasis following Starch Ingestion in Adults. *Journal of Nutrition, 142*(5), pp. 853–858.

Mandel, A. L., des Gachons, C. P., Plank, K. L., Alarcon,S. and Breslin, P. A. S. (2010) Individual Differences in AMY1 Gene Copy Number, Salivary alpha-Amylase Levels, and the Perception of Oral Starch. *PLoS ONE, 5*(10).

Marsh, P. D. (2004) Dental plaque as a microbial biofilm. *Caries Research, 38*(3), pp. 204–211.

Matsui, M., Chosa, N., Shimoyama, Y., Minami, K., Kimura, S. and Kishi, M. (2014) Effects of tongue cleaning on bacterial flora in tongue coating and dental plaque: a crossover study. *BMC Oral Health, 14*, p. 4.

Matsuo, R. (2000) Role of saliva in the maintenance of taste sensitivity. *Critical Reviews in Oral Biology & Medicine, 11*(2), pp. 216–229.

Matsuo, R. and Yamamoto, T. (1992) Effects of inorganic constituents of saliva on taste reponse of the rat chorda tympani nerve *Brain Research, 583*(1-2), pp. 71–80.

Mese, H. and Matsuo, R. (2007) Salivary secretion, taste and hyposalivation. *Journal of Oral Rehabilitation, 34*(10), pp. 711–723.

Moore, L. M., Nielsen, C. R. and Mistretta, C. M. (1982) Sucrose taste thresholds: age-related differences. *Journal of Gerontology, 37*(1), pp. 64–69.

Morales-Bozo, I., Urzua-Orellana, B., Dominguez, P., Aguilera, S. and Lopez-Solis, R. (2006) Patterns and variability in electrophoretic polypeptide profiles of human saliva in a healthy population. *Journal of Physiology and Biochemistry, 62*(3), pp. 179–188.

Moro, I., Umemura, S., Crago, S. S. and Mestecky, J. (1984) Immunohistochemical distribution of immunoglobulins, lactoferrin, and lysozyme in human minor salivary glands. *Journal of Oral Pathology, 13*(2), pp. 97–104.

Morris-Wiman, J., Sego, R., Brinkley, L. and Dolce, C. (2000) The effects of sialoadenectomy and exogenous EGF on taste bud morphology and maintenance. *Chemical Senses, 25*(1), pp. 9–19.

Morzel, M., Chabanet, C., Schwartz, C., Lucchi, G., Ducoroy, P. and Nicklaus, S. (2014a) Salivary protein profiles are linked to bitter taste acceptance in infants. *European Journal of Pediatrics, 173*(5), pp. 575–582.

Morzel, M., Jeannin, A., Lucchi, G., Truntzer, C., Pecqueur, D., Nicklaus, S., Chambon, C. and Ducoroy, P. (2012) Human infant saliva peptidome is modified with age and diet transition. *Journal of Proteomics, 75*(12), pp. 3665–3673.

Morzel, M., Palicki, O., Chabanet, C., Lucchi, G., Ducoroy, P., Chambon, C. and Nicklaus, S. (2011) Saliva electrophoretic protein profiles in infants: Changes with age and impact of teeth eruption and diet transition. *Archives of Oral Biology, 56*(7), pp. 634–642.

Morzel, M., Tai, S., Brignot, H. and Lherminier, J. (2014b) Immunocytological detection of salivary mucins (MUC5B) on the mucosal pellicle lining human epithelial buccal cells. *Microscopy Research and Techniques, 77*(6), pp. 453–457.

Mounayar, R., Morzel , M., H., B., Tremblay-Franco, M., Canlet, C., Lucchi, G., Ducoroy, P., Feron, G. and Neyraud, E. (2014) Salivary markers of taste sensitivity to oleic acid: a combined proteomics and metabolomics approach. *Metabolomics, 10*, pp. 688–696.

Nassar, M., Hiraishi, N., Islam, M. S., Otsuki, M. and Tagami, J. (2014) Age-related changes in salivary biomarkers. *Journal of Dental Sciences, 9*(1), pp. 85–90.

Nater, U. M., La Marca, R., Florin, L., Moses, A., Langhans, W., Koller, M. M. and Ehlert, U. (2006) Stress-induced changes in human salivary alpha-amylase activity-associations with adrenergic activity. *Psychoneuroendocrinology, 31*(1), pp. 49–58.

Nayak, A. and Carpenter, G. H. (2008) A physiological model of tea-induced astringency. *Physiology & Behavior, 95*(3), pp. 290–294.

Neyraud, E., Sayd, T., Morzel, M. and Dransfield, E. (2006) Proteomic analysis of human whole and parotid salivas following stimulation by different tastes. *Journal of Proteome Research, 5*(9), pp. 2474–2480.

Neyraud, E., Tremblay-Franco, M., Gregoire, S., Berdeaux, O. and Canlet, C. (2013) Relationships between the metabolome and the fatty acid composition of human saliva; effects of stimulation. *Metabolomics, 9*(1), pp. 213–222.

Oh, H. I. and Hoff, J. E. (1986) Effect of condensed grape tannins on the in vitro activity of digestive proteases and activation of their zymogens *Journal of Food Science, 51*(3), pp. 577–580.

Ohno, T., Uematsu, H., Nozaki, S. and Sugimoto, K. (2003) Improvement of taste sensitivity of the nursed elderly by oral care. *Journal of Medical and Dental Sciences, 50*(1), pp. 101–107.

Oppenheim, F. G., Salih, E., Siqueira, W. L., Zhang, W. M. and Helmerhorst, E. J. (2007) Salivary proteome and its genetic polymorphisms. *Oral-Based Diagnostics, 1098*, pp. 22–50.

Otagiri, K., Nosho, Y., Shinoda, I., Fukui, H. and Okai, H. (1985) Studies on a model of bitter peptides including arginine, proline and phenylalanine residues. 1. Bitter taste of dipeptides and tripeptides, and bitterness increase of the model peptides by extension of the peptide-chain. *Agricultural and Biological Chemistry, 49*(4), pp. 1019–1026.

Ovalle, W. K. and Nahirney, P. C. (2013) *Netter's Essential Histology*, Elsevier, Saunders.

Oxford, G. E., Nguyen, K. H., Alford, C. E., Tanaka, Y. and Humphreys-Beher, M. G. (1998) Elevated salivary EGF levels stimulated by periodontal surgery. *Journal of Periodontology, 69*(4), pp. 479–484.

Ozdener, M. H. and Rawson, N. E. (2012) Exogenous EGF and BDNF treatment enhances differentiation and proliferation of cultured taste cells. *Turkish Journal of Medical Sciences, 42*(2), pp. 179–192.

Paran, N., Mattern, C. F. and Henkin, R. I. (1975) Ultrastructure of the taste bud of the human fungiform papilla. *Cell and Tissue Research, 161*(1), pp. 1–10.

Park, N. J., Li, Y., Yu, T. W., Brinkman, B. M. N. and Wong, D. T. (2006) Characterization of RNA in saliva. *Clinical Chemistry, 52*(6), pp. 988–994.

Passilly-Degrace, P., Chevrot, M., Bernard, A., Ancel, D., Martin, C. and Besnard, P. (2014) Is the taste of fat regulated? *Biochimie, 96C*, pp. 3–7.

Pedersen, A. M., Bardow, A., Jensen, S. B. and Nauntofte, B. (2002) Saliva and gastrointestinal functions of taste, mastication, swallowing and digestion. *Oral Diseases, 8*(3), pp. 117–129.

Perry, G. H., Dominy, N. J., Claw, K. G., Lee, A. S., Fiegler, H., Redon, R., Werner, J., Villanea, F. A., Mountain, J. L., Misra, R., Carter, N. P., Lee, C. and Stone, A. C. (2007) Diet and the evolution of human amylase gene copy number variation. *Nature Genetics, 39*(10), pp. 1256–1260.

Piludu, M., Lantini, M. S., Cossu, M., Piras, M., Oppenheim, F. G., Helmerhorst, E. J., Siqueira, W. and Hand, A. R. (2006) Salivary histatins in human deep posterior lingual glands (of von Ebner). *Archives of Oral Biology, 51*(11), pp. 967–973.

Pramanik, R., Osailan, S. M., Challacombe, S. J., Urquhart, D. and Proctor, G. B. (2010) Protein and mucin retention on oral mucosal surfaces in dry mouth patients. *European Journal of Oral Sciences, 118*(3), pp. 245–253.

Quirynen, M., Avontroodt, P., Soers, C., Zhao, H., Pauwels, M. and van Steenberghe, D. (2004) Impact of tongue cleansers on microbial load and taste. *Journal of Clinical Periodontology, 31*(7), pp. 506–510.

Rautemaa, R. and Ramage, G. (2011) Oral candidosis--clinical challenges of a biofilm disease. *Critical Reviews in Microbiology, 37*(4), pp. 328–336.

Redl, B., Merschak, P., Abt, B. and Wojnar, P. (1999) Phage display reveals a novel interaction of human tear lipocalin and thioredoxin which is relevant for ligand binding. *FEBS Letters, 460*(1), pp. 182–186.

Reich, M., Kümmerer, K., Al-Ahmad, A. and Hannig, C. (2013) Fatty Acid Profile of the Initial Oral Biofilm (Pellicle): an In-Situ Study. *Lipids, 48*(9), pp. 929–937.

Riggio, M. P., Lennon, A., Rolph, H. J., Hodge, P. J., Donaldson, A., Maxwell, A. J. and Bagg, J. (2008) Molecular identification of bacteria on the tongue dorsum of subjects with and without halitosis. *Oral Diseases, 14*(3), pp. 251–258.

Riva, A., Loffredo, F., Puxeddu, R. and Testa Riva, F. (1999) A scanning and transmission electron microscope study of the human minor salivary glands. *Archives of Oral Biology, 44 Suppl 1*, pp. S27–S31.

Riva, A. and Tandler, B. (2000) Three-dimensional structure of oncocyte mitochondria in human salivary glands: a scanning electron microscope study. *Ultrastructural Pathology, 24*(3), pp. 145–150.

Riva, F. T., Cossu, M., Lantini, M. S. and Riva, A. (1985) Fine structure of human deep posterior lingual glands. *Journal of Anatomy, 142*, pp. 103–115.

Rosing, C. K. and Loesche, W. (2011) Halitosis: an overview of epidemiology, etiology and clinical management. *Brazilian Oral Research, 25*(5), pp. 466–471.

Ruhl, S., Rayment, S. A., Schmalz, G., Hiller, K. A. and Troxler, R. F. (2005) Proteins in whole saliva during the first year of infancy. *Journal of Dental Research, 84*(1), pp. 29–34.

Sakashita, S., Takayama, K., Nishioka, K. and Katoh, T. (2004) Taste disorders in healthy "carriers" and "non-carriers" of Candida albicans and in patients with candidosis of the tongue. *Journal of Dermatology, 31*(11), pp. 890–897.

Salles, C., Chagnon, M. C., Feron, G., Guichard, E., Laboure, H., Morzel, M., Semon, E., Tarrega, A. and Yven, C. (2011) In-mouth mechanisms leading to flavor release and perception. *Critical Reviews in Food Science and Nutrition, 51*(1), pp. 67–90.

Sbarbati, A., Crescimanno, C. and Osculati, F. (1999) The anatomy and functional role of the circumvallate papilla/von Ebner gland complex. *Medical Hypotheses, 53*(1), pp. 40–44.

Schlueter, N., Ganss, C., Potschke, S., Klimek, J. and Hannig, C. (2012) Enzyme activities in the oral fluids of patients suffering from bulimia: a controlled clinical trial. *Caries Research, 46*(2), pp. 130–139.

Schmale, H., Ahlers, C., Blaker, M., Kock, K. and Spielman, A. I. (1993) Perireceptor events in taste. *Ciba Found Symposium, 179*, pp. 167–80; discussion 180-185.

Schmale, H., Holtgrevegrez, H. and Christiansen, H. (1990) Possible role for salivary-gland protein in taste reception indicated by homology to lipophilic-ligand carrier proteins. *Nature, 343*(6256), pp. 366–369.

Scinska-Bienkowska, A., Wrobel, E., Turzynska, D., Bidzinski, A., Jezewska, E., Sienkiewicz-Jarosz, H., Golembiowska, K., Kostowski, W., Kukwa, A., Plaznik, A. and Bienkowski, P. (2006) Glutamate concentration in whole saliva and taste responses to monosodium glutamate in humans. *Nutritional Neuroscience, 9*(1-2), pp. 25–31.

Seneviratne, C. J., Zhang, C. F. and Samaranayake, L. P. (2011) Dental plaque biofilm in oral health and disease. *The Chinese Journal of Dental Research, 14*(2), pp. 87–94.

Sengupta, A., Valdramidou, D., Huntley, S., Hicks, S. J., Carrington, S. D. and Corfield, A. P. (2001) Distribution of MUC1 in the normal human oral cavity is localized to the ducts of minor salivary glands. *Archives of Oral Biology, 46*(6), pp. 529–538.

Sevenhuysen, G. P., Holodinsky, C. and Dawes, C. (1984) Development of Salivary Alpha-Amylase in Infants from Birth to 5 Months. *American Journal of Clinical Nutrition, 39*(4), pp. 584–588.

Shatzman, A. R. and Henkin, R. I. (1980) Metal-binding characteristics of the parotid salivary protein gustin. *Biochimica et Biophysica Acta, 623*(1), pp. 107–118.

Shatzman, A. R. and Henkin, R. I. (1981) Gustin Concentration Changes Relative to Salivary Zinc and Taste in Humans. *Proceedings of the National Academy of Sciences of the United States of America, 78*(6), pp. 3867–3871.

Siqueira, W. L. and Dawes, C. (2011) The salivary proteome: challenges and perspectives. *Proteomics Clinical Application, 5*(11-12), pp. 575–579.

Siqueira, W. L., Salih, E., Wan, D. L., Helmerhorst, E. J. and Oppenheim, F. G. (2008) Proteome of human minor salivary gland secretion. *Journal of Dental Research, 87*(5), pp. 445–450.

Slomiany, B. L., Zdebska, E., Murty, V. L. N., Slomiany, A. and Mandel, I. D. (1984) Lipid composition of human pellicle. *Journal of Dental Research, 63*, pp. 271.

Smith, C. H., Boland, B., Daureeawoo, Y., Donaldson, E., Small, K. and Tuomainen, J. (2013) Effect of Aging on Stimulated Salivary Flow in Adults. *Journal of the American Geriatrics Society, 61*(5), pp. 805–808.

Smith, F. R., Henkin, R. I. and Dell, R. B. (1976) Disordered gustatory acuity in liver disease. *Gastroenterology, 70*(4), pp. 568–571.

Smith, L. J., Kaminsky, S. and Dsouza, S. W. (1986) Neonatal fat digestion and lingual lipase. *Acta Paediatrica Scandinavica, 75*(6), pp. 913–918.

Soares, R. V., Lin, T., Siqueira, C. C., Bruno, L. S., Li, X., Oppenheim, F. G., Offner, G. and Troxler, R. F. (2004) Salivary micelles: identification of complexes containing MG2, sIgA, lactoferrin, amylase, glycosylated proline-rich protein and lysozyme. *Archives of Oral Biology, 49*(5), pp. 337–343.

Soares, S., Vitorino, R., Osorio, H., Fernandes, A., Venancio, A., Mateus, N., Amado, F. and de Freitas, V. (2011) Reactivity of human salivary proteins families toward food polyphenols. *Journal of Agriculture and Food Chemistry, 59*(10), pp. 5535–5547.

Solemdal, K., Sandvik, L., Willumsen, T., Mowe, M. and Hummel, T. (2012) The impact of oral health on taste ability in acutely hospitalized elderly. *PLoS One*, 7(5), p. e36557.

Spielman, A. I. (1990) Interaction of saliva and taste. *Journal of Dental Research, 69*(3), pp. 838–843.

Spielman, A. I., D'Abundo, S., Field, R. B. and Schmale, H. (1993) Protein analysis of human von Ebner saliva and a method for its collection from the foliate papillae. *Journal of Dental Research, 72*(9), pp. 1331–1335.

Sreebny, L. M. (2000) Saliva in health and disease: an appraisal and update. *International Dental Journal, 50*(3), pp. 140–161.

Sreebny, L. M. and Valdini, A. (1988) Xerostomia. Part I: Relationship to other oral symptoms and salivary gland hypofunction. *Oral Surgery Oral Medicine Oral Pathology, 66*(4), pp. 451–458.

Sreebny, L. M., Valdini, A. and Yu, A. (1989) Xerostomia. Part II: Relationship to nonoral symptoms, drugs, and diseases. *Oral Surgery Oral Medicine Oral Pathology, 68*(4), pp. 419–427.

Tabak, L. A., Levine, M. J., Mandel, I. D. and Ellison, S. A. (1982) Role of salivary mucins in the protection of the oral cavity. *Journal of Oral Pathology, 11*(1), pp. 1–17.

Takahashi, N., Washio, J. and Mayanagi, G. (2010) Metabolomics of supragingival plaque and oral bacteria. *Journal of Dental Research, 89*(12), pp. 1383–1388.

Takeda, I., Stretch, C., Barnaby, P., Bhatnager, K., Rankin, K., Fu, H., Weljie, A., Jha, N. and Slupsky, C. (2009) Understanding the human salivary metabolome. *Nmr in Biomedicine, 22*(6), pp. 577–584.

Takeda, M. and Hoshino, T. (1975) Fine structure of taste buds in the rat. *Archivum Histologicum Japonicum, 37*(5), pp. 395–413.

Tancredi, T., Pastore, A., Salvadori, S., Esposito, V. and Temussi, P. A. (2004) Interaction of sweet proteins with their receptor - A conformational study of peptides corresponding to loops of brazzein, monellin and thaumatin. *European Journal of Biochemistry, 271*(11), pp. 2231–2240.

Temussi, P. A. (2011) Determinants of sweetness in proteins: a topological approach. *Journal of Molecular Recognition, 24*(6), pp. 1033–1042.

Thatcher, B. J., Doherty, A. E., Orvisky, E., Martin, B. M. and Henkin, R. I. (1998) Gustin from human parotid saliva is carbonic anhydrase VI. *Biochemical and Biophysical Research Communications, 250*(3), pp. 635–641.

Vacca Smith, A. M. and Bowen, W. H. (2000) In situ studies of pellicle formation on hydroxyapatite discs. *Archives of Oral Biology, 45*(4), pp. 277–291.

Valdez, I. H. and Fox, P. C. (1991) Interactions of the salivary and gastrointestinal systems. *I. the role of saliva in digestion. Digestive diseases, 9*, pp. 125–132.

Veerman, E. C., van den Keybus, P. A., Valentijn-Benz, M. and Nieuw Amerongen, A. V. (1992) Isolation of different high-Mr mucin species from human whole saliva. *The Biochemical Journal, 283*(3), pp. 807–811.

Ventre, G., Colonna, C., Smith, J., Alfano, D. and Moldow, R. (2013) Salivary VIP concentrations are elevated in humans after acute stress. *Peptides, 49*, pp. 27–31.

Vitkov, L., Hannig, M., Krautgartner, W. D. and Fuchs, K. (2002a) Bacterial adhesion to sulcular epithelium in periodontitis. *FEMS Microbiology Letters, 211*(2), pp. 239–246.

Vitkov, L., Hannig, M., Nekrashevych, Y. and Krautgartner, W. D. (2004) Supramolecular pellicle precursors. *European Journal of Oral Sciences, 112*(4), pp. 320–325.

Vitkov, L., Krautgartner, W. D., Hannig, M. and Fuchs, K. (2001) Fimbria-mediated bacterial adhesion to human oral epithelium. *FEMS Microbiology Letters, 202*(1), pp. 25–30.

Vitkov, L., Krautgartner, W. D., Hannig, M., Weitgasser, R. and Stoiber, W. (2002b) Candida attachment to oral epithelium. *Oral Microbiology and Immunology, 17*(1), pp. 60–64.

Wada, O. H., Yamamori, T., Shimazaki, N., Ishibashi, K. and Seino, K. (2010) Function of salivary histatin 5 for bitter substances. in *88th General Session and Exhibition of the IADR (International Association for Dental Research)*, Barcelona.

Watanabe, I. S., Ogawa, K., Cury, D. P., Dias, F. J., Sosthenes, M. C., Issa, J. P. and Iyomasa, M. M. (2013) Fine structure of bacterial adhesion to the epithelial cell membranes of the filiform papillae of tongue and palatine mucosa of rodents: a morphometric, TEM, and HRSEM study. *Microscopy Research and Techniques, 76*(12), pp. 1226–1233.

Watanabe, M., Asatsuma, M., Ikui, A., Ikeda, M., Yamada, Y., Nomura, S. and Igarashi, A. (2005) Measurements of several metallic elements and matrix metalloproteinases (MMPs) in saliva from patients with taste disorder. *Chemical Senses, 30*(2), pp. 121–125.

Weiffenbach, J. M., Fox, P. C. and Baum, B. J. (1986) Taste and Salivary Function. *Proceedings of the National Academy of Science USA, 83*(16), pp. 6103–6106.

Witt, M. and Miller, I. J., Jr., (1992) Comparative lectin histochemistry on taste buds in foliate, circumvallate and fungiform papillae of the rabbit tongue. *Histochemistry, 98*(3), pp. 173–182.

Xiao, J., Ni, X., Kai, G. and Chen, X. (2013) A review on structure-activity relationship of dietary polyphenols inhibiting alpha-amylase. *Critical Reviews in Food Science and Nutrition, 53*(5), pp. 497–506.

Yan, Q. Y. and Bennick, A. (1995) Identification of Histatins as Tannin-Binding Proteins in Human Saliva. *The Biochemical Journal, 311*, pp. 341–347.

Zimmerman, J. N., Custodio, W., Hatibovic-Kofman, S., Lee, Y. H., Xiao, Y. and Siqueira, W. L. (2013) Proteome and peptidome of human acquired enamel pellicle on deciduous teeth. *International Journal of Molecular Sciences, 14*(1), pp. 920–934.

CHAPTER 5

Basic physiology of the intranasal trigeminal system

Florence Gingras-Lessard[1,3] & Johannes Frasnelli[1,2]
[1] CÉAMS, Research Center, Sacré-Cœur Hospital, Montréal, Québec, Canada
[2] Department of Anatomy, Université du Québec à Trois-Rivières, Trois-Rivières, Québec, Canada
[3] Faculté de Médecine, Université de Montréal, Montréal, Québec, Canada

5.1 Introduction

The senses of taste and of smell are both interconnected, to produce a wide variety of information about what we are eating or smelling. However, we analyze our chemical environment with an additional chemosensory system. From an anatomical point of view, it is located on the trigeminal nerve, the fifth and thickest cranial nerve. The trigeminal nerve is responsible for the somatosensory innervation of the whole face including the mucosa of the mouth and the nose. Here it can also be stimulated by certain chemical substances, which give rise to sensations such as burning, cooling, scratching etc. This chemosensory portion of the trigeminal nerve together with the central nervous processing units is called the chemosensory trigeminal system, or, shorter, the trigeminal system. Together with smell and taste, it plays a major role in the perception of food, and classic examples of the effect of the trigeminal system are the cold sensation of peppermint and the burning evoked by jalapeño peppers. It is important to note that at a high concentration, almost every odorant activates the trigeminal system (Doty et al. 1978) underlining the intimate connection between the trigeminal and the olfactory system.

In this chapter, the interested reader will find information on anatomy and the physiology of the trigeminal system. We will detail the methods used to assess the function of this system, and describe the interaction between olfaction and the trigeminal system.

Flavour: From food to perception, First Edition.
Edited by Elisabeth Guichard, Christian Salles, Martine Morzel, and Anne-Marie Le Bon.

5.2 Neuroanatomy of the trigeminal system

5.2.1 General neuroanatomy of the trigeminal system

The trigeminal nerve is the main structure allowing us to feel touch on our face (Figure 5.1). The trigeminal nerve is the fifth cranial nerve (CNV) and is divided into three branches (therefore trigeminal from latin tri: three and geminus: twin; here triplet), each one responsible for a specific region of the face. The first branch (V_1) is the ophthalmic nerve, the second one (V_2) is the maxillary nerve and the third one (V_3) is the mandibular nerve. They provide the sensation for respectively the upper, the middle and the lower portions of the face. The trigeminal nerve is also responsible for a few motor innervations, as for the muscles of mastication and the tensor tympani muscle (Blumenfeld 2010).

From the periphery to central, the three nerves aggregate in the Gasserian ganglion (or semilunar ganglion, or trigeminal ganglion) situated in Meckel's cave (or trigeminal cave), near the tip of the petrous part of the temporal bone. The cell bodies of the neurons (pseudo-unipolar neurons) are located in this ganglion. It contains approximately 27,000 neurons (Borsook et al. 2003). From here, the nerve fibers enter the central nervous system at the ventrolateral pons. In the following, the fibers split to reach two different parts of the trigeminal nucleus: the chief trigeminal sensory nucleus for the light touch (epicritic sensation), and the spinal nucleus for the deep touch, the sensation of pain and of temperature (protopathic sensation). In these nuclei the first synaptic relay takes places. The axons of the postsynaptic neurons form the trigeminal lemniscus and the trigeminothalamic tract, respectively. Both tracks decussate to reach the contralateral thalamus, in the ventro-posteriormedial nucleus (VPM), where the second synapse is located (Blumenfeld 2010). Finally, the sensitive fibers reach the primary somatosensory cortex, where the information reaches conscience and the individual gets aware of being touched.

Because it conveys pain, the trigeminal nerve is also connected to the pain matrix (Mouraux et al. 2011), an ensemble of central structures integrating noxious stimuli. The pain matrix comprises the primary and secondary somatosensory cortex, the insular cortex, the anterior cingulate cortex and the prefrontal cortex. Each of these structures plays a distinct role in the perception of pain: the primary and secondary somatosensory cortex together with the anterior part of the insular cortex are involved in the processing of the sensory aspects of the noxious stimulus. On the other hand, the posterior part of the insular cortex provides the insight, the memory and the emotions related to the stimulus. The anterior cingulate cortex integrates the emotions and also plays a role in the cognition. Finally, the prefrontal cortex serves to evaluate the intensity of the painful stimulus (Iannilli et al. 2008).

One interesting characteristic about the trigeminal nerve is the somatotopy, which is preserved from the periphery to the cortex. The term somatotopy describes the organization of the neurons in their course through the central

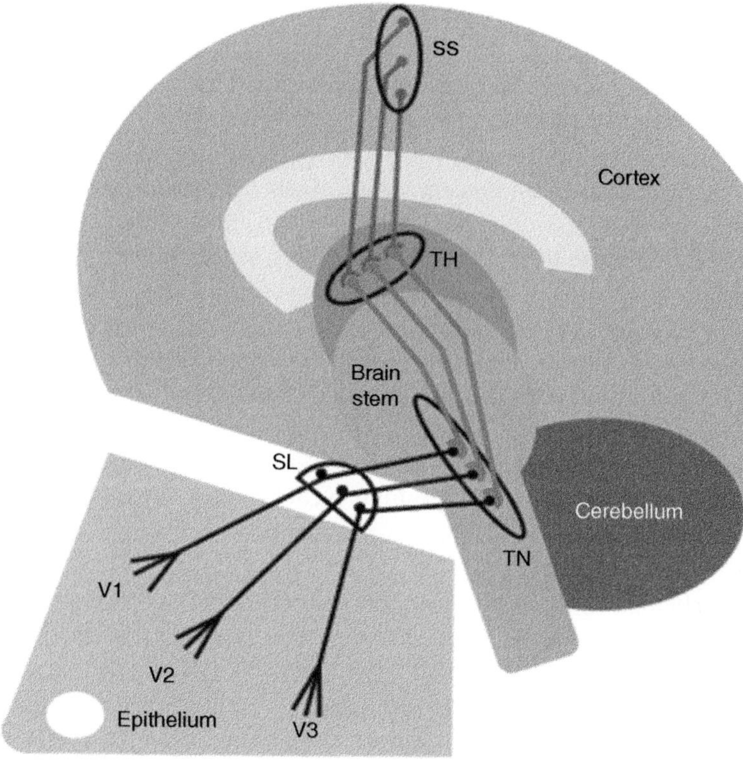

Figure 5.1 Overview of the trigeminal pathway. Neurons within sensory fibers from V1 (ophthalmic nerve), V2 (maxillary nerve) and V3 (mandibular nerve) have their cell bodies within the semilunar ganglion in Meckel's cave (SL); their first relay station is in the trigeminal nucleus of the brain stem (TN). Second order neurons project to the ventro-posteriormedial nucleus of the thalamus (TH). Third order neurons project to the somatosensory cortex (SS) in the parietal cortex. Note the somatotopy which is preserved from the periphery to the central nervous system. A colored version of this figure can be found in the online version of this chapter.

nervous system. Fibers from V_1 terminate in ventral portions of the trigeminal nucleus, whereas V_3 fibers end dorsally; V_2 lies between V_1 and V_2. The central part of the face (containing the nose and the mouth) is therefore represented rostrally while the peripheral is located in the caudal part of the nucleus (Borsook et al. 2003). Similarly, the somatosensory cortex is organized according to the somatosensory homunculus (Penfield and Rasmussen 1950). Within the area representing the face, the fibers of the ophthalmic nerve are projecting to the caudal part of the primary somatosensory cortex, while the sensitive territory of the maxillary nerve is located in the rostral part. The area of the mandibular nerve lays between these two areas (DaSilva et al. 2002).

5.2.2 Specific neuroanatomy of the chemosensory trigeminal system

The chemosensory portion of the trigeminal nerve has some specific anatomical particularities. The inner nose is innervated by branches of the maxillary nerve and by the ethmoidal and the infraorbital nerves, both branches of the ophthalmic nerve, whereas the oral mucosa is exclusively innervated by branches of the mandibular nerve (Frasnelli et al. 2014). Here, trigeminal fibers can react to molecules that are inhaled or ingested; thus, all three branches of the trigeminal nerve contribute to the trigeminal chemosensory system. In fact, the receptive structures of the trigeminal system are located throughout the nasal and oral mucosa. Nevertheless, sensitivity towards trigeminal stimuli is not equal between different regions. This has most prominently been shown for the nasal mucosa. The nasal septum and the antero-superior areas of the nasal cavity exhibit highest susceptibility and therefore probably highest receptor densities, while the posterior part of the nasal cavity is less sensitive (Scheibe et al. 2006). These data suggest that receptive structures are not distributed homogeneously throughout the oral and nasal mucosa.

Trigeminal chemosensory stimulation leads to two kinds of nervous fibers, namely the A-delta and the C fibers (Scheibe et al. 2008). Both fiber types are thin; A-delta fibers are slightly myelinated whereas C fibers are not myelinated at all. Accordingly, they convey the information relatively slowly. If the trigeminal nerve is stimulated by chemosensory stimuli, the central activation patterns are distinct from "classical" somatosensory trigeminal activation patterns. Trigeminal chemosensory stimulation activates the cerebral zones associated with somatosensory stimulation of the trigeminal nerve, such as the somatosensory cortex. In contrast to somatosensory stimulation, however, trigeminal chemosensory stimulation leads to additional activations. Typically, these regions include the piriform cortex and the orbitofrontal cortex, two regions usually associated with the sense of smell (Boyle et al. 2007b).

Interestingly, stimulation of the left and right nostril leads to different activation patterns, with larger activation in the right side of the brain (Hummel et al. 2009a). This may explain why some studies observed participants to more easily localize trigeminal stimuli presented to the right nostril than to the left nostril (Frasnelli et al. 2009). Although higher order somatosensory processing takes place contralaterally, the trigeminal chemosensory stimuli seem to predominantly activate the ipsilateral cortex, as do olfactory stimuli. For example, painful chemosensory stimulation activates the ipsilateral thalamus to a larger extent than the contralateral thalamus, which is in striking contrast to mechanical or electrical stimulation of the trigeminal nerve (Iannilli et al. 2008). The exact neuroanatomical underpinnings of these findings are yet unknown. Another interesting fact is the difference found between the central processing of men and women. Women tend to be more sensitive than men, with significantly lower CO_2 thresholds (Frasnelli et al. 2006). A recent review summarizes that irritation perception is greater in women (Ohla and Lundstrom 2013) and suggests that the differences between both sexes are driven by the central trigeminal processing. Further, trigeminal sensitivity declines with age, especially beyond 60 years (Wysocki et al. 2003).

5.3 Physiology of the trigeminal system

5.3.1 Specific roles of the trigeminal system

The trigeminal nerve has different functions. First, it is well known that the trigeminal nerve plays the main role in the sensitive innervation for both, epicritic and protopathic information from the face. It therefore also allows the perception of temperature and texture of objects within the oral (such as food) and less frequently nasal cavity. In addition to this, the trigeminal system provides information on the chemical composition of substances we are exposed to. It therefore contributes to increase the complexity of odour perception. Within this framework, the trigeminal system constitutes the afferent part of several reflexes protecting the mucosa of both the upper and the lower airways (Shusterman 2007). In response to some irritants, it induces a greater production of mucus, a change in respiration patterns and congestion in the erectile tissue of the nose. Those changes therefore protect the mucosa from toxic products, since a prolonged exposition to trigeminal stimuli can lead to damage as far as in the lungs (Doty et al. 1978). The trigeminal system also allows for the localization of an odor source. In fact, the trigeminal nerve has the ability to integrate the difference of time between the right and the left nostrils as much as the difference of concentration between those two. This seems to be impossible *via* the olfactory system (Kobal et al. 1989).

5.3.2 Receptors and perception

The trigeminal system is activated when its fibers within the nasal and the oral mucosa are stimulated. For long time, the scientific community thought that trigeminal chemoreception is the result of an unspecific interaction between chemical substances with free nerve endings of these fibers. Only over the last two decades, specific receptors on trigeminal fibers have been detected which allow for a specific interaction between the chemical stimulus and the neuron (Gerhold and Bautista 2009). The best-known group of trigeminal receptors belongs to the transient receptor potential (TRP) family. These receptors are ion channels with sensitivity to both changes in temperature and chemical substances (Bandell et al. 2004). Activation of these receptors leads to the sensation of the according temperature change, even in an environment with stable temperature, a phenomenon well known to lovers of Indian or Mexican cuisine. In fact, the heat evoked by a Vindaloo dish or by Jalapeño peppers is due to capsaicin, the spicy ingredient of hot peppers. Capsaicin activates the TRPV1 receptor (originally named Vanilloid Receptor - VR1), the first TRP receptor to be discovered, just as noxious temperatures above 43 degrees do. As a consequence, both activation with capsaicin and noxious heat evoke the same painful burning sensation. TRPV1's sensitivity is potentiated by a low pH (Caterina et al. 1997), explaining why carbonated drinks with a low pH such as beer or soda do not provide any relieve when eating hot and spicy dishes. With long term exposure, capsaicin acts as an analgesic by evoking desensitization. Again, it is well-known by lovers of spicy food that the painful burning will diminish after a couple of bites. Long-term exposure to capsaicin further induces lysis of the TRP-containing neurons (Caterina et al. 1997), which may explain habituation to spicy food in the long run as tolerance for capsaicin increases with the consumption of spicy food. Many other chemicals have been discovered which activate TRPV1 and provide burning sensations of different degrees when in contact with the mucosa, such as eugenol (the spicy ingredient of cloves; Bandell et al. 2004, Saunders et al. 2013, Yang et al. 2003), gingerol (the spicy ingredient of ginger; Bandell et al. 2004), and allicin (the spicy ingredient of garlic; Macpherson et al. 2005).

The next TRP receptor to be discovered covers the opposite range of physiological temperatures in the nasal and oral cavity. The TRPM8 receptor (originally coined Cold and Menthol Receptor - CMR1) is sensitive to cold temperatures, menthol, and several other substances (e.g., eucalyptol, icilin). These stimuli have in common to produce a cooling sensation. TRPM8 reacts to temperature from 8 to approximately 25°C. Similar to TRPV1, TRPM8 also desensitizes in few minutes (McKemy et al. 2002).

Some of the TRP receptors are also present on keratinocytes and therefore in the skin (Sondersorg et al. 2014). Specifically, the TRPV3 receptor has been discovered there. TRPV3 is nevertheless also present in mucosal fibers and activated by warm temperatures above 39°C (Peier et al. 2002). It is further activated by chemical substances such as thymol (a main component of Mediterranean herbs such as oregano and thyme; Vogt-Eisele et al. 2007, Xu et al. 2006). Activation of TRPV3 is associated with the sensation of warmth (Xu et al. 2006).

Another TRP receptor which is sensitive towards the range of noxious cold is the TRPA1 channel. It is activated by cold temperatures below 17°C (Bandell et al. 2004, Jordt et al. 2004), as well as chemicals such as allyl thioisocyanate (the spicy ingredient of mustard; Bandell et al. 2004, Jordt et al. 2004). TRPA1 stimulation gives rise to a dull, painful sensation as one can experience when eating wasabi.

Other, non TRP receptors are known, but less well explored, such as the acid-sensing ion receptor (ASIC), which is involved in acid evoked nociception and probably also in tasting (Waldmann et al. 1997). The burning of nicotine is probably evoked by the stimulation of cholinergic receptors in the mucosa (Thuerauf et al. 1999). Finally, solitary chemosensory cells, which have been found in the nasal cavities of mammals although not in humans may also be involved in trigeminal perception (Finger et al. 2003).

5.4 Assessment of the trigeminal system

The main challenge when testing the trigeminal system is the fact that most trigeminal stimuli activate the olfactory system as well, and they do so in lower concentrations than the ones needed for trigeminal activation (Doty et al. 1978). Therefore, trigeminal sensations will almost inevitably be contaminated by some olfactory component. It is therefore (nearly) impossible to assess a trigeminal threshold with the same methods as it is done for the olfactory system, because one cannot rule out that the result is based on olfactory sensations rather than trigeminal perceptions. Another challenge is the high degree of desensitization and adaptation in the trigeminal system which forces researchers to use very long inter-stimulus intervals (in the range of more than 30 seconds) rendering assessments quite time consuming.

The different methods to assess the trigeminal system can be subdivided into specific and unspecific methods.

5.4.1 Specific methods

Specific methods are techniques, which are based on unique features of the trigeminal system. For instance, the lateralization task is based on the fact that we can only determine which of both nostrils received an odorous stimulus if the stimulus also activated the trigeminal system. The Negative Mucosal Potential on the other hand allows assessment of peripheral activation directly from the mucosa.

5.4.1.1 Lateralization task

The presence of two nostrils in human beings has pushed researchers to think that, as for audition and vision, we are able to localize a stimulus with our nose. The proof of this phenomenon has been made by a study in which pure olfactory and trigeminal stimuli, that is, stimuli which activate the olfactory system or the trigeminal system exclusively, were presented to participants, but only to one nostril. Only the trigeminal stimuli could be localized better than by chance (Kobal et al. 1989). This topic remains somehow controversial as some groups claim that localization of pure odours is possible for some animals (including rats (Rajan et al. 2006) and sharks (Gardiner and Atema 2010) or, under certain circumstances even for humans (Porter et al. 2005). It is however very difficult to rule out a possible trigeminal activation evoked by the supposedly pure olfactory stimuli used in these studies (Frasnelli et al. 2011). To summarize the evidence provided by these studies, localization of pure olfactory stimuli is impossible or virtually impossible, whereas localization of trigeminal stimuli is a very easy task (Croy et al. 2014, Frasnelli et al. 2009, Kleemann et al. 2009, Negoias et al. 2013). Therefore, by providing a dichotomous stimulus, we can assess the trigeminal system, without any interference from the olfactory system. To do so, stimuli can be presented manually with squeeze bottles or mechanically by a computer controlled olfactometer. The usual paradigm of the lateralization task is to give an odorous stimulus in one of the nostril while the other nostril receives only air, in order to avoid lateralization based on somatosensory information from the unilateral air puff (Keita et al. 2013). Usually a semi-quantitative approach is used to evaluate trigeminal sensitivity, for example, by counting the number of correct responses in a series of trials (Hummel et al. 2003), but recently researchers suggested a technique to determine trigeminal detection thresholds by using a localization task (Naka et al. 2014).

5.4.1.2 Negative mucosal potential

The Negative Mucosal Potential (NMP) is an electrophysiological signal recorded from the nasal epithelium. It represents the peripheral activation of the trigeminal system, on the level of the mucosa (Thürauf et al. 1991, Thürauf et al. 2002). In fact, it measures the sum of generator potentials caused by the depolarization of chemosensory nociceptive receptors (Frasnelli et al. 2014). The NMP is recorded with a silver-chlorided silver electrode coated with ringer-agar filled Teflon™ tubing. This technique is specific to the trigeminal system as it is recordable from the respiratory epithelium of the nasal cavity where no olfactory receptor neurons can be found (Kobal 1985). The NMP signal is a negative wave with a relatively long latency of approximately one second. The measurement of NMP requires an olfactometer – a device allowing for automated stimulus delivery – assuring the absence of any thermal or mechanical co-stimulation and the averaging of several individual responses in order to minimize background activity and electrical noise. The amplitude of the NMP is correlated to both, stimulus concentration and perceived intensity (Frasnelli et al. 2007a).

5.4.1.3 Trigeminal reflexes

Finally, the evaluation of nasal reflexes may provide indirect information about the trigeminal nerve. For example, breathing patterns may change in the presence of irritants, which may even evoke apnea. One can measure the strength and the speed of these reflexes to evaluate the trigeminal system (Walker et al. 2001), but one has to keep in mind that olfactory information also alters breathing patterns (Johnson et al. 2006). One may also count the stimulation of the cornea into this category. The cornea – the transparent front of the eye – responds to chemical stimulation with only one type of perception, namely pain. Thus, an additional method to investigate the trigeminal system is based on the fact that the eye's cornea is responsive to trigeminal stimulation, without any concomitant olfactory interference (de Wijk et al. 1998, Vesaluoma et al. 2000). Typically, participants respond with a blink reflex when stimulated above threshold.

5.4.2 Unspecific methods

Unspecific methods are techniques which are also used in other sensory domains and thus are not specific for the trigeminal system. In order to avoid contamination from other sensory systems – mainly olfaction – one has to make sure to either use stimuli specific to the trigeminal system (e.g., carbon dioxide) or rely on subjects which lack olfactory function (such as individuals with congenital anosmia).

First, carbon dioxide is one of the few molecules which in humans (as opposed to amphibians for example; Coates 2001) seem to exclusively activate the trigeminal system. Such a purely trigeminal stimulus allows for the evaluation of the trigeminal system, for example, by assessing perception thresholds (Frasnelli et al. 2010) or by using other techniques such as event related potentials (see below). Major drawback of this technique is the fact that one cannot completely rule out olfactory stimulation by carbon dioxide or possible contaminations within the gas tanks.

A second method to rule out olfactory interference is the investigation of individuals with anosmia. Because their olfactory nerve is defective, these individuals have to rely completely on the trigeminal system in order to detect volatiles. By comparing their results to those of individuals with normal olfactory function, one can assess the trigeminal contribution to the experience of smelling different volatile compounds. Anosmia is a relatively common condition, with a prevalence of approximately 5% in the general population, while hyposmia, that is, a reduced olfactory function, affect another 15% (Bramerson et al. 2004, Landis et al. 2004). While this promising approach has provided significant insight into the mechanisms of the trigeminal system, one has however to be careful when interpreting results obtained in individuals with smell loss as olfactory dysfunction per se has been shown to affect trigeminal function. In fact, individuals with anosmia are less sensitive to trigeminal stimuli than people with a normal sense of smell (Hummel et al. 2003).

A third method to assess the trigeminal system is to train healthy participants with a normal sense of smell to only focus on the trigeminal sensation. Researches have shown that the results of individuals with anosmia and the trained participants are similar at low concentration. However, this method is less reliable if high concentrations are used, because the olfactory nerve could therefore influence the perception of the stimuli (Doty et al. 1978). Again, this technique is imperfect, as one cannot completely rule out interference form the olfactory system.

In order to evaluate the trigeminal system, typically one of the aforementioned approaches is used together with the methods described in the following.

5.4.2.1 Event-related potentials

Event-related potentials (ERP) are electroencephalogram (EEG)-based signals used to evaluate the cortical activation specific to sensory stimulation. ERP are used in all sensory domains, and can therefore also be measured for trigeminal stimuli. EEG is recorded by means of electrodes placed on the scalp. This is typically done by using a standardized system (such as the 10-20 system; Khazi et al. 2012). In order to record trigeminal ERP, one needs to use a stimulus, which specifically activates the trigeminal system, such as carbon dioxide (Ohla and Lundstrom, 2013). Because the electrical activity of the brain is never zero when the participant is alive, event-related potentials must be extracted from the background noise. To do so, multiple stimulations are averaged which deletes the random background noise (Hummel and Kobal 1992). While this can easily be achieved for auditory and visual stimuli, it is more challenging for the trigeminal system, as the high degree of habituation requires the stimulations to be separated by at least 30 to 40 seconds. In order not to eternalize measurements and to keep the subjects' vigilance stable, trigeminal ERP are obtained after averaging relatively few stimulations. While in vision or audition research ERP are obtained after averaging hundreds of stimuli, trigeminal ERP usually rely on the averaging of a dozen or so stimulations. Unfortunately this implies a relative poor noise-to-signal ratio (Boesveldt et al. 2007).

An ERP is a wave signal which typically consists of several positive and at least one negative peak. The maximal signal can typically be obtained at the vertex (Hummel and Kobal 1992). The first positive wave (P1) is followed by a larger negative one (N1) and subsequently by a complex of positive peaks (P2, P3, or late positive complex) (Frasnelli et al. 2003). The exact neurological sources of these peak activities are still unknown (Hari et al. 1997). Typically, one can measure the latency and the amplitude of the peaks. The early components of the ERP (P1, N1) are mainly the results of the stimulus effects; as such they largely depend on stimulus parameters such as concentration and duration. The late positive complex on the other hand is largely influenced by the psychological state of the participant. For example, it reflects habituation as repeated delivery of an identical stimulus results in a great reduction of the late positive complex but not of the early components (Laudien et al. 2008).

In summary, measuring trigeminal ERP allows to evaluate trigeminal processing at different levels and this in a more objective way than other methods do. Biggest advantage of recording trigeminal ERP is the very high temporal resolution; in turn its disadvantages are a poor spatial resolution and the need for an olfactometer, similar to NMP.

5.4.2.2 Neuroimaging

If one aims to evaluate the trigeminal system with high spatial resolution, different neuroimaging methods are available. The predominantly used technique is functional Magnetic Resonance Imaging (MRI). Neuronal activity induces a change in blood oxygenation of the respective brain area. This induces alterations in the magnetic property of the area in question as oxygenated hemoglobin has other magnetic susceptibility than desoxygenated hemoglobin (Holdsworth and Bammer 2008). This difference can be visualized by functional MRI by measuring the blood-oxygen-level dependent (BOLD) signal. When comparing the BOLD signal in a particular brain region during a task and in baseline condition, one can determine the degree of oxygen consumption in this region. Oxygen consumption is an indirect measure of neuronal activity; thus the BOLD signal allows inference about the underlying neuronal activity.

Activation patterns are obtained by comparing different conditions to each other. Functional MRI has a typical voxel (the 3D equivalent of a pixel) size of 3*3*3mm or less. This spatial resolution is superior to other functional imaging techniques, explaining the widespread use of this technique. Still, one has to keep in mind that one voxel contains thousands of neurons which do not all have the exactly same function and thus activity patterns. The major disadvantages of functional MRI are its price, a certain discomfort to the participants and a relatively poor temporal resolution.

Another interesting but less frequently used technique is the positron emission tomography (PET) scan. Here, the patient receives an injection of a slightly radioactive molecule, which can then be detected by a scanner. When investigating the brain, usually a radioactive atom such as ^{18}F is added as a marker to glucose resulting in 2-deoxy-2-fluoro-D-glucose. The marker accumulates in tissue with high energy consumption which is associated with neuronal activity allowing for determining the activity of different areas of the brain (Otte and Halsband 2006). Spatial and temporal resolution of PET is weaker than functional MRI; further it leads to a certain exposure to radioactivity (Hummel et al. 2009b).

5.5 Interactions between trigeminal system and olfaction

Even if they are two well-separated structures sustained by two different cranial nerves, the trigeminal system and olfaction are closely related. From the level of stimuli to central processing, both systems interact with each other. This will be discussed in the present section.

5.5.1 Stimuli

Almost all of the molecules we inhale will activate both the olfactory system and, at higher concentrations, the trigeminal system. In fact, even the few components which were for long time considered to be purely olfactory stimuli such as phenyl ethanol or vanillin (Doty et al. 1978) appear to stimulate the trigeminal system at higher concentrations (Frasnelli et al. 2011). Similarly, only few stimuli appear to be purely trigeminal, including carbon dioxide and capsaicin. Thus, most stimuli stimulate both the olfactory and the trigeminal system and are therefore called mixed olfactory trigeminal stimuli. They lend further support to the notion of an intimate connection between both sensory systems.

5.5.2 Peripheral interaction

Olfactory input seems to modulate the trigeminal response already at the level of the mucosa. In fact, the olfactory nerve appears to have an inhibitive effect on the peripheral trigeminal response as NMP are larger in patients suffering from anosmia than in matched controls (Frasnelli et al. 2007a, Frasnelli et al. 2007b). This effect appears not be straightforward as recovery from anosmia tends to further increase these measures of peripheral susceptibility (Frasnelli et al. 2007a). The exact neural underpinnings of this are unclear. Some researchers support the notion that some trigeminal fibers terminating within the olfactory bulb may explain how olfactory input modulates the trigeminal response. From their origin in the trigeminal nucleus, these fibers travel with the other trigeminal fibers to the nasal epithelium, but then re-enter the central nervous system and terminate within the olfactory bulb (Schaefer et al. 2002).

5.5.3 Central interaction

At a central level, the interaction between the trigeminal system and olfaction is well established. Although information from both sensory systems reaches central areas *via* different and independent channels, they share several common processing areas (Lombion et al. 2009). In fact, trigeminal stimulation leads to activation of both somatosensory areas and olfactory regions, specifically the piriform and the orbitofrontal cortex. In fact, simultaneous stimulation of both nostrils with a trigeminal and an olfactory stimulus, respectively, leads to mutual inhibition and enhancement, suggesting central interactions (Cain and Murphy 1980). One study shed light on this interaction, as a mixed stimulus led to larger activation in the integration centers and in the orbitofrontal cortex than the mathematical sum of its components (Boyle et al. 2007a).

Some more evidence stems from individuals with sensory loss: patients with olfactory dysfunction show weaker results in a lateralization task (Hummel et al. 2003). In fact, ERP measures reveal that central activity in response to trigeminal stimuli is reduced in patients with anosmia (Frasnelli et al. 2007a). These effects seem to fade off with longer lasting olfactory dysfunction (Hummel et al. 2003) and are absent in individuals with congenital anosmia (Frasnelli et al. 2007a). This lends further support to the notion of a highly plastic interaction between both sensory systems.

Much less evidence is available from patients with loss of trigeminal sensitivity. Still, a case study of a patient suffering from meningeoma affecting the left trigeminal nerve without touching the olfactory nerve showed a largely decreased olfactory sensitivity on the affected side (Husner et al. 2006). This finding suggests that the interaction between both sensory systems is indeed mutual.

5.6 Conclusion

The trigeminal system is a third chemosensory system, next to olfaction and gustation. As smell and taste, trigeminal sensations arise from the interactions of chemicals with specific chemoreceptors. The chemical senses allow us to explore our outer and, as far as it concerns the oral and the nasal cavity, inner environment. While it is less sensitive that the olfactory system, at higher concentrations most odorants will also activate the trigeminal system. Generally speaking, the olfactory and the trigeminal system are intimately connected and interact on several levels.

Future research will deepen our understanding of the interaction between stimuli of different trigeminal receptors, at receptor levels, within nerve fibers and in higher order processing centers. Further, researchers will aim to understand the interaction of the trigeminal system and other sensory systems, including but not limited to the chemical senses of smell and taste. An interesting direction is the investigation of the pleasure we perceive when ingesting trigeminal stimuli such as herbs and spices in the context of local and exotic foods. A better comprehension may then be used in clinical research where scientists will investigate to what extend trigeminal stimuli could replace olfactory stimulation in individuals with anosmia and hyposmia. These widespread conditions are associated with reduced enjoyment of food, and some affected individuals have the tendency to increase sugar and/or salt intake in order to perceive at least gustatory information – which has deleterious effects in a society which struggles from overweight, diabetes and related diseases.

Acknowledgement

Johannes Frasnelli holds the Université du Québec à Trois-Rivières (UQTR) Research Chair in Chemosensory Neuroanatomy. He is supported by NSERC and installation grants from the Research Centre of Sacré-Coeur Hospital and from UQTR.

References

Bandell, M., Story, G. M., Hwang, S. W., Viswanath, V., Eid, S. R., Petrus, M. J., Earley, T. J. and Patapoutian, A. (2004) Noxious cold ion channel TRPA1 is activated by pungent compounds and bradykinin. *Neuron, 41*(6), pp. 849–857.

Blumenfeld, H. (2010) *Neuroanatomy through Clinical Cases,* Sunderland (MA, USA): Sinauer Associates.

Boesveldt, S., Haehner, A., Berendse, H. W. and Hummel, T. (2007) Signal-to-noise ratio of chemosensory event-related potentials. *Clinical Neurophysiology, 118*(3), pp. 690–695.

Borsook, D., DaSilva, A.F.M., Ploghaus, A. and Becerra, L. (2003) Specific and Somatotopic Functional Magnetic Resonance Imaging Activation in the Trigeminal Ganglion by Brush and Noxious Heat. *The Journal of Neuroscience, 23*(21), pp. 7897–7903.

Boyle, J. A., Frasnelli, J., Gerber, J., Heinke, M. and Hummel, T. (2007a) Cross-modal integration of intranasal stimuli: a functional magnetic resonance imaging study. *Neuroscience, 149*(1), pp. 223–231.

Boyle, J. A., Heinke, M., Gerber, J., Frasnelli, J. and Hummel, T. (2007b) Cerebral activation to intranasal chemosensory trigeminal stimulation. *Chemical Senses, 32*(4), pp. 343–353.

Bramerson, A., Johansson, L., Ek, L., Nordin, S. and Bende, M. (2004) Prevalence of olfactory dysfunction: the skovde population-based study. *Laryngoscope, 114*(4), pp. 733–737.

Cain, W. S. and Murphy, C. L. (1980) Interaction between chemoreceptive modalities of odour and irritation. *Nature, 284*(5753), pp. 255–257.

Caterina, M. J., Schumacher, M. A., Tominaga, M., Rosen, T. A., Levine, J. D. and Julius, D. (1997) The capsaicin receptor: a heat-activated ion channel in the pain pathway. *Nature, 389*(6653), pp. 816–823.

Coates, E. L. (2001) Olfactory CO_2 chemoreceptors. *Respiration Physiology, 129,* pp. 219–229.

Croy, I., Schulz, M., Blumrich, A., Hummel, C., Gerber, J. and Hummel, T. (2014) Human olfactory lateralization requires trigeminal activation. *Neuroimage, 98,* pp. 289–295.

DaSilva, A. F. M., Becerra, L., Makris, N., Strassman, A. M., Gonzalez, R. G., Geatrakis, N. and Borsook, D. (2002) Somatotopic Activation inthe Human Trigeminal Pain Pathway. *The Journal of Neuroscience, 22*(18), pp. 8183–8192.

de Wijk, R., Cain, W. S. and Pilla-Caminha, G. (1998) Human psychophysical and neurophysiological measurements on ethanol. *Chemical Senses, 23,* p. 586.

Doty, R. L., Brugger, W. E., Jurs, P. C., Orndorff, M. A., Snyder, P. J. and Lowry, L. D. (1978) Intranasal Trigeminal Stimulation from Odorous Volatiles : Psychometric Responses from Anosmic and Normal Humans. *Physiology & Behavior, 20,* pp. 175–185.

Finger, T. E., Bottger, B., Hansen, A., Anderson, K. T., Alimohammadi, H. and Silver, W. L. (2003) Solitary chemoreceptor cells in the nasal cavity serve as sentinels of respiration. *Proceedings of the National Academy of Sciences of the United States of America, 100*(15), pp. 8981–8986.

Frasnelli, J., Charbonneau, G., Collignon, O. and Lepore, F. (2009) Odor localization and sniffing. *Chemical Senses, 34*(2), pp. 139–144.

Frasnelli, J., Hummel, T., Berg, J., Huang, G. and Doty, R. L. (2011) Intranasal localizability of odorants: influence of stimulus volume. *Chemical Senses, 36*(4), pp. 405–410.

Frasnelli, J., Hummel, T. and Shusterman (2014) Clinical Disorders of the Trigeminal System. in Welge-Luessen, A. and Hummel, T. (eds.) *Management of Smell and Taste Disorders : A practical Guide for Clinicians,* Stuttgart (Germany): Thieme Medical Publishers.

Frasnelli, J., Lotsch, J. and Hummel, T. (2003) Event-related potentials to intranasal trigeminal stimuli change in relation to stimulus concentration and stimulus duration. *Journal of Clinical Neurophysiology, 20*(1), pp. 80–86.

Frasnelli, J., Schuster, B. and Hummel, T. (2007a) Interactions between olfaction and the trigeminal system: what can be learned from olfactory loss. *Cerebral Cortex, 17*(10), pp. 2268–2275.

Frasnelli, J., Schuster, B. and Hummel, T. (2007b) Subjects with congenital anosmia have larger peripheral but similar central trigeminal responses. *Cerebral Cortex, 17*(2), pp. 370–377.

Frasnelli, J., Schuster, B. and Hummel, T. (2010) Olfactory dysfunction affects thresholds to trigeminal chemosensory sensations. *Neuroscience Letters, 468*(3), pp. 259–263.

Frasnelli, J., Schuster, B., Zahnert, T. and Hummel, T. (2006) Chemosensory specific reduction of trigeminal sensitivity in subjects with olfactory dysfunction. *Neuroscience, 142*(2), pp. 541–546.

Gardiner, J. M. and Atema, J. (2010) The Function of Bilateral Odor Arrival Time Differences in Olfactory Orientation of Sharks. *Current Biology, 20*(13), pp. 1187–1191.

Gerhold, K. A. and Bautista, D. M. (2009) Molecular and cellular mechanisms of trigeminal chemosensation. *Annals of the New York Academy of Sciences, 1170*, pp. 184–189.

Hari, R., Portin, K., Kettenmann, B., Jousmäki, V. and Kobal, G. (1997) Right-hemisphere preponderance of responses to painful CO_2 stimulation of the human nasal mucosa. *Pain, 72*, pp. 145–151.

Holdsworth, S. J. and Bammer, R. (2008) Magnetic resonance imaging techniques: fMRI, DWI, and PWI. *Seminars in Neurology, 28*(4), pp. 395–406.

Hummel, T., Futschik, T., Frasnelli, J. and Hüttenbrink, K.-B. (2003) Effects of olfactory function, age, and gender on trigeminally mediated sensations: a study based on the lateralization of chemosensory stimuli. *Toxicology Letters, 140-141*, pp. 273–280.

Hummel, T., Iannilli, E., Frasnelli, J., Boyle, J. and Gerber, J. (2009a) Central processing of trigeminal activation in humans. *Annals of the New York Academy of Sciences, 1170*, pp. 190–195.

Hummel, T. and Kobal, G. (1992) Differences in human evoked potentials related to olfactory or trigeminal chemosensory activation. *Electroencepalography and Clinical Neurophysiology, 84*, pp. 84–89.

Hummel, T., Oehme, L., van den Hoff, J., Gerber, J., Heinke, M., Boyle, J. A. and Beuthien-Baumann, B. (2009b) PET-based investigation of cerebral activation following intranasal trigeminal stimulation. *Human Brain Mapping, 30*(4), pp. 1100–1104.

Husner, A., Frasnelli, J., Welge-Lussen, A., Reiss, G., Zahnert, T. and Hummel, T. (2006) Loss of trigeminal sensitivity reduces olfactory function. *Laryngoscope, 116*(8), pp. 1520–1522.

Iannilli, E., Del Gratta, C., Gerber, J. C., Romani, G. L. and Hummel, T. (2008) Trigeminal activation using chemical, electrical, and mechanical stimuli. *Pain, 139*(2), pp. 376–388.

Johnson, B. N., Russell, C., Khan, R. M. and Sobel, N. (2006) A comparison of methods for sniff measurement concurrent with olfactory tasks in humans. *Chemical Senses, 31*(9), pp. 795–806.

Jordt, S. E., Bautista, D. M., Chuang, H. H., McKemy, D. D., Zygmunt, P. M., Hogestatt, E. D., Meng, I. D. and Julius, D. (2004) Mustard oils and cannabinoids excite sensory nerve fibres through the TRP channel ANKTM1. *Nature, 427*(6971), pp. 260–265.

Keita, L., Frasnelli, J., La Buissonniere-Ariza, V. and Lepore, F. (2013) Response times and response accuracy for odor localization and identification. *Neuroscience, 238*, pp. 82–86.

Khazi, M., Kumas, A. and Didya, M. J. (2012) Analysis of EEG Using 10:20 Electrode system. *International Journal od Innovative Research in Science, Engineering and Technology, 1*(2), pp. 185–191.

Kleemann, A. M., Albrecht, J., Schopf, V., Haegler, K., Kopietz, R., Hempel, J. M., Linn, J., Flanagin, V. L., Fesl, G. and Wiesmann, M. (2009) Trigeminal perception is necessary to localize odors. *Physiology & Behavior, 97*(3-4), pp. 401–405.

Kobal, G. (1985) Pain-Related Electrical Potentials of the Human Nasal Mucosa Elicited by Chemical Stimulation. *Pain, 22*(2), pp. 151–163.

Kobal, G., Van Toller, S. and Hummel, T. (1989) Is there directional smelling? *Experientia, 45*(2), pp. 130–132.

Landis, B. N., Konnerth, C. G. and Hummel, T. (2004) A study on the frequency of olfactory dysfunction. *Laryngoscope, 114*(10), pp. 1764–1769.

Laudien, J. H., Wencker, S., Ferstl, R. and Pause, B. M. (2008) Context effects on odor processing: an event-related potential study. *Neuroimage, 41*(4), pp. 1426–1436.

Lombion, S., Comte, A., Tatu, L., Brand, G., Moulin, T. and Millot, J. L. (2009) Patterns of Cerebral Activation During Olfactory and Trigeminal Stimulations. *Human Brain Mapping, 30*(3), pp. 821–828.

Macpherson, L. J., Geierstanger, B. H., Viswanath, V., Bandell, M., Eid, S. R., Hwang, S. and Patapoutian, A. (2005) The pungency of garlic: Activation of TRPA1 and TRPV1 in response to allicin. *Current Biology, 15*(10), pp. 929–934.

McKemy, D. D., Neuhausser, W. M. and Julius, D. (2002) Identification of a cold receptor reveals a general role for TRP channels in thermosensation. *Nature, 416*(6876), pp. 52–58.

Mouraux, A., Diukova, A., Lee, M. C., Wise, R. G. and Iannetti, G. D. (2011) A multisensory investigation of the functional significance of the "pain matrix". *Neuroimage, 54*(3), pp. 2237–2249.

Naka, A., Wolf, A., Renner, B. and Mueller, C. A. (2014) A Novel Device for the Clinical Assessment of Intranasal Trigeminal Sensitivity. *Annals of Otology, Rhinology & Laryngology, 123*(6), pp. 428–433.

Negoias, S., Aszmann, O., Croy, I. and Hummel, T. (2013) Localization of odors can be learned. *Chemical Senses, 38*(7), pp. 553–562.

Ohla, K. and Lundstrom, J. N. (2013) Sex differences in chemosensation: sensory or emotional? *Frontiers in Human Neuroscience, 7*, p. 607.

Otte, A. and Halsband, U. (2006) Brain imaging tools in neurosciences. *Journal of Physiology-Paris, 99*(4-6), pp. 281–292.

Peier, A. M., Reeve, A. J., Andersson, D. A., Moqrich, A., Earley, T. J., Hergarden, A. C., Story, G. M., Colley, S., Hogenesch, J. B., McIntyre, P., Bevan, S. and Patapoutian, A. (2002) A heat-sensitive TRP channel expressed in keratinocytes. *Science, 296*(5575), pp. 2046–2049.

Penfield, W. and Rasmussen, T. (1950) *The Cerebral Cortex of Man. A Clinical Study of Localization of Function*, New York (USA): Macmillan.

Porter, J., Anand, T., Johnson, B., Khan, R. M. and Sobel, N. (2005) Brain mechanisms for extracting spatial information from smell. *Neuron, 47*(4), pp. 581–592.

Rajan, R., Clement, J. P. and Bhalla, U. S. (2006) Rats smell in stereo. *Science, 311*(5761), pp. 666–670.

Saunders, C., Li, W. Y., Patel, T. D., Muday, J. A. and Silver, W. L. (2013) Dissecting the role of TRPV1 in detecting multiple trigeminal irritants in three behavioral assays for sensory irritation. *F1000Research, 2*,p. 74.

Schaefer, M. L., Bottger, B., Silver, W. L. and Finger, T. E. (2002) Trigeminal collaterals in the nasal epithelium and olfactory bulb: a potential route for direct modulation of olfactory information by trigeminal stimuli. *Journal of Comparative Neurology., 444*, pp. 221–226.

Scheibe, M., van Thriel, C. and Hummel, T. (2008) Responses to trigeminal irritants at different locations of the human nasal mucosa. *Laryngoscope, 118*(1), pp. 152–155.

Scheibe, M., Zahnert, T. and Hummel, T. (2006) Topographical differences in the trigeminal sensitivity of the human nasal mucosa. *Neuroreport, 17*, pp. 1417–1420.

Shusterman, D. (2007) Trigeminally-mediated health effects of air pollutants: sources of inter-individual variability. *Human & Experimental Toxicology, 26*(3), pp. 149–157.

Sondersorg, A. C., Busse, D., Kyereme, J., Rothermel, M., Neufang, G., Gisselmann, G., Hatt, H. and Conrad, H. (2014) Chemosensory Information Processing between Keratinocytes and Trigeminal Neurons. *Journal of Biological Chemistry, 289*(25), pp. 17529–17540.

Thuerauf, N., Kaegler, M., Dietz, R., Barocka, A. and Kobal, G. (1999) Dose-dependent stereoselective activation of the trigeminal sensory system by nicotine in man. *Psychopharmacology, 142*, pp. 236–243.

Thürauf, N., Friedel, I., Hummel, C. and Kobal, G. (1991) The mucosal potential elicited by noxious chemical stimuli with CO_2 in rats: is it a peripheral nociceptive event? *Neuroscience Letters, 128*(2), pp. 297–300.

Thurauf, N., Gunther, M., Pauli, E. and Kobal, G. (2002) Sensitivity of the negative mucosal potential to the trigeminal target stimulus CO_2. *Brain Research, 942*(1-2), pp. 79–86.

Vesaluoma, M., Muller, L., Gallar, J., Lambiase, A., Moilanen, J., Hack, T., Belmonte, C. and Tervo, T. (2000) Effects of oleoresin capsicum pepper spray on human corneal morphology and sensitivity. *Investigative Ophthalmology & Visual Science, 41*(8), pp. 2138–2147.

Vogt-Eisele, A. K., Weber, K., Sherkheli, M. A., Vielhaber, G., Panten, J., Gisselmann, G. and Hatt, H. (2007) Monoterpenoid agonists of TRPV3. *British Journal of Pharmacology, 151*(4), pp. 530–540.

Waldmann, R., Champigny, G., Bassilana, F., Heurteaux, C. and Lazdunski, M. (1997) A proton-gated cation channel involved in acid sensing. *Nature, 386*, pp. 173–177.

Walker, J. C., Kendal-Reed, M., Hall, S. B., Morgan, W. T., Polyakov, V. V. and Lutz, R. W. (2001) Humain Responses to Proprionic Acid. II. Quantification of Breathing Responses and their Relationship to Perception. *Chemical Senses, 26*, pp. 351–358.

Wysocki, C. J., Cowart, B. J. and Radil, T. (2003) Nasal trigeminal chemosensitivity across the adult life span. *Perception & Psychophysics, 65*(1), pp. 115–122.

Xu, H., Delling, M., Jun, J. C. and Clapham, D. E. (2006) Oregano, thyme and clove-derived flavors and skin sensitizers activate specific TRP channels. *Nature Neuroscience, 9*(5), pp. 628–635.

Yang, B. H., Piao, Z. G., Kim, Y. B., Lee, C. H., Lee, J. K., Park, K., Kim, J. S. and Oh, S. B. (2003) Activation of vanilloid receptor 1 (VR1) by eugenol. *Journal of Dental Research, 82*(10), pp. 781–785.

CHAPTER 6

Characterization of aroma compounds: structure, physico-chemical and sensory properties

Henryk Jeleń & Anna Gracka
Faculty of Food Science and Nutrition, Poznań University of Life Sciences, Poznań, Poland

6.1 Introduction: volatile vs. aroma compounds in food

Over 10000 volatile compounds have been detected in foods. Among them, only a few hundred (5–10%) are aroma compounds. Moreover, in a recent meta-analysis of 5642 publications on food aroma from 1980 to 2013 and 949 other documents, Hofmann and co-workers (Dunkel et al. 2014) identified 226 key food odorants from the analysis of a total of 119 publications dealing with 227 food samples, based on their odour activity value (OAV>1). Ranking the 226 key odorants in order of their abundance they found 16 compounds (termed "generalists") that are detected as key odorants in ≥25% of analysed samples. The most frequently occurring ones are methional (53.7% of samples, boiled potatoes odour), 2- and 3-methyl butanal (50.7%, malty), butane-2,3-dione (41.9%, buttery), (E,E)-2,4-decadienal (41.4%, fried, fatty) and furaneol (40.5%, caramel), followed by sotolon (seasoning), 1-octene-3-one (mushroom), acetic acid (vinegar), acetaldehyde (fruity, fresh), ethyl 2-and 3-methylbutanoate (fruity), (E)-2-nonenal (cardboard-like), vanillin (vanilla-like), 2-acetyl-1-pyrroline (roasted, popcorn), 2- and 3-methylbutanoic acid (sweaty) and butanoic acid (sweaty). Another set of 57 key odorants ("intermediaries") contributed to the aroma of 5-25% of analysed food samples, while the third group consisted of 151 compounds ("individualists") present in less than 5% of foods.

The key aroma compounds have odour thresholds in a very broad range of values (from mg/L down to pg/L). When odour impact of compounds in foods are discussed, two features, odour threshold and compound concentration, must

Flavour: From food to perception, First Edition.
Edited by Elisabeth Guichard, Christian Salles, Martine Morzel, and Anne-Marie Le Bon.

be taken into consideration. The odour impact of compounds could be approximated by their odour activity value (OAV), which is a ratio of compound's concentration to its odour threshold. Compounds play significant role in overall product aroma when, despite their relatively high odour thresholds, they occur in high concentrations. On the other hand, compounds with extremely low odour thresholds, even when present in very low concentrations, may be key odorants. In an OAV concept, gas chromatography-olfactometry (GC-O) enables the determination of most odour active compounds using the aroma extract dilution analysis (AEDA) approach in which serial dilutions of aroma extract are sniffed, or other approaches that provide quantitation of aroma sensation. Odorants with low OAVs (typically ≤ 1) are generally considered to be unimportant in global sensory perception. However, determination of aroma profile by GC-O is not enough to characterise the global aroma. Compounds present at the subthreshold level that cannot be detected by GC-O can also contribute to the global aroma when present in mixture with supra-thresholds compounds due to the existence of perceptual interactions (reconstitution discrepancy) (Bult et al. 2001). There is evidence for the influence of compounds at or below their threshold levels on the perceived aroma of food indicating that addition or synergy occurs between compounds (Ryan et al. 2008). To fully characterise aroma *via* odorants identification, reconstitution tests followed by omission tests should be carried out.

6.2 Food odorants: structure, aroma and physicochemical properties

Food odorants are low molecular mass (<300 Da) compounds which belong to many chemical classes, mainly alcohols, esters, carbonyl compounds, terpenoids and heterocycles. The physicochemical properties of these compounds play a major role in their sensory perception.

Depending on compound/matrix interactions, chemical composition and structure, odorants may have different odour thresholds (OT). As it is impossible to measure directly a concentration of aroma compound interacting at human receptor level, odour thresholds measured in different matrices can be used as an indicator of their odour activity. There are examples of aroma compounds with extremely low odour thresholds (ng/L) as well as others with high thresholds (mg/L). Although relationships between molecular structure and sensory qualities have been investigated for a long time and much progress has been done recently on the perception mechanisms for odorants at the olfactory receptor level, it is hard to establish a universal structure—activity theory, and prediction of odour thresholds from a chemical structure fails in most cases. Odour thresholds for particular compounds published in the literature can vary in a broad range. As examples, odour thresholds determined in water for 3-methylbutanal (malty odour) range from 0.2 to 170 μg/L, for ethanol from 25000 to 900000 μg/L, for ethyl butanoate (fruity odour) from 0.0032 to 450 μg/L or for β-ionone (violet-like) from 0.007 to 23 μg/L (Czerny et al. 2008). For active amyl alcohol (2-methyl-1-butanol), the odour detection thresholds compiled by van Gemert (2011) from various publications are 4.15, 1.9, 0.25-0.42, 0.3, 50-60, 32.5, 1.2, 3.7 and 0.0159 mg/kg. With such variability, it is often hard to estimate an average OT value based on literature data. Such differences can be attributed to different levels of experience of the sensory panel members, to the size of the panel, as well as to contamination of the reference compounds with other odour active compounds, sometimes closely related isomers (Czerny et al. 2008).

Odour thresholds and odour qualities depend on structural parameters such as chain length, unsaturation degree, double bonds position, functional groups position and molecular characteristics and stereochemistry of molecules. Physicochemical properties of volatile compounds, such as solubility, volatility, viscosity, odour etc., vary as a function of chain length. Short-chain saturated alcohols are polar solvents highly soluble in water, but when the impact of the hydroxyl group on a given alcohol structure decreases due to increasing the hydrophobic chain length (higher than C5), the ability to dissolve in water decreases; such alcohols become nonpolar. Chain length mostly affects odour threshold and aroma properties. In the case of aldehydes, compounds with low molecular weight (<150 Da) have unpleasant odours, which disappear for compounds with higher molecular weight which have sweet, fruity aromas. Also, viscosity of alcohols increases with the increasing chain length. For the 4-alkyl substituted γ-lactones an increase in alkyl chain length causes decrease in coconut aroma and an increase in fruity sweet notes. For these compounds (R)-enantiomers have more intense odours than the corresponding (S)-enantiomers (Cooke et al. 2009). Odour descriptors of R-enantiomers of γ-lactones are shown in Figure 6.1.

Odour threshold (OT) is a parameter which depends not only on the length of the chain, but also on the structure of the bonds and functional groups. In saturated aldehydes OT decreases from C5 to C8, then increases. However no clear relationship has been found between chain length or bond structure and OT (Belitz et al. 2009). For straight chain acetates found as odorants in some fruits (Takeoka et al. 1996), ethyl acetate and propyl acetate have relatively high OTs (13500 and 2000 μL/10^3L respectively). Butyl and pentyl acetate have low OTs (respectively 66 μL/10^3L and 43 μL/10^3L) which rise with increased chain length, to reach higher values (190 μL/10^3L for octyl acetate). For branched esters, higher OTs were noted for unsaturated compared to saturated esters. The double bond position and the configuration also influenced odour threshold. The addition of double bond in the C-2 position of acid resulted in higher OTs, whereas the addition of double bond further away from carbonyl carbon had a smaller effect on OT (Takeoka et al. 1998). Addition of double bond to a structure of 2-acetyl-1-pyrroline and changing positions of double bonds, which results in a structurally similar 2-acetyl-pyrrole, caused a drastic difference in

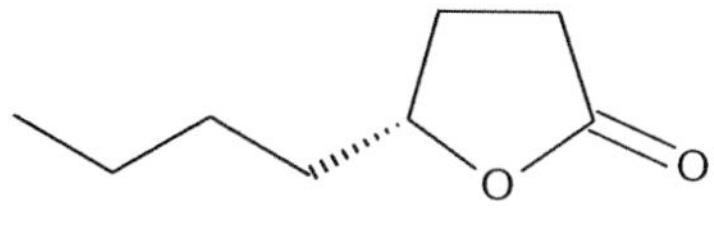

[1]

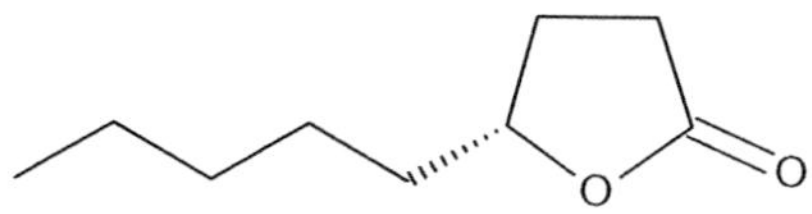

[2]

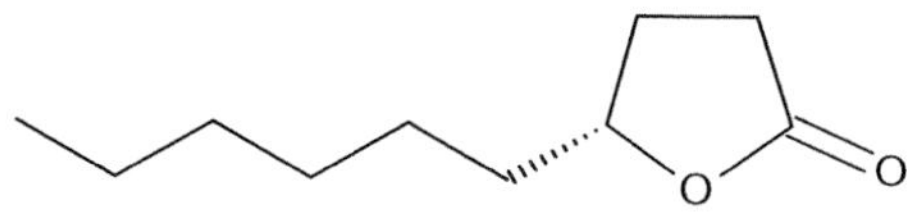

[3]

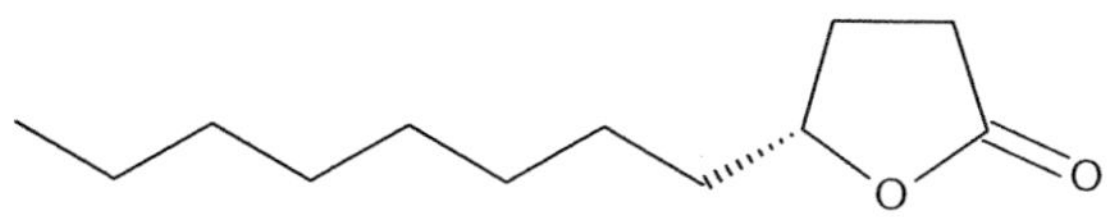

[4]

Figure 6.1 Structures of alkyl substituted γ-lactones with different odour descriptors. [1] (R)-γ-octalactone (coconut notes with almond notes, spicy green); [2] (R)-γ-nonalactone (soft coconut with fatty-milky aspects, strong, sweet); [3] (R)-γ-decalactone (caramel, fatty sweet fruity note, soft coconut); [4] (R)-γ-dodecalactone (bloomy notes with aldehyde and woody aspects, strong, fruit-sweet). Source: Adapted from Cooke et al. (2009).

OTs of these compounds as for 2-acetyl-1-pyroline it was estimated at 0.1 μg/L, whereas for 2-acetylpyrrole it was 170 000 μg/L (Buttery 1999).

A change of the double bond configuration or functional group's position influences the odour threshold or aroma quality of a given compound, but these properties depend also on the size of the molecule. For example, changing the bond of a molecule from saturated to unsaturated may result in modifying aroma quality in small molecules, as illustrated by *trans*-2-hexenal which has a herbal green-like aroma while saturated hexanal has a fruity, strawberry-like odour. On the other hand, in larger molecules, these changes affect only the odour threshold of the compound. The effect of removing or changing the hydroxyl group in aliphatic or aromatic alcohols often leads to the disappearance of odour (Reineccius 2005). Comparing odour thresholds between *cis* and *trans* isomers of alka-1,5-dien-3-ones and alka-1,5-dien-ols shows that few differences are observed for long chain (C14) homologues whereas, for octadienones and octadienols significant differences exist. (Z)-octa-1,5-dien-3-one (geranium-like metallic odour) has an odour threshold of 0.003 ng/L (air), whereas its corresponding (E)-octa-1,5-dien-3-one (geranium-like odour) has an odour threshold of 1.4 ng/L (air). Similarly, (Z)-octa-1,5-dien-3-ol (geranium-like) has an odour threshold of 1.3 ng/L, whereas its (E) isomer (geranium-like, green odour) has an odour threshold of 36.2 ng/L (Lorber et al. 2014).

Odour thresholds of compounds of similar structures with different functional groups locations can differ substantially. As shown in Figure 6.2, 2-Ethyl-3,5-dimethyl pyrazine and 2-ethyl-3,6-dimethyl pyrazine differ in position of one of the methyl groups only, which makes them similar in retention on nonpolar GC columns, however their odour thresholds in water differ over 200 times: 8.6 μg/L for 2-ethyl-3,6-dimethylpyrazine, and 0.04 μg/L for 2-ethyl-3,5-dimethylpyrazine (Buttery and Ling 1997a). An additional alkylation on 4-methylphenol in ortho- and para- position respectively modifies the odour threshold of the corresponding molecules to 27 ng/L for 2,4-dimethylphenol and 143 ng/L for 3,4-dimethylphenol which corresponds to an increase by factors of 225 and 1190 respectively. Trimethyl phenols are also characterised by lower odour thresholds than dimethylphenols (Czerny et al. 2011). Other examples of large differences in odour thresholds of structurally similar earthy/musty smelling pyrazines are given in Figure 6.2. 2-Ethenyl-3-ethyl-5-methylpyrazine has an odour threshold of 0.014 ng/L in air, whereas 3-ethenyl-2-ethyl-5-methylpyrazine's odour threshold in air

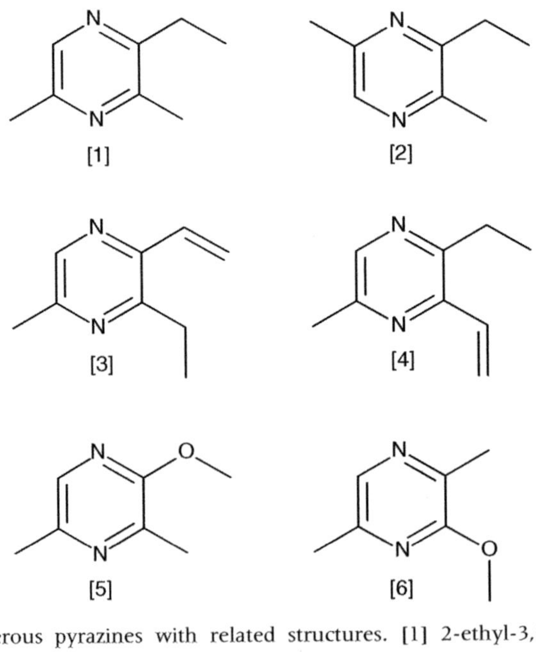

Figure 6.2 Odoriferous pyrazines with related structures. [1] 2-ethyl-3,5-dimethylpyrazine (OT 0.04 μg/L in water); [2] 2-ethyl-3,6-dimethylpyrazine (OT 8.6 μg/L in water); [3] 2-ethenyl-3-ethyl-5-methylpyrazine (OT 0.014 ng/L in air); [4] 3-ethenyl-2-ethyl-5-methylpyrazine (OT 109.5 ng/L in air); [5] 2-methoxy-3,5-dimethylpyrazine (OT 0.01ng/L in air); [6] 3-methoxy-2,5-dimethylpyrazine (OT 56 ng/L in air).

is 109.5 ng/L (Czerny et al. 1996). Also 2-methoxy-3,5-dimethylpyrazine (earthy smelling) and 3-methoxy-2,5-dimethylpyrazine (pea-like smelling) have very different odour thresholds of 0.001 ng/L (air) and 56 ng/L respectively (Molyneux and Schieberle 2007). From the above mentioned examples of pyrazines it is evident that trace amounts of another more potent isomer can drastically influence the odour threshold of analysed compound and that the great chemical similarity of isomers with different odour threshold is an additional challenge for the separation of these compounds by GC and in consequence also by GC-O.

Chirality influences both odour quality and odour threshold of aroma compounds (Brenna et al. 2003). In the field of food flavours, there are well known examples of different odour notes of enantiomers: carvone (caraway (+) and spearmint (−)), nootkatone (grapefruit (+) and woody/spicy (−)), limonene (orange (+) and turpentine (−)), menthol (vegetable/mild mint (+) and strong mint/cooling (−)), or linalool (sweet/petitgrain (+) and woody/lavender (−)). The enantiomers of whisky (or quercus) lactones (Figure 6.3), which are extracted from oak barrels in which alcoholic beverages are kept, have been described with different odour notes (Brown et al. 2006). Their natural enantiomeric excesses are 77% for (3S, 4S)-lactone and 23% for (3S, 4R)-lactone. Odour descriptors for (3S, 4S) enantiomer are weak coconut, earthy, mouldy, hay, whereas for (3S, 4R) they are spicy, celery, weak coconut, distinct green walnut. The remaining enantiomers have sweet, fresh, bright, coconut notes (3R, 4R), and an intense coconut note with an after odour reminiscent of celery

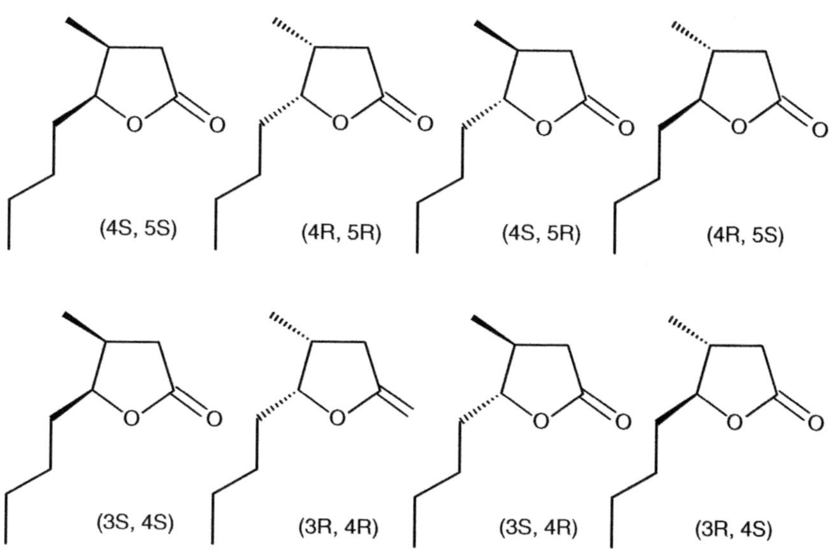

Figure 6.3 Whisky lactones (quercus lactones) enantiomeric structures with different odour notes.

(3R, 4S). The odour thresholds in white wine of naturally occurring isomers, (3S,4S) cis-oak lactone and (3S, 4R) trans-oak lactone, are 20 and 140 µg/L respectively, whereas those of the two non-naturally occurring isomers, (3R, 4R)-cis- and (3R 4S)-trans, are 132 and 330 µg/L respectively (Brown et al. 2006). Guth (1996) identified in white wines a sweet, coconut smelling wine lactone (3a,4,5,7a-tetrahydro-3,6-dimethylbenzofuran-2(3H)-one). The configuration of this naturally occurring wine lactone was established as (3S,3aS,7aR) enantiomer. All possible enantiomers were synthesized and odour thresholds estimated for these compounds (Figure 6.4). The lowest odour threshold (0.02 pg/L in air, range of 0.01-0.04 pg/L) was noted for the naturally occurring isomer while the others exhibited odour thresholds ranging from 0.007 to >1000 ng/L in air (Guth 1996).

Aroma perception is not only due to the amount of odorant but also influenced by the matrix from which aroma compounds are released into the vapour phase. Interactions between the components matrix and odorants (discussed in detail in Chapter 9) influence odorants solubility and as a result their vapour/matrix partition coefficients. For example for apple odorants, the ratio of odour thresholds in juice vs. in water can vary from 5.2 (determined for butyl butanoate) to

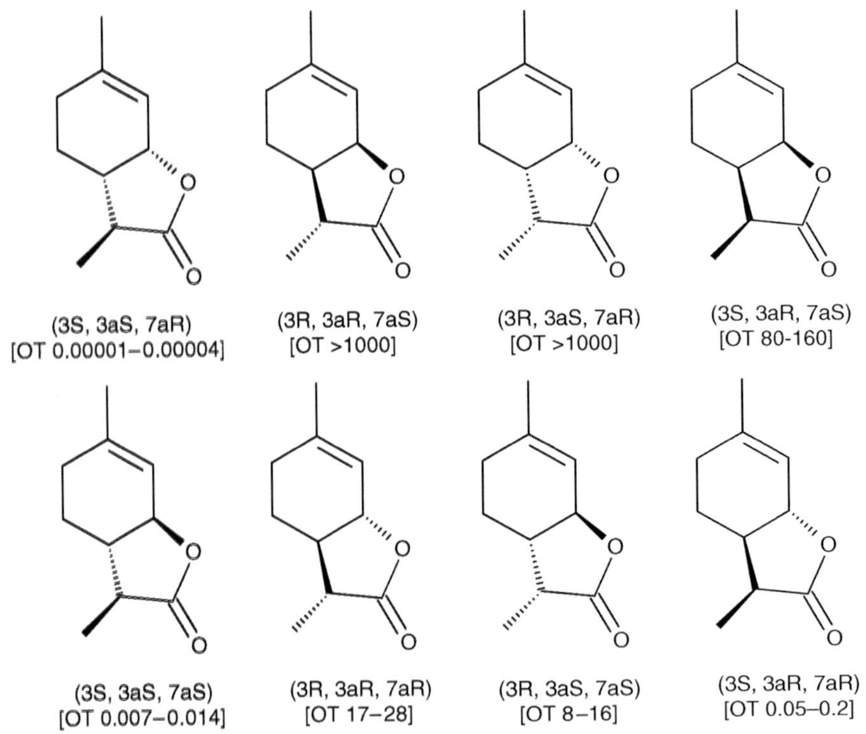

Figure 6.4 Wine lactones enantiomers with their odour thresholds (ng/L in air).
Source: Adapted from Guth (1996).

600 for ethyl 2-methyl butanoate (Cliff et al. 2011). Oak lactones odour thresholds in wine differ between white and red wines: in white wine (3S, 4S) *cis* and (3S, 4R) *trans* isomers have odour thresholds of 20 and 140 μg/L respectively, whereas in red wine they are 54 and 370 μg/L respectively (Brown et al. 2006). As summarised by Mazzoleni and Maggi (2007), odour thresholds for trichloroanisole (2,4,6-TCA) differ substantially from 0.03 ng/L in water, 0.8 ng/L in mineral water, 4 ng/L in dry white wine to 22 ng/L in red wine.

6.3 Main classes of food odorants

Aroma compounds influencing flavour of foods belong practically to all classes of chemical organic compounds and more precisely alcohols, acids and esters, phenolic compounds, carbonyl compounds, sulphur and nitrogen aliphatic compounds, heterocyclic aroma compounds, terpenes and norisoprenoids.

6.3.1 Alcohols, acids and esters

Alcohols are important flavour compounds which are widely present in nature. Their hydroxyl group determines most of their properties. When the hydroxyl group is substituted, the odour of the compound may change. Alcohols are present in many fruits. Among saturated alkyl alcohols, which occur widely in nature, the most known is ethanol. It is an important component of spirits, but does not have a significant impact on food flavour. However, during fermentation processes in which ethanol is produced, significant amounts of higher alcohols are formed as by-products. In this fraction (fusel oils), compounds such as 2-methyl-1-propanol and isoamyl alcohol often prevail. Isoamyl alcohol is a mixture of 2-methylbutan-1-ol and 3-methylbutan-1-ol, and contributes often to fruity aroma in food and may be used in flavourings to elicit such notes (Vandamme and Soetaert 2002). The higher alcohols described above are products of the Ehrlich pathway. Most straight chain C4-C9 alcohols also have a fruit-like aroma. Alcohols above C9 have a more fatty and unpleasant oily odour note. With the increasing length of the chain, the viscosity of alcohols increases. Unsaturated alcohols have a greater impact on food flavour than saturated alcohols. Alcohols with low molecular weight and a double bond near the hydroxyl group have an unpleasant, harsh aroma (Reineccius 2005). They are also precursors for synthesis of aldehydes and esters (Surburg et al. 2006). C6-alcohols such as *cis*-3-hexenol and *cis*-6-hexenol (leaf alcohol), as well as C6-aldehydes which are formed in plants, are important for the flavour industry because of their characteristic odour of freshly cut grass (Figure 6.5). *Cis*-3-hexenol is also a precursor for the synthesis of 2-*trans*-6-*cis*-nonadien-1-ol (violet leaf alcohol) and 2-*trans*-6-*cis*-nonadien-1-al. These alcohols, and also some other C6 and C9 aliphatic aldehydes are formed in plants as a result

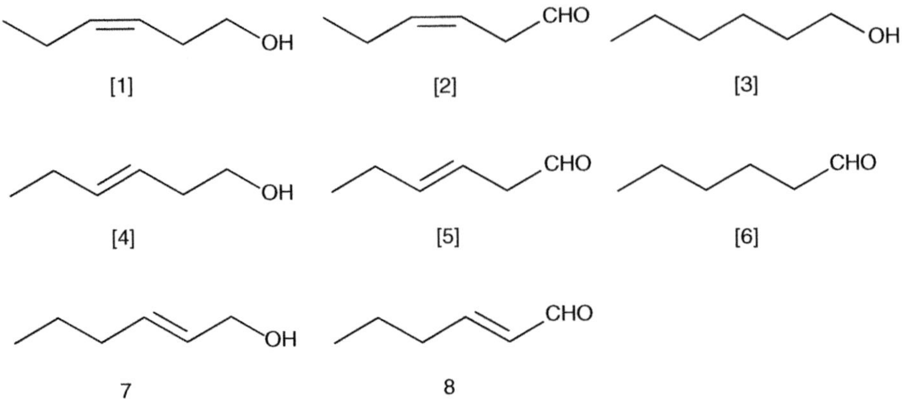

Figure 6.5 Fresh green odour components from green leaves. [1] – (3Z)-hexenol (leaf alcohol), [2] – (3Z)-hexenal, [3] - n-hexanol, [4] – (3E)-hexenol, [5] – (3E)-hexenal, [6] – n-hexanal. [7] – (2E)-hexenol. [8] – (2E)-hexenal (leaf aldehyde). Source: Adapted from Hatanaka (1996).

of enzymatic oxidation of linoleic and linolenic acid by lipooxygenases and hydroperoxide lyases (Jeleń and Wąsowicz 2012). In the case of *cis*-3-hexenol, the reaction with linoleic acid gives unstable *cis*-3-hexenal, which is transformed into *cis*-3-hexenol. Finally, as a result of its reaction with dehydrogenases, proper alcohols are formed. The synthesis of "leaf alcohols" is caused by the damage of plant cells by enzymes, whereby the enzymes have the ability to react with substrates. One of them is (R)-1-octen-3-ol, an unsaturated aliphatic alcohol with a mushroom-like odour, widely used as an earthy odorant.

Almost all aromatic alcohols have desirable pleasant aromas. The most common aromatic alcohol is 2-phenylethanol with a pleasant rose-honey-like and floral odour (Li et al. 2014). This compound contributes to rose notes in food (Majcher et al. 2014). It has been found mainly in fermented foods, such as beer, cheese, wine, bread, coffee and soy sauce (Etschmann et al. 2002). 2-Phenylethanol can by formed by microbiological metabolism. From 2-phenylethanol, aroma-active 2-phenethyl esters, such as 2-phenethyl acetate, 2-phenethyl butyrate, 2-phenethyl isobutyrate and 2-phenethyl propionate with its powerful rose-like aroma can be formed (Li et al. 2014). Benzyl alcohol after esterification forms important flavour compounds. In cinnamaladehyde reduction, cinnamyl and dihydrocinnamyl alcohols are formed. Cinnamyl alcohol exists in *cis* and *trans* forms, the latter more frequent in nature. Both have a floral, hyacinth and cinnamon aroma. Cinnamyl alcohol can be formed by the reduction of cinnamaldehyde (Berger 2007).

Acids often increase and enhance characteristic aroma profiles of food. For example, aliphatic C3-C8 acids enhance fruity notes, especially 2-methyl-2-pentenoic acid, used in strawberry flavours. C4, C6-C12 acids (e.g. trans-2-butenoic acid) also enhance cheese flavour. The strength of aroma decreases with acid chain length. Simple acids have high odour thresholds. If used as odorants, most of them have a "pungent" odour. Diluted 2- and 3-methylbutyric acids have a fruity aroma, and acetic acid has the characteristic sharp odour of vinegar. Long-chain acids are odourless, but—when added to food—they increase its acidity. Other acids widely used in flavour industry are propionic acid, butyric acid with its strong buttery aroma, and 2-methylpropanoic acid. Some enantiomers of acids are characterized by different aromas depending on their orientation: *trans*- 2-methylbutanoic acid has a sweaty, cheesy aroma, and *cis*-2-methylbutanoic acid has a fruity aroma (Tachihara et al. 2006). Unsaturated acids generally have sharper and stronger aromas than saturated ones. Aromatic acids, with the exception of phenyl acetic acid, are unfrequently found as food odorants. Phenyl acetic acid has a sweet, fruity, honey-like odour, and is mainly present in honey.

Acids are used also as substrates for ester formation, which has a significant role in flavour development. Esters with low molecular weight are present mainly in fruit, flowers and fermented beverages. Most of them are formed from the corresponding alcohols and acids with an even number of carbon atoms. Those with simple structures have characteristic fruity aromas; others have fruity aromas as well, but often nonspecific. Aroma of many esters depends on their molecular structure and conformation (Brenna et al. 2003). Short-chain esters have a fruity odour, but as the length of the chain increases, their smell can get more fatty, soapy or metallic. The strength of aroma depends on whether the ester structure is saturated or unsaturated (unsaturated esters have stronger aroma) and decreases as volatility decreases. *Trans*-2-hexenyl acetate has a green, fresh-fruity aroma, and 2-propenyl hexanoate has a pineapple odour. The most common esters in food are ethyl esters and acetates. Isoamyl acetate is one of the key aroma compounds of banana aroma (Berger 2007). Esters contribute mainly to the flavour of fruits and have often "fruity" aroma descriptors (trans-2-hexenyl acetate—fresh fruit, green apple; butyl 2-methylbutanoate—fruity, green apple; butyl acetate —fruity, apple; butyl butanoate—fruity, berry; ethyl 2-methyl butanoate—fruity, green apple, sweet; hexyl acetate—fruity, solvent-like) (Cliff et al. 2011). As demonstrated by several authors (Takeoka et al. 1996, Cliff et al. 2011), the addition of a methyl group in the 2- or 3- position to butyl acetate decreases odour threshold, and methylation decreases thresholds for some esters (ethyl-, propyl) but increases them for others (pentyl, hexyl).

6.3.2 Phenolic compounds

Phenols, particularly alkylated phenols, are often related to off-odours in food. They are typically described as having a smoky, phenolic, medicinal odour, leather-like or ink-like (Czerny et al. 2011). A compilation of data on the orthonasal odour thresholds for alkylated phenols shows high discrepancies in reported values: from 2.7 to 200 μg/L (water) for 4-methylphenol, from 21 to 600 μg/L for 4-ethylphenol, from 1.7 to 800 μg/L for 3-ethyl phenol and from 31 to 10000 μg/L for phenol (Czerny et al. 2011). Some volatile phenolic compounds (4-ethylphenol, 4-ethylguaiacol, 4-vinylphenol and 4-vinylguaiacol) are associated with the "Brett" off-flavour of wine. Even though the odour threshold of 4-ethylphenol is quite high in wine (0.440 mg/L) (Heresztyn 1986), thresholds in water range from 0.013 to 0.6 mg/kg and vary very broadly in other media: 0.004 mg/kg in 7% acetic acid, 0.2 mg/kg in sunflower oil, 0.6 mg/kg in 20% ethanol but also up to 140 mg/kg in 14% ethanolic solution (Van Gemert 2011). Sensory odour threshold of 4-ethyl guaiacol in water ranges from 0.0044 (Czerny et al. 2008) to 0.089 mg/kg (Giri et al. 2010) and odour thresholds of p-vinyl phenol and p-vinyl guaiacol in water are 0.022 and 0.012 ppm respectively (Fallico et al. 1996). *Brettanomyces* spp., the causative microorganisms responsible for Brett off-flavour, produce many odour-active compounds including lactones, fatty acids, esters or lipid oxidation products. Other microorganisms have also the ability to convert phenolic acids to volatile phenols, which results in off-flavours as demonstrated with strawberries infested with *Phythophthora cactorum* (Jeleń et al. 2005).

Phenolic volatile compounds are also abundant compounds associated with smoking processes and detected in smoked meat or fish (Varlet et al. 2006), but also in smoke affected grapes (Parker et al. 2012). Cresol isomers and guaiacol odour thresholds range from 20 to 64 μg/L depending on the isomer (Parker et al. 2012). Volatile phenols form also an important group of coffee odorants (Flament and Bessière-Thomas 2002).

6.3.3 Carbonyl compounds

The aroma of aliphatic aldehydes changes with the length of the carbon chain and aldehydes with more than 13 carbon atoms are characterised by a weak odour. Acetaldehyde has an apple-like odour, and is widely used in food flavours, not only to provide smell, but also to contribute to freshness and juiciness of food (Burdock and Fenaroli 2010; Goodrich et al. 1998). As the number of carbon atoms rises, aldehydes have an increasingly fatty odour. Aldehydes such as isobutyraldehyde, isovaleraldehyde, and 2-methylbutyraldehyde have fruity and roasted aromas (Surburg et al. 2006). Hexanal, depending on the concentration, can have a fatty-green odour or, in low concentration, an "unripe fruit" aroma. Its synthesis is *via* decomposition of linoleic acid in a lipid oxidation pathway (Ozcan and Barringer 2011). Among 2-alkenals, the most common is *trans*-2-hexenal, called "leaf aldehyde" because of its green odour (Hatanaka 1996). Along with *cis*-3-hexenal, *trans*-2-hexenal is one of the main compounds responsible for green, strawberry and fresh flavour in food (Myung et al. 2006; Németh et al. 2004; Ozcan and Barringer 2011). *Cis*-4-heptenal is also widely used in the food flavour industry as an odorant with a powerful fatty, fish-like, creamy aroma (Surburg et al. 2006). It has been shown that unsaturated aldehydes with the double bond at C2, C3, C4 and (E) or (Z) geometry have odour thresholds 100 to 1000 times lower than their corresponding unsaturated alcohols (Hatanaka 1996). The most common polyunsaturated aldehyde is *trans*-2-*trans*-4-decadienal with a fatty/citrus/chicken aroma, one of the key volatile compounds in chicken meat. The second most common is *trans*-2-*cis*-6-nonadienal (violet leaf aldehyde), present in cucumber. C6-aldehydes are formed from unsaturated fatty acids. Unsaturated fatty acids such as linoleic and linolenic acid are broken down forming shorter fatty acids, which in turn react with lipoxygenase and hydroperoxide lyase to form hexanal and *cis*-3-hexanal. In case of the occurrence of *cis*-3/*trans*-2-isomerase, *cis*-3-hexenal is isomerised into *trans*-2-hexenal (Stone and Kazeniac 1975). Benzaldehyde has a bitter almond and cherry-like aroma and is produced on an industrial scale as flavouring component by the hydrolysis of benzyl chloride in the presence of a catalyst and serves as a substrate for the synthesis of other flavour compounds. Therefore, this aromatic aldehyde is one of the most significant aromatic aldehydes in food. Besides benzaldehyde, cinnamaldehyde and its α-amyl- and α-hexyl- homologues are found as food odorants. Acetals, derived from aliphatic aldehydes in reaction with alcohols (ethanol), are present in alcoholic beverages mostly. Aldehydes are vulnerable to oxidation and to the formation of Schiff's bases with amines, but the formation of acetals counteracts their further decomposition. The aroma of acetals depends on aldehyde precursors (Rowe 2005; Surburg et al. 2006). One of the most widely detected acetals in alcoholic beverages is acetaldehyde diethyl acetal, with a nutty aroma.

In the ketone class, odd-numbered methyl ketones C7, C9, C11, being formed *via* β-oxidation and decarboxylation of corresponding fatty acids, are well known food odorants because of their nut-like aroma. Some monoketones, such as octan-3-one and alkan-2-ones, have been identified in cheese (Hassan et al. 2012). Heptan-2-one and nonan-2-one are also present in blue-veined cheeses produced using *Penicillium roqueforti*. Straight and branched chain ketones up to C8 contribute to caramel notes in food (Qian et al. 2002). Among diketones, the most common is butane-2,3-dione (diacetyl). Together with 3-hydroxy-2-butanone (acetoin), a partial oxidation product of 2,3-butanediol, they are responsible for the buttery flavour in dairy products and margarine (Jyoti et al. 2003). Butanedione is also an indicator of the fermentation stage in beer production and contributes to the "green" notes of freshly fermented beers, so a decrease usually to < 100 ppb is required in production process for light Pils type of beers. Ketones have also been found in fruit, coffee and various other types of food. 4-Hydroxyphenyl-2-butanone, called the raspberry ketone, is the most important of all aromatic ketones, and one of the key aroma compounds of raspberry (Fischer et al. 2001). Due to the high cost of production (in raspberries, there is less than 4 mg kg^{-1} of the natural flavour, and the total production cost of a kilogram is about \$1000), the synthetic raspberry ketone is the one most often used, but also biotechnology opens perspectives for its production from natural sources i.e. from betuloside obtained from birch (*Betula pendula*) (Serra et al. 2005).

6.3.4 Sulfur and nitrogen aliphatic compounds

Sulfur compounds comprise mercaptans (thiols), thioesters, sulfides, polysulfides, dithiazins, oxathiazines, thiophenes and thiazoles. Compared with other volatile compounds, they are present in food in very low concentrations and most of them are chemically labile. However, because of low odour thresholds, they still constitute important food odorants. Allyl isothiocyanate, the main aroma compound of mustard oil, is the sulfur odorant naturally occurring there in largest amounts (Surburg et al. 2006). It is partly responsible for the aroma of spoilt vegetables, but also for the aroma of heated food. Fruits, vegetables and dairy products contain aliphatic thiols mostly. The edible plant richest in sulfur volatile compounds is garlic, which has more than half of all the sulfur volatile compounds responsible for aroma. Some sulfur compounds can also be responsible for off-flavours, particularly when present in high amounts. Hydrogen sulfide, which is responsible for the "rotten egg" aroma in wines, may also have an impact on the development of a fresh, "citrus" aroma. Dimethyl sulphide which in high concentration has a cabbage aroma, in low levels is responsible for undesirable notes in canned citrus juices, as well as the aroma of heated milk (McGorrin 2011; Mussinan and Keelan 1994). Sulfur compounds are mainly related to onion-like, cabbage-like, skunky aromas, but they are also present in tropical fruits, e.g. 3-mercaptohexanol and 3-methylthiohexanol. Most of these compounds have an impact on the overall flavour of foods, but do not have a characteristic aroma of their own. These sulfur compounds can be identified as having the same organoleptic properties as particular foods (character-impact compounds) (Boelens and Van Gemert 1993).

Thiols are compounds which have a very low odour threshold. They are present in fermented food and drinks mostly, such as cheese and wine. Methanethiol is one of the most common sulfur compounds in food, not only because of its odour, but because of its being a substrate for other volatile sulfur compounds' synthesis, for example, the formation of dimethyl sulfide by oxidation. In high concentration, it has a cabbage-like odour, such as in Camembert (Qian et al. 2002), whereas in other cheeses, it is responsible for flavour intensity. Methanethiol is a product of the degradation of methionine, and it can be oxidized to other sulfur compounds such as thioesters and sulfides. Methionol has a cabbage-like aroma, and it is responsible for the desirable aroma of wines. Methionol can be synthesized from methional (Landaud et al. 2008). 2-Furanmethanethiol, which is present in low concentrations in some wines, contributes to their "roasted coffee" aroma. Other thiols have a fruity aroma. The most common is 1-ρ-menthene-8-thiol, known as grapefruit mercaptan, one of key odorants in grapefruit juice, with an OT in water of 0.0001 μg/L (Buettner and Schieberle 2001). Thiols are a significant group of odorants, which can react with other volatile compounds. Thioesters have stronger aromas than their corresponding thiols, but the strength of the aroma decreases from acetates to hexanoates (Engel 1999). 4-Mercapto-4-methyl-pentan-2-one ("cat ketone") is one of the most important aroma compounds in wine, especially in Sauvignon Blanc. This compound is frequently described as having a "blackcurrant" or, at higher doses, "cat urine" aroma (Landaud et al. 2008). 1-Methoxy-3-methyl-3-mercaptobutane, because of its fruity, blackcurrant aroma, is used in juice flavourings. Some sulfur compounds are strictly related to off-flavours. Among them, the most common are 3-methyl-2-butene-1-thiol with a skunky odour, an indicator of the "light-struck" defect in beer, which appears as a result of improper storage (McGorrin 2011). Methional can be formed from methionine in Strecker degradation, and it may be oxidized to methanethiol. It has a characteristic aroma of boiled potatoes. Methional may appear in oxidized white wine and UHT-milk (Mussinan and Keelan 1994). It has been observed that enantiomeric forms of alcohols and esters (acetates, butanoates, hexanoates) may differ with regard to the quality of aroma or its intensity. Only the (R)-esters of 3-mercaptohexanol have a fruity aroma, while the (S)-esters have a distinctly sulfurous aroma (Engel 1999).

Sulfides are mainly present in alliaceous vegetables such as onion, leek, and garlic. These sulfides can be either symmetrical (with identical hydrocarbon groups) or unsymmetrical (with different groups). Dimethyl sulfide, a compound with a very low odour threshold, has a sulfurous aroma, also described as boiled cabbages, asparagus, or corn. Depending on its concentration in wine, dimethyl sulfide can affect the wine's aroma positively or negatively. Other sulfides, such as dimethyl disulfide, have a typical garlic aroma, which is important in cheeses (Landaud et al. 2008). Butyl sulfide is one of the few characteristic sulfur compounds with floral notes.

From all aliphatic nitrogen containing compounds, only amines, isothiocyanates and aminophenols are important odorants in food. Amino acids, in Maillard reaction, are substrates for thermal reactions with reducing sugars, which result in a series of reactions whose products are significant odour compounds. Amines are mainly quality indicators in fish. Dimethylamine and trimethylamine for example increase as a result of fish putrefaction (Gram and Dalgaard 2002). Amines with low molecular weight have a strong odour and an ammonia-like aroma at low concentration. Amines with higher molecular weight have a fish-like aroma, which is stronger in secondary and tertiary amines, but as the molecular weight increases, the odour gets weaker until the amines become odourless. Some amines are Strecker degradation products, and their odour depends on their pH. Aromatic amines with the amino group in the nucleus have a very weak odour or are completely odourless, while aromatic amines with the amino group on the side chain have generally an ammoniac aroma.

Isothiocyanates have the characteristic –N=C=S group, which contributes to pungent notes in vegetables. These compounds occur naturally as glucosinolate conjugates, for example in cauliflower, broccoli, turnip or cabbage (Berger 2007). Benzyl isothiocyanate, also known as mustard-oil-aroma, is present in cress as a pungent odorant, and in unripe papaya at high levels as a pungent-sour odorant (Burdock and Fenaroli 2010; Pino 2014). It has been discovered that, in low concentration, benzyl isothiocyanate has a fruity aroma (Pino 2014). 3-Butenyl isothiocyanate is one of the most important odorants in horseradish. It is also present in wasabi and mustard, giving off a strong, penetrating aroma (Burdock and Fenaroli 2010).

6.3.5 Heterocyclic aroma compounds

Heterocyclic aroma compounds share few ring structures containing sulfur, nitrogen or oxygen atoms in a ring. Examples of heterocyclic aroma compounds are tiophenes, pyrrolines, thiazoles, furanones, pyrones, pyrazines or pyridines (Figure 6.6).

Though the compounds shown on this figure rarely appear in food as aroma compounds in unsubstituted form, their substituted derivatives belong to compounds with often very low odour thresholds. Heterocyclic aroma compounds are formed from relatively simple precursors, sugars and amino acids, at high temperatures. Most of them are formed during the Maillard reaction, the most important reaction in flavour chemistry of thermally processed foods. Maillard reaction determines the flavour of roasted products (coffee, cocoa, nuts) occurring during baking, boiling and steaming (Cerny 2008). Heterocyclic aroma compounds are formed also from thermal degradation of carbohydrates and as products resulting from interaction of Maillard reaction products with lipid oxidation products.

Thermal reactions result in the formation of a great number of products from a limited number of precursors, even a single amino acid and single sugar. In a reaction of ribose and cysteine, sulfur containing aroma

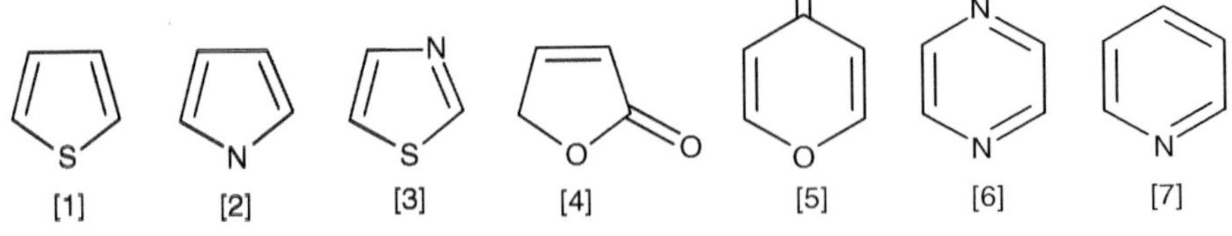

Figure 6.6 Basic heterocyclic structures of aroma compounds in food. [1] tiophene; [2] 1-pyrroline; [3] thiazole; [4] 2-furanone; [5] 4-pyrone; [6] pyrazine; [7] pyridine.

molecules such as 2-methylfuran, thiazole, 2-methylthiophene, furaldehyde, 2-methyl-3-furanthiol, 2-furfurylthiol, 2-methyl-3-(methylthio)furan, 3-thiophenethiol, bis(2-methyl-3-furyl)disulfide, or bis-(2-furfuryl)disulfide are formed (Cerny and Davidek 2003). In a model roasting reaction of serine, threonine and sucrose, around 350 volatiles were detected including more than 100 monocyclic furans, 8 bicyclic furans and 13 furanones (Baltes and Bochmann 1987a). Numerous sulfur aroma heterocycles, mainly thiophenes, thiazoles and alicyclic compounds can be formed from L-ascorbic acid and cysteine in Maillard reaction (Yu et al. 2012). Sensory evaluation of glucose – amino-acids interactions products under acidic conditions at 100°C revealed flowery, fragrant aromas from proline, rose aroma from phenylalanine and tyrosine, flowery and fruity odour from alanine. Glycine, lysine, threonine and valine produced caramel-like odours, whereas cysteine and methionine yielded savoury, meaty and soy sauce-like aromas (Wong et al. 2008). In a reaction of glutathione with glucose and fructose, 2,5-dimethylthiophene (metallic, sulfurous odour) and 5-methyl-2-furfural (almond-like, caramel-like odour) were predominant aroma compounds, based on their flavour dilution (FD), a measure of intensity of particular odorant in food obtained by sniffing the serially diluted extract of volatile/flavour compounds (Lee et al. 2010).

Among heterocycles compounds contributing to caramel aroma (Paravisini et al. 2012), 21 oxygenated compounds were detected including furans, furanones, pyranones and furans with oxygenated substitutes. The main odorants detected by GC-O were furfural (almond odour), acetylformoin, 2,4-dihydroxy-2,5-dimethyl-3(2H)-furanone (toasty, caramel flavour), 2-methyl-2(3H)-furanone (roasted, burnt odour), 2,3-dihydro-5-hydroxy-6-methyl-(4H)-pyran-4-one (dihydromaltol) and maltol (caramel, cotton candy odour) (Paravasini et al. 2012).

Coffee is a rich source of heterocyclic compounds, and with the introduction of capillary gas chromatography numerous, new heterocycles were identified, among them 27 oxazoles, 25 thiazoles, 18 pyrazines, 4 pyrroles, 3 pyridines and 3 indoles (Vitzthum and Werkhoff 1974). In a coffee roasting model experiment with serine, threonine and sucrose as precursors 123 mono and bicyclic pyrazines were detected (Baltes and Bochmann 1987b). Alkylpyrazines detected in coffee, cocoa, peanuts, chips have mainly roasted, nut-like flavour. At present approximately 200 pyrazines, 80 pyridines and 150 pyrroles have been detected in various food products. Although pyrazines are usually associated with roasting processes and alkylated pyrazines form one of most odoriferous groups of heterocycles (Figure 6.2), 3-alkyl-2 methoxy pyrazines formed in enzymatic processes are also an important group of heterocyclic odorants, which contribute to aroma of green peas and bell peppers. 3-Isobutyl-2-methoxypyrazine (IBMP), 3-sec-butyl-2-methoxypyrazine (SBMP) and 3-isopropylo-2-methoxypyrazine (IPMP) influence aroma of some wines (Cabernet Sauvignon, Sauvignon Blanc, Merlot), contributing to "veggie" aromas. Odour thresholds for these pyrazines in water are in the order of 1-2 ng/L, whereas in wines they are higher, in a range from 10 to 16 ng/L (Sala et al. 2002). Lactic acid bacteria were found responsible for formation of N-heterocycles causing mousy off-flavour in wine, 2-ethyltetrahydropyridine, 2-acetyltetrahydropyridine, but also 2-acetyl-1-pyrroline (Costello et al. 2001, Costello and Henschke 2002). 2-Acetyl-1-pyrroline is an important nitrogen containing compound identified in popcorn, cooked rice, basmati rice, wheat bread crust, rye bread, pearl millet, roasted mango seeds and has an odour threshold of 0.01 μg/L (water) (Demyttenaere et al. 2002, Buttery et al. 1997). Its thio-analogue, 2-acetylthiazoline has an odour threshold of 2 μg/L, whereas the corresponding aromatic analogues have much higher odour threshold values of 170 ppm for 2-acetylpyrrole and 10 ppb for 2-acetylthiazole.

Sulfur containing compounds of high importance in formation of food aromas are aliphatic thiols, sulfides and di, tri-sulfides, but a majority of odoriferous sulfur containing compounds are heterocycles, containing sulfur and sometimes another heteroatom. They include thiophenes, tiophenones, thiazoles, thiazolines, and poly-sulfur heterocycles. Sulfur containing heterocycles along with nitrogen compounds are the main odorants in roasted peanuts, fried meats, boiled/cooked meats, cocoa, coffee, black tea, bread crumb. Though their amount in food is usually low, they play a significant role in the aroma perception of thermally treated foods (Blank 2002).

Thermal degradation of thiamine is a source of numerous aroma compounds. As thiamine is present in various foods, such as meat, bread, rice, peanuts, yeast extract in a broad range of concentrations from 0.05mg/100g in beef to 3.10 mg/100 g in yeast extract (Mottram and Mottram 2002) it can be a source of numerous sulfur-containing aroma compounds when subjected to heating, such as furanthiols and thiophenes with savoury aromas. Among these compounds, bis(2-methyl-3-furyl)-disulfide is known to be an important contributor to cooked meaty aromas having extremely low odour threshold of 0.02 ng/kg. 2-Methyl-3-(methylthio)-furan identified in beef was reported to have an OT of 0.05 μg/kg, with a detectable meaty aroma at concentrations below 1 μg/kg. 2-Furfurylthiol has an odour threshold of 0.05-0.12 μg/L and its odour description changes from roasted coffee at low concentrations to sulfurous at higher concentrations. Thiazoles, alkyl substituted have odour thresholds ranging from 1 to 1000 μg/L and have various odour notes like cereal, bready, roasty (2-acetylthiazole), hazelnut, earthy (4-acetylthiazole), green, tomato leaf (isobutylthiazole). They are formed in a reaction of α-dicarbonyls with ammonia and hydrogen sulfide. Thiazoles are detected in various meats. Alkyl thiazolines and alkylthiazoles were detected in cooked beef and lamb as Maillard reaction products with a potential contribution to fatty, fried aromas of meat, but with a low influence on flavour (Elmore and Mottram 1997; Mottram 1998).

Table 6.1 shows selected heterocyclic aroma compounds in foods with mainly meaty, roasty odour notes and also sweet, cotton candy, caramel odours. Their

Table 6.1 Selected heterocyclic compounds of high sensory importance in foods.

Formula	Odorant	Odour sensation	OT water [μg/kg]	OT air [μg/m³]
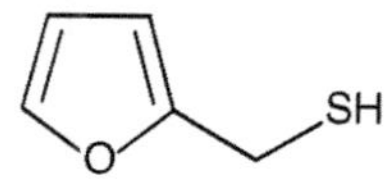	2-furfurylthiol	Sulfury, coffee	0.001 – 0.13	0.004 – 0.02
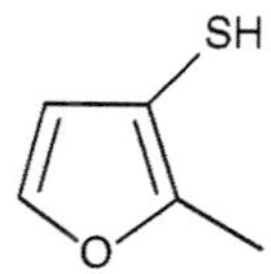	2-methyl-3-furanthiol	Sulfury, meaty	0.0004 – 0.01	0.001 – 0.01
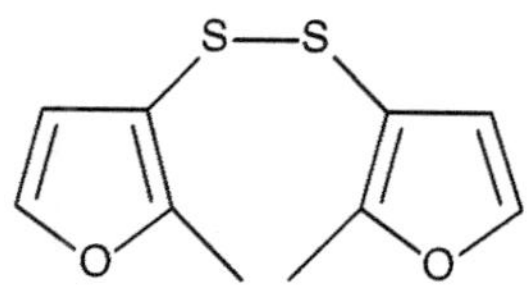	bis-(2-methyl-3-furyl)-disulfide	Sulfury, meaty	0.00002 – 0.00032	0.0006 – 0.057
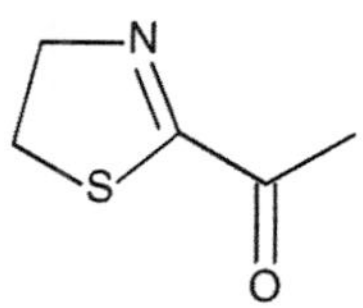	2-acetyl-2-thiazoline	Roasty, popcorn	0.12 – 1.3	0.016 – 0.08
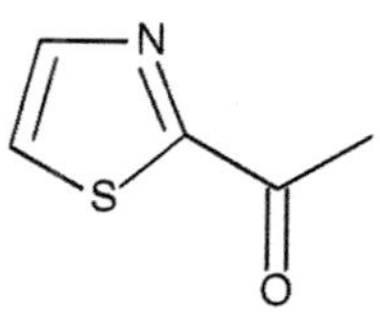	2-acetylthiazole	Sulfury, roasty	3 – 10	4 – 4.1

(continued overleaf)

Table 6.1 (*continued*)

Formula	Odorant	Odour sensation	OT water [μg/kg]	OT air [μg/m³]
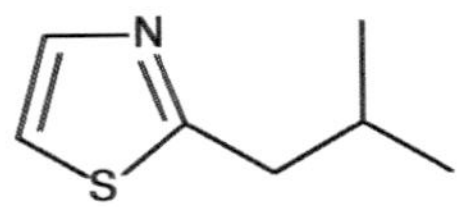	2-isobutylthiazole	Green tomato	3 - 29	nd
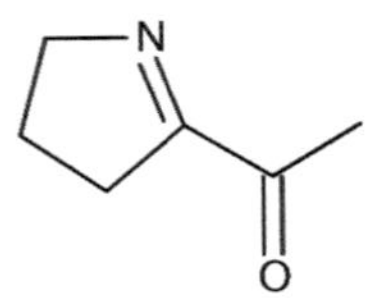	2-acetyl-1-pyrroline	Popcorn, toasty	0.053 – 0.12	0.02 – 0.04
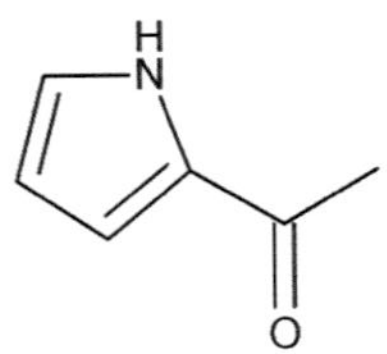	2-acetylpyrrole	Musty, nutty	58600 - 170000	>2000
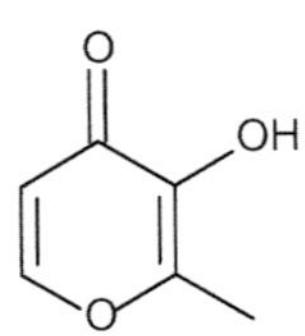	3-hydroxy-2-methyl-4H-pyran-4-one (maltol)	Cotton candy, caramel, sweet	210 – 35000	nd

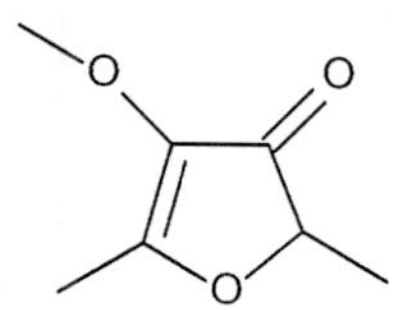	2,5-dimethyl-4-methoxy-3(2)H-furanone (mesifurane)	Musty, sweet, maple, fruity	0.03 - 3400	nd
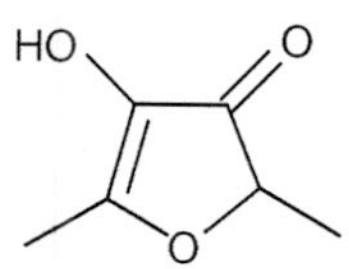	2,5-dimethyl-4-hydroxy-3(2H)-furanone (Furaneol®)	Cotton candy, caramel, strawberry	0.04 - 60	0.5 - 2
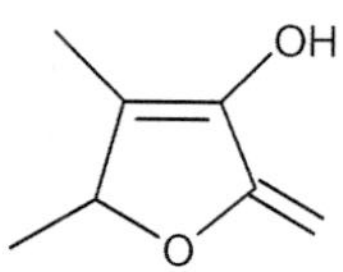	3-hydroxy-4,5-dimethyl-2(3H)-furanone (sotolone)	Sweet, strong burnt sugar, seasoning	0.3 - 5804	0.01 – 1.2

Source: Odour threshold information adapted from van Gemert (2011)

common characteristic is a very low odour threshold but relatively high ranges in odour threshold values reported in the literature. Popcorn, roasted notes are usually associated with nitrogen heterocycles mainly pyrrolines, pyrroles but also thiazolines and thiazoles (Buttery 1997a, Buttery et al. 1999). Furanones listed in Table 6.1 often contribute to sweet, caramel, cotton candy odour notes to food. Furaneol has been found as an odorant in corn products, and various foods such as strawberry, raspberry, mango, tomato, bread crust, roasted coffee, beef and wines (Buttery and Ling 1997b, Buttery 1999). Sotolon with odour notes ranging from sweet, burnt sugar to seasoning has been confirmed a key odorant in many foods such as roasted coffee, roasted beef or chicken.

6.3.6 Terpenes and norisoprenoids

Terpenes are secondary metabolites originating form mevalonate pathway and formed from five carbon (isoprene) units joined "head–to tail" to form linear chains or rings. The term "terpenes" refers sometimes to hydrocarbons only, whereas terpenoids include also their oxygenated forms. Though a diverse array of compounds is synthesized in this way, from the point of view of flavour chemistry, mainly monoterpenes and to a lesser extent sesquiterpenes play a role as odorants. The main sources of monoterpenes are essential oils, especially citrus essential oils, but also other fruits. Monoterpenes formed from geranyl diphosphate have 10 carbon atoms, whereas sesquiterpenes formed from farnesyl diphosphate have 15 carbon atoms (Croteau et al. 2000), which influences their volatility and sensory properties. Due to their unsaturated character, terpenes are prone to reactions and transformations both in plants and in food products such as wines. In plants (peppermint) limonene is transformed into (−)-isopiperitenone, then (+)-pulegone and finally to (−)-menthol and related compounds: (+)-isomenthol, (+)-neoisomenthol and (+)-neomenthol. In spearmint (−)-limonene is converted to (−)-carvone in a more simplified pathway. It results in differences in flavour notes of peppermint and spearmint (Croteau et al. 2000). Wines, especially from Muscat grape varieties, are rich in monoterpene alcohols, mainly nerol, linalool, α-terpineol and geraniol. Terpene compounds are relatively labile and susceptible to changes caused by acidity of the medium, time and temperature of storage. The examples include transformation of nerol and linalool to α-terpineol, limonene and terpinolene, or linalool to linalyl oxides, which can be in their *cis* or *trans* pyranyl or furanyl forms (Riberau-Gayon et al. 1975). Transformation of terpenes influences aroma of wines due to modification of the key odorants as demonstrated by Castino and Di Stefano (1981). Odour thresholds of most of the monoterpene alcohols are relatively high in wine with values of 20 μg/L for citronellol, 130 μg/L for geraniol, 110 μg/L for hotrienol, 100 μg/L for linalool or 400 μg/L for nerol (Marais 1983). As monoterpene hydrocarbons are nonpolar with weak solubility in water, their odour thresholds in water are relatively low, OT data provided for limonene vary from 0.004 to 3.5 mg/L (van Gemert 2011). Sesquiterpene hydrocarbons usually have a lower impact on flavour of food products than monoterpenes. Odour threshold in water for key sesquiterpnes in hops, such as β-caryophyllene and humulene, vary from 1.5 to 13 mg/kg and from 0.12 to 0.39 mg/kg respectively (van Gemert 2011).

Carotenoid pigments are tetraterpenes frequently occurring in plants. Over 400 carotenoids were identified in higher plants, algae and microorganisms. Among the most prominent tetraterpenes present in nature are lycopene, carotenes (α, β and γ) and lutein. The cleavage of C9-C10, C8-C9, C7-C8 and C6-C7 bonds of the polyene chain results in formation of cyclic aroma compounds containing 13, 11, 10 and 9 carbons respectively during oxidation process. The carotenoid derived aroma compounds are called norisoprenoids (or norterpenoids). Examples of important norisoprenoids are shown in Figure 6.7. Damascenone is one of the best known norisoprenoids, which has been identified for the first time in Bulgarian rose essential oil (*Rosa damascena*) and reported in alcoholic beverages, mainly wines, brandies and whiskeys and beers, but also in fruits (grapes, blackberry, black currant, nectarine) and in tomatoes, both fresh and processed, coffee, malt, molasses and tea (Sefton et al. 2011). Odour detection threshold reported for damascenone range from 0.75 ng/L to 23.7 ng/L in water and higher values in alcoholic media (up to 10000 ng/L) (Sefton et al.

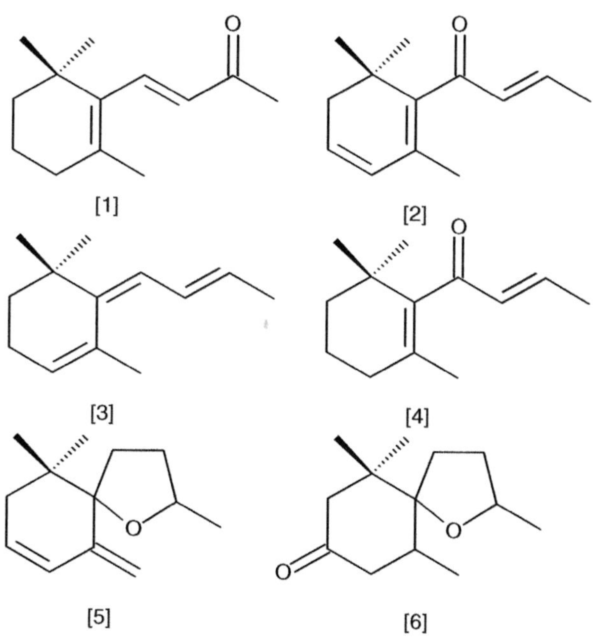

Figure 6.7 Carotenoid degradation products found as food odorants. [1] – β-ionone, [2] – β-damascenone, [3] – (E,E)-megastigma-4,6,8-triene, [4] – β-damascone, [5] – vitispirane, [6] – theaspirone.

2011). Apart from damascenone and its derivatives, ionones, ionols and their related compounds were detected, often in their bound forms as glycosides, in nectarines (Aubert et al. 2003). Abundant norisoprenoids are present in orange juice, where they contribute to its floral aroma (Mahattanatawee et al. 2005), in quince, passionfruit, honey (Winterhalter and Ruseff 2002) and in wines, in which other odorant compounds such as vitispirane and TDN (1,1,6-trimethyl-1,2-dihydronaphtalene) are often detected (Vinholes et al. 2009).

6.4 Conclusion

Among the great number of volatile compounds a relatively small percentage significantly contributes to the aroma of particular foods. Odoriferous compounds influencing aroma of particular foods—key odorants—belong to different chemical classes and present different polarity, volatility and stability. Their aroma and odour thresholds depend strongly on their stereochemistry. An additional issue is the coexistence of aroma compounds in substantially different concentrations in one product and possible interactions of odorants with the food matrix. These factors make the characterisation of key odorants a demanding analytical task, in which the main challenges are related to reliable quantitation of odorants, identification of closely related compounds and reliable quantitation of sensory sensations and determination of odour thresholds.

References

Aubert, C., Ambid, C., Baumes, R., Günata, Z. (2003) Investigation of bound aroma constituents of yellow-fleshed nectarines (Prunus persica L. Cv. Springbright). Changes in bound aroma profile during maturation. *Journal of Agricultural and Food Chemistry*. *51*(21), pp. 6280–6286.

Baltes W., Bochmann G. (1987a) Model ractions on roast aroma formation. II. Mass spectrometric identification of furans and furanones from the reaction of serine and threonine with sucrose under the conditoins of coffee roasting. *Zeitschrift für Lebensmittel-Untersuchung und -Forschung*. *184*(3), pp. 179–186.

Baltes, W., Bochmann, G. (1987b) Model reactions on roast aroma formation. IV Mass spectrometric identification of pyrazines from the reaction of serine and threonine with sucrose under the condition of coffee roasting. *Zeitschrift für Lebensmittel-Untersuchung und –Forschung*. *184*(6), pp. 485–493.

Belitz, H.-D., Grosch, W., Schieberle, P. (2009) *Food Chemistry*, Berlin (Germany): Springer-Verlag Berlin Heidelberg.

Berger, R.G. (ed.) (2007) *Flavours and Fragrances: Chemistry, Bioprocessing and Sustainability*, Berlin (Germany): Springer Science & Business Media, Place.

Blank, I. (2002) Sensory relevance of volatile organic sulfur compounds. In: Reineccius, G. A., Reineccius, T. A. *Heteroatomic aroma compounds*. ACS Symposium Series 826. American Chemical Society, Washington, DC, (USA), pp. 25–54.

Boelens, M.H., Van Gemert, L.J. (1993) Volatile character-impact sulfur compounds and their sensory properties. *Perfumer & Flavorist* *18*(3), pp. 29–39.

Brenna, E., Fuganti, C., Serra, S. (2003) Enantioselective perception of chiral odorants. *Tetrahedron: Assymetry 14*(1), pp. 1–42.

Brown, R. C., Sefton, M. A., Taylor, D. K., Elsey, G. M. (2006) An odour detection threshold determination of all four possible stereoisomers of oak lactone in white and red wine. *Australian Journal of Grape and Wine Research 12*(2), pp. 115–118.

Buettner, A., Schieberle, P. (2001) Evaluation of key aroma compounds in hand squeezed grapefruit juice (Citrus paradisi Macfayden) by quantitation and flavor reconstitution experiments. *Journal of Agricultural and Food Chemistry 49*(3), pp. 1358–1363.

Bult, J. H. F., Schifferstein, H. N. J., Roozen, J. P., Voragen, A. G. J., Kroeze, J. H. A. (2001) The influence of olfactory concept on the probability of detectiong sub- and peri-treshold components in a mixtrue of odorants. *Chemical Senses 26*(5), pp. 459–469.

Burdock, G.A., Fenaroli, G. (2010) *Fenaroli's handbook of flavor ingredients*, Boca Raton, FL (USA): CRC Press.

Buttery, R., Ling, L. C. (1997a) 2-ethyl-3,5-dimethylpyrazine and 2-ethyl-3,6-dimethylpyrazine: odor thresholds in water solution. *Lebensmittel-Wissenschaft & Technologie 30*(1), pp. 109–110.

Buttery. R. G., Ling, L. C., Stern, D. J. (1997) Studies on popcorn aroma and flavor volatiles. *Journal of Agricultural and Food Chemistry 45*(3), pp. 837–843.

Buttery, R. G., Ling, L.C. (1997b) 2,5-dimethyl-4-hydroxy-3(2H)-furanone in corn products *Journal of Agricultural and Food Chemistry 45*(4), pp. 1306–1308.

Buttery, R. G., Orts, W. J., Takeoka, G. R., Nam, Y. (1999) Volatile flavor components of rice cakes. *Journal of Agricultural and Food Chemistry 47*(10), pp. 4353–4356.

Buttery, R. G. (1999) Flavor Chemistry and odor thresholds. in: Teranishi, R., Wicks, E.L., Horstein, I. *Flavor Chemistry. Thirty Years of Progress*, New York (USA): Kluver Academic/Plenum Publkishers, pp. 353–367.

Castino M., Di Stefano R. (1981) Effetti della temperatura di conservazione sulle caratteristiche dell'Asti Spumante. *Rivista di Viticoltura e di Enologia 34*, pp. 106–119.

Cerny C. (2008) The aroma side of the Maillard reaction. *Annals of the New York Academy of Sciences 1126*(1), pp. 66–71.

Cerny, C., Davidek, T. (2003) Formation of aroma compounds from ribose and cysteine during Maillard reaction. *Journal of Agricultural and Food Chemistry 51*(9), pp. 2714–2721.

Cliff, M., Stanich, K., Trujillo, J. M., Toivonen, P, Forney, C. F. (2011) Determination and prediction of odor thresholds for odor active volatiles in a neutral apple juice matrix. *Journal of Food Quality 34*(3), pp. 177–186.

Costello, P. J., Lee, T.H., Henschke, P. A. (2001) Ability of lactic acid bacteria to produce N-heterocycles causing mousy off-flavour in wine. *Australian Journal of Grape and Wine Research 7*(3), pp. 160–167.

Costello, P. J., Henschke, P.A. (2002) Mousy off-flavor in wine: precursors and biosynthesis of the causative N-heterocycles 2-ethyltetrahydropyridine, 2-acetyltetrahydropyridine and 2-acetyl-1-pyrroline by lactobacillus hilgardii DSM 20176. *Journal of Agricultural and Food Chemistry 50*(24), pp. 7079–7089.

Cooke (neé Brown), R. C., van Leeuven, K. A., Capone, D. L., Gawel, R., Elsey, G.M., Sefton, M. (2009) Odor detection thresholds and enantiomeric distribution of several 4-alkyl substituted γ-lactones in Australian red wine. *Journal of Agricultural and Food Chemistry 57*(6), pp. 2462–2467.

Croteau, R., Kutchan, T. M., Lewis, N. G. (2000) Natural products (secondary metabolites). in: Buchanan, B., Gruissem, W., Jones, R. (Eds.) *Biochemistry & Molecular Biology of Plants*, Rockville (USA): American Society of Plant Pathologists. pp. 1250–1318.

Czerny, M., Wagner, R., Grosch, W. (1996) Detection of odor-active ethenylalkylpyrazines in roasted coffee. *Journal of Agricultural and Food Chemistry 44*(10), pp. 3268–3272.

Czerny, M., Christlbauer, M., Christlbauer, M., Fischer, A., Granvogl, M., Hammer, C., Hartl, C., Hernandez, N. M, Schieberle, P. (2008) Re-investigation on odour thresholds of key aroma compounds and development of an aroma language based on odour qualities of defined aqueous odorant solutions. *European Food Research and Technology 228*(2), pp. 265–273.

Czerny, M., Brueckner, R., Kirchoff, E., Schmitt, R., Buettner, A. (2011) The influence of molecular structure on odor qualities and odor detection thresholds of volatile alkylated phenols. *Chemical Senses 36*(6), pp. 539–553.

Demyttenaere, J., Tehrani, K. A., De Kimpe, N. (2002) The chemistry of the most important Maillard flavor compounds of bread and cooked rice. In: Reineccius, G. A., Reineccius, T. A. *Heteroatomic aroma compounds*. Washington, DC, (USA): ACS Symposium Series 826. American Chemical Society, pp. 150–166.

Dunkel, A., Steinhaus, M., Kotthoff, M., Nowak, B., Krautwurst, D., Schieberle, P., Hofmann, T. (2014) Nature's chemical signatures in human olfaction: a foodborne perspective for future biotechnology. *Angewandte Chemie International Edition 53*(28), pp. 7124–7143.

Elmore, J. S., Mottram, D. S. (1997) Novel thiazoles and 3-thiazolines in cooked beef aroma. *Journal of Agricultural and Food Chemistry 45*(9), pp. 3603–3607.

Engel, K.-H. (1999) The importance of sulfur-containing compounds to fruit flavors. in Teranishi R., Wick E.L., Hornstein I. (eds.) *Flavor Chemistry*, Boston, MA (USA): Springer US, pp. 265–273.

Etschmann, M.M., Bluemke, W., Sell, D., Schrader, J. (2002) Biotechnological production of 2-phenyloethanol. *Applied Microbiology and Biotechnology 59*(1), pp. 1–8.

Fallico, B., Lanza, M. C., Maccarone, E., Asmundo, C. N., Rapisarda, P. (1996) Role of hydroxycinnamic acids and vinylphenols in the flavor alteration of blood orange juices. *Journal of Agricultural and Food Chemistry 44*(9), pp. 2654–2657.

Fischer, M., Boker, A., Berger, R. G. (2001) Fungal formation of raspberry ketone differs from the pathway in plant cell culture. *Food Biotechnology 15*(3), pp. 147–155.

Flament, I., Bessière-Thomas, Y. (2002) *Coffee Flavor Chemistry*, West Sussex, England: John Wiley & Sons.

Giri, A., Osato, K., Ohshima, T. (2010) Identification and chracterization of headspace volatiles of fish miso, a japanese fish meat based fermented paste, with special emphasis on effect of fish species and meat washing. *Food Chemistry 120*(2), pp. 621–631.

Goodrich, R. M., Braddock, R. J., Parish, M. E., Sims, C. A. (1998) Bioconversion of Citrus Aroma Compounds by Pichia pastoris. *Journal of Food Science 63*(3), pp. 445–449.

Gram, L., Dalgaard, P., (2002) Fish spoilage bacteria – problems and solutions. *Current Opinion in Biotechnology 13*(3), pp. 262–266.

Guth, H. (1996) Determination of the configuration of wine lactone. *Helvetica Chimica Acta 79*(6), pp. 1559–1571.

Hatanaka, A. (1996) The fresh green odors emitted by plants. *Food Reviews International 12*(3), pp. 303–351.

Hassan, F.A., Abd El-Gawad, M.A., Enab, A.K. (2012) Flavour compounds in cheese. *International Journal of Academic Research 2*(2), pp. 15–29.

Heresztyn, T. (1986) Metabolism of volatile phenolic compounds from hydroxycinnamic acids by Brettanomyces yeasts. *Archives of Microbiology 146*(1), pp. 96–98.

Jeleń, H. H., Krawczyk, J., Larsen, T.O., Jarosz, A., Gołębniak, B. (2005) Main compounds responsible for off-odour of strawberries infected by Phythophthora cactorum. *Letters in Applied Microbiology 40*(4), pp. 255–259.

Jeleń, H., Wąsowicz, E. (2012) Lipid derived flavor compounds. in: Jeleń H. (ed.) *Food Flavors. Chemical, Sensory and Technological Properties*, Boca Raton, FL (USA): CRC Press, Taylor & Francis Group, pp. 65–93.

Jyoti, B.D., Suresh, A.K., Venkatesh, K.V., 2003. Diacetyl production and growth of lactobacillus rhamnosus on multiple substrates. *World Journal of Microbiology and Biotechnology 19*, pp. 509–514.

Landaud, S., Helinck, S., Bonnarme, P. (2008) Formation of volatile sulfur compounds and metabolism of methionine and other sulfur compounds in fermented food. *Applied Microbiology and Biotechnology* 77(6), pp. 1191–1205.

Lee, S. M., Jo, Y.-J., Kim, Y.-S. (2010) Investigation of the aroma active compounds formed in the Maillard reaction between glutathione and reducing sugars. *Journal of Agricultural and Food Chemistry 58*(5), pp. 3116–3124.

Li, C., Sun, J., Li, T., Liu, S.-Q., Huang, D. (2014) Chemical and enzymatic synthesis of a library of 2-phenethyl esters and their sensory attributes. *Food Chemistry 154*, pp. 205–210.

Lorber, K., Schieberle, P., Buettner, A. (2014) Influence of the chemical structure on odor qualites and odor thresholds in homologous series of alka-1,5-dien-3-ones, alken-3-ones, alka-1,5-dien-3-ols and alk-1-en-3ols. *Journal of Agricultural and Food Chemistry 62*(5), pp. 1025–1031.

Majcher, M.A., Myszka, K., Kubiak, J., Jeleń, H.H. (2014) Identification of key odorants of fried cottage cheese and contribution of *Galactomyces geotrichum* MK017 to the formation of 2-phenylethanol and related rose-like aroma compounds. *International Dairy Journal 39*(2), pp. 324–329.

Mahattanatawee, K., Rouseff, R., Valim, M. F., Naim, M. (2005) Identification and aroma impact of norisoprenoids in orange juice. *Journal of Agricultural and Food Chemistry 53*(2), pp. 393–397.

Marais J. (1983) Terpenes in the aroma on grape and wine: a review. *South African Journal of Enology and Viticulture 4*(2), pp. 49–58.

Mazzoleni, V., Maggi, L. (2007) Effect of wine style on the perception of 2,4,6-trichloroanisole, a compound related to cork taint in wine. *Food Research International 40*(6), pp. 694–699.

McGorrin, R.J. (2011) The significance of volatile sulfur compounds in food flavors: an overview. in: Qian, M.C., Fan, X., Mahattanatawee, K. (eds.) *Volatile sulfur compounds in food*, Washington, DC (USA): American Chemical Society, pp. 3–31.

Molyneux, R. J., Schieberle, P. (2007) Compound identification: A Journal of Agricultural and Food Chemistry perspective. *Journal of Agricultural and Food Chemistry 55*(12), pp. 4625–4629.

Mottram, D. S. (1998) Flavor formation in meat and meat products: a review. *Food Chemistry 62*(4), pp. 415–424.

Mottram, D. S., Mottram, H. R. (2002). An overview of the contribution of sulfur containing compounds to the aroma of heated foods. in: Reineccius, G. A., Reineccius, T. A. *Heteroatomic aroma compounds*, Washington, DC, (USA): ACS Symposium Series 826. American Chemical Society, pp. 73–93.

Mussinan, C.J., Keelan, M.E. (1994) Sulfur compounds in foods: an overview, in: Mussinan, C.J., Keelan, M.E. (eds.) *Sulfur compounds in foods*, Washington, DC (USA): American Chemical Society, pp. 1–6.

Myung, K., Hamilton-Kemp, T.R., Archbold, D.D. (2006) Biosynthesis of trans-2-hexenal in response to wounding in strawberry fruit. *Journal of Agricultural and Food Chemistry 54*(4), pp. 1442–1448.

Németh, Á.S., Márczy, J.S., Samu, Z., Háger-Veress, Á., Szajáni, B. (2004) Biocatalytic production of 2(E)-hexenal from hydrolysed linseed oil. *Enzyme Microbial Technology 34*(7), pp. 667–672.

Ozcan, G., Barringer, S. (2011) Effect of enzymes on strawberry volatiles during storage, at different ripeness level, in different cultivars, and during eating. *Journal of Food Science 76*(2), pp. 324–333.

Paravisini, L., Gourrat-Pernin, K., Gouttefangeas, C., Moretton, C., Nigay, H., Dacremont, C., Guichard, E. (2012) Identification of compounds responsible for the odorant properties of aromatic caramel. *Flavor and Fragrance Journal 27*(6), pp. 424–432.

Parker, M., Osidacz, P., Baldock, G. A., Hayasaka, Y., Black, C. A., Pardon, K. H., Jeffery, D.. W., Geue, J. P., Herderich, M. J., Francis, I. L. (2012) Contribution of several volatile phenols and their glycoconjugates to smoke related sensory properties of red wine. *Journal of Agricultural and Food Chemistry 60*(10), pp. 2629–2637.

Pino, J.A. (2014). Odour-active compounds in papaya fruit cv. Red Maradol. *Food Chemistry 146*, pp. 120–126.

Qian, M., Nelson, C., Bloomer, S. (2002) Evaluation of fat-derived aroma compounds in Blue cheese by dynamic headspace GC/olfactometry-MS. *Journal of the American Oil Chemists' Society 79*(7), pp. 663–667.

Reineccius, G. (2005). *Flavor Chemistry and Technology*, Boca Raton, FL (USA): CRC Press, Place.

Ribereau-Gayon P., Boidron J.N., Terrier A., (1975) Aroma of Muscat grape varietes. *Journal of Agricultural and Food Chemistry 23*(6), pp. 1042–1047.

Rowe, D.J. (2005) *Chemistry and Technology of Flavors and Fragrances*, Oxford (England): Blackwell Publishing Ltd.

Ryan, D., Prenzler, P. D., Saliba, A. J., Scollary, G. R. (2008) The significance of low impact odorants in global odour perception. *Trends in Food Science and Technology 19*(7), pp. 383–389.

Sala, C., Mestres, M., Marti, M. P., Busto, O., Guasch, J. (2002) Headspace solid phase microextraction analysis of 3-alkyl-2-methoxypyrazines in wines. *Journal of Chromatography A 953*(1), pp. 1–6.

Sefton, M. A., Skouroumounis, G. K., Elsey, G. M., Taylor, D. K. (2011) Occurrence, sensory impact, formation and fate of damascenone in grapes, wines, and other foods and beverages. *Journal of Agricultural and Food Chemistry 59*(18), pp. 9717–9746.

Serra, S., Fuganti, C., Brenna, E. (2005) Biocatalytic preparation of natural flavours and fragrances. *Trends in Biotechnology 23*, pp. 193–198.

Stone, E.J., Kazeniac, S.J. (1975) Formation of aldehydes and alcohols in tomato fruit from U-14 C labeled linolenic and linoleic acids. *Journal of Food Science 40*(6), pp. 1138–1141.

Surburg, H., Panten, J., Bauer, K. (2006) *Common fragrance and flavor materials: preparation, properties and uses*, Weinheim (Germany): John Wiley.

Tachihara, T., Hashimoto T., T., Ishizaki, S., Komai, T., Fujita, A., Ishikawa, M., Kitahara, T. (2006) Microbial resolution of 2-methylbutyric acid and its application to several chiral flavour compounds in: Bredie WLP, Petersen MA (eds) Flavour science: recent advances and trends. *Developments in Food Science 43*, Amsterdam (Netherlands): Elsevier, pp. 97–104.

Takeoka, G., Buttery, R. G., Ling, L. (1996) Odour thresholds of various branched and straight chain acetates. *Lebensmittel-Wissenschaft & Technologie 29*(7), pp. 677–680.

Takeoka, G. R., Buttery, R. G., Ling, L. C., Wong, R. Y., Dao, L. T., Edwards, R. H., Berrios, J. de J. (1998) Odor thresholds of various unsaturated branched esters. *Lebensmittel-Wissenschaft & Technologie 31*(5), pp. 443–448.

Van Gemert, L.J. (2011) *Odour thresholds. Compilation of odour threshold values in air, water and other media*, Netherlands: Oliemans Punter & Partners BV, pp. 1–486.

Vandamme, E.J., Soetaert, W. (2002) Bioflavours and fragrances via fermentation and biocatalysis. *Journal of Chemical Technology and Biotechnology 77*(12), pp. 1323–1332.

Varlet, V., Knockaert, C., Prost, C., Serot, T. (2006) Comparison of odor-active volatile compounds of fresh and smoked salmon. *Journal of Agricultural and Food Chemistry 54*(9), pp. 3391–3401.

Vinholes, J., Coimbra M. A., Rocha, S. M. (2009) Rapid tool for assessment of C13 norisoprenoids in wines. *Journal of Chromatography A 1216*(47), pp. 8398–8403.

Vitzthum, O. G., Werkhoff, P. (1974) Neu entdeckte stickstoffheterocyclen im kaffee-aroma. *Zeitschrift für Lebensmitteluntersuchung und Forschung 156*(5), pp. 300–307.

Winterhalter. P., Rouseff, R. L. (2002) Carotenoid derived aroma compounds: an introduction. in: Winterhalter. P., Rouseff, R. L. (eds.) *Carotenoid derived compounds*, ACS Symposium Series 802, Washington, DC (USA): American Chemical Society, pp. 1–17.

Wong, K. H., Aziz, S. A., Mohamed, S. (2008) Sensory aroma from Maillard reaction of individual and combinations of aminoacids with glucose in acidic conditions. *International Journal of Food Science and Technology 43*(9), pp. 1512–1519.

Yu, A.-N., Tan, Z. –W., Wang, F.-S. (2012) Mechanism of formation of sulfur aroma compounds from L-ascorbic acid and L-cysteine during the Maillard reaction. *Food Chemistry 132*(3), pp. 1316–1323.

CHAPTER 7

Characterization of taste compounds: chemical structures and sensory properties

Anni Laffitte, Fabrice Neiers & Loïc Briand
Center for Taste and Feeding Behaviour, CNRS, INRA, University of Bourgogne Franche-Comté Dijon, France

7.1 Introduction

Taste is one of the most important senses involved in food choice and in the regulation of food intake in humans. The sense of taste informs us on the presence of fundamental nutrients or harmful compounds such as toxic molecules. Humans, like many living organisms, are able to perceive and discriminate five primary taste qualities including sweet, salty, sour, bitter, and umami (delicious in Japanese, the taste of some amino acids). Taste compounds are detected by taste buds located mainly on the surface of the tongue within taste papillae. Sapid molecules dissolved in saliva are perceived through the stimulation of specific chemo-detectors (reviewed in Chapter 3 of this book) by taste compounds present in foods. In addition to the five fundamental taste qualities, a number of other taste sensations are still matter of debate. Among non-canonical sensations, calcium taste (Tordoff 2001), kokumi (mouthfulness and thickness in Japanese) taste (Maruyama et al. 2012) and fat taste (Khan and Besnard 2009) have been proposed. Finally, trigeminal and astringent compounds also contribute to the non-olfactory flavour of food, but are not considered here as they are not considered as tastants.

In this chapter, we will review the characteristics of the main known tastants that are important contributors to the five basic tastes. Recently, new tastants or taste modulators have been generated using classical approaches (fractionation of food extract or structure-activity relationship) combined with novel screening technologies based on taste receptors. These compounds will also be described.

Flavour: From food to perception, First Edition.
Edited by Elisabeth Guichard, Christian Salles, Martine Morzel, and Anne-Marie Le Bon.

7.2 Sweet-tasting compounds

Sweet taste is involved in the estimation of the caloric content of food and especially in identifying carbohydrate-rich compounds. For ages, honey has constituted the main source of sweetness in most parts of the world. In the eighteenth century, the extraction of sucrose, or table sugar from sugar cane (*Saccharum officinale*), grew rapidly. During the Napoleonic wars, sugar beet (*Beta vulgaris*) production was developed to become the major source of sugar in Europe. Today, global sugar production has reached 174 million metric tons, and western populations face major health issues directly linked to the consumption of large quantities of sucrose such as tooth decay, type II diabetes and obesity. It has thus become evident that alternatives to high-calorie sweeteners are necessary, and the demand for low-calorie sweeteners keeps rising. Many artificial and natural sweeteners have been identified since the early 1800s. Some of these compounds have been approved for human consumption in the United States and/or Europe. However, the use of these substances has been subject to some controversy due to health risks associated with their consumption. At the beginning of the 2000s, a major breakthrough came from the identification of the receptor responsible for sweet taste recognition (Nelson et al. 2001). It has been shown that T1R2 (taste receptor type 1, member 2) assembles with T1R3 (taste receptor type 1, member 3) to form a functional sweet taste receptor able to detect all sweet-tasting compounds. Recent new screening technologies based on the high-throughput screening of taste receptors have allowed the discovery of new sweeteners and sweetness enhancers. These compounds will be described in section 6 of this chapter.

In this section, natural and artificial sweeteners belonging to many various chemical classes will be presented, and their threshold and organoleptic properties will be outlined (Table 7.1). These compounds include natural sugars (sucrose), sugar alcohols (xylitol), dipeptides (aspartame), sulfonyl amide (saccharin) and terpenoid glycosides (stevioside). Some amino acids, including glycine, L-alanine, L-threonine, and L-phenylalanine, generate a sweet taste in humans. However, because their detection threshold is in the millimolar range and because of their sweetness potencies, they have not been considered as potential sweeteners. In addition, a more surprising class of proteins originating from plants with sweet-tasting or taste-modifying properties will be reviewed.

7.2.1 Natural sweeteners

7.2.1.1 Natural carbohydrates: sucrose, glucose and fructose

Sucrose, known as table sugar, is the disaccharide generally associated with sweetness and is chemically known as α-D-glucopyranosyl-(1→2)-β-D-fructofuranose (Figure 7.1). This carbohydrate, composed of the monosaccharides glucose and fructose linked by an ether bond is the reference substance for

Table 7.1 Sweet-tasting compounds, relative sweetness compared to sucrose and aftertaste.

Sweetener	Relative sweetness	Aftertaste	Reference
Sucrose	1	–	(Schiffman and Gatlin 1993)
D-Fructose	0.75	–	(Schiffman and Gatlin 1993)
Stevioside	300	Bitter, liquorice	(Lindley 2012)
Rebaudioside A	450	Liquorice	(Lindley 2012)
Mogroside V	250-400	Liquorice, cooling effect	(Kinghora et al. 1986)
Glycyrrhizic acid	90-170	Strong liquorice	(Kim and Kinghorn 2002)
NHDC	250-1800	Liquorice, menthol	(Borrego et al. 1995)
Saccharin	150-300	Bitter, metallic	(Bassoli et al. 2002)
Cyclamate	30-50	Metallic, salt	(Schiffman and Gatlin 1993)
Acesulfame-K	119	Bitter	(Schiffman and Gatlin 1993)
Aspartame	160-220	Bitter	(Bassoli et al. 2002)
Neotame	11 000	Liquorice	(Bassoli et al. 2002)
Sucralose	755	Bitter	(Bassoli et al. 2002)
Thaumatin	100 000	Liquorice	(De Vos et al. 1985)
Monellin	3000	Lingering	(Kohmura et al. 1990)
Brazzein	2000	Liquorice	(Poirier et al. 2012)
Mabinlin	100	Bitter, astringent	(Li et al. 2008)
Miraculin	400 000	n.d.	(Gibbs et al. 1996)
Neoculin	20 000	n.d.	(Gibbs et al. 1996)

n.d. = not described

psychophysical evaluations to which all other sweet-tasting substances are compared. Sucrose taste is described as pure and clean (Schiffman and Gatlin 1993). Other common disaccharides used in food include lactose (derived from galactose and glucose), maltose (formed by two units of glucose) and trehalose (formed from two glucose units joined by a 1-1 alpha bond). Lactose is the most abundant sugar found in milk (2-8% depending on the species) and milk-derived products (Scrimshaw and Murray 1988). The sweetening power of lactose is approximately four times lower than that of sucrose. Maltose is the disaccharide formed in breaking down starch by amylase, and it is found in germinating seeds such as barley and in beer for example. Maltose is approximately 30% as sweet as sucrose. Finally, trehalose is produced by a large panel of organisms, from invertebrates to plants and fungi. Trehalose is implicated in anhydrobiosis and helps cells withstand long periods of desiccation. Trehalose has approximately 45% the sweetness of sucrose. Among monosaccharides, glucose (Figure 7.1) elicits a sweet taste and is approximately 75% as sweet as sucrose. Fructose (Figure 7.1) occurs naturally in fruits, some root vegetables and honey and is the sweetest of the natural sugars (Schiffman and Gatlin 1993). Carbohydrates are not only used as sweet agents; they are also sometimes added to food products because of their texturing capacities. Corn syrup for example, that contains maltose and higher oligosaccharides, is widely used in many products because in addition to being sweet, it can also be a thickener and a humectant. Corn

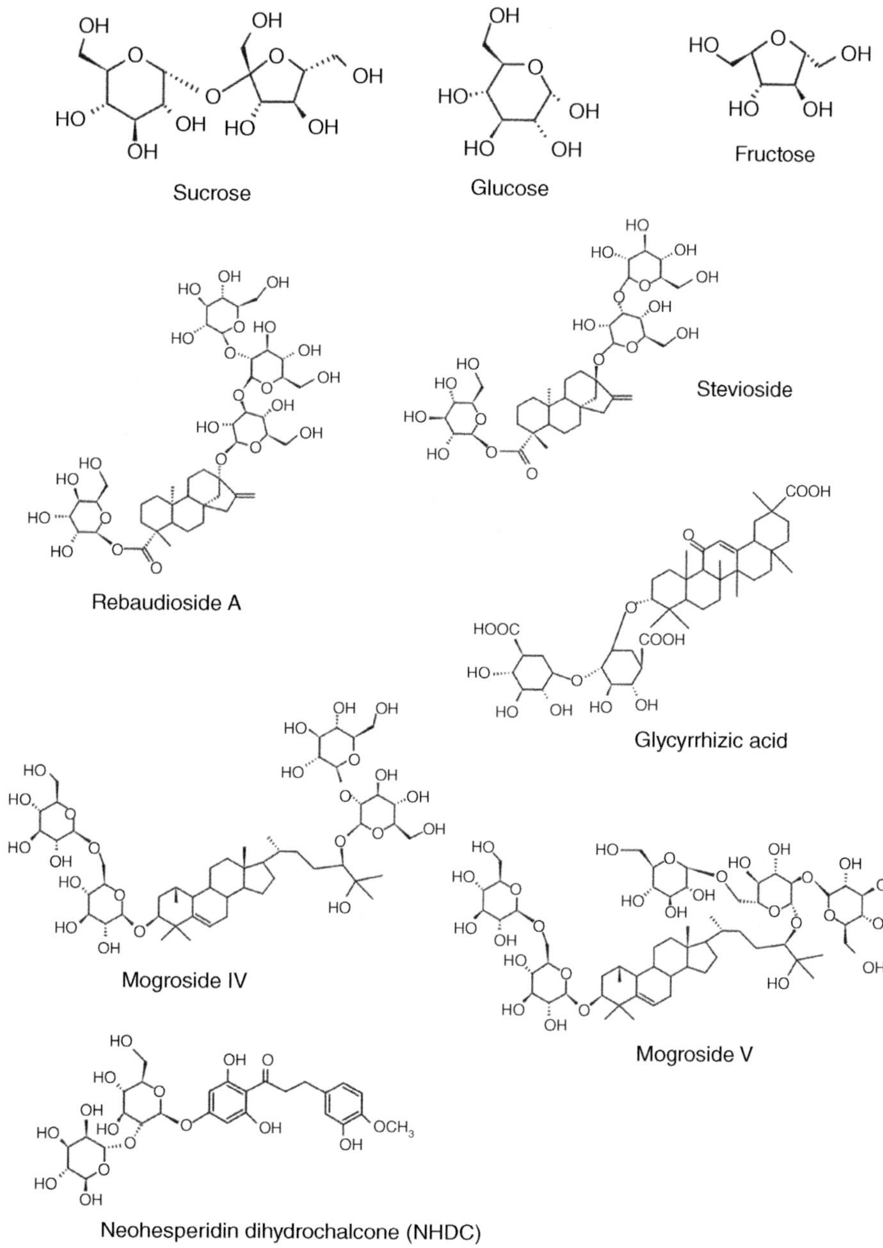

Figure 7.1 Chemical structure of natural sweet-tasting compounds.

syrup is produced from corn starch using an enzyme called glucoamylase, which is able to hydrolyse D-glucose (dextrose) from starch. Often mixed with regular corn syrup is high fructose corn syrup; a much sweeter and cheaper alternative to sugar, this type of syrup has become the predominant sweetener used in the US in processed foods. High fructose corn syrup is obtained from corn syrup using the enzyme glucose isomerase, which is able to convert D-glucose to the sweeter D-fructose (Inglett 1984). Due to health issues that have risen from the modern high consumption of sugar and natural carbohydrates, there is a constant increase in consumer demand for natural, low-calorie sweeteners. Plant species containing natural sweeteners are currently the subject of intensive research to identify compounds with high-intensity sweetening properties.

7.2.1.2 Sweet terpenoids: steviosides, glycyrrhizin and mogrosides

Stevia rebaudiana is a plant from Paraguay and Brazil that has been known to have sweet-tasting leaves for centuries by indigenous populations. Extracted from the leaves of this plant, a mixture of terpenoids called steviosides has recently attracted attention for their organoleptic properties. From the 9 diterpenes present in the plant, two molecules, rebaudioside A and stevioside (Figure 7.1), are able to elicit a sweet taste, with a potency that makes them attractive as sweeteners (300 times and 450 times sweeter than sucrose, respectively) (Lindley 2012). Rebaudioside A generates a cleaner sweet taste similar to that of sucrose compared to stevioside, which delivers a strong bitter aftertaste in addition to its sweet taste. Rebaudioside A and stevioside are used as food additives or dietary supplements in several countries, and their use is particularly widespread in Japan and Korea (Lindley 2012). Rebaudioside A has been approved by the FDA and has a generally recognised as safe (GRAS) status (notice n°467, E960); furthermore, the European Union has approved its use as a food additive and has established the recommended daily intake for the sweetener (4 mg/kg bw/day). However, the safety of steviosides is still a matter of debate.

Glycyrrhizin or glycyrrhizic acid (E958) is another sweet terpenoid (Figure 7.1) and constitutes the well-known sweet component of liquorice root (*Glycyrrhiza glabra*). Glycyrrhizic acid is approximately 90-170 times sweeter than sucrose (Kim and Kinghorn 2002). Its temporal profile in the mouth is slow compared to that of sucrose. Its well-known liquorice aftertaste, its poor organoleptic and pharmacological properties limit the use of glycyrrhizic acid as a sweetener (Schiffman and Gatlin 1993). With respect to regulatory status, glycyrrhizic acid has been accepted as a flavouring agent only in the US, and its use as a sweetener is not approved in Europe.

Mogrosides are a family of terpenoids extracted from a Chinese plant locally known as Lo Han Guo or Monkfruit (*Siraitia grosvenorii*). Two mogrosides (mogrosides IV and V) have been isolated from fruits and are well described (Figure 7.1). They are described as being 250-400 times sweeter than sucrose depending on the concentration (Kinghora et al. 1986). The sweetness of mogrosides has a slow onset profile, typical of natural high-potency sweeteners (Lindley 2012). Although these natural sweeteners are promising for food applications, their liquorice aftertaste coupled with a cooling taste effect may limit their use. Extracts of Lo Han Guo have been granted GRAS status and are commercially available in some countries such as Japan (Lindley 2012).

7.2.1.3 Neohesperidin dihydrochalcone

Neohesperidin dihydrochalcone or NHDC (E959) is a semi-synthetic sweetener obtained from the alkali hydrogenation of nehesperidin, a flavonone isolated from the peel of Seville oranges (Figure 7.1). NHDC is 250-1800 times sweeter than sucrose and generates a slower onset for sweet taste than sucrose with a lingering menthol-like aftertaste (Borrego et al. 1995). NHDC has the ability to synergise with other bulk sweeteners such as saccharin, sucralose and aspartame because its binding site is located in the transmembrane domain of the T1R3 subunit (Winnig et al. 2007) of the sweet taste receptor (see Chapter 3 for details on the receptor). NHDC is also used as a flavour enhancer because it has the ability to mask the bitter aftertaste of other sweeteners. Its use has been approved in the EU for several foodstuffs. However, NHDC has not obtained GRAS status in the US and is not currently accepted as a sweetener there.

7.2.2 Artificial sweeteners

Since the late nineteenth century, it has been known that artificial sweet-tasting molecules other than sugars and naturally sweet molecules can be generated through chemical synthesis. The use of these artificial sweeteners was the first answer to health problems associated with sugar consumption, such as diabetes. Even if they have been consumed for a long time, some concerns about the safety of these artificial sweeteners, such as aspartame, acesulfame-K and saccharin, have appeared. In this section, some of the most common artificial sweeteners are described. Their sensory properties are reviewed, and their chemical structures are presented in Figure 7.2.

7.2.2.1 Saccharin

Also known as *o*-benzoic sulfimide, saccharin (E954) was the first artificial sweetener, discovered in 1879 by Fahlberg (Cohen 1986). Saccharin is 300 times sweeter than sucrose and generates a slightly bitter and metallic aftertaste. Saccharin is commercially available today as sodium and calcium salts. The origin of its bitter off-taste has been explained by its capacity to activate the human bitter taste receptor TAS2R44 (Meyerhof et al. 2010) (see chapter 3). Saccharin is readily soluble in water, and its temporal profile is similar to that of sucrose. Saccharin has been observed to synergise with other sweeteners such as aspartame and NHDC, as previously mentioned. At low concentrations,

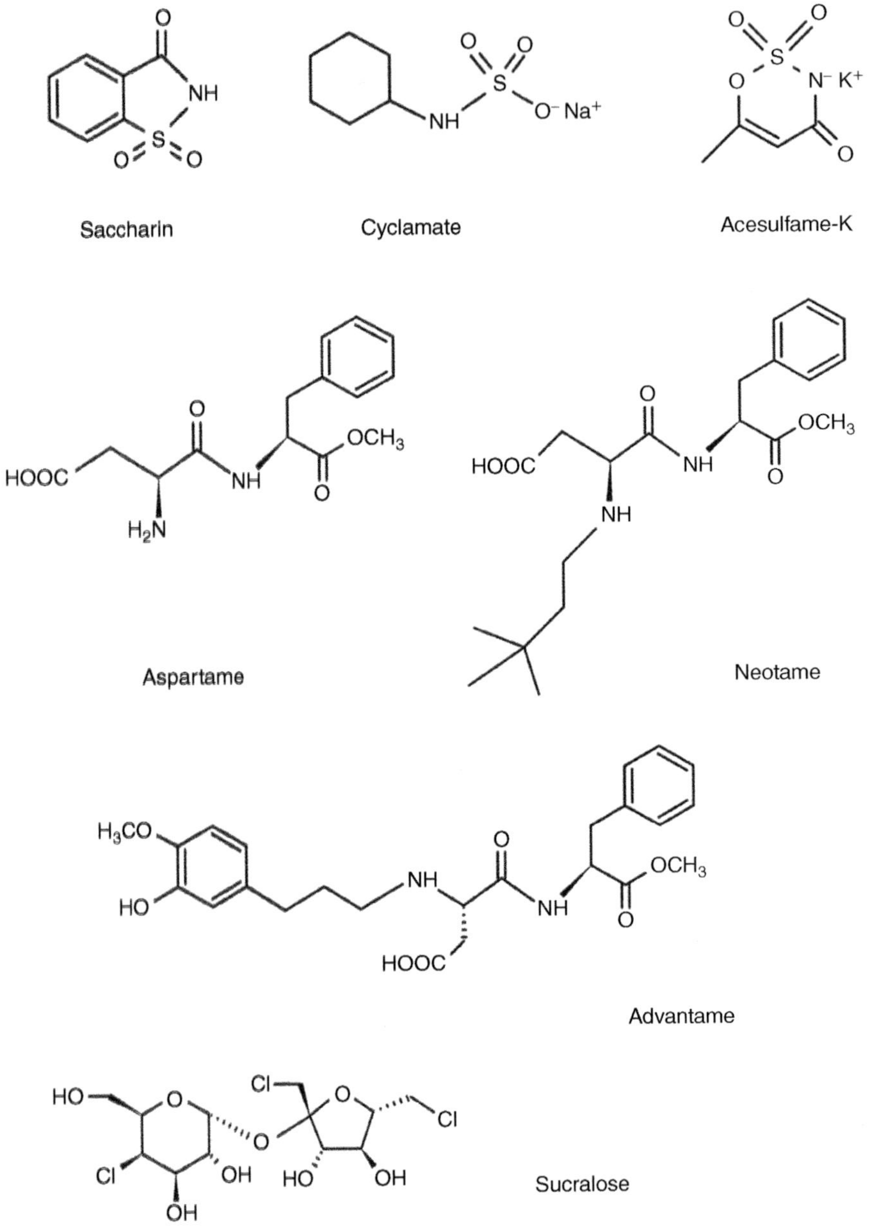

Figure 7.2 Chemical structure of artificial sweet-tasting compounds.

saccharin can be used as a sweetener, but when used at high concentration saccharin elicits a strong bitter aftertaste. A remarkable gustatory sensation elicited by water following the removal of high concentrations of saccharin from the mouth is called "sweet-water" taste. The mechanism of action of saccharin has been elucidated, and the presence of an allosteric inhibitory binding site for saccharin in the transmembrane domain of T1R3 subunit has been identified (Galindo-Cuspinera et al. 2006).

7.2.2.2 Cyclamate

The sweetener cyclamate (salts of cyclohexylsulfamic acid) was discovered by Sveda in 1937 (Audrieth and Sveda 1944). These salts are commonly referred to as cyclamic acid salts or simply cyclamates. Cyclamate belongs to a class of sweet molecules known as sulfamates and it is commonly used in the food industry as salts of sodium or calcium (E952). Its taste closely resembles that of sucrose; however, a weak metallic and salty aftertaste is perceivable (Schiffman and Gatlin 1993). Cyclamate is mostly used as a blend sweetener, and it has high sweetening power. Its aftertaste is thus less of a problem in food-beverage applications. Cyclamate has a temporal profile similar to that of sucrose, and it has been observed to synergise with most known sweeteners. Indeed, cyclamate, like NHDC, has been demonstrated to bind to the transmembrane domain of the T1R3 subunit of the sweet receptor (Jiang et al. 2005). Cyclamate is approved as a commonly used sweetener in most major countries, including the EU, but due to health concerns raised by a study performed on rodents in the 1960s, it has since been banned in the US and is currently pending FDA approval.

7.2.2.3 Acesulfame-K

Acesulfame-K (E950) is an intense sweetener that belongs to the class of dihydro-oxathiazinone dioxides. Acesulfame-K was discovered in 1967 by the German company Hoechst Co. by accident, and its development was undertaken when its sweetness was discovered (Clauss and Jensen 1973). The potassium salt of acesulfame was found to be sweeter than its sodium salt analogue, and it was considerably easier to produce and purify than the calcium salt analogue. Acesulfame-K is approximately 200 times sweeter than sucrose, and its temporal profile is similar to that of sucrose. At low concentrations, acesulfame-K does not elicit any unpleasant aftertaste, and it is often used as a blend sweetener to mask the unpleasant aftertaste of other sweeteners. Acesulfame-K readily synergises with cyclamate and aspartame (Schiffman and Gatlin 1993), and these blends are among the most commonly used by the food-beverage industry. The use of acesulfame-K as a sweetener is approved in all major countries, including the United States and the EU.

7.2.2.4 Aspartame, neotame and advantame

Serendipitously discovered in 1965 (Mazur et al. 1969), the sweet dipeptide L-aspartyl-L-phenylalanine methyl ester (Figure 7.2) called aspartame (E951) is 160-220 times sweeter than sucrose. Its taste closely resembles that of sucrose, with no distinct aftertaste. Aspartame is known to synergise with many bulk and intense sweeteners at different levels, depending on the contents and concentration of the blend. The most common mixtures used in food industry are with acesulfame-K and saccharin. Under these conditions, the sweetness intensity synergy can be as high as 30% (O'Donnell 2007). Aspartame has a small energetic content of 4 kcal/g, but because very little is used to sweeten foods and beverages, there is no real caloric gain from products sweetened with aspartame. Aspartame is commonly used in many low-calorie products such as sodas, sweets and dairy preparations (Schiffman and Gatlin 1993). Aspartame has received bad press about possible health risks associated with its metabolism after consumption, but it is generally accepted that its use at current levels has no health risks associated. However, because of the bad press it has received, aspartame has become difficult to use in many products. Other peptide derivatives of aspartame have thus been developed and studied, such as neotame or recently advantame (Figure 7.2). Neotame (E961) is a derivative of aspartame, approximately 11 000 and 50 times sweeter than sucrose and aspartame, respectively, thus making it one of the most potent sweeteners. The taste of neotame is quite clean and resembles that of sucrose; however, at high concentration, a liquorice-like aftertaste becomes noticeable. Neotame and aspartame have been shown to bind the N-terminal domain of T1R2 (Xu et al. 2004). The synergy of neotame with other sweeteners is limited, and the only noticeable synergy is with saccharin (Schiffman and Gatlin 1993). When blended with other sweeteners, the liquorice-like aftertaste of neotame becomes less dominant. Its use as a commercial sweetener has been approved since 2002 in the US (GRAS status) while the EU has banned its use in European markets. Advantame is a sweetener even more potent than neotame, 20 000 times sweeter than sucrose. Its use has very recently been approved in the US by the FDA and by the European commission, but very little data concerning this sweetener are available due to its very recent discovery.

7.2.2.5 Sucralose

Sucralose is a semi-synthetic sweetener that resulted from intense research conducted by the Tate & Lyle Company in the 1970s on the properties of sucrose. It was discovered in 1976 that a chemical modification of sucrose by the addition of a halogenated derivative leads to increased sweetening power. The most intense sweet taste was elicited by a derivative in which three hydroxyl groups on sucrose were replaced by chlorine atoms (4'-,1'- and 6' positions) (Schiffman and Gatlin 1993). Sucralose was identified for its high sweetening power and resemblance to sucrose, along with its high chemical/thermal stability and low toxicity. Sucralose is approximately 600 times sweeter than sucrose, with a slight bitter aftertaste. Sucralose synergises with acesulfame-K, cyclamate and saccharin (Schiffman and Gatlin 1993), and its use as a sweetener in foodstuff is approved in many countries including the US and EU. Sucralose has been shown to bind the N-terminal domains of T1R2 and T1R3 (Maitrepierre et al. 2012, Nie et al. 2005).

7.2.3 Sweet-tasting proteins and taste-modifying proteins

A special class of sweet-tasting molecules has attracted interest for industrial use as natural sweeteners and sugar substitutes, sweet-tasting proteins. In general, molecules with a high molecular weight such as proteins have no taste. However, a limited number of proteins can elicit a strong sweet taste. Originating from tropical plants, some of these proteins such as brazzein and thaumatin have been consumed for centuries by indigenous population, thus reducing concern about their toxicity. To date, seven intensely sweet-tasting proteins have been identified: thaumatin, monellin, brazzein, pentadin, mabinlin, neoculin and miraculin. It should be noted that hen egg white lysozyme is classified by some authors as a sweet-tasting protein (Maehashi and Udaka 1998). However, the very low sweetness potency of lysozyme, approximately 20 times sweeter than sucrose, limits its applications in the food-industry (Masuda et al. 2001). Therefore, its sensory properties will not be described here.

Although they have properties in common, such as their isoelectric point, which is alkaline for all proteins except for brazzein, the sweet-tasting proteins differ in their structure and surface charge. Indeed, they share no sequence homology or structural identity (Figure 7.3), and they originate from plants that are phylogenetically different.

7.2.3.1 Thaumatin

Described as the sweet constituent of the berries of the West African plant *Thaumatococcus daniellii*, also called katemfe berry, thaumatin is the common designation of what is actually a mixture of two intensely sweet-tasting proteins, thaumatin I and thaumatin II. These two proteins are monomeric proteins composed of 207 amino acids (22 kDa), and the two differ by approximately 3% in their amino acid composition. The three-dimensional structure of thaumatin II (Masuda et al. 2011) is stabilised by eight disulphide bridges, making it a very stable protein with respect to pH and heat denaturation (Figure 7.3). According to the SCOPe classification (http://scop.berkeley.edu), thaumatin is part of the osmotin thaumatin-like superfamily. Site-directed mutagenesis studies have demonstrated that alkaline residues such as lysine and arginine present on the surface of thaumatin are crucial for its sweet-tasting properties (Ohta et al. 2011). Thaumatin is the only sweet-tasting protein to have actually been commercialised in the 1970s as an extract from *T. daniellii* under the name Talin (E957), still available today from Naturex (Avignon, France). Thaumatin

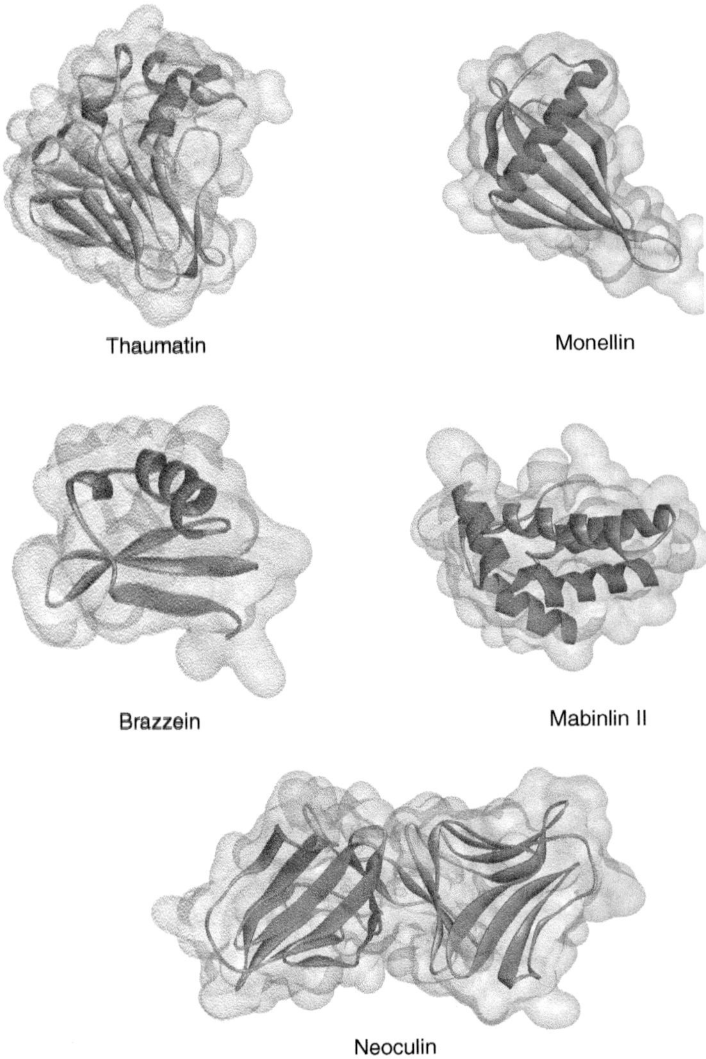

Figure 7.3 Three-dimensional structure of sweet-tasting proteins.

has received approval from both the FDA and the European Commission to be used as a sweetener in foodstuff. The sweetness potency of thaumatin is approximately 1 600 times higher than that of sucrose on a weight basis and 100 000 times higher on a molar basis. A lingering sweet taste has been attributed to thaumatin, including a liquorice-like aftertaste. The production of recombinant thaumatin has been described in many expression systems, including *Escherichia coli, Bacillus subtilis, Saccharomyces cervisiae, Pichia pastoris* and *Aspergillus niger*. Recently, heterologous expression of thaumatin in plants such as tobacco and cucumber has been reported (Pham et al. 2012, Szwacka et al. 2009).

7.2.3.2 Monellin

Monellin is the sweet-tasting component of the plant *Dioscoreophyllum cumminsii*. The name monellin comes from the research centre where the protein was identified, the Monell Chemical Senses Center (USA). Monellin is a heterodimeric protein with a molecular weight of 11.4 kDa (Figure 7.3), is composed of two chains A and B made of 42 and 50 amino acid residues, respectively (Morris et al. 1973) and is approximately 3 000 times sweeter than sucrose (Kohmura et al. 1990). Monellin is related to the cystatin family of protease inhibitors (Murzin 1993). The two chains are held together only by weak forces, making the protein relatively unstable to high temperature and to acidic pH. When heated to temperatures above 50°C, monellin dissociates, thus losing its sweetness properties (Kim et al. 1989). To prevent the dissociation of the two chains and to facilitate monellin heterologous expression, a recombinant single-chain monellin (SCM) was successfully designed by Kim et al. (1989) based on the crystal structure of the natural protein (Hobbs et al. 2007, Ogata et al. 1987). In SCM, the C-terminal residue of the chain B and the first N-terminal residues of the chain A were fused. Later, another recombinant single chain monellin called MNE I was designed (Tancredi et al. 1992), in which a Gly-Phe dipeptide links the two chains together. These recombinant versions of monellin are as sweet as natural monellin but are far more stable with respect to pH and heat denaturation (Kim et al. 1989, Tancredi et al. 1992). Monellin is approximately 3000 times as sweet as sucrose on a weight basis and 90 000 times sweeter on a molar basis. Monellin has been produced in various expression systems, including bacteria and the yeasts *Saccharomyces cerevisiae* and *Candida utilis* (Aghera and Udgaonkar 2011, Chen et al. 2005, Kondo et al. 1997).

7.2.3.3 Brazzein and pentadin

Brazzein is a small (6.4 kDa) sweet-tasting protein (Ming and Hellekant 1994) expressed in the seed pulp of the fruit of the West African plant *Pentadiplandra brazzeana*. Brazzein is composed of a single polypeptide chain of 54 amino acid residues. Brazzein's three-dimensional structure (Figure 7.3) is composed of a short alpha helix and three antiparallel beta-sheets stabilised by four disulfide bridges (Assadi-Porter et al. 2000, Nagata et al. 2013). On a structural basis, brazzein can be classified as a scorpion toxin-like protein, and closely resembles the defensin superfamily (Wintjens et al. 2011). Brazzein's compact structure makes it very stable to heat and pH denaturation. Thus brazzein can withstand temperatures up to 80°C for several hours and pH levels in the range of 2.5 to 8. Sensory analyses have revealed that brazzein is approximately 2 000 times sweeter than a 2% sucrose solution on a weight basis. Its taste has a lingering quality and is described as being more similar to that of sucrose than to that of thaumatin. Although not yet commercialised, brazzein heterologous expression has been described in the yeast *Pichia pastoris* expression systems (Poirier et al. 2012), with physico-chemical and sensory properties identical to those of the natural protein. Interestingly, it has been shown that the cysteine-rich domain of T1R3 is a major determinant for the response of the T1R2/T1R3 sweet taste receptor towards brazzein (Jiang et al. 2004), but the mechanism of sweet taste receptor activation remains to be elucidated.

Interestingly, a second sweet-tasting protein called pentadin has been extracted from the seed pulp of *Pentadiplandra brazzeana* (Van der Wel et al. 1989). There is very little information available on pentadin, except that its molecular weight is approximately 12 kDa and its potency is 500 times that of sucrose on a weight basis. Further studies on its structure would be necessary to determine its chemical properties.

7.2.3.4 Mabinlin

In 1982, from the seeds of the Chinese subtropical plant *Capparis masaikai,* a group of four sweet-tasting proteins named mabinlins I, II, III, and IV was identified (Nirasawa et al. 1994). The seeds have long been used by locals of the Yunnan province in China as traditional medicine and are known to generate a long-lasting sweet sensation after chewing, accompanied by a bitter aftertaste and an astringent sensation. The two extracted proteins were called mabinlin I and mabinlin II after the local name of the plant, mabinlang. Although both have the same sweetening potency, mabinlin I is much less heat-stable than II (Nirasawa et al. 1994), making it a less interesting candidate for commercial use. Mabinlin II was estimated to be 375 times sweeter than sucrose on a molar basis and it is a heterodimer of two chains composed of 33 and 72 amino acids with molecular weights of 4.6 and 5.2 kDa, respectively. Mabinlin II is remarkably stable when heated at high temperature (Li et al. 2008). The three-dimensional structure of natural mabinlin II (Figure 7.3) was solved, showing that the cross-linking of the two chains, A and B by four disulfide bridges may account for its extreme thermostability (Guan et al. 2000, Li et al. 2008). Using cellular-based assays, natural mabinlin II has been shown to activate the sweet taste T1R2/T1R3, but the mechanism of receptor activation is still unknown. To the best of our knowledge, no successful heterologous expression of mabinlin has been reported. Currently, mabinlin is not commercially used as a sweetener, but its stability at very high temperature makes it an interesting candidate for food application.

7.2.3.5 Miraculin and neoculin

In addition to the previously described sweet-tasting proteins, two unique proteins called miraculin and neoculin have taste-modifying properties. Miraculin is isolated from the berries of the West African plant *Synespalum dulcificum,* whose unique properties have long been recognised by indigenous populations. The fruit, also named the "miracle fruit", contains miraculin, which transforms a sour taste into a sweet taste. Miraculin, first isolated in 1968 (Kurihara and Beidler 1968), is a tetramer and dimer of a 25 kDa protein composed of 191 amino acids, and both the tetramer and the dimer have taste-modifying properties (Igeta et al. 1991). The dimer is stabilised by a covalent bond formed by a disulphide bridge. In contrast to all sweet-tasting proteins, miraculin is glycosylated on two N-glycosylation sites. Miraculin is homologous with soybean trypsin inhibitor (Selvakumar et al. 2011). The effect of miraculin has been reported to last for approximately one hour (Kurihara and Beidler 1969). It is very difficult to compare the sweetness of miraculin to that of other sweet-tasting proteins because of its unique mechanism. However, miraculin has been estimated to be as much as 400 000 times sweeter than sucrose (Gibbs et al. 1996) on a molar basis. Although the three-dimensional structure of miraculin is not known, its mechanism of action has been recently elucidated (Koizumi et al. 2011). At neutral pH, miraculin interacts with the sweet taste receptor T1R2/T1R3; however, it does not activate the receptor and partially suppresses the response to other sweeteners. At acidic pH, miraculin changes its conformation and activates the sweet taste receptor. The natural extract of miraculin is commercially available on a small scale. Heterologous expression of miraculin has been described in the yeast *Pichia pastoris* (Ito et al. 2011), transgenic plants (Kato et al. 2010, Sun et al. 2007) and in *E. coli* (Matsuyama et al. 2009). The unique properties of miraculin make this protein interesting for use as a sweetener or for food applications.

Neoculin (previously named curculin) is the only protein known to elicit sweet taste and also to exhibit sweet-taste modifying properties. The heterodimeric neoculin is isolated from the fruits of the Malaysian plant *Curculigo latifolia*, known for centuries by locals for its sensory properties. The reason why this protein is sometimes referred to as either curculin or neoculin is the confusion regarding its structure. When the recombinant protein was first produced, to the general surprise of everyone, it did not taste sweet. To determine the reason for this, in 2004, two groups isolated a second gene 77% identical to the gene of curculin from the plant and produced a panel of homo and heterodimers (Shirasuka et al. 2004, Suzuki et al. 2004). It was found that the sweet taste was elicited only by a curculin1 and curculin2 heterodimer, renamed neoculin. The crystal structure of the heterodimer extracted from the plant shows two chains with molecular weights of 13 and 11 kDa connected by four disulfide bonds (Figure 7.3). Neoculin has been produced as a recombinant protein in *E. coli* (Suzuki et al. 2004) and in the filamentous fungus *Aspergillus oryzae* (Nakajima et al. 2006). Neoculin is the first protein to be found to have both taste-modifying and sweet-taste-eliciting properties, making it an interesting and unique sweetener or taste modifier for future use in food applications.

7.3 Bitter tastants

Bitter taste is generated by probably more than tens of thousands of different bitter molecules. Interestingly, all of these bitter compounds are perceived in humans by only 25 different taste receptors (TAS2Rs) (Meyerhof et al. 2010). Whereas some TAS2R receptors respond to only few bitter compounds, some of them are broadly tuned and recognise a broad range of different bitter molecules (Meyerhof 2005, Shi and Zhang 2009). The response to bitter molecules is innate and well known to provoke a stereotypical aversion. This aversion level depends on the diet of each species. Thus, human beings present an intermediate threshold level between carnivore species and herbivore species. It is commonly accepted that bitter aversion is an evolutionary result of animal protection against toxic molecules, including bacterial products originating from spoiled foods. However, no real correlation between the toxicity and the bitter taste of a molecule has been established. Indeed, some toxic molecules are not perceived as bitter, and many bitter molecules are not toxic and are commonly used for human consumption. In addition, the threshold detection of toxic molecules with a bitter taste is not correlated with toxicity level (Drewnowski and Gomez-Carneros 2000). In parallel, some food molecules perceived as bitter may be of special interest in view of their beneficial effects on human health. For instance, phenethyl isothiocyanate and benzyl isothiocyanate (Figure 7.4), two bitter-tasting compounds found in cruciferous vegetables, such as broccoli, have been shown to inhibit carcinogenesis (Hecht 2000).

The majority of bitter-tasting molecules found in human food have plant origins. The bitter compounds are often molecules which take part in plant defences against herbivorous animals by making the plant unpalatable. A recent study conducted on a large panel of animal species demonstrated that the number of bitter taste receptors present in a species correlates with the fraction occupied by plant-based food in their diet (Li and Zhang 2014). However, it has been shown that herbivores may tolerate some bitter molecules, with this tolerance varying from one species to another (Nolte et al. 1994). Interestingly, most of the molecules perceived as bitter by human beings have been observed to be aversive for most animal species, including worms and insects, suggesting that plants have evolved to produce bitter compounds as a defence against pests and herbivores.

It is known that individual bitter sensitivity is highly variable in humans and can be partially explained by genetic polymorphisms of some of the bitter taste receptors. The difference in perception of thiourea molecules, more specifically phenylthiocarbamide (PTC) and 6-*n*-propylthiouracil (PROP) (Figure 7.4), is a good illustration of the effect of gene receptor polymorphism on bitter recognition. The sensitivity of these two molecules can be easily tested through the filter paper method, which has been used in many studies for more than 70 years. PTC and PROP are not perceived by 30% of the population and are perceived as being moderately to highly bitter by the remaining 70%. PTC sensitivity has been directly linked to the genetic polymorphism of a single TAS2R receptor (TAS2R38) (U. K. Kim et al. 2003, Prodi et al. 2004). The sensitivity for these

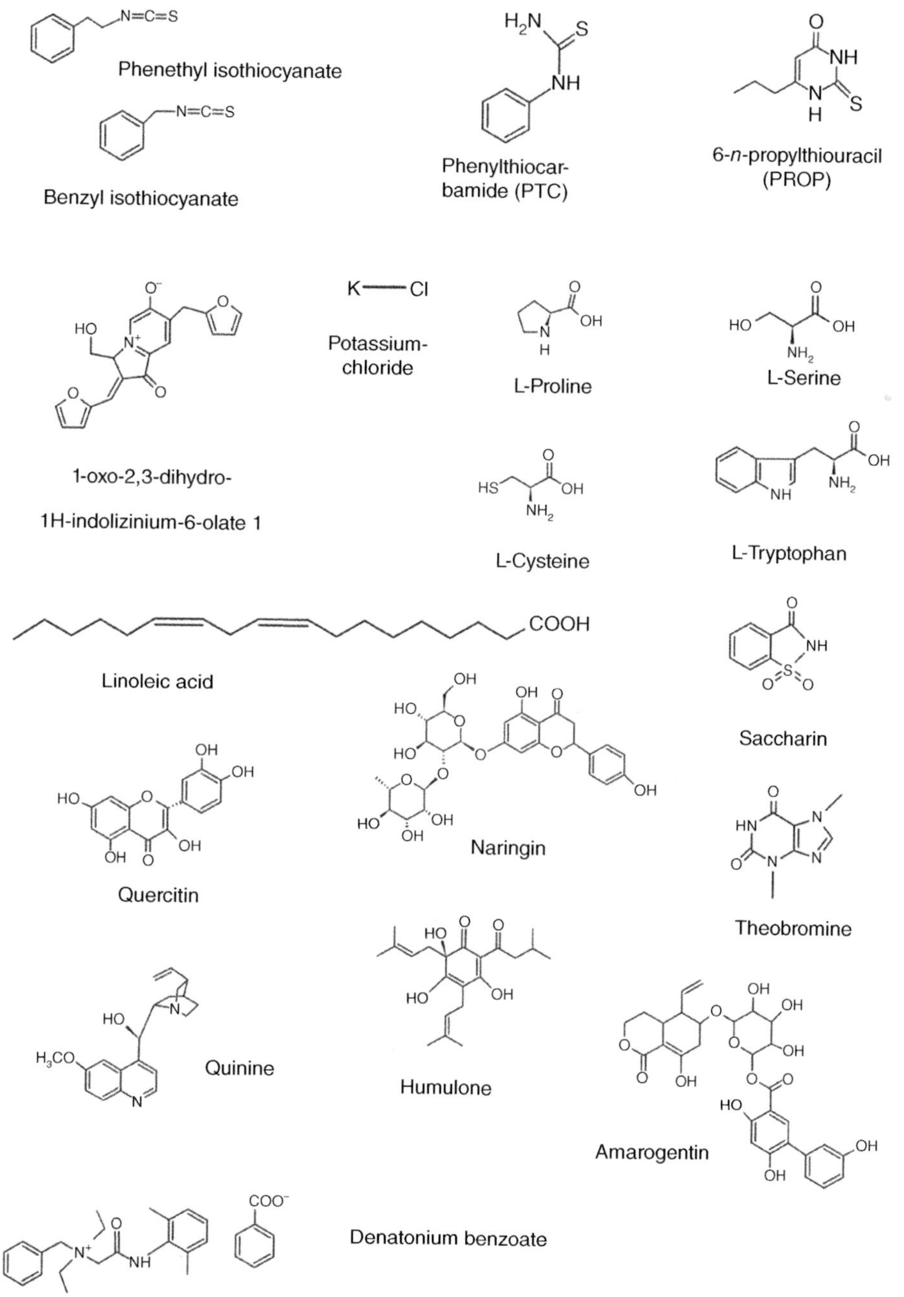

Figure 7.4 Chemical structure of some typical bitter-tasting molecules.

two molecules has also been linked to other inter-individual factors such as gender or ethnicity. For example, people from Asia or Africa are more sensitive than Caucasian populations. Concerning gender, women are more sensitive to these bitter molecules than men (Drewnowski et al. 2001). An age relation has also been observed, with the sensitivity to these two molecules declining with ageing. This decline in the sensitivity threshold with age has also been observed for quinine (Figure 7.4), another bitter molecule (Cowart et al. 1994). Thus, the inter-individual differences and the TAS2R polymorphism may account for the considerable divergence of bitterness perception in and between populations.

7.3.1 Origin of bitter molecules

Plants represent the main source of bitter molecules in human food. Therefore, many vegetables such as chicory, chard, endives, artichokes, broccoli, or asparagus have a bitter taste. Some drinks prepared from plants are known to possess bitterness, such as coffee, green tea, wine, beer or sugar-free chocolate. Some bitter compounds can also originate from food processing, for example through bacterial activity. Different bacteria are known to generate bitter molecules during food processing, for example, those used in the production of beer, wine, cheese, soy sauce, or other fermented foods (Hagedorn and Kaphammer 1994). Proteolytic and lypolytic bacteria may cause bitterness in milk by digesting proteins such as casein, generating bitter-tasting peptides. In cheddar cheese, the bitterness is provided by the starter proteolytic activity from *Lactococcus lactis* (Broadbent et al. 2002). The bitterness may also result from the production of bitter molecules from chemical reactions occurring in food processing, such as the Maillard reaction induced by thermal treatment. For instance, the most intense bitter compounds produced during the simple Maillard reaction between xylose, rhamnose and L-alanine are 1-oxo-2,3-dihydro-1H-indolizinium-6-olates 1-5, with 1 being the most bitter (Frank et al. 2003). This compound (Figure 7.4) presents a bitter taste threshold value of 0.25 μM. A bitter taste can also appear in food after cold storage through oxidation; for example, bitterness in linseed oil appears after cold storage mainly due to the formation of a methionine-oxidised peptide (Bruhl et al. 2007).

7.3.2 Bitter molecules

The molecules perceived as bitter are chemically very diverse (Figure 7.4), as are their origins and their detection thresholds (Table 7.2). No relationship between structure and bitterness has been established, likely due to the relatively large number of taste receptors involved in their detection. Nevertheless, based on the chemical structure of a molecule, it is possible in some cases to anticipate its bitterness by analogy with the structure of known bitter molecules. Bitter molecules can be found in many chemical classes, including salts, flavonoids, alkaloids, amino acids, peptides and lipids. Due to the large number of bitter

Table 7.2 Bitter-tasting compounds, chemical classes, examples of food source and detection thresholds.

Compound	Chemical class	Food source and concentration	Detection threshold
KCl	Salt	Goat cheese (5–20 g/kg)	5 mM
L-Trp	Amino acid	Wheat (70 g/kg)	4–6 mM
NALPE	Peptide	Soybean (n.d.)	74 μM
Linoleic acid	Fatty acid	Poppy seed (n.d)	4–6 mM
Saccharin	Artificial sweetener	Sweet drinks (20–270 mg/L)	0.3 mM
Quercetin	Flavonol	Capers (2.3 g/kg) Onion (0.32 g/kg)	1.7 mM
Naringin	Flavanone	Grapefruit juice (300–750 mg/L) Oroblanco juice (346–489 mg/L)	0.64 mM
Theobromine	Alkaloid	Cocoa beans (20–30 g/kg)	56 μM
Quinine	Alkaloid	Tonic drinks (80 g/L)	25 μM
Caffeine	Alkaloid	Coffee (10–20 g/L)	0.8 – 1.2 mM
Catechin	Phenolic flavonoid	Red wine (1–3.5 mg/L)	N.D.
Genistein	Isoflavone	Soybeans (24–40 mg/kg) Tofu (29–78 mg/L)	14 mM
Limonin	Limonoid aglycone	Lemon juice (12.2 mg/L) Orange juice (9.7 mg/L)	8 μM
Sinigrin		Brussels sprouts (110–1560 mg/kg)	0.300 – 12.5 mM
Humulone		Beer (15–100 mg/L)	14 μM

n.d. = not described

molecules that exist, we will only review representative molecules of the main chemical classes.

7.3.2.1 Salts

Many simple salts are perceived as bitter, for example potassium, magnesium and calcium chloride. Despite its bitterness, potassium chloride (Figure 7.4) has been shown to be a good substitute of sodium chloride, for example, in cheddar-style cheese (Grummer et al. 2012). Calcium chloride (E509) is a widely used salt as an additive in food, mainly as a firming agent, stabiliser or thickener. It is also used in sport drinks as an electrolyte. Interestingly, magnesium and potassium chloride are present in high concentrations in many vegetables such as spinach, in which the salts contribute to the bitterness of the foods.

7.3.2.2 Amino acids and peptides

Amino acids are known to elicit complex tastes, such as umami, sweetness and bitterness. Nevertheless, some amino acids primarily have a bitter taste. Bitter amino acids can be found among both series D and L. Originally, the D-amino acids were considered to be sweet, whereas the L-amino acids were described as bitter. However, this simple view was amended (Kawai et al. 2012, Schiffman et al. 1981) to consider all the hydrophobic or branched amino acids as bitter. Indeed, some branched amino acids as L-threonine are not bitter but sweet. Interestingly, L-proline, L-serine and L-cysteine (Figure 7.4) taste bitter in their crystal form but not in solution, partially explaining the divergence between the different studies (Asao et al. 1987). The contribution of free amino acids to the bitterness of food can be substantial. For example, in whole wheat bread, L-tryptophan (Figure 7.4) appears to be one of the compounds exhibiting the highest bitterness intensities (threshold value of 4-6 mM) (Jiang and Peterson 2013). In addition, the chemical modification of amino acids can dramatically change their taste. For instance, the enzymatic industrial γ-glutamylation of L-phenylanaline abolishes its bitterness (Suzuki et al. 2002), and a very intense bitterness is observed after L-proline and sucrose roasting. Many examples of bitter peptides generated during food processing have been reported in the literature. Some peptides are known to generate bitterness in soy sauce, cheese or sake (Maehashi and Huang 2009). For instance, two peptides from Gouda cheese, Tyr-Pro-Phe-Pro-Gly-Pro-Ile-His-Asn-Ser and Leu-Val-Tyr-Pro-Phe-Pro-Gly-Pro-Ile-His-Asn, generate a bitter taste. A non-bitter peptide can acquire bitterness after hydrolysis, as previously mentioned for casein during cheese ripening. For example, the 11S glycinin protein present in soybean is cleaved into two bitter peptides by trypsin (I. M. R. Kim et al. 2003). Another peptide, Asn-Ala-Leu-Pro-Glu (also referred to as NALPE), was isolated from soybean, presenting an intense bitter taste (threshold value of 74 μM) (Kim et al. 2008). Although the bitterness of peptides has not been shown to be related to their amino acid content, it is interesting to note that an empirical relationship between peptide hydrophobicity and bitterness has been observed. Recently, it has been shown that the detection of bitter peptides is not mediated by a single broadly tuned TAS2R receptor but rather by a set of sensitive TAS2Rs (Kohl et al. 2013).

7.3.2.3 Lipids

Rather surprisingly, some free fatty acids (for instance oleic acid and linoleic acid) are perceived as bitter by humans. The bitterness of fatty acids can explain the bitterness of some foodstuff. For example, linoleic acid (Figure 7.4) contributes to the bitter taste of poppy seeds (Grosch and Laskawy 1984) (threshold value of 4-6 mM). The transformation of lipids may also be the origin of molecules that generate bitterness. Thus, soybean lecithin, used in various products (ice cream, bakery, etc.), presents a bitter taste after hydrolysis. The same effect is observed with neutral-tasting lipids that become bitter after their enzymatic hydrolysis in salmon after freezing (Refsgaard et al. 2000).

7.3.2.4 Sweet molecules with bitter off-taste

In addition to being sweet, many low-calorie sweeteners, natural or synthetic, generate a bitter off-taste. Sensitivity to this off-taste has been shown to differ between individuals (Bartoshuk 1979) and could be correlated to bitter taste receptor polymorphism. Some sweeteners such as saccharin (Figure 7.4) and acesulfame-K share a common mechanism of interaction to the same two bitter receptors (TAS2R43 and TAS2R44) (Horne et al. 2002, Kuhn et al. 2004). The interaction with the bitter receptors can also vary from one artificial sweetener to another, explaining the possibility of adding another sweetener to increase the sweet taste without reaching the bitter threshold. It is also interesting to note that some sweeteners such as tagatose and xylitol or sweet-tasting proteins such as brazzein do not present any bitterness (Fujimaru et al. 2012, Poirier et al. 2012).

7.3.2.5 Other bitter molecules

Numerous chemical families contain molecules with bitter taste. The main ones are lactones, flavonoids, pyranosides, terpenes, esters, amides, lactames, sulfimides, flavonols, alkaloids and glucosides. These molecules are found in food; for example, flavanone and flavonols are found in grapefruits and in Brussels sprouts, and glucosides are found in mustard greens. Moreover, most of these molecules are the major bitterness compounds in said foods. Quercitin (Figure 7.4) is a flavonol and is the major bitter molecule of immature apples (Bravo 1998), and naringin (Figure 7.4), a flavanone, is a major bitterness molecule found in orange juice. The alkaloid family contains the largest number of bitter molecules present in human food, including theobromine (Figure 7.4), which is found in chocolate, tea or cola, and caffeine, which is found in coffee. Another bitter alkaloid molecule, quinine (Figure 7.4) (detection threshold value of 25 μM), is naturally found in the bark of the cinchona tree. Quinine is added to some drinks (e.g., tonic drinks contain approximately 80 g/L) and used as a bitter reference compound for sensory evaluations. Most food contains a combination of bitter molecules. Beer, for example, contains more than 90 bitter molecules. Alpha-acids such as humulone (Figure 7.4) are the major contributors to its bitterness. One of the most potent bitter molecules found in nature is a glycoside, amarogentin (Figure 7.4) (threshold value is lower than 30 μM), which is found in gentian roots and used in some beverages. In parallel, one of the most potent artificial bitter molecules obtained by chemical synthesis is denatonium benzoate (Figure 7.4) (threshold value of 0.5 nM), which is currently used as an aversive agent to prevent accidental ingestion of household products or to denature alcohol.

7.3.3 Bitterness masking and bitterness inhibitors

Reducing bitterness is a challenge for both the food industry and the drug industry. In the drug industry, the main difficulty is the bitterness of the majority of drugs, which reduces patient compliance in taking medicine. This problem of bitter perception is accentuated with young children. In the food industry, bitterness acceptance depends on the product and becomes more problematic with various new "healthy products", which generally contain reduced amounts of fat and/or sugar. Indeed, a simple way to reduce or mask bitterness is to add sweetness. In the drug industry, a common method is based on the coating of pills, physically limiting the diffusion of a bitter drug towards the corresponding receptors. Another bitter-masking technique used in both industries is the addition of sweet congruent flavours. For example, a strawberry flavour is often added to medicinal syrups for children. Another food industry method for reducing the concentration of bitter molecules in food products is simply by removing the molecules. The solution that is implemented depends both on the product and the origin of the bitter molecule. Plant selection has been used to reduce bitter molecule concentrations in various vegetables such as cucumber, lettuce or spinach. Bitter compounds can also be removed during food processing. In orange juice, the two major bitter compounds, naringin and limonin, can be removed using specific resins. However, removing bitter molecules may lead to the loss of health-benefit properties, such as the anti-carcinogenicity of isothiocyanates. Finally, another strategy for bitterness reduction is based on bitter taste inhibition using, for example, zinc salt (E650), which suppresses the bitter taste of molecules such as quinine or denatonium benzoate. Another strategy used by the food industry is to add sodium chloride to food. It has been shown that salt enhances flavour by selectively inhibiting bitterness (Breslin and Beauchamp 1997). Recent developments in the study of bitter-blockers based on the screening of bitter taste receptors using cellular assays have led to the discovery of new inhibitors. These new compounds, which may be useful for some food applications, will be reviewed in section 6 of this chapter.

7.4 Umami and kokumi compounds

7.4.1 Basic properties of umami substances, history

It is currently well established that umami is one of the five primary taste qualities. Its physiological role is believed to be the signalling of the presence of free amino acids or proteins in food. For humans, umami is the typical taste induced by monosodium glutamate (MSG) (Figure 7.5) found naturally in many protein-rich foods, such as seafood, meat, cheese and certain fruits and vegetables. Although numerous molecules can generate an umami taste in humans, the umami sensation is one of the most difficult to describe. Umami taste is also one of the most complex fundamental tastes. This complexity is mainly due to the synergistic effects observed between different umami-tasting molecules and the fact that umami tastants, which possess a pure umami taste, are rare. For instance, the anion of MSG (glutamate) induces the umami taste,

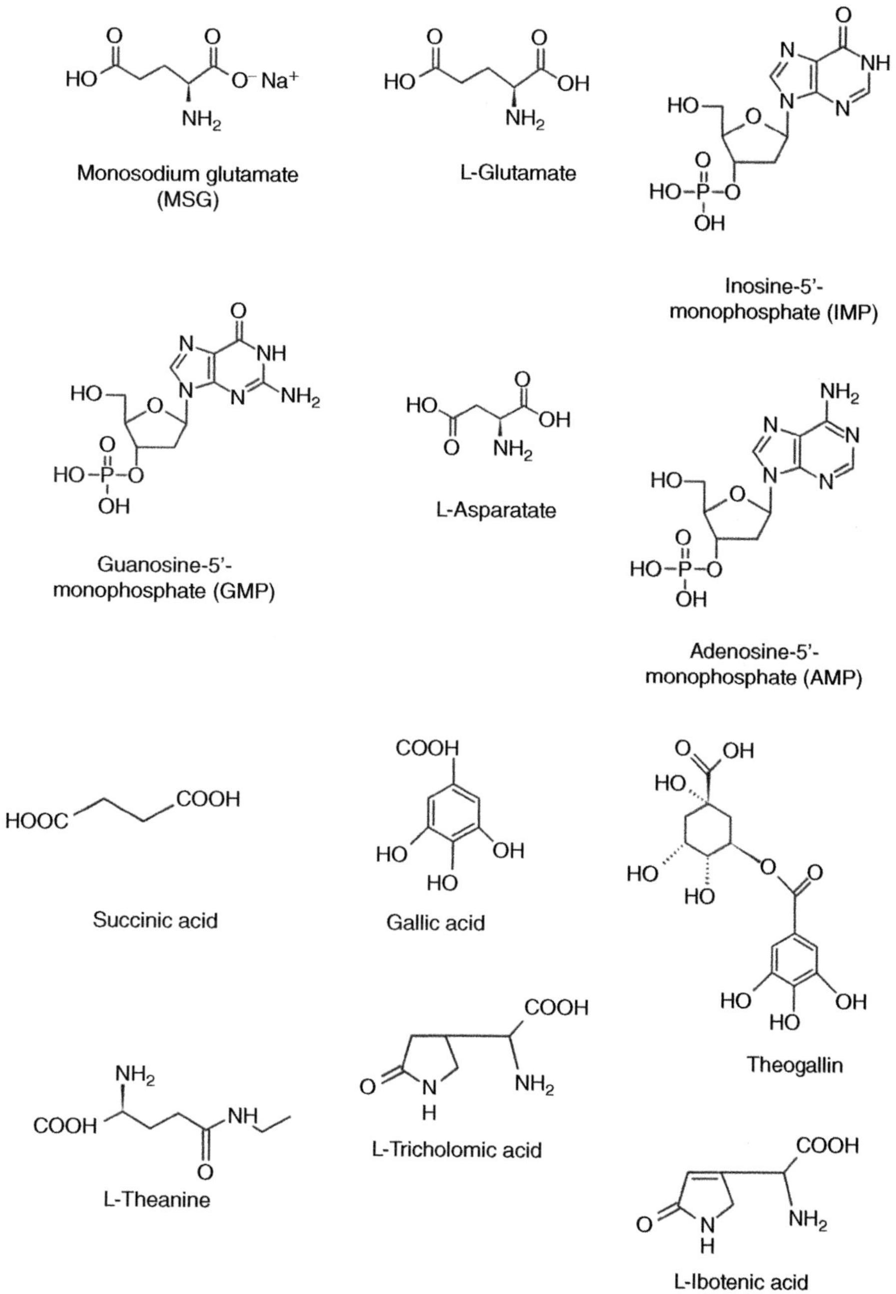

Figure 7.5 Chemical structure of some umami compounds.

whereas the cation (sodium) is responsible for the salty taste of MSG. Although the existence of a fifth taste has challenged researchers for a well over a century, it is known that ancient Romans 2000 years ago enjoyed numerous dishes containing significant concentrations of umami molecules (Curtis 2009). In his treatise "La physiologie du goût" (The physiology of taste) published in 1826, Jean Anthelme Brillat-Savarin identified a specific taste for the part of meat soluble in water. Glutamic acid (Figure 7.5) itself was isolated and chemically described in 1866 by the German chemist Karl Heinrich Ritthausen, but the umami taste concept was proposed in 1908 by Kikunae Ikeda, a physical chemistry professor at the University of Tokyo. Professor Ikeda isolated MSG from seaweed (dried kelp) and was the first to describe its unique taste properties, which were distinct from those of the other fundamental taste qualities (sweet, bitter, sour and salty). There is no English word for umami. Umami (うま味) is a Japanese word that literally translates to delicious, meaty or savoury taste. In 1913, Professor Shintaro Kodama, who was a collaborator of Professor Ikeda, discovered that the constituents of katsuobushi (dried, fermented and smoked skipjack tuna) contained another umami tastant, a 5′ ribonucleotide (Figure 7.5), inosine-5′-monophosphate (IMP). Guanosine-5′-monophosphate (GMP), another important umami molecule (Figure 7.5b), was identified many years later from the biochemical degradation of yeast RNA (Kuninaka 1960, Kuninaka 1964, Sakaguchi et al. 1958) and was found to be the main umami component in shiitake mushrooms. Another important discovery was the potentiation of MSG by IMP and GMP (Kuninaka 1960, Kuninaka 1964, Yamaguchi 1991). This synergy is accepted as the most unique hallmark of umami taste. In the 1980s and 1990s, taste responses to umami molecules were investigated in humans and animals using electrophysiology and behavioural experiments. Unfortunately, the use of animal models is of limited use because the responses towards amino acids vary from one species to another. Therefore, only two amino acids, L-glutamate and L-aspartate (Figure 7.5), are umami-tasting to humans, whereas behavioural and electrophysiological experiments have revealed that mice are able to perceive a large range of amino acids (Maruyama et al. 2006). This observation has been correlated to the ability of the recombinant mouse T1R1/T1R3 receptor to function as a broadly tuned amino acid detector (Nelson et al. 2002, Toda et al. 2013). It should be noted that umami detection is stereospecific because the enantiomers D-glutamic and D-aspartic acids have no umami taste. The identification of a GPCR specifically involved in the detection of umami molecules that reflects the properties of umami taste was a major breakthrough in the early 2000s. These receptors are reviewed in chapter 3 of this book. Briefly, it was demonstrated that the T1R1 (taste receptor type 1, member 1) and T1R3 (taste receptor type 1, member 3) GPCR heterodimer is involved in umami taste perception in humans, whereas T1R2 (taste receptor type 1, member 2) and T1R3 assemble to form a single taste receptor sensitive to several sweet-taste-eliciting chemicals

(Li et al. 2002, Nelson et al. 2002, Nelson et al. 2001). The identification of the T1R1/T1R3 receptor formally established umami taste as the fifth taste. In addition, metabotropic glutamate receptor types 1 and 4 and their truncated forms, namely taste-mGluR1 (San Gabriel et al. 2009, San Gabriel et al. 2007, Toyono et al. 2002) and taste-mGluR4 (Chaudhari et al. 2000, Chaudhari et al. 1996, Yang et al. 1998), have been proposed to contribute to umami tastant detection. However, the physiological involvement of these taste-mGluRs in umami molecule detection in humans is still a matter of debate because these receptors are activated by L-glutamate and analogues but are not sensitive to 5′ ribonucleotides.

7.4.2 Natural umami compounds

L-Glutamic acid is one of the major amino acids constituting plant and animal food proteins. In addition, free L-glutamic acid is naturally present in numerous foods, such as seafood, meat, poultry, fruits, vegetables, mushrooms and green tea (Table 7.3). Interestingly, L-glutamic acid is the most abundant free amino acids (more than 50% of the free amino acid content) in human breast milk. L-glutamic acid, with a concentration of 190 mg/L (1.65 mM), may account for the taste acceptability of milk to human infants. The ripening of cheese or the maturation of fermented foods also release free L-glutamic acid by protein proteolysis, such as in fish or vegetable sauces used to generate umami flavour in daily cooking. The two 5′ ribonucleotides IMP and GMP are also naturally occurring in many foods. IMP has been mostly observed in meat, whereas GMP is most abundant in fish, shellfish and mushrooms (Table 7.4). The umami-tasting 5′ ribonucleotide AMP (Figure 7.5) has been found primarily in fish and seafoods (Table 7.4). In addition to L-glutamic acid and 5′ ribonucleotides, other rare umami tastants have been identified in various food sources. Certain di- or tri-peptides rich in acidic amino acid residues, such as pGlu-Pro-Ser or Asp-Glu-Ser resulting from vegetable or fish protein proteolysis, have been reported to generate a weak umami taste, which could be significantly enhanced by IMP (Noguchi et al. 1975, Schlichtherle-Cerny and Amado 2002, Winkel et al. 2008). An octapeptide, Lys-Gly-Asp-Glu-Glu-Ser-Leu-Ala, isolated from beef hydrolysate generated by a papain treatment has been reported to have a "delicious" umami taste (Yamasaki and Maekawa 1978). However, we have to be cautious about the sensory properties of these umami peptides since they have been questioned since (van den Oord and van Wassenaar 1997). In Japanese green tea, in addition to L-glutamic acid and succinic acid (Figure 7.5), Kaneko and co-workers (Kaneko et al. 2006), identified gallic acid, theogallin and the amino acid derivative N-ethyl glutamine (also named theanine) (Figure 7.5) as key contributors to the umami taste of green tea leaves. Recently, using cellular assays, L-theanine was demonstrated to elicit umami taste and to synergise with IMP *via* T1R1/T1R3 taste activation (Narukawa et al. 2014).

Table 7.3 Concentration of Free L-Glutamic Acid in Foods

Food	Free L-glutamic acid (mg/100g)	Reference
Cheese		
Sheep milk cheese	5115	(Ghirri and Bignetti 2012)
Parmesan cheese	1200	(Kato et al. 1974)
Cheddar cheese	182	(Ghirri and Bignetti 2012)
Meat and poultry		
Salami	4944	(Ghirri and Bignetti 2012)
Chicken	22	(Yamaguchi and Ninomiya 2000)
Beef	10	(Kato et al. 1974)
Pork	9	(Yamaguchi and Ninomiya 2000)
Fish and seafood		
Fresh tuna fish	3397	(Ghirri and Bignetti 2012)
Fish sauce (Japan)	1383	(Yamaguchi and Ninomiya 2000)
Scallop	140	(Yamaguchi and Ninomiya 2000)
Prawn	43	(Kato et al. 1974)
Vegetable		
Mushroom	180	(Kato et al. 1974)
Tomato	140	(Kato et al. 1974)
Corn	106	(Yamaguchi and Ninomiya 2000)
Green peas	106	(Yamaguchi and Ninomiya 2000)
Potatoes	102	(Kato et al. 1974)
Onion	51	(Yamaguchi and Ninomiya 2000)
Cabbage	50	(Yamaguchi and Ninomiya 2000)
Green asparagus	49	(Yamaguchi and Ninomiya 2000)
Spinach	48	(Yamaguchi and Ninomiya 2000)
Fruit		
Kiwi	176	(Ghirri and Bignetti 2012)
Grapes	143	(Ghirri and Bignetti 2012)
Banana	110	(Ghirri and Bignetti 2012)
Avocado	18	(Yamaguchi and Ninomiya 2000)
Milk		
Human breast milk	19	(Yamaguchi and Ninomiya 2000)
Bovine whole milk	1	(Yamaguchi and Ninomiya 2000)
Seaweed		
Kelp	2240	(Kato et al. 1974)
Dried laver	1378	(Yamaguchi and Ninomiya 2000)
Miscellaneous		
Fermented beans (China)	476	(Yamaguchi and Ninomiya 2000)
Green tea	668	(Kato et al. 1974)
Soy sauce (Japan)	782	(Yamaguchi and Ninomiya 2000)
Wheat flour (whole)	3529	(Ghirri and Bignetti 2012)
Pine nut	6740	(Ghirri and Bignetti 2012))

Table 7.4 Concentration of 5′ ribonucleotides in foods.

Food	IMP (mg/100g)	GMP (mg/100g)	AMP(mg/100g)
Beef	70	4	8
Pork	200	2	9
Chicken	201	5	13
Squid	n.d.	n.d.	184
Tuna	286	n.d.	6
Snow crab	5	4	32
Scallop	n.d.	n.d.	172
Tomato	n.d.	n.d.	21
Green peas	n.d.	n.d.	2
Dried Shiitake mushroom	n.a.	150	n.a.
Dried porcini fungi	n.d.	10	n.a.
Oyster mushroom	n.d.	10	n.a.

Source: Adapted from Yamaguchi and Ninomiya (2000). IMP = inosine 5′monophosphate; GMP = guanosine 5′monophosphate; AMP = adenosine 5′monophosphate; n.d. = not detected; n.a. = not analysed

7.4.3 Synergy between umami compounds

One of the most interesting features of umami taste is the synergy observed between umami tastants, particularly between MSG and 5′ ribonucleotides (IMP and GMP). The detection threshold for MSG is approximately 0.12 g/L (0.7 mM) while, in the presence of IMP, this threshold is reduced more than 60-fold (0.0019 g/L; 0.011 mM). Typical umami tastants can be roughly divided into two classes. One class of tastants is related to L-glutamic acid, whereas the other includes 5′ ribonucleotide derivatives. Several compounds have a sensory threshold lower than that of MSG, such as L-tricholomic acid or L-ibotenic acid (Figure 7.5), which have been discovered in mushrooms. Hence, the umami tastes of L-tricholomic acid and L-ibotenic acid are 4 and 25 times more intense than the umami taste of MSG, respectively. Interestingly, IMP and GMP have an umami taste on their own. Functional cellular assays have demonstrated that human T1R1/T1R3 is activated by L-glutamate. These responses are enhanced by the presence of IMP (Li et al. 2002, Nelson et al. 2002, Nelson et al. 2001, Zhao et al. 2003), further demonstrating that T1R1/T1R3 is the major candidate receptor for umami detection in humans. Because IMP or GMP alone are not able to generate a receptor response (Li et al. 2002), it has been proposed that 5′ ribonucleotides may synergise with L-glutamate naturally present at low concentration in saliva. The molecular mechanism of synergy between L-glutamate and IMP at the molecular level has been recently elucidated by combining site-directed mutagenesis and cellular assay (Zhang et al. 2008). It should be noted that lactisole, a potent sweetness inhibitor that binds human

T1R3, is able to inhibit umami taste as well (Galindo-Cuspinera and Breslin 2006). Similarly, the sweetener cyclamate, which binds T1R3, does not activate T1R1/T1R3 by itself (cyclamate has no umami taste) but potentiates the receptor response to L-glutamate (Galindo-Cuspinera and Breslin 2006, Xu et al. 2004). All together, these data strongly suggest that the T1R1/T1R3 receptor is the main receptor for MSG detection.

7.4.4 Effects of umami substances on the taste of food

There is a widespread agreement that umami makes a variety of foods more palatable. Since ancient Roman times, ingredients rich in umami compounds such as fish or soy sauces have been used to improve foods in Western and Asian cooking. Interestingly, unlike sweet taste, the taste of pure MSG is generally not judged as pleasant. The palatability enhancement of umami may have been learned during breast-feeding because human milk contains a high concentration of free L-glutamate. Since the 1960s, L-glutamate and 5′ ribonucleotides have been introduced as taste enhancers into many industrial foods such as soups, noodles, potato chips and sauce bases. It should be noted that yeast extracts are used as taste enhancers in the food industry because they contain L-glutamate and natural 5′ribonucleotides such as GMP and AMP originating from yeast RNA degradation during yeast autolysis.

7.4.5 Kokumi taste and kokumi molecules

In addition to the five basic tastes, a taste sensation named Kokumi taste was proposed in the 1990′s (Ueda et al. 1990, Ueda et al. 1997). In Japanese, kokumi (コク味) literally means "mouthfulness and thickness". Kokumi compounds have been used for many years in traditional Japanese cuisine. Kokumi molecules exhibit taste-enhancing properties. They have no taste themselves, but they are able to enhance the sweet, salty and umami tastes. Three types of flavour sensation are attributed to kokumi molecules: mouthfulness, thickness and long-lasting savoury profile. The tripeptide glutathione (Glu-Cys-Gly, GSH (Figure 7.6)) is a typical kokumi taste compound. GSH is tasteless, but in the presence of umami compounds such as MSG or IMP, it reinforces the compounds′ taste and increases the long-lasting taste sensation (Ueda et al. 1990, Ueda et al. 1997). Isolated from water extracts of garlic and onions, the first kokumi compounds (Ueda et al. 1990) were sulfur-containing peptides derivatives such as γ-glutamyl-*trans*-(+)-S-propenyl-L-cysteine sulfoxide (γ-Glu-PeCSO (Figure 7.6)). From matured Gouda cheese and from the aqueous extract of edible beans, γ-L-glutamyl peptides such as γ-Glu-Val and γ-Glu-Leu were identified as key kokumi molecules. Sensory analyses indicated that these molecules enhanced mouthfulness and induced a longer-lasting savoury taste on the tongue (Dunkel et al. 2007, Toelstede et al. 2009). Recently, it has been demonstrated that the kokumi taste involves the calcium-sensing

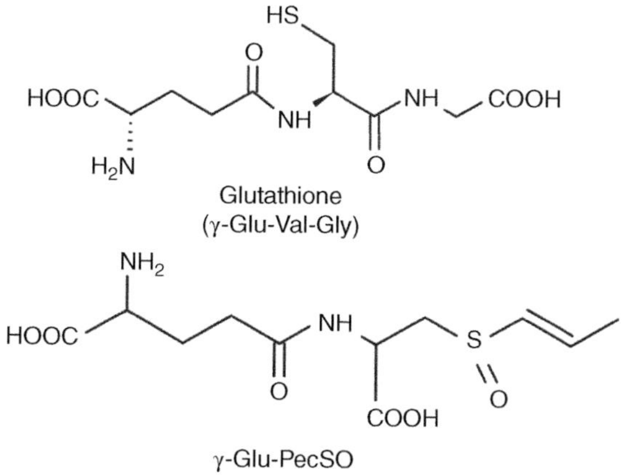

Figure 7.6 Chemical structure of some kokumi compounds.

receptor (CaSR), a close relative of the class C GPCRs T1R1, T1R2, and T1R3 taste receptors (Maruyama et al. 2012, Ohsu et al. 2010). Cellular assays have revealed that some CaSR agonists such as calcium, L-histidine, protamine or poly-lysine are indeed able to produce a kokumi taste in humans, further demonstrating the function of CaSR in taste perception. Interestingly, Oshu and co-workers, screening libraries of di- and tri-peptides for CaSR activities, identified the tripeptide γ-Glu-Val-Gly as a potent kokumi compound (Ohsu et al. 2010).

7.5 Salty and sour compounds

Salty taste is the taste elicited by Na^+ and other cations. Na^+ is the major cation in our body and is an essential compound of extracellular fluids. The need to detect cations such as Na^+ comes from the significant loss of Na^+ in sweat and urine and from our inability to store these ions. In food, Na^+ (NaCl) is the main salty compound, but it is known that other cations, such as Ca^{2+}, K^+, Li^+ and $NH_4{}^+$ have a salty taste. However, it should be noted that Na^+ is the only cation to produce a clean salt taste for humans. For instance, K^+ generates a bitter or a metallic taste, which considerably limits its application as a salt substitute in food. Interestingly, basic dipeptides, including ornithyl-taurine taste like salt and have been proposed as salt substitutes. Salt can produce two divergent behavioural responses in humans and animals depending on its concentration: sodium is attractive at low concentrations (<100 mM NaCl) and is aversive at higher concentrations. This aversion may be due to the harmful effect of sodium at high concentration, which may lead to health problems such as hypertension and kidney disease. In rodents, two mechanisms of salt detection have been described. One involved in

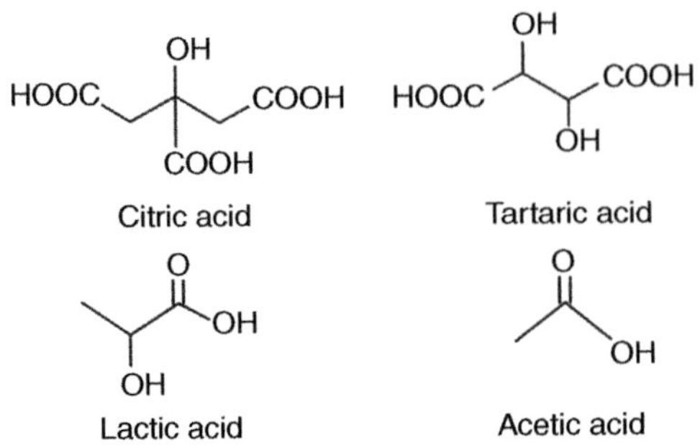

Figure 7.7 Chemical structure of some sour molecules.

the detection of low concentration of salt is sensitive to amiloride, whereas the other involved in the detection of high sodium concentration (>100 mM NaCl) is amiloride-insensitive. In humans, salt taste perception is amiloride-insensitive. Whereas the molecular detector of high salt concentrations remains unknown, knock-out studies conducted in mice have identified epithelial Na^+ channels (ENaCs) as the main low-salt receptor (Chandrashekar et al. 2010).

Sourness is the taste specialised in the detection of protons, that is, acidity. This taste may act as a warning signalling the presence of immoderate concentrations of acids (which may be dangerous for the body) and cautioning humans to avoid the consumption of unripe fruits or spoiled foods contaminated by acid-producing microorganisms. Numerous acids, including citric acid, tartaric acid, lactic acid and acetic acid (Figure 7.7) generate sourness. These acids are naturally present in sour foods such as citrus fruit (lemon, lime, orange and grapefruit), or are produced though fermentation. The sourness of acids is rated relatively to diluted hydrochloric acid. Whereas the intensity of sourness is proportional to the proton concentration, it has been observed that a low correlation exists between the intensity of sour taste and the pH of stimuli, especially for organic acids. This poor correlation could be attributed to different rates of acid-induced salivary secretions with buffering properties. However, pH of stimuli and perceived sourness have been correlated with strong inorganic acids such as hydrochloric and nitric acids (Ganzevles and Kroeze 1987). Interestingly, a low level of sour taste is attractive to humans and animals in some foods and drinks such as oranges, lemon juice and cola but is aversive in unripe fruits or spoiled food.

7.6 Development of new tasting compounds

The identification of new tastants has been traditionally achieved by the psychophysiological evaluation of purified fractions from existing flavour extracts or by the chemical derivation of known tasting molecules. The discovery of taste

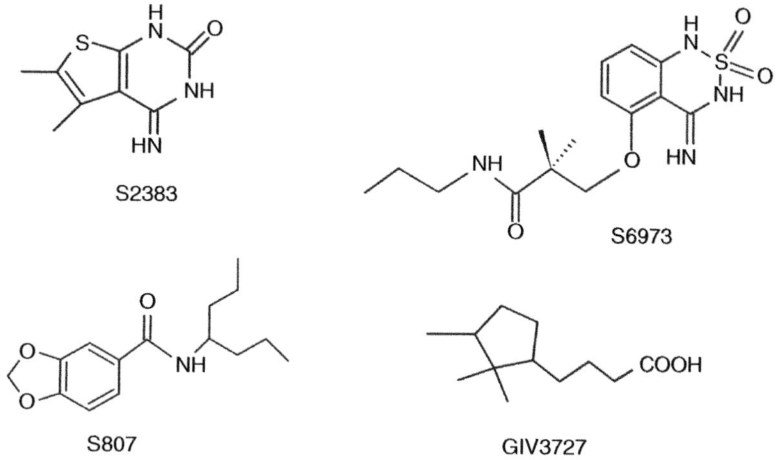

Figure 7.8 Chemical structure of new tasting molecules.

receptors has ushered a new era of novel screening technologies based on taste receptors. Recently, Senomyx Inc., a biotechnology company located in in San Diego (California, USA) that uses an innovative high-throughput screening system, has identified new sweet-tasting compounds and taste modulators (Servant et al. 2010). S2383 (Figure 7.8) is the first commercially available sweet taste modulator able to increase the perception of the artificial sweetener sucralose. This compound received GRAS status in 2008. More recently, Senomyx, combining sweet-taste receptor screening and optimisation, identified a new taste modulator named S6973 (Figure 7.8), which received GRAS status in 2009. Interestingly, S6973 molecule has no sweet taste of its own, but it enhances the sweetness of natural sugars, providing an opportunity for its use in a wide variety of applications, from beverages to foods.

Furthermore, there is an interest in searching for alternatives to MSG. Increasing the umami taste may help to reduce NaCl levels, leading to beneficial health effects. Senomyx recently discovered an umami molecule, named S807 (Figure 7.8) (Zhang et al. 2008). This new tastant generates a strong umami taste with a sensory threshold in the micromolar range, making this molecule one the most potent umami molecules. Interestingly, it has been shown that S807 binds to the transmembrane domain of the T1R1 subunit (Zhang et al. 2008), explaining its capability to synergise with L-glutamate.

There is also great interest in finding bitter blockers able to reduce the bitterness of foods, beverages and pharmaceuticals. In collaboration with Givaudan, the group of Professor Meyerhof, using a high-throughput screening approach, identified the GIV3727 (Figure 7.8) compound as a new bitter receptor blocker (Slack et al. 2010). This molecule was observed to effectively inhibit activation of the human bitter taste receptor TAS2R31 by the two artificial sweeteners acesulfame-K and saccharin. Interestingly, this compound was also proven able to inhibit five other bitter taste receptors, opening the opportunity for its use as a general bitter blocker. There is also great interest in finding salt taste enhancers or salt substitutes. It is likely that these screening approaches will help identify novel flavour ingredients allowing for the reduction of sodium content in foods without altering their organoleptic properties.

Acknowledgements

This work was supported by grants from the Institut National de la Recherche Agronomique and the Burgundy Regional Council (PARI 2010-2011-2012, AGRALE1 Project), the Institut Benjamin Delessert, as well as a fellowship from the Burgundy Regional Council (AL).

References

Aghera, N. and Udgaonkar, J. B. (2011) Heterologous expression, purification and characterization of heterodimeric monellin. *Protein Expression and Purification, 76*(2), pp. 248–253.

Asao, M., Iwamura, H., Akamatsu, M. and Fujita, T. (1987) Quantitative structure-activity relationships of the bitter thresholds of amino acids, peptides, and their derivatives. *Journal of Medicinal Chemistry, 30*(10), pp. 1873–1879.

Assadi-Porter, F. M., Aceti, D. J. and Markley, J. L. (2000) Sweetness determinant sites of brazzein, a small, heat-stable, sweet-tasting protein. *Archives of Biochemistry and Biophysics, 376*(2), pp. 259–265.

Audrieth, L. F. and Sveda, M. (1944) Preparation And Properties Of Some N-substituted Sulfamic Acids. *The Journal of Organic Chemistry, 09*(1), pp. 89–101.

Bartoshuk, L. M. (1979) Bitter taste of saccharin related to the genetic ability to taste the bitter substance 6-n-propylthiouracil. *Science, 205*(4409), pp. 934–935.

Bassoli, A., Drew, M. G., Merlini, L. and Morini, G. (2002) General pseudoreceptor model for sweet compounds: a semiquantitative prediction of binding affinity for sweet-tasting molecules. *Journal of Medicinal Chemistry, 45*(20), pp. 4402–4409.

Borrego, F., Canales, I. and Lindley, M. (1995) Neohesperidin dihydrochalcone: state of knowledge review. *Zeitschrift für Lebensmittel-Untersuchung und Forschung, 200*(1), pp. 32–37.

Bravo, L. (1998) Polyphenols: chemistry, dietary sources, metabolism, and nutritional significance. *Nutrition Reviews, 56*(11), pp. 317–333.

Breslin, P. A. and Beauchamp, G. K. (1997) Salt enhances flavour by suppressing bitterness. *Nature, 387*(6633), p. 563.

Broadbent, J. R., Barnes, M., Brennand, C., Strickland, M., Houck, K., Johnson, M. E. and Steele, J. L. (2002) Contribution of Lactococcus lactis cell envelope proteinase specificity to peptide accumulation and bitterness in reduced-fat Cheddar cheese. *Applied and Environmental Microbiology, 68*(4), pp. 1778–1785.

Bruhl, L., Matthaus, B., Fehling, E., Wiege, B., Lehmann, B., Luftmann, H., Bergander, K., Quiroga, K., Scheipers, A., Frank, O. and Hofmann, T. (2007) Identification of bitter off-taste compounds in the stored cold pressed linseed oil. *Journal of Agricultural and Food Chemistry, 55*(19), pp. 7864–7868.

Chandrashekar, J., Kuhn, C., Oka, Y., Yarmolinsky, D. A., Hummler, E., Ryba, N. J. and Zuker, C. S. (2010) The cells and peripheral representation of sodium taste in mice. *Nature, 464*(7286), pp. 297–301.

Chaudhari, N., Landin, A. M. and Roper, S. D. (2000) A metabotropic glutamate receptor variant functions as a taste receptor. *Nature Neurosciences, 3*(2), pp. 113–119.

Chaudhari, N., Yang, H., Lamp, C., Delay, E., Cartford, C., Than, T. and Roper, S. (1996) The taste of monosodium glutamate: membrane receptors in taste buds. *The Journal of Neuroscience, 16*(12), pp. 3817–3826.

Chen, Z., Cai, H., Lu, F. and Du, L. (2005) High-level expression of a synthetic gene encoding a sweet protein, monellin, in Escherichia coli. *Biotechnology Letters, 27*(22), pp. 1745–1749.

Clauss, K. and Jensen, H. (1973) Oxathiazinone dioxides—A new group of sweetening agents. *Angewandte Chemie International Edition in English, 12*(11), pp. 869–876.

Cohen, S. M. (1986) Saccharin: past, present, and future. *Journal of the American Dietetic Association, 86*(7), pp. 929–931.

Cowart, B. J., Yokomukai, Y. and Beauchamp, G. K. (1994) Bitter taste in aging: compound-specific decline in sensitivity. *Physiology & Behavior, 56*(6), pp. 1237–1241.

Curtis, R. I. (2009) Umami and the foods of classical antiquity. *The American Journal of Clinical Nutrition, 90*(3), pp. 712s–718s.

De Vos, A. M., Hatada, M., van der Wel, H., Krabbendam, H., Peerdeman, A. F. and Kim, S. H. (1985) Three-dimensional structure of thaumatin I, an intensely sweet protein. *Proceedings of the National Academy of Sciences of the United States of America, 82*(5), pp. 1406–1409.

Drewnowski, A. and Gomez-Carneros, C. (2000) Bitter taste, phytonutrients, and the consumer: a review. *The American Journal of Clinical Nutrition, 72*(6), pp. 1424–1435.

Drewnowski, A., Kristal, A. and Cohen, J. (2001) Genetic taste responses to 6-n-propylthiouracil among adults: a screening tool for epidemiological studies. *Chemical Senses, 26*(5), pp. 483–489.

Dunkel, A., Koster, J. and Hofmann, T. (2007) Molecular and sensory characterization of gamma-glutamyl peptides as key contributors to the kokumi taste of edible beans (Phaseolus vulgaris L.). *Journal of Agricultural and Food Chemistry, 55*(16), pp. 6712–6719.

Frank, O., Jezussek, M. and Hofmann, T. (2003) Sensory activity, chemical structure, and synthesis of Maillard generated bitter-tasting 1-oxo-2,3-dihydro-1H-indolizinium-6-olates. *Journal of Agricultural and Food Chemistry, 51*(9), pp. 2693–2699.

Fujimaru, T., Park, J. H. and Lim, J. (2012) Sensory characteristics and relative sweetness of tagatose and other sweeteners. *Journal of Food Science, 77*(9), pp. S323–S328.

Galindo-Cuspinera, V. and Breslin, P. A. (2006) The liaison of sweet and savory. *Chemical Senses, 31*(3), pp. 221–225.

Galindo-Cuspinera, V., Winnig, M., Bufe, B., Meyerhof, W. and Breslin, P. A. (2006) A TAS1R receptor-based explanation of sweet 'water-taste'. *Nature, 441*(7091), pp. 354–357.

Ganzevles, P. G. and Kroeze, J. H. (1987) Effects of adaptation and cross-adaptation to common ions on sourness intensity. *Physiology & Behavior, 40*(5), pp. 641–646.

Ghirri, A. and Bignetti, E. (2012) Occurrence and role of umami molecules in foods. *International Journal of Food Science and Nutrition, 63*(7), pp. 871–881.

Gibbs, B., Alli, I. and Mulligan, C. (1996) Sweet and Taste Modifying Proteins: A Review. *Nutrition Research 16*(9), pp. 1619–1630.

Grosch, W. and Laskawy, G. (1984) [Contribution of linoleic acid to the bitter taste of poppy seeds (Papaver somniferum)]. *Zeitschrift fur Lebensmittel-Untersuchung und -Forschung, 178*(4), pp. 257–259.

Grummer, J., Karalus, M., Zhang, K., Vickers, Z. and Schoenfuss, T. C. (2012) Manufacture of reduced-sodium Cheddar-style cheese with mineral salt replacers. *Journal of Dairy Science, 95*(6), pp. 2830–2839.

Guan, R. J., Zheng, J. M., Hu, Z. and Wang, D. C. (2000) Crystallization and preliminary X-ray analysis of the thermostable sweet protein mabinlin II. *Acta Crystallographica Section D-Biological Crystallography, 56*(Pt 7), pp. 918–919.

Hagedorn, S. and Kaphammer, B. (1994) Microbial biocatalysis in the generation of flavor and fragrance chemicals. *Annual Review of Microbiology, 48*, pp. 773–800.

Hecht, S. S. (2000) Inhibition of carcinogenesis by isothiocyanates. *Drug Metabolism Reviews, 32*(3-4), pp. 395–411.

Hobbs, J. R., Munger, S. D. and Conn, G. L. (2007) Monellin (MNEI) at 1.15 A resolution. *Acta Crystallographica Section F Structural Biology Communications, 63*(Pt 3), pp. 162–167.

Horne, J., Lawless, H. T., Speirs, W. and Sposato, D. (2002) Bitter taste of saccharin and acesulfame-K. *Chemical Senses, 27*(1), pp. 31–38.

Igeta, H., Tamura, Y., Nakaya, K., Nakamura, Y. and Kurihara, Y. (1991) Determination of disulfide array and subunit structure of taste-modifying protein, miraculin. *Biochimica Biophysica Acta, 1079*(3), pp. 303–307.

Inglett, G. (1984) Sweeteners: An Overall Perspective. in Stegink, F., (ed.) *Aspartame Physiology and Biochemistry*: Food Science (Marcel Dekker, Inc.). pp. 11–25.

Ito, K., Sugawara, T., Koizumi, A., Nakajima, K., Shimizu-Ibuka, A., Shiroishi, M., Asada, H., Yurugi-Kobayashi, T., Shimamura, T., Asakura, T., Misaka, T., Iwata, S., Kobayashi, T. and Abe, K. (2011) Cysteine-to-serine shuffling using a Saccharomyces cerevisiae expression system improves protein secretion: case of a nonglycosylated mutant of miraculin, a taste-modifying protein. *Biotechnology Letters, 33*(1), pp. 103–107.

Jiang, D. and Peterson, D. G. (2013) Identification of bitter compounds in whole wheat bread. *Food Chemistry, 141*(2), pp. 1345–1353.

Jiang, P., Cui, M., Zhao, B., Snyder, L. A., Benard, L. M., Osman, R., Max, M. and Margolskee, R. F. (2005) Identification of the cyclamate interaction site within the transmembrane domain of the human sweet taste receptor subunit T1R3. *Journal of Biological Chemistry, 280*(40), pp. 34296–34305.

Jiang, P., Ji, Q., Liu, Z., Snyder, L. A., Benard, L. M., Margolskee, R. F. and Max, M. (2004) The cysteine-rich region of T1R3 determines responses to intensely sweet proteins. *Journal of Biological Chemistry, 279*(43), pp. 45068–45075.

Kaneko, S., Kumazawa, K., Masuda, H., Henze, A. and Hofmann, T. (2006) Molecular and sensory studies on the umami taste of Japanese green tea. *Journal of Agricultural and Food Chemistry, 54*(7), pp. 2688–9264.

Kato, H., Rhue, M. R. and Nishimura, T. (1974) *Role of Free Amino Acids and Peptides in Food Taste*. ACS Symposium Series.

Kato, K., Yoshida, R., Kikuzaki, A., Hirai, T., Kuroda, H., Hiwasa-Tanase, K., Takane, K., Ezura, H. and Mizoguchi, T. (2010) Molecular breeding of tomato lines for mass production of miraculin in a plant factory. *Journal of Agricultural and Food Chemistry, 58*(17), pp. 9505–9510.

Kawai, M., Sekine-Hayakawa, Y., Okiyama, A. and Ninomiya, Y. (2012) Gustatory sensation of (L)- and (D)-amino acids in humans. *Amino acids, 43*(6), pp. 2349–2358.

Khan, N. A. and Besnard, P. (2009) Oro-sensory perception of dietary lipids: new insights into the fat taste transduction. *Biochimica Biophysica Acta, 1791*(3), pp. 149–155.

Kim, I. M. R., Kawamura, Y. and Lee, C. H. (2003) Isolation and Identification of Bitter Peptides of Tryptic Hydrolysate of Soybean 11S Glycinin by Reverse-phase High-performance Liquid Chromatography. *Journal of Food Science, 68*(8), pp. 2416–2422.

Kim, M. R., Yukio, K., Kim, K. M. and Lee, C. H. (2008) Tastes and structures of bitter peptide, asparagine-alanine-leucine-proline-glutamate, and its synthetic analogues. *Journal of Agricultural and Food Chemistry, 56*(14), pp. 5852–5858.

Kim, N.-C. and Kinghorn, A. D. (2002) Highly sweet compounds of plant origin. *Archives of Pharmacal Research, 25*(6), pp. 725–746.

Kim, S. H., Kang, C. H., Kim, R., Cho, J. M., Lee, Y. B. and Lee, T. K. (1989) Redesigning a sweet protein: increased stability and renaturability. *Protein Engineering, 2*(8), pp. 571–575.

Kim, U. K., Jorgenson, E., Coon, H., Leppert, M., Risch, N. and Drayna, D. (2003) Positional cloning of the human quantitative trait locus underlying taste sensitivity to phenylthiocarbamide. *Science, 299*(5610), pp. 1221–1225.

Kinghora, A. D., Soejarto, D. D. and Inglett, G. E. (1986) Sweetening agents of plant origin. *Critical Reviews in Plant Sciences, 4*(2), pp. 79–120.

Kohl, S., Behrens, M., Dunkel, A., Hofmann, T. and Meyerhof, W. (2013) Amino acids and peptides activate at least five members of the human bitter taste receptor family. *Journal of Agricultural and Food Chemistry, 61*(1), pp. 53–60.

Kohmura, M., Nio, N. and Ariyoshi, Y. (1990) Complete amino acid sequence of the sweet protein monellin. *Agricultural and Biological Chemistry, 54*(9), pp. 2219–2224.

Koizumi, A., Tsuchiya, A., Nakajima, K., Ito, K., Terada, T., Shimizu-Ibuka, A., Briand, L., Asakura, T., Misaka, T. and Abe, K. (2011) Human sweet taste receptor mediates acid-induced sweetness of miraculin. *Proceedings of the National Academy of Sciences of the United States of America, 108*(40), pp. 16819–16824.

Kondo, K., Miura, Y., Sone, H., Kobayashi, K. and Iijima, H. (1997) High-level expression of a sweet protein, monellin, in the food yeast Candida utilis. *Nature Biotechnology, 15*(5), pp. 453–457.

Kuhn, C., Bufe, B., Winnig, M., Hofmann, T., Frank, O., Behrens, M., Lewtschenko, T., Slack, J. P., Ward, C. D. and Meyerhof, W. (2004) Bitter taste receptors for saccharin and acesulfame K. *The Journal of Neuroscience, 24*(45), pp. 10260–10265.

Kuninaka, A. (1960) Study of taste ribonucleic acid derivatives. *Journal of Agricultural Chemistry Society Japan, 34*, pp. 487–492.

Kuninaka, A. (1964) The Nucleotides, a Rationale of Research on Flavor Potentiation. *Symposium on Flavor Potentiation*, pp. 4–9.

Kurihara, K. and Beidler, L. M. (1968) Taste-modifying protein from miracle fruit. *Science, 161*(3847), pp. 1241–1243.

Kurihara, K. and Beidler, L. M. (1969) Mechanism of the action of taste-modifying protein. *Nature, 222*(5199), pp. 1176–1179.

Li, D.-F., Jiang, P., Zhu, D.-Y., Hu, Y., Max, M. and Wang, D.-C. (2008) Crystal structure of Mabinlin II: A novel structural type of sweet proteins and the main structural basis for its sweetness. *Journal of Structural Biology, 162*(1), pp. 50–62.

Li, D. and Zhang, J. (2014) Diet shapes the evolution of the vertebrate bitter taste receptor gene repertoire. *Molecular Biology and Evolution, 31*(2), pp. 303–309.

Li, X., Staszewski, L., Xu, H., Durick, K., Zoller, M. and Adler, E. (2002) Human receptors for sweet and umami taste. *Proceedings of the National Academy of Sciences of the United States of America, 99*(7), pp. 4692–4966.

Lindley, M. (2012) Natural High-Potency Sweeteners. in *O'Donnell, K.* and Kearsley, M. W., (eds.) Sweeteners and Sugar Alternatives in Food Technology John Wiley & Sons, Ltd. pp. 185–212.

Maehashi, K. and Huang, L. (2009) Bitter peptides and bitter taste receptors. *Cellular and Molecular Life Sciences, 66*(10), pp. 1661–1671.

Maehashi, K. and Udaka, S. (1998) Sweetness of lysozymes. *Bioscience, Biotechnology and Biochemistry, 62*(3), pp. 605–606.

Maitrepierre, E., Sigoillot, M., Le Pessot, L. and Briand, L. (2012) Recombinant expression, in vitro refolding, and biophysical characterization of the N-terminal domain of T1R3 taste receptor. *Protein Expression and Purification, 83*(1), pp. 75–83.

Maruyama, Y., Pereira, E., Margolskee, R. F., Chaudhari, N. and Roper, S. D. (2006) Umami responses in mouse taste cells indicate more than one receptor. *The Journal of Neuroscience, 26*(8), pp. 2227–2234.

Maruyama, Y., Yasuda, R., Kuroda, M. and Eto, Y. (2012) Kokumi substances, enhancers of basic tastes, induce responses in calcium-sensing receptor expressing taste cells. *PLoS One, 7*(4), p. e34489.

Masuda, T., Ohta, K., Mikami, B. and Kitabatake, N. (2011) High-resolution structure of the recombinant sweet-tasting protein thaumatin I. *Acta Crystallographica Section F Structural Biology Communications, 67*(Pt 6), pp. 652–658.

Masuda, T., Ueno, Y. and Kitabatake, N. (2001) Sweetness and enzymatic activity of lysozyme. *Journal of Agricultural and Food Chemistry, 49*(10), pp. 4937–4941.

Matsuyama, T., Satoh, M., Nakata, R., Aoyama, T. and Inoue, H. (2009) Functional expression of miraculin, a taste-modifying protein in Escherichia coli. *Journal of Biochemistry, 145*(4), pp. 445–450.

Mazur, R. H., Schlatter, J. M. and Goldkamp, A. H. (1969) Structure-taste relationships of some dipeptides. *Journal of the American Chemical Society, 91*(10), pp. 2684–2691.

Meyerhof, W. (2005) Elucidation of mammalian bitter taste. *Reviews of Physiology, Biochemistry and Pharmacology, 154,* pp. 37–72.

Meyerhof, W., Batram, C., Kuhn, C., Brockhoff, A., Chudoba, E., Bufe, B., Appendino, G. and Behrens, M. (2010a) The molecular receptive ranges of human TAS2R bitter taste receptors. *Chemical Senses, 35*(2), pp. 157–170.

Ming, D. and Hellekant, G. (1994) Brazzein, a new high-potency thermostable sweet protein from Pentadiplandra brazzeana B. *FEBS Letters, 355*(1), pp. 106–108.

Morris, J. A., Martenson, R., Deibler, G. and Cagan, R. H. (1973) Characterization of monellin, a protein that tastes sweet. *Journal of Biological Chemistry, 248*(2), pp. 534–539.

Murzin, A. G. (1993) Sweet-tasting protein monellin is related to the cystatin family of thiol proteinase inhibitors. *Journal of Molecular Biology, 230*(2), pp. 689–694.

Nagata, K., Hongo, N., Kameda, Y., Yamamura, A., Sasaki, H., Lee, W. C., Ishikawa, K., Suzuki, E. and Tanokura, M. (2013) The structure of brazzein, a sweet-tasting protein from the wild African plant Pentadiplandra brazzeana. *Acta Crystallographica Section D-Biological Crystallography, 69*(Pt 4), pp. 642–647.

Nakajima, K., Asakura, T., Maruyama, J., Morita, Y., Oike, H., Shimizu-Ibuka, A., Misaka, T., Sorimachi, H., Arai, S., Kitamoto, K. and Abe, K. (2006) Extracellular production of neoculin, a sweet-tasting heterodimeric protein with taste-modifying activity, by Aspergillus oryzae. *Applied and Environmental Microbiology, 72*(5), pp. 3716–3723.

Narukawa, M., Toda, Y., Nakagita, T., Hayashi, Y. and Misaka, T. (2014) L-Theanine elicits umami taste via the T1R1 + T1R3 umami taste receptor. *Amino Acids, 46*(6), pp. 1583–1587.

Nelson, G., Chandrashekar, J., Hoon, M. A., Feng, L., Zhao, G., Ryba, N. J. and Zuker, C. S. (2002) An amino-acid taste receptor. *Nature, 416*(6877), pp. 199–202.

Nelson, G., Hoon, M. A., Chandrashekar, J., Zhang, Y., Ryba, N. J. and Zuker, C. S. (2001) Mammalian sweet taste receptors. *Cell, 106*(3), pp. 381–390.

Nie, Y., Vigues, S., Hobbs, J. R., Conn, G. L. and Munger, S. D. (2005) Distinct contributions of T1R2 and T1R3 taste receptor subunits to the detection of sweet stimuli. *Current Biology, 15*(21), pp. 1948–1952.

Nirasawa, S., Nishino, T., Katahira, M., Uesugi, S., Hu, Z. and Kurihara, Y. (1994) Structures of heat-stable and unstable homologues of the sweet protein mabinlin. The difference in the heat stability is due to replacement of a single amino acid residue. *European Journal of Biochemistry, 223*(3), pp. 989–995.

Noguchi, M., Arai, S., Yamashita, M., Kato, H. and Fujimaki, M. (1975) Isolation and identification of acidic oligopeptides occurring in a flavor potentiating fraction from a fish protein hydrolysate. *Journal of Agricultural and Food Chemistry, 23*(1), pp. 49–53.

Nolte, D. L., Russell Mason, J. and Lewis, S. L. (1994) Tolerance of bitter compounds by an herbivore,Cavia porcellus. *Journal of Chemical Ecology, 20*(2), pp. 303–308.

O'Donnell, K. (2007) Aspartame and Neotame. in *Sweeteners and Sugar Alternatives in Food Technology*: Blackwell Publishing Ltd. pp. 86–102.

Ogata, C., Hatada, M., Tomlinson, G., Shin, W. C. and Kim, S. H. (1987) Crystal structure of the intensely sweet protein monellin. *Nature, 328*(6132), pp. 739–742.

Ohsu, T., Amino, Y., Nagasaki, H., Yamanaka, T., Takeshita, S., Hatanaka, T., Maruyama, Y., Miyamura, N. and Eto, Y. (2010) Involvement of the calcium-sensing receptor in human taste perception. *Journal of Biological Chemistry, 285*(2), pp. 1016–1022.

Ohta, K., Masuda, T., Tani, F. and Kitabatake, N. (2011) Introduction of a negative charge at Arg82 in thaumatin abolished responses to human T1R2-T1R3 sweet receptors. *Biochemical and Biophysical Research Communications, 413*(1), pp. 41–45.

Pham, N. B., Schafer, H. and Wink, M. (2012) Production and secretion of recombinant thaumatin in tobacco hairy root cultures. *Biotechnology Journal, 7*(4), pp. 537–545.

Poirier, N., Roudnitzky, N., Brockhoff, A., Belloir, C., Maison, M., Thomas-Danguin, T., Meyerhof, W. and Briand, L. (2012) Efficient production and characterization of the sweet-tasting brazzein secreted by the yeast Pichia pastoris. *Journal of Agricultural and Food Chemistry, 60*(39), pp. 9807–9814.

Prodi, D. A., Drayna, D., Forabosco, P., Palmas, M. A., Maestrale, G. B., Piras, D., Pirastu, M. and Angius, A. (2004) Bitter taste study in a sardinian genetic isolate supports the association of phenylthiocarbamide sensitivity to the TAS2R38 bitter receptor gene. *Chemical Senses, 29*(8), pp. 697–702.

Refsgaard, H. H., Brockhoff, P. M. and Jensen, B. (2000) Free polyunsaturated fatty acids cause taste deterioration of salmon during frozen storage. *Journal of Agricultural and Food Chemistry, 48*(8), pp. 3280–3285.

Sakaguchi, K., Kibi, M. and Kuninaka, M. (1958) Japanese patent application. *SN 11586*.

San Gabriel, A., Nakamura, E., Uneyama, H. and Torii, K. (2009) Taste, visceral information and exocrine reflexes with glutamate through umami receptors. *The Journal of Medical Investigation, 56* Suppl, pp. 209–217.

San Gabriel, A. M., Maekawa, T., Uneyama, H., Yoshie, S. and Torii, K. (2007) mGluR1 in the fundic glands of rat stomach. *FEBS Letters, 581*(6), pp. 1119–1123.

Schiffman, S. S. and Gatlin, C. A. (1993) Sweeteners: State of knowledge review. *Neuroscience & Biobehavioral Reviews, 17*(3), pp. 313–345.

Schiffman, S. S., Sennewald, K. and Gagnon, J. (1981) Comparison of taste qualities and thresholds of D- and L-amino acids. *Physiology & Behavior, 27*(1), pp. 51–59.

Schlichtherle-Cerny, H. and Amado, R. (2002) Analysis of taste-active compounds in an enzymatic hydrolysate of deamidated wheat gluten. *Journal of Agricultural and Food Chemistry, 50*(6), pp. 1515–1522.

Scrimshaw, N. S. and Murray, E. B. (1988) The acceptability of milk and milk products in populations with a high prevalence of lactose intolerance. *The American Journal of Clinical Nutrition, 48*(4 Suppl), pp. 1079–1159.

Selvakumar, P., Gahloth, D., Tomar, P. P., Sharma, N. and Sharma, A. K. (2011) Molecular evolution of miraculin-like proteins in soybean Kunitz super-family. *Journal of Molecular Evolution, 73*(5-6), pp. 369–379.

Servant, G., Tachdjian, C., Tang, X. Q., Werner, S., Zhang, F., Li, X., Kamdar, P., Petrovic, G., Ditschun, T., Java, A., Brust, P., Brune, N., DuBois, G. E., Zoller, M. and Karanewsky, D. S. (2010) Positive allosteric modulators of the human sweet taste receptor enhance sweet taste. *Proceedings of the National Academy of Sciences of the United States of America, 107*(10), pp. 4746–4751.

Shi, P. and Zhang, J. (2009) Extraordinary diversity of chemosensory receptor gene repertoires among vertebrates. *Results and problems in cell differentiation, 47*, pp. 1–23.

Shirasuka, Y., Nakajima, K., Asakura, T., Yamashita, H., Yamamoto, A., Hata, S., Nagata, S., Abo, M., Sorimachi, H. and Abe, K. (2004) Neoculin as a new taste-modifying protein occurring in the fruit of Curculigo latifolia. *Bioscience, Biotechnology and Biochemistry, 68*(6), pp. 1403–1407.

Slack, J. P., Brockhoff, A., Batram, C., Menzel, S., Sonnabend, C., Born, S., Galindo, M. M., Kohl, S., Thalmann, S., Ostopovici-Halip, L., Simons, C. T., Ungureanu, I., Duineveld, K., Bologa, C. G., Behrens, M., Furrer, S., Oprea, T. I. and Meyerhof, W. (2010) Modulation of bitter taste perception by a small molecule hTAS2R antagonist. *Current Biology, 20*(12), pp. 1104–1109.

Sun, H. J., Kataoka, H., Yano, M. and Ezura, H. (2007) Genetically stable expression of functional miraculin, a new type of alternative sweetener, in transgenic tomato plants. *Plant Biotechnology Journal, 5*(6), pp. 768–777.

Suzuki, H., Kajimoto, Y. and Kumagai, H. (2002) Improvement of the bitter taste of amino acids through the transpeptidation reaction of bacterial gamma-glutamyltranspeptidase. *Journal of Agricultural and Food Chemistry, 50*(2), pp. 313–318.

Suzuki, M., Kurimoto, E., Nirasawa, S., Masuda, Y., Hori, K., Kurihara, Y., Shimba, N., Kawai, M., Suzuki, E. and Kato, K. (2004) Recombinant curculin heterodimer exhibits taste-modifying and sweet-tasting activities. *FEBS Letters, 573*(1-3), pp. 135–138.

Szwacka, M., Siedlecka, E., Zawirska-Wojtasiak, R., Wisniewski, L. and Malepszy, S. (2009) Expression pattern of the pre-prothaumatin II gene under the control of the CaMV 35S promoter in transgenic cucumber (Cucumis sativus L.) flower buds and fruits. *Journal of Applied Genetics, 50*(1), pp. 9–16.

Tancredi, T., Iijima, H., Saviano, G., Amodeo, P. and Temussi, P. A. (1992) Structural determination of the active site of a sweet protein. A 1H NMR investigation of pMNEI. *FEBS Letters, 310*(1), pp. 27–30.

Toda, Y., Nakagita, T., Hayakawa, T., Okada, S., Narukawa, M., Imai, H., Ishimaru, Y. and Misaka, T. (2013) Two distinct determinants of ligand specificity in T1R1/T1R3 (the umami taste receptor). *Journal of Biological Chemistry, 288*(52), pp. 36863–36877.

Toelstede, S., Dunkel, A. and Hofmann, T. (2009) A series of kokumi peptides impart the long-lasting mouthfulness of matured Gouda cheese. *Journal of Agricultural and Food Chemistry, 57*(4), pp. 1440–1448.

Tordoff, M. G. (2001) Calcium: taste, intake, and appetite. *Physiological Reviews, 81*(4), pp. 1567–1597.

Toyono, T., Seta, Y., Kataoka, S., Harada, H., Morotomi, T., Kawano, S., Shigemoto, R. and Toyoshima, K. (2002) Expression of the metabotropic glutamate receptor, mGluR4a, in the taste hairs of taste buds in rat gustatory papillae. *Archives of Histology and Cytology, 65*(1), pp. 91–96.

Ueda, Y., Sakaguchi, M., Hirayama, K., Miyajima, R., Kimizuka and A. (1990) *Characteristics flavor constituents in water extract of garlic*, Tokyo, JAPON: Agricultural Chemical Society of Japan.

Ueda, Y., Yonemitsu, M., Tsubuku, T., Sakaguchi, M. and Miyajima, R. (1997) Flavor characteristics of glutathione in raw and cooked foodstuffs. *Bioscience, Biotechnology and Biochemistry, 61*(12), pp. 1977–1980.

van den Oord, A. H. A. and van Wassenaar, P. D. (1997) Umami peptides: assessment of their alleged taste properties. *Zeitschrift für Lebensmitteluntersuchung und -Forschung A, 205*(2), pp. 125–130.

Van der Wel, H., Larson, G., Hladik, A., Hladik, C. M., Hellekant, G. and Glaser, D. (1989) Isolation and characterization of pentadin, the sweet principle of Pentadiplandra brazzeana Baillon. *Chemical Senses, 14*(1), pp. 75–79.

Winkel, C., de Klerk, A., Visser, J., de Rijke, E., Bakker, J., Koenig, T. and Renes, H. (2008) New developments in umami (enhancing) molecules. *Chemistry & Biodiversity, 5*(6), pp. 1195–1203.

Winnig, M., Bufe, B., Kratochwil, N. A., Slack, J. P. and Meyerhof, W. (2007) The binding site for neohesperidin dihydrochalcone at the human sweet taste receptor. *BMC Structural Biology*, 7, p. 66.

Wintjens, R., Viet, T. M., Mbosso, E. and Huet, J. (2011) Hypothesis/review: the structural basis of sweetness perception of sweet-tasting plant proteins can be deduced from sequence analysis. *Plant Science*, *181*(4), pp. 347–354.

Xu, H., Staszewski, L., Tang, H., Adler, E., Zoller, M. and Li, X. (2004) Different functional roles of T1R subunits in the heteromeric taste receptors. *Proceedings of the National Academy of Sciences of the United States of America*, *101*(39), pp. 14258–14263.

Yamaguchi, S. (1991) Basic properties of umami and effects on humans. *Physiology & Behavior*, *49*(5), pp. 833–841.

Yamaguchi, S. and Ninomiya, K. (2000) Umami and food palatability. *Journal of Nutrition*, *130*(4S Suppl), pp. 921s–926s.

Yamasaki, Y. and Maekawa, K. (1978) A Peptide With Delicious Taste. *Agricultural and Biological Chemistry*, *42*(9), pp. 1761–1765.

Yang, S. N., Wu, J. N., Liu, D. and Tung, C. S. (1998) Metabotropic glutamate receptors are involved in calcium-induced LTP of AMPA and NMDA receptor-mediated responses in the rat hippocampus. *Brain Research Bulletin*, *46*(6), pp. 505–512.

Zhang, F., Klebansky, B., Fine, R. M., Xu, H., Pronin, A., Liu, H., Tachdjian, C. and Li, X. (2008) Molecular mechanism for the umami taste synergism. *Proceedings of the National Academy of Sciences of the United States of America*, *105*(52), pp. 20930–20934.

Zhao, G. Q., Zhang, Y., Hoon, M. A., Chandrashekar, J., Erlenbach, I., Ryba, N. J. and Zuker, C. S. (2003) The receptors for mammalian sweet and umami taste. *Cell*, *115*(3), pp. 255–266.

CHAPTER 8

Sensory characterization of compounds with a trigeminal effect for taste modulation purposes

Isabelle Cayeux & Christian Starkenmann
Firmenich Sa Research & Development Department, Geneva, Switzerland

8.1 Introduction

The food industry is on a continual quest to deliver new and healthier products to consumers. Creating taste enhancers and ingredient substitutes for high sugar, high salt, and high fat foods has started decades ago. Yet consumers often do not like these products because of their poor flavour. This creates a need within the food industry to innovate by testing natural sources for novel compounds that show interesting trigeminal or taste effects. Thus, the challenge for flavour companies today is to provide "natural" flavouring solutions to reduce the use of sugar, salt, and fat in food while maintaining the same flavour perception. How can we find these new molecules and charactesrize their potential interesting sensory effects?

In humans, the olfactory and gustatory systems are the principal chemosensory systems, but the trigeminal somatosensory system also plays a fundamental role in chemosensation and the overall flavour perception of food and beverages (Viana 2011). Although flavour perception arises from the integration of multiple sensory inputs, it is possible for consumers to distinguish between them, especially when their attention is drawn to specific sensory characteristics. Small and Prescott (2005) proposed that flavour represents a functional sensory system for feeding with inputs from somatosensation, gustation, and olfaction, the key to the system being the meaning of the sensation rather than the particular organ of transduction. Considering flavour perception as a functional integrated sensory system could allow flavourists to compensate for the taste of reduced sugar in a food product, for example, by using another sensory input such as trigeminal perception.

Flavour: From food to perception, First Edition.
Edited by Elisabeth Guichard, Christian Salles, Martine Morzel, and Anne-Marie Le Bon.

8.2 Trigeminal perception

Trigeminal perception is particularly rich in sensory qualities (Table 8.1), but is also characterized by the evolution of sensations over time: delay, sensitization, desensitization (Lawless 1984, Green 1989, Green and Lawless 1991, Prescott 1999, Green 2001). Trigeminal sensation is perceived through the sensory endings of the trigeminal (V cranial) nerve as described in Chapter 5. These endings can be activated by physical stimuli (mechanical forces and temperature) and by chemical agents (chemesthesis), and evoke sensations of touch, temperature and pain (Viana 2011). Oral chemesthesis explains the pungent or sharp feel of many different foods and spices such as chili pepper, horseradish, wasabi roots and Szechuan pepper; the coolness of peppermint; the tingle of carbonated drinks; and the irritation produced by substances such as garlic extract (Viana 2011). Most of the chemical stimuli can produce more than one sensation and can be described more precisely with specific terms (Table 8.1). The concentration of the compound determines to some extent the perceived sensation. For example, menthol is pure cooling at low concentration while burning is also perceived at high concentration. In recent years, tremendous progress has been made in our understanding of how different chemicals are detected by trigeminal endings as transient receptor potential (TRP) channels.

8.3 Interactions between taste and trigeminal perception

The perceptual integration between olfaction and taste and the cognitive mechanisms involved have been widely studied and are addressed in Chapter 14. The trigeminal sensation is also often associated to the sensation of smell as it is described in Chapter 5. But the interactions between taste and trigeminal perception are important and are somehow less described (Delwiche 2004, Trotier et al. 2012). Spence (2014) proposed that trigeminal inputs could influence flavour perception by tactile, thermal, painful and/or kinesthetic effects and that this effect has been underestimated for many years. Many psychological studies have emphasized the suppressive or masking effect of chemesthetic compounds on taste perception, but not for all tastes (Lawless and Stevens 1984, Cowart 1987, Simons et al. 2002, Delwiche 2004). Koskinen et al. (2003) showed that menthol decreased the sweetness and increased the sourness of lemon-flavoured yoghurt. Similarly, capsaicin reduced the sweetness in food (soup and flavoured mixes), but had no effect on saltiness or sourness (Prescott and Stevenson 1995). One study showed that panellists rated a sodium chloride (NaCl) solution as being less salty when irritation was reduced by desensitization (Gilmore and Green, 1993), suggesting that the saltiness rating

Table 8.1 Sensory qualities associated to different types of trigeminal stimuli. The dark grey indicates the main sensation at usual concentrations in food products.

	Menthol	Shanshool	Spinanthol	Eugenol	Capsaicin	Piperine	Allyl isothiocyanate	Gingerol	Shoagol	CO_2	Ethanol	Cinnamaldehyde	Acids, polyphenols	Cuminaldehyde	Temperature	Product viscosity	Product body	Softness of surfaces	Mouth after feel
Cooling	■																		
Numbing	▒	■	■	■				▒	▒	▒									
Pungent, sharp					■	■	▒			▒	▒	▒		▒					
Burning	▒				■	■	▒	■	■		■	▒		■					
Tingling, tickling	▒				▒	▒		■	■	■	▒	▒		■					
Foamy, bubbly										■									
Stinging, biting	▒				▒	▒	■			▒	▒	▒							
Itching						▒													
Warmth												■							
Astringency, puckering													■						
Hot, Cold															■				
Thick, thin																■			
Watery, heavy, light																▒	■		
Smooth, pulpy																▒	▒	■	
Clean, lingering, mouth coating																			■
Food Product	Peppermint, spearmint	Szechuan pepper	Jambu fruit	Cloves, cinnamon	Chili pepper	Pepper	Wasabi, mustard, horseradish	Ginger				Cinnamon	Citrus, beverage	Cumin	All	Dairy product, beverage, soup, sauce			
	Chemesthesis														T°c	Tactile			

(■ = dark grey; ▒ = light grey)

Source: Adapted from Viana (2011) and Lawless and Heymann (2010)

could be affected by irritation intensity. Some chemicals such as spilanthol are claimed to enhance saltiness (Miyazawa et al. 2006).

Concerning the cooling perception, some authors (Le Calvé et al. 2010, Petit et al. 2007, Saint-Eve et al. 2010) showed that the trigeminal sensations elicited by carbonation or cooling agents could impact the flavour perception of model beverages, depending on the congruency between trigeminal sensation and flavour. Starkenmann et al. (2011a) showed that trigeminal sensations could also induce new sensations, which help to maintain global taste and flavour perception of products that have reduced sugar, salt, or fat.

Green (2001) also described different types of interactions between taste and texture in the mouth. Sediva et al. (2004) specified how viscosity impacts the perception of sourness. Other investigators demonstrated the importance of temporal synchrony in perceptual interaction mechanisms among olfaction, taste, and texture (Labbe et al. 2008). Recently, Slocombe et al. (2015) showed that the evaluation of taste components can also be influenced by the tactile quality of the food. They found that the roughness or smoothness of the foodstuff itself significantly influenced perception: food was rated as significantly more sour if it had a rough (versus smooth) surface.

8.4 Taste compounds with trigeminal sensory qualities

Many studies showed that conventional taste compounds such as salts and acids can elicit oral chemesthesis (Stevens and Lawless 1986, Gilmore and Green 1993, Dessirier et al. 2001). At high concentration, NaCl and KCl were found to have properties of both saltiness and irritation (Green and Gelhard 1989). The irritation property of NaCl at high concentration can enhance the burning sensation of capsaicin (Prescott et al. 1984). The irritation elicited by NaCl and capsaicin may be mediated by similar capsaicin-sensitive trigeminal nociceptors in the oral epithelium, or by different cellular transduction mechanisms (Dessirier et al. 2001).

Again, this type of modulation may be peripheral (Simons and Noble 2003) and/or cognitive (Cerf-Ducastel et al. 2001) and has the potential to be powerful. Perceptions of the quality and the time course of these trigeminal compounds, however, create difficulties for practical use in flavour optimization.

Flavour companies are interested in evaluating products that stimulate trigeminal perception in the oral cavity and can potentially modulate taste (e.g., increased sweetness and saltiness), which improves the overall flavour. The sensory methods presented here allow us to screen and to assess the performances of these products.

8.4.1 Scope of the sensory studies

In this review, we describe the sensory strategy and methods that we use with human panels to discover and to characterize uncommon trigeminal-active compounds. The use of human panels allows us to assess the whole trigeminal performance of novel compounds and their potential taste modulation effects.

Following the isolation and characterization of novel compounds that potentially elicit an interesting mouth effect, we perform detailed sensory tests with a trained panel to evaluate the parameters of this effect (Starkenmann et al. 2011a). The sensory aspects currently assessed are determined in two steps. The first step allows identifying types of trigeminal and modulation taste effects, for example, cooling sensation, enhanced saltiness in an NaCl solution or masking of bitterness and metallic off-note effects in a KCl solution, by comparing model solutions with or without the novel compound. The second step focuses on the sensory characterization of the novel compound by using both quantitative measures, such as detection thresholds, dose-response curves or intensity profile over time, and qualitative measures to define the most appropriate terms to describe the products. The sensory characterization of these new compounds allows comparing their overall performance with that of benchmark products that have previously been characterized.

Our approach, which involves sourcing, screening and sensory characterization steps, is outlined in Figure 8.1. Here, we focus on screening and characterization methods that are based on sensory processes and human panels.

As the sensory process with human panels is a low throughput screening method compared with that of a cell-based assay, sourcing must be selective. Product selection is based on knowledge gained from the botanical literature,

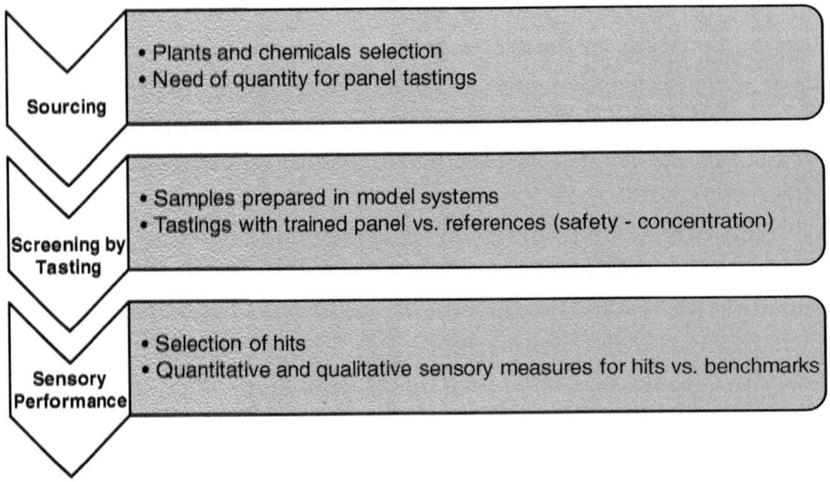

Figure 8.1 Approach to screening and characterization of samples with potential taste modulation and trigeminal effects.

ethnobotanists, chefs, recipes and common plant uses. These products must be delivered at a given quantity to allow panel tastings at an acceptable concentration.

8.4.2 Screening by tasting

The screening of compounds having trigeminal effects and/or potential taste modulation is achieved with a trained panel of 15 subjects. Tastings are organized on a weekly basis using water and water-based model systems (Figure 8.2). Panellists taste the sample from a black cup with or without using a nose clip in order to focus on taste and avoid the influence of vision (since some samples are coloured). All samples are first tasted in water and then in the water-based model system.

For sample tasting in water, the sensory task of the panellist is to describe the sample by selecting appropriate attributes (Table 8.2). For the sample evaluation in model systems, the subject has to first taste the model system alone as a reference and to indicate the perceived intensity of the taste qualities on a linear scale from not intense (0) to very intense (10). For example, a solution of sucrose will be evaluated for its sweetness only, but a solution of stevia extract (SG95) will be scored for both sweetness and liquorice taste (the attribute defined with the panel for this specific effect of stevia-like products). The panellist may also select to add comments to the same list as proposed for the evaluation in water.

The treatment of data for the sample evaluation in water is simply the number of attributes cited by the panellists. For the sample evaluated in different model systems, a student t-test (one-tailed paired test) is performed to compare the systems with the sample and those without it (reference).

The concentrations of the compound used to set up the different model systems (Table 8.3) are defined by using dose-responses curves (perceived intensity as a function of the concentration) in order to propose a moderate taste quality as a reference.

The example presented in Figure 8.3 concerns an extract of alligator pepper from Africa tasted at 400 ppm by 15 trained panellists. The left column provides (in brackets) the number of subjects who chose the attribute. The extract has a clear hot/pungent trigeminal effect, which is similar to the effects produced by capsaicin or piperine. In the right column, the differences in scores between the reference and the sample are indicated for each water-based model system. The number of asterisks corresponds to the significance level of the t-test (* = 95%, ** = 99%, and *** = 99.9% significant difference from the reference). Interestingly, this extract has a significant salt-enhancing effect.

This example demonstrates that some products having trigeminal properties could have an interesting modulation effect on taste. This was also the case for some products that induced a tactile sensation in the mouth and increased sweetness.

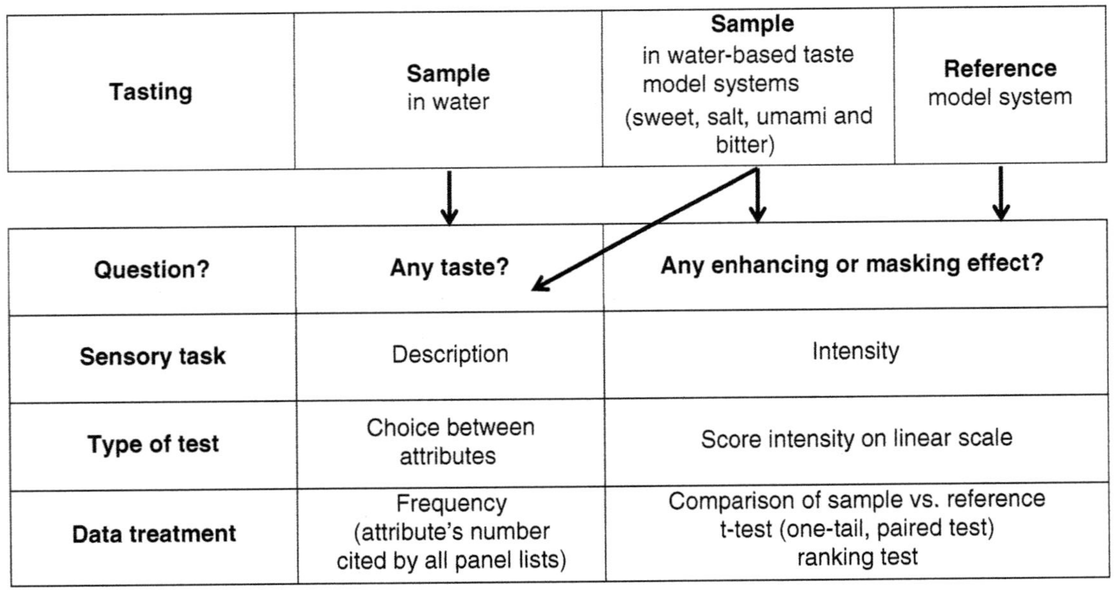

Figure 8.2 Sample evaluation procedure in water and water-based model systems.

Table 8.2 Terms used in the sensory characterization of taste compounds with trigeminal sensory qualities.

Without nose clip			
Trigeminal qualities		*Taste qualities*	
Cooling	Astringent	Bitter	"Liquorice" Taste
Burning	Fatty	Salty	Metallic
Pungent	Mouth coating	Sour	
Tingling	Thick	Sweet	
Salivating		Umami	
Without nose clip			
Smell qualities			
Earthy			
Fruity			
Green			
Herbal			
Mushroom			
Nutty			
Tea			
Woody			
Others			

Table 8.3 Examples of water-based model systems to screen trigeminal and taste modulation effects.

Taste	Taste compound	Concentration (%)
Water	Mineral water henniez	Mineralization 581 mg/L
Sweet	Sucrose	4
Sweet	Stevia extract	0.02
Bitter	Paracetamol	0.0875
	Quinine HCl	0.000675
Salty	NaCl	0.25
Salty/Bitter	KCl	0.25
Umami	MSG	0.05

The next steps are to identify the chemical(s) that elicits these interesting mouth effects as cooling, burning, tingling, tactile and/or taste modulation and then to characterize the sensory effects in comparison with benchmark products.

8.4.3 Sensory characterization of new trigeminal compounds

In this part, we present examples of compounds having interesting trigeminal effects, which can be used for taste modulation and more generally to enhance the global flavour of food products or beverages. For some of these compounds, their sensory characteristics are compared to benchmark products.

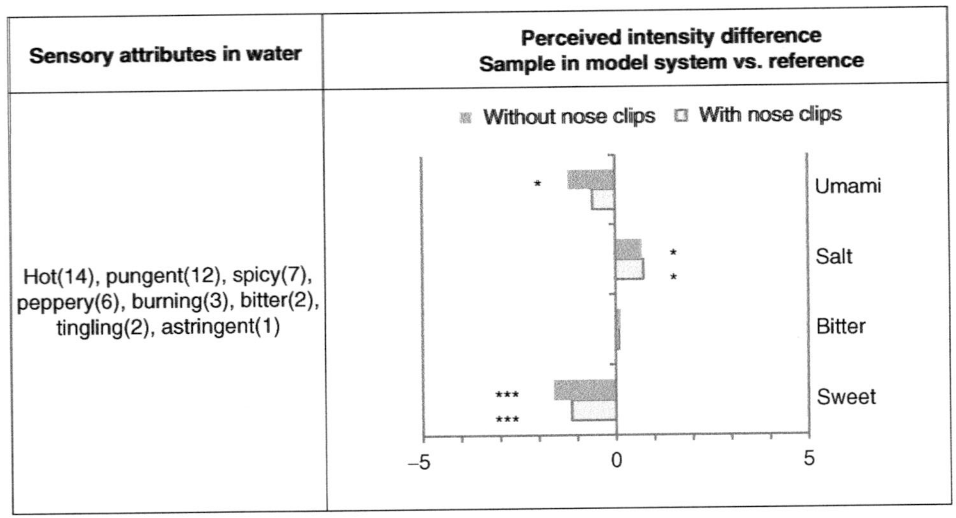

Figure 8.3 Sensory attributes and taste modulation properties of alligator pepper extract tasted in water-based model systems (Test performed by 15 trained panellists).

8.4.3.1 Mouthfeel effect of (R)-strombine from dried scallop

The mouthfeel of seafood is quite different from that of chicken broth, for example, which is fattier, saltier, and more umami. We investigated the taste-active compounds in dried scallop and discovered that one fraction was clearly described as sweet with fullness in the mouth. (R)-strombine and glycine contribute significantly to this specific taste (Starkenmann et al. 2009).

To assess the sensory performance of (R)-strombine and glycine, trained panellists received a 20 mL water solution. They were asked to take the entire sample in their mouth. They spat out the sample after 5 seconds. The sensory characterization was combined with threshold measurements (ASTM E679 Ascending Concentration Series Method of Limits) in one session per product, with choices of sensory attributes. The panellists were asked to choose an attribute from among sweet, salty, bitter, umami, acid, mouthfeel (Figure 8.4). (R)-strombine was perceived as salty, umami, and contributing to mouthfeel by most panellists. The multisensory aspect of this compound is interesting, especially for savoury applications. The taste component of glycine is unimodal and clearly sweet.

8.4.3.2 Hot/Pungent effect of polygodial from laksa *persicaria odorata* (lour.) sojak

Laksa *Persicaria odorata* (Lour.) Sojak, a coriander plant commonly used in Thai cuisine, displays a pungency that is quite different from the pungency or tingling of known compounds such as piperine, capsaicin, gingerol or even spilanthol (Rentmeister-Bryant and Green 1997, Szallasi and Blumberg 1999, Calixto et al. 2005). It resembles the pungency of galangal acetate, which occurs in *Alpinia galangal* (Zingiberaceae family). Extraction and analysis of the leaves of laksa led to the discovery of the known polygodial, a compound also present in mountain pepper *Tasmannia lanceolata* (Poir.) from the family Winteraceae (Starkenmann et al. 2006). Polygodial is a good example of the potential uses of pungent molecules in different applications. It can be used for example in wasabi paste to enhance the pungent effect of allyl isothiocyanate without the associated rubbery olfactive note, to enhance the freshness of toothpaste (Hisashi et al. 1996) or even to improve the sensory acceptability of artificial sweeteners (Kang et al. 1999).

The sensory characterization of polygodial, piperine, and capsaicin was performed by a panel of 25 subjects. Threshold measurements were achieved with the 3-alternative forced choice method (based on the ASTM E679 Ascending Concentration Series Method of Limits) in one session per product. According to average threshold values (Figure 8.5), polygodial is twice as strong as piperine, but 40 times weaker than capsaicin. The slightly wider distribution range for the perception of polygodial than for the other two compounds means that there are more differences between subjects for this compound. This characterization allows determining relevant concentrations of this compound for dedicated applications.

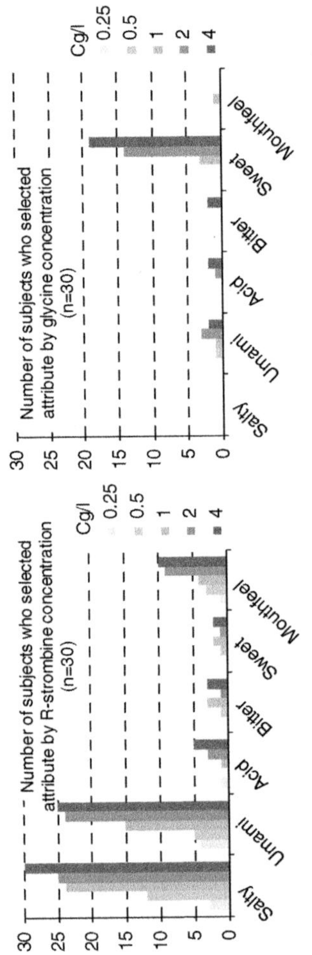

Figure 8.4 Sensory characterization of (R)-strombine and glycine at various concentrations: Number of subjects (Out of 30 panellists) who selected the proposed sensory attributes.

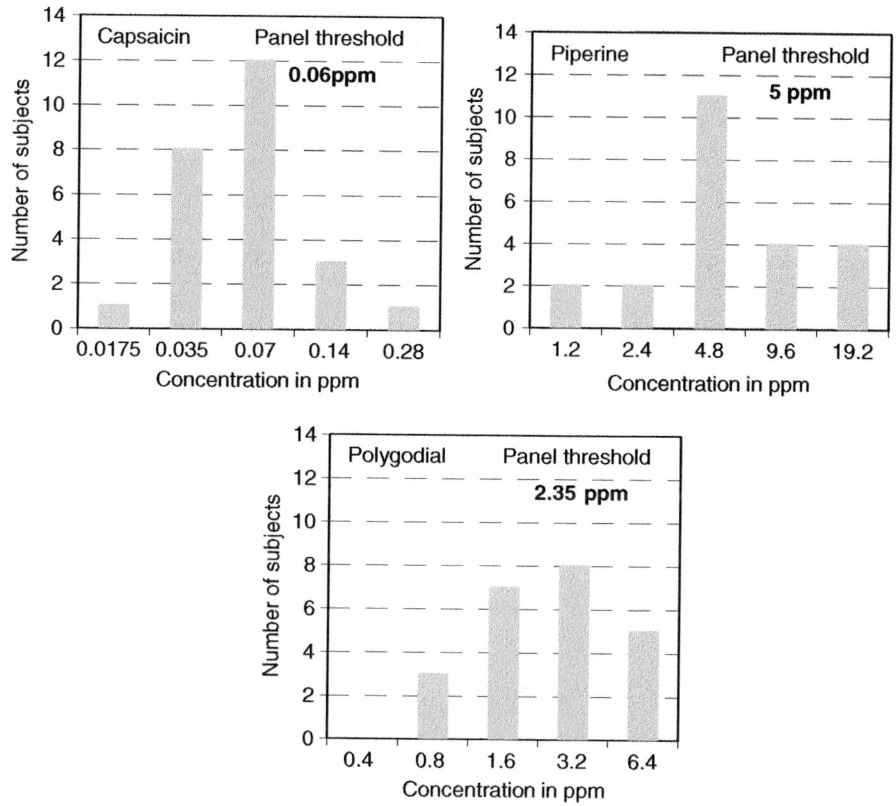

Figure 8.5 Detection thresholds for polygodial in comparison with those of capsaicin and piperine.

8.4.3.3 Cooling effect of dihydroumbellulol from the leaves of *umbellularia california* nutt

When the leaves of *Umbellularia californica* Nutt. are crushed, smelled or tasted, a clear trigeminal effect is perceived. As a result of this observation, this plant was studied and umbellulone isolated as a major constituent. Dihydroumbellulol was obtained from the umbellulone (Starkenmann et al. 2011b). The cooling strength of this compound was then evaluated by a trained panel. The objective of the sensory evaluations was first to confirm the cooling effect of dihydroumbellulol, and then to compare its performance with that of (-)-menthol.

Thirty trained panellists evaluated the perceived cooling intensity of the two samples, dihydroumbellulol and (-)-menthol, over time (one sample per session, at 50 mg/L). Each subject received a 30 mL sample and was asked to place it in the mouth and to start a timer at the same time. Subjects spat out the sample after 5 seconds. Panellists evaluated the time at which they started to perceive the cooling sensation (Tbegin), the maximum perceived cooling intensity and its corresponding time (Imax and Tmax), the time when the

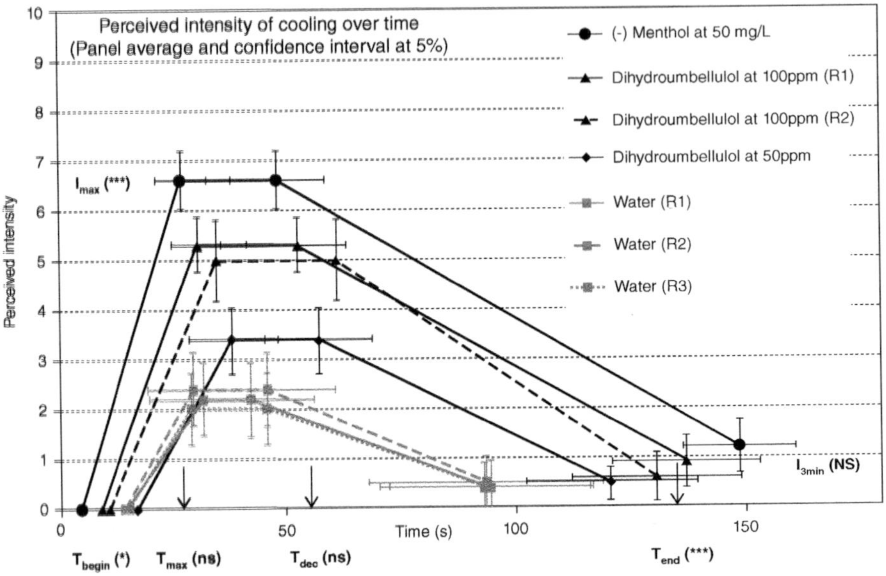

Figure 8.6 Intensity of cooling effect over time for menthol, dihydroumbellulol and water.

cooling sensation started to decrease (Tdec), and finally the perceived cooling intensity 3 min after they first tasted the sample (Iend) or the time when the cooling sensation disappeared, if it was less than 3 min (Tend). Each evaluation was rated on a linear scale from not perceived (=0) to very intense (=10). We performed student t-tests (two-tailed, paired) on each parameter (Tbegin, Imax, Tmax, Tdec, Iend, and Tend) to compare data obtained from the two cooling compounds. The probabilities obtained for each of these tests indicate whether the compounds have been perceived as significantly different or not (significance was defined as $p < 0.05$) for the parameter under consideration. This sensory evaluation (Figure 8.6) confirmed that dihydroumbellulol has a similar cooling profile over time to that of (-)-menthol, but that overall, it is less powerful. The Imax values showed that dihydroumbellulol is weaker than (-)-menthol (significant difference for Imax) and its persistent trigeminal effect is shorter, with a more pronounced delay. Although this new natural cooling compound is less powerful than menthol, it has a similar cooling profile without the minty note, which makes it interesting for flavouring solutions especially citrus ones.

8.5 Conclusion

We have demonstrated that sensory screening and characterization of potential plant extracts/fractions or chemicals by a trained panel is a key analytical step in the discovery and selection of molecules that contribute to the taste of foods. This is a robust and powerful method for identifying interesting trigeminal and taste modulators. Sensory characterization of the hits can be compared with those of benchmark products for further development of new taste modulators, or for use in synergy with other compounds or other applications.

Naturally derived compounds that elicit an interesting taste have considerable potential as innovative flavours. However, with the large volume of cheap commodity compounds such as monosodium glutamate (MSG) being the market benchmark, the principal hurdle is the cost of using these types of compounds.

In terms of launching commercial products, more potential can be seen in molecules that produce trigeminal effects. Mouth refreshing and fat-, salt-, and sweet-enhancing products will always be molecules of interest. Our recent results have clearly shown that it is still possible to discover new molecules in plants that elicit trigeminal effects, which motivates us to continue the search, despite the fact that such molecules have been the subject of intense research over many years.

Furthermore, this sensory knowledge is a useful tool for flavourists in order to magnify the taste of existing flavours, to optimize the flavour of healthy products, and to develop new products. It is also a powerful motivating tool for our chemists in developing new compounds according to structure-activity relationships.

Acknowledgements

The authors are grateful for their precious advices and their theoretical and technical competences to Nadine Gaudreau, Bénédicte Le Calvé, Christian Margot and Christine Saint-Léger, from Human Perception and Bioresponses Department of the Research and Development Division of Firmenich, SA.

References

Calixto, J.B., Kassuya, C.A.L., Andre, E., and Ferreira, J. (2005) Contribution of natural products to the discovery of the transient receptor potential (TRP) channels family and their functions. *Pharmacology & Therapeutics, 106*, pp. 179–208.

Cowart, B.J. (1987) Oral chemical irritation: does it reduce perceived taste intensity? *Chemical Senses, 12*, pp. 467–479.

Delwiche, J. F. (2004) The impact of perceptual interactions on perceived flavour. *Food Quality and Preference, 15*, pp. 137–146.

Dessirier, J.-M., O'Mahony, M., Iodi-Carstens, M., Yao, E., and Carstens, E. (2001) Oral irritation by sodium chloride: sensitization, self-desensitization, and cross-sensitization to capsaicin. *Physiology & Behavior, 72*, pp. 317–324.

Cerf-Ducastel B., Van de Moortele P.F., MacLeod P., Le Bihan D., Faurion A. (2001) Interaction of gustatory and lingual somatosensory perception at the cortical level in humans: an fMRI study. *Chemical Senses, 26*, pp. 371–383.

Gilmore, M.M. and Green, B.G. (1993) Sensory irritation and taste produced by NaCl and citric acid: effects of capsaicin desensitization. *Chemical Senses, 18,* pp. 257–272.

Green, B.G. and Gelhard, B. (1989) Salt as an oral irritant. *Chemical Senses, 14,* pp. 259–271.

Green, B.G. (1989) Capsaicin sensitization and desensitization on the tongue produced by brief exposures to a low concentration. *Neuroscience Letters, 107,* pp. 173–178.

Green, B. G. and Lawless H.T. (1991) The psychophysics of somatosensory chemoreceptors in the nose and in the mouth. in Doty et al. (eds) *Smell and Taste in Health and Disease,* New-York: Haven press. pp. 235–253.

Green, B. G. (2001) Psychophysical measurement of oral chemesthesis. in Simon S. A. and Nicolelis M. A. (Eds) *Methods in chemosensory research,* Boca Raton, FL: CRC Press. pp. 3–20.

Hisashi, I., Suzuki, A.S., and Shigeki, I. (1996) *U.S. Patent 5,523,105.*

Kang, L.L., Zyzak, T., and Nakasu , (1999) U.S. Patent 5,948,460.

Koskinen S., Kälviäinen N., Tuorila H. (2003) Perception of chemosensory stimuli and related responses to flavoured yogurts in the young and elderly. *Food Quality and Preference, 14*(8), pp. 623–635.

Labbe, D., Gilbert, F., and Martin, N. (2008) Impact of olfaction on taste, trigeminal and texture perceptions. *Chemosensory Perception, 4,* pp. 217–226.

Lawless, H.T. (1984) Oral chemical irritation, psychophysical properties. *Chemical Senses, 9,* pp. 143–155.

Lawless, H.T. and Stevens, D.A. (1984) Effects of oral chemical irritation on taste. *Physiology & Behavior, 32,* pp. 995–998.

Lawless H.T. and Heymann H. (2010) Chemesthesis. in Menary, R.C. (ed.) *Sensory evaluation of food principles and practices,* Second edition, New-York: Springer. pp. 39–46.

Le Calvé B., Goichon H., and Cayeux I. (2010) CO_2 perception and its influence on flavour. In Blank I., Wüst M., and Yeretzian C., (eds) *Expression of Multidisciplinary Flavour Science,* Winterthur, CH: ZHAW. pp. 55–58.

Miyazawa T., Matsuda T., Muranishi S., Miyake K. (2006) Salty taste enhancers containing spinanthol, flavouring materials containing them, foods and beverages containing the flavouring materials, and taste-enhancing methods. JP 2006 296,357, Japan: Ogawa and Co. Ltd.

Petit, C.E.F., Hollowood, T.A., Wulfert, F., and Hort, J. (2007) Colour-coolant-aroma interactions and the impact of congruency and exposure on flavour perception. *Food Quality & Preference, 18*(6), pp. 880–889.

Prescott, J., Allen, S., and Stephens, L. (1984) Interactions between oral chemical irritation, taste and temperature. *Chemical Senses, 18*(4), pp. 389–404.

Prescott, J. and Stevenson, R.J. (1995) Effects of oral chemical irritation on tastes and flavours in frequent and infrequent users of chili. *Physiology & Behavior, 58,* pp. 1117–1127.

Prescott J., (1999) The Generalizability of Capsaicin Sensitization and Desensitization. *Physiology & Behavior, 66,* pp. 741–749.

Rentmeister-Bryant, H. and Green, B.G. (1997) Perceived irritation during ingestion of capsaicin or piperine: comparison of trigeminal and non-trigeminal areas. *Chemical Senses, 22,* pp. 257–266.

Saint-Eve, A., Deleris, I., Feron, G., Ibarra, D., Guichard, E., and Souchon, I. (2010) How trigeminal, taste and aroma perceptions are affected in mint-flavoured carbonated beverages. *Food Quality and Preference, 21,* pp. 1026–1033.

Sediva A., Panovská Z., Pokorný J. (2004). Effect of viscosity on the perceived intensity of acid taste. *Czech Journal of Food Science, 22,* pp. 143–150.

Simons, C.T., O'Mahony, M., and Carstens, E. (2002). Taste suppression following lingual capsaicin pre-treatment in humans. *Chemical Senses, 27*(4), pp. 353–365.

Simons, C. T., and Noble, A. C. (2003). Challenges for the sensory sciences from the food and wine industries. *Nature Reviews Neuroscience, 4,* pp. 599–605.

Slocombe B.G., Carmichael D.A., Simner J. (2015) Cross-modal tactile-taste interactions in food evaluations. *Neuropsychologia*. 2015 Jul 10. pii: *S0028-3932*(15)30096–8. doi: 10.1016/j.neuropsychologia.2015.07.011

Small, D. and Prescott, J. (2005) Odor/taste integration and the perception of flavour. *Experimental Brain Research, 166*, pp. 345–357.

Spence, C. (2014) Confusing tastes with flavours. in Stokes D. et al. (eds) *Perception and its modalities*, Oxford, UK: Oxford University Press. pp. 26–37.

Starkenmann, C., Luca, L., Niclass, Y., Praz, E., and Roguet, D. (2006) Comparison of volatile constituents of *Persicaria odorata* (Lour.) Sojak (*Polygonum odoratum* Lour.) and *Persicaria hydropiper* L. Spach (*Polygonum hydropiper* L.). *Journal of Agricultural and Food Chemistry, 54*, pp. 3067–3071.

Starkenmann, C., Cayeux, I., Brauchli, F., and Mazenyet, F. (2009) Taste contribution of (R)-strombine to dried scallop. *Journal of Agricultural and Food Chemistry, 57*, pp. 7938–7943.

Starkenmann, C., Cayeux, I., and Birkbeck, A. (2011a) Exploring natural products for new taste sensations. *Chimia, 65*, pp. 1–4.

Starkenmann, C., Cayeux, I., Brauchli, F., and Mazenyet, F. (2011b) Hemisynthesis of dihydroumbellulols from umbellulone: new cooling compounds. *Journal of Agricultural and Food Chemistry, 59*, pp. 677–683.

Stevens, D.A. and Lawless, H.T. (1986) Putting out the fire: effects of tastants on oral chemical irritation. *Perception & Psychophysics, 39*, pp. 346–350.

Szallasi, A. and Blumberg, P.M. (1999) Vanilloid receptor (capsaicin) and mechanisms. *Pharmacological Reviews, 51*, pp. 159–211.

Trotier, D., Ishii-Foret, A., Djoumoi, A., Bourdonnais, M., Chéruel, F., Faurion, A. (2012) La sensibilité trigéminale chimique. in Salesse R. et Gervais R., (eds) *Odorat et Goût : de la neurobiologie des sens chimiques aux applications*, Versailles, France : Quae. pp. 215–223.

Viana, F. (2011) Chemosensory properties of the trigeminal system. *ACS Chemical Neurosciences Review, 2*, pp. 38–50.

CHAPTER 9
Interactions between aroma compounds and food matrix

Laurianne Paravisini[1,2] & Elisabeth Guichard[1]
[1] *Center for Taste and Feeding Behaviour, CNRS, INRA, University of Bourgogne Franche-Comté Dijon, France*
[2] *Department of Food Science and Nutrition, University of Minnesota, Saint Paul, Minnesota, United States*

9.1 Introduction

Flavour perception is a multimodal sensation involving olfactory, taste and trigeminal stimulations generated by food components. Aroma compounds are small molecules characterized by physico-chemical properties which make them easily released in the gas phase, such as low molecular weight (<400 Da) and high vapour pressure. They belong to different chemical classes and are responsible for various odorant notes, as described in details in Chapter 6. During food consumption, these compounds will activate the olfactory receptors localized in the olfactory epithelium through the retro and ortho-nasal pathways and generate the olfactory stimulus. Among the non-volatile compounds of the food matrix, some are responsible for taste and trigeminal sensations, as described in chapters 7 and 8. However, the food matrix is also composed of macromolecules which are able to modify aroma perception due to specific interactions with aroma compounds. Depending on their molecular weight and/or their specific chemical properties, for example, high hydrophobicity, non-volatile compounds can modify the partitioning of aroma compounds between the matrix and the gas phase and thus impact the release of these compounds in the nasal cavity. This chapter describes the current state-of-the-art regarding the physico-chemical interactions occurring in foods between aroma compounds and macromolecules. After a general overview of the physico-chemical characteristics of aroma compounds which can influence binding and release phenomena, this chapter will present a general overview of the interactions between aroma compounds and not only the main classes of macromolecules (proteins, lipids and carbohydrates) but also other natural compounds such as polyphenols, Maillard melanoidins and artificial sweeteners. As food is a mixture of different ingredients, some examples of more complex food systems will then be discussed. Finally, the

Flavour: From food to perception, First Edition.
Edited by Elisabeth Guichard, Christian Salles, Martine Morzel, and Anne-Marie Le Bon.

last part presents new trends in different modelling approaches of binding and release studies.

9.2 Physico-chemical characteristics of aroma compounds influencing binding and release

The different chemical classes and the sensory properties of aroma compounds are described extensively in Chapter 6. Table 9.1 gives some physico-chemical characteristics of aroma compounds from different chemical classes. Most of the aroma compounds, except maltol, present a high volatility, expressed by their saturated vapour pressure at 25°C. They are also few soluble in water, except diacetyl and maltol. Their distribution between oil and water can be estimated by their log P value, which corresponds to the logarithm of their octanol/water partition coefficient. A log P value below 1 (diacetyl and maltol) means that the compound is more soluble in water than in oil.

In order to be perceived, the aroma compounds have to be transferred from the food matrix to the gas phase. The air/matrix partition coefficient K describes the distribution of volatile compounds between the matrix and the gas phase:

$$K = \frac{C_G}{C_M}$$

Table 9.1 Main physico-chemical characteristics of aroma compounds.

Compounds	MW (g.mol-1)	log P	Saturated vapor pressure (25°C) (Pa)	Solubility in water (25°C) (g.L-1)	Odour properties
diacetyl	86	−1.3	8306	200	Butter
hexanal	100	1.85	1491	5.6	Fruity, green
heptanal	114	2.31	465	1.3	Fatty, pungent
octanal	128	2.7	155	0.56	Green, soap, lemon
2-heptanone	114	1.97	151	4.3	Fruity
2-octanone	138	2.5	187	1.3	Fatty, green, cheese
2-nonanone	142	3.1	59	0.17	Hot milk, soap, green
ethyl butanoate	116	1.49	1885	4.3	Fruity, pineapple
ethyl pentanoate	130	1.95	640	2.2	Apple, nut
ethyl hexanoate	144	2.41	215	0.6	Fruity
ethyl heptanoate	158	2.86	91	0.03	Fruity, apple
1-hexanol	102	1.88	122	6	Sweet, alcohol
3-methyl-1-butanol	88	1.2	400	20	Burnt
linalool	154	2.74	21	1.59	Flower, lavender
limonene	136	4.3	220	0.36	Lemon, orange
maltol	126	0.1	0.07	10.9	Caramel

Source: (http://pubchem.ncbi.nlm.nih.gov/search/#collection=compounds and http://www.flavornet.org/flavornet.html)

with C_G, the concentration of the volatile compound in the headspace and C_M, the concentration of the volatile compound in the matrix. In the literature, these coefficients can be expressed using the activity coefficient (γ_ι), mass fraction (k_m), the molar fraction (K_i) or the molar concentration (k_i). Therefore, to compare values from the literature, a conversion in the same unity is often necessary. The parameters used for conversion are either standard constants (R, T, P, M_a, M_{liq} and d_{liq}) or the saturation pressure of the pure compound at the corresponding temperature (p_i^s).

Conversion of partition coefficient k_i expressed in molar concentration, or in mass concentration to K_i expressed in molar fraction

$$K_i = k_i \times \frac{RTd_{liq}}{PM_{liq}} \times 10^6$$

Conversion of partition coefficient k_m expressed in mass fraction to K_i expressed in molar fraction

$$K_i = k_m \times \frac{M_a}{M_{liq}}$$

Conversion of partition coefficient K_i expressed in molar fraction to γ_ι activity coefficient

$$\gamma_\iota = K_i \times \frac{P}{p_i^s}$$

Conversion of partition coefficient k_i expressed in molar concentration to γ_ι activity coefficient

$$\gamma_\iota = k_i \times \frac{RTd_{liq}}{p_i^s M_{liq}} \times 10^6$$

with :
P - pressure (Pa)
T - temperature (K)
R - gas constant (8.314 J mol^{-1} K^{-1})
d_{liq} - liquid density (kg L^{-1})
M_{liq} - molar mass of liquid (g mol^{-1})
M_a - molar mass of air (28.8 g mol^{-1})
p_i^s- saturated vapour pressure of compound i (Pa)

These conversions were used to compare air/water partition coefficients available in the literature for ethyl butanoate at different temperatures (Kopjar et al. 2010). Despite the differences in calculated values which could be explained by the different methodologies used, the air/water partition coefficients obtained at different temperatures were used to calculate the enthalpy of affinity. Air/matrix partition coefficients are given in Tables 9.2 and 9.3 for selected aroma compounds in different media. For the same aroma compound, a comparison of the partition coefficients calculated in different matrices gives an indication of the retention/release behaviour.

Table 9.2 Air/Matrix partition coefficients (w/v) of aroma compounds in systems containing lipids.

Compound	37°C			37°C, Artificial saliva		25°C	
	Van Ruth et al. (2002a)			Van Ruth et al. (2002b)		Guichard (2002)	
	Air/Sunflower oil	Air/Emulsion (Oil-40/Water-60)	Air/Emulsion (Oil-60/Water-40)	Air/Emulsion (Oil-40/Water-60)	Air/Emulsion (Oil-60/Water-40)	Olive oil / Water	Sunflower oil / Water
diacetyl	0.00274	0.00213	0.00231	0.00204	0.00223	0.412	0.418
hexanal	0.0004	0.00096	0.00064	0.00147	0.00137		
heptanal	0.00014	0.00035	0.00023	0.00075	0.00068		
octanal	0.00006	0.00015	0.0001	0.00026	0.00024		
2-heptanone	0.00019	0.00046	0.00031	0.00089	0.00081		
2-octanone	0.00007	0.00017	0.00011	0.00062	0.00056		
2-nonanone						644	687
ethyl butanoate	0.00058	0.00138	0.00093	0.000201	0.000177		
ethyl hexanoate						380	406
1-hexanol	0.00014	0.00031	0.00022	0.00104	0.00102		
3-methyl-1-butanol	0.00053	0.00079	0.00068	0.00117	0.00121		

Table 9.3 Air/Matrix partition coefficients (w/v) of aroma compounds in water solutions.

	37°C		37°C, pH6		30°C	
	Bylaite et al. (2004)		Van Ruth et al. (2002c)		Ayed et al. (2014)	
Compound	**Air/Water**	**Air/Lambda-carrageenan (0.5%)**	**Air/Water**	**Air/Beta-lactoglobulin (2%)**	**Air/Water**	**Air/Pectin (0.8%)**
diacetyl	0.0017	0.0033	0.00186	0.00171		
hexanal	0.0257	0.0238	0.01681	0.01034	0.00623	0.00550
heptanal	0.0394	0.0328	0.01982	0.00908	0.00763	0.00679
octanal	0.0675	0.0513	0.01936	0.00341	0.00924	0.00875
2-heptanone			0.00908	0.00642	0.00456	
2-octanone			0.01251	0.00672		
2-nonanone	0.0434	0.0178				
ethyl butanoate	0.0299	0.033	0.02016	0.01565	0.00729	0.00725
ethyl pentanoate	0.039	0.0444			0.00831	0.00712
ethyl hexanoate	0.062	0.0731			0.01031	0.00949
ethyl heptanoate	0.1467	0.0896			0.01249	0.01041
1-hexanol	0.002	0.0021	0.00048	0.00045	0.00070	0.00082
3-methyl-1-butanol	0.0013	0.0013	0.00118	0.00109		
linalool					0.00200	0.00204

9.3 Physico-chemical interactions between aroma compounds and different classes of non-volatile compounds

Most of the literature on the interactions between aroma compounds and food matrix relates to simple systems. The main effects are observed for lipids, proteins and carbohydrates, but other constituents such as polyphenols, melanoidins, artificial sweeteners, alcohol, salt can also affect aroma release from water solution.

9.3.1 Lipids

In fat-containing foods, the retention/release behaviour of aroma compounds mainly depends on their solubility in the matrix. First, most of the aroma compounds are more soluble in oil than in water, these compounds are considered as hydrophobic (log $P > 1$) and are thus less released in the vapour phase from oil than from water (lower air/oil than air/water partition coefficient). The less hydrophobic or more hydrophilic compounds (log $P < 1$), such as diacetyl (or 2,3-butanedione, log $P = -1.34$), present quite similar air/water and air/oil partition coefficients (Tables 9.2 and 9.3). In systems containing lipids, aroma compounds are distributed between the lipid and the aqueous phase following the physical laws of partition. However, the oil/water partition coefficients of aroma compounds depend on the nature of the oil as illustrated in Table 9.2 for hydrophobic compounds (ethyl hexanoate and 2-nonanone). Sunflower oil/water partition coefficients are higher than olive oil/water partition coefficients. This can be explained by the higher amount of unsaturated fatty acids in sunflower oil (C18:2) which leads to a higher solubility of aroma compounds (Guichard 2002; Roudnitzky et al. 2003). Aroma compounds are also more soluble in short chain than in long chain saturated triglycerides (Maier 1975). As aroma compounds are only soluble in liquid fat, a high amount of solid fat reduces their solubility and thus increases their release in the vapour phase (Roudnitzky et al. 2003).

9.3.2 Proteins

Food proteins have little flavour of their own, but are known to bind and trap aroma compounds, which reduces their release in the air phase (Tromelin et al. 2006). The mechanisms of aroma compounds binding by proteins depend both on protein structure and nature of aroma compound (aldehyde, alcohol, ketone, ester) (Heng et al. 2004, Semenova et al. 2002a). Most of the studies related to aroma-protein interactions have been done using milk proteins and few others used proteins from leguminous plants.

9.3.2.1 Milk proteins

Among the milk proteins, β-lactoglobulin has been the subject of many studies (Hansen 1997, Lubbers et al. 1998, O'Neill 1996). This protein is able to bind aroma compounds by hydrophobic interactions. Within the same chemical class of aroma compounds, the strength of the interaction increases with the chain length (Jouenne and Crouzet 2000, O'Neill and Kinsella 1987, Sostmann et al. 1997), except for terpenes (Reiners et al. 2000). Indeed, 3-Dimensional Quantitative Structure Activity Relationships (3-D QSAR) molecular modelling studies have shown the existence of two groups of ligands, confirming the presence of at least two binding sites on β-lactoglobulin and putting forward the role of hydrogen bonding (Tromelin and Guichard 2003, Tromelin and Guichard 2004). The binding within the central cavity occurs for aroma compounds that have an elongated structure whereas the binding onto the protein surface in a site located between strand β G, α helix, and strand β I of β-lactoglobulin occurs for aroma compounds that have or adopt a compact structure (Tavel et al. 2008). In response to a change in environment during processing (heating, mechanical stress ...), this globular protein may shift from a "native" state to a "denatured" state, which will change the nature of the interactions (Famelart et al. 2004, O'Neill 1996). As an example, interactions between two aroma compounds (β-ionone and guaiacol) and β-lactoglobulin in a molten globule state were studied by 2-Dimensional Nuclear Magnetic Resonance (2-D NMR) spectroscopy. The less tightly packed structure of the molten globule favoured ligand binding, in particular within the central cavity. The greater flexibility of the calyx entrance and the conformational change of loop EF induced an easier access of the central cavity after the thermal treatment (Tavel et al. 2010). The pH and ionic strength also affect the strength of the binding by modifying the structure of the protein (Tromelin et al. 2006). As an example, the retention of hydrophobic aroma compounds increased from pH 3 to 9, due to modification of flexibility of the protein which allows better accessibility to the primary or the secondary hydrophobic sites. Conversely, a dramatic decrease was observed at pH 11 which could be the consequence of an alkaline denaturation of the protein (Jouenne and Crouzet 2000). The addition of salts modifies the ionic force of the medium which may modify the interactions between proteins and flavour compounds. For example, the retention of 2-octanone by β-lactoglobulin increased with increasing sodium chloride concentration (Jouenne and Crouzet 1997), whereas the amount of benzaldehyde bound by β-lactoglobulin decreased with the addition of NaCl (Andriot et al. 1999). The different behaviour between the two aroma compounds may be explained by the presence of two different binding sites or by the lower sensibility of 2-octanone to the ionic environment near the binding site.

Caseins are among the most abundant milk proteins. The caseins are unfolded proteins but more flexible than the typical globular proteins (Swaisgood 1993). All caseins exhibit an amphipathic character (Wong et al. 1996), they preferentially locate at hydrophobic interfaces but are more hydrophilic than almost all globular proteins (Holt et al. 2013). There are two fundamental functions of caseins: the effective transport of Ca^{2+} and the self-associations that lead to the colloid state (Farrell et al. 2002). The association of bovine casein with small hydrophobic molecules involves an inclusion mechanism within the hydrophobic interior of micelle-like protein associates. In fat-free dairy matrices, a high amount of caseins increases the retention of limonene, while the less hydrophobic ethyl hexanoate and the hydrophilic diacetyl are not affected (Heilig et al. 2011). The addition of sodium caseinate in model wine induces a decrease in the activity coefficients of aroma compounds in the following order: β-ionone > hexanol > ethyl hexanoate = isoamyl acetate (Voilley et al. 1991). The authors explain the retention of hexanol by the formation of hydrogen bonds and that of esters by hydrophobic effect. This retention effect was also associated with a decrease in odour intensity.

The interactions between milk proteins and aroma compounds mainly involve hydrogen bonding or hydrophobic effect (Tromelin et al. 2006). However, aldehydes and sulfur compounds are able to interact with proteins by irreversible binding as was demonstrated between trans-2-hexenal and whey proteins with the formation of fluorescent compounds. (Meynier et al. 2004).

9.3.2.2 Leguminous proteins

Proteins of leguminous plants are one of the most promising materials for formulating new forms of food products. These proteins are isolated from soy, peas, and beans and may be used as functional additives. The role of the protein structure in binding and release of aroma was investigated in the case of hexyl acetate and revealed that the native leguminous molecules possess the higher binding affinity for this compound (Semenova et al. 2002b). Denaturating proteins by heat results in intensive aggregation of protein basic chains and causes an increase in the number of binding sites altogether with a decrease in the value of intrinsic affinity constant. Acid denaturation (pH 3) dramatically alters the native protein structure, producing a loss of hexyl acetate bound to the protein. Competitive binding of aroma compounds with 11S glycinin suggested that some structures of aroma compounds are more suitable than others for the binding to the protein (Semenova et al. 2002a).

9.3.3 Carbohydrates

Carbohydrate compounds are ubiquitous in food products and usually play a key role in both taste, especially sweetness, and texture. Three categories of carbohydrates can be considered regarding their molecular weight: simple sugars such as glucose and sucrose (mono- and disaccharides), oligosaccharides (2–10 monosaccharides) such as oligofructose and polysaccharides (>10 monosaccharides) such as starch (Delarue and Giampaoli 2006). Their impact on aroma release is quite difficult to predict since they are able to induce both retention and release effects, depending on the conditions and on the molecules. In one way, they can induce small physicochemical (pH, ionic strength, water activity ...) and textural changes in the matrix and thus indirectly modify aroma release. In another way, carbohydrates can directly trap the aroma compounds and induce a retention effect. In any way, sugars-induced effects are highly dependent on the intrinsic properties of the food matrix such as the type of carbohydrates, their concentration (Nawar 1971) and also on the properties of aroma compounds such as steric hindrance, polarity and relative volatility, as already reviewed by Goubet et al. (1998).

9.3.3.1 Simple sugars: mono- and disaccharides

Among all simple sugars, sucrose is likely the most abundant. It occurs naturally as a polymerization product of glucose and fructose by glycosidic linkage. In most cases, the effect of sucrose on aroma release is dependent on the level of concentration added. Indeed, a dynamic flavour release study from sucrose solutions showed that the rate-limiting factor determining the initial flavour release was the hydration of sucrose which depends on the molarity of sucrose (Rabe et al. 2003). In a simple model system, Ebeler et al. (1988) showed that the addition of sucrose up to 40% induced only a slight increase in the headspace concentration of menthone and isoamyl acetate, whereas a significant increase in the concentrations of acetone and 1-octanol in the vapour phase was observed by addition of increasing amounts of sugar (from 1 to 50%) (Voilley et al. 1977). In a soft drink model system, an increase in the concentration of sucrose (from 20 to 60% w/w) was shown to significantly increase the release of *tutti-frutti* aroma compounds such as esters and terpenes (Hansson et al. 2001). This phenomenon can be explained by the so-called salting-out effect due to the hydration of the sucrose molecules by water which leads to decrease the water activity of the matrix. This usually tends to lower the solubility of aroma compounds in water and thus increases their concentration in the gas phase. However, the addition of sucrose also makes the solvent character of the solution more hydrophobic which reduces or reverses the salting-out effect for the more hydrophobic compounds. In viscous solutions, sucrose can also induce a retention effect which may be explained by steric bulk in link with an increase in the viscosity or by the existence of specific interactions between sucrose and aroma compounds (Roberts et al. 1996). In water solutions, an addition of 35% sucrose induced a decrease in self-diffusion coefficients measured by DOSY NMR for ethyl butanoate which was not associated with differences in chemical shifts of aroma peaks in the different media, but only to an increase in viscosity (Savary et al. 2006). Nahon et al. (2000) attempted to model aroma release from aqueous solutions containing sucrose. At lower sucrose concentrations, the partition coefficient primarily controls aroma release, whereas at higher sucrose concentrations, the influence of the mass transfer coefficient is more important.

Among monosaccharides, fructose and glucose are reducing sugars commonly used as food ingredients themselves but also occur naturally from the degradation of sucrose. Lactose is the main carbohydrate of milk and thus widely present in dairy products. In coffee beverages, Piccone et al. (2012) showed that the addition of the same sugar in different matrices containing no sugar (Ready-to-Drink or Expresso coffee) could induce different effects on aroma release. In Expresso coffee beverage, the addition of sucrose, fructose or lactose caused a significant higher release of some furan compounds and a lower release of pyrazines. No significant effect was observed between the different sugars. On the contrary, in Ready-to-Drink beverage, the presence of sugars induced either no change or a retention effect depending on the sugar type. Lactose for example was the sugar that mostly caused a high retention. However, while a significant higher release could be explained by a salting-out effect, the authors did not explain the retention effect. In this case, the effects could also be the results of interactions between other non-volatile compounds and aroma compounds. For example, the non-volatile matrix of coffee contains up to 30% of brown polymers called melanoidins known to interact with aroma compounds (See section 3.5).

9.3.3.2 Oligo- and polysaccharides

Polysaccharides are commonly used in food for their textural properties as thickeners. Polysaccharides are also widely used as encapsulating agents to trap aroma compounds and prevent aroma loss during processing, for example during thermal treatment. In low-fat food, natural polysaccharides such as starch can be used as fat replacers (Jorgensen et al. 2012). In its native state, starch is composed of a well-organized network of amylose and amylopectin. Both are made up of 1,4-linked α-D-anhydroglucopyranoses but they differ in the degree of polymerization and ramification. Amylose is an essentially linear polymer, whereas amylopectin is much larger and is branched. The ratio between amylose and amylopectin will determine the physical properties of starch. As shown in model systems, amylose binds flavour compounds by formation of inclusion complexes (Jouquand et al. 2006, Rutschmann and Solms 1990a, Rutschmann and Solms 1990b, Rutschmann and Solms 1990c). The type of complex between aroma and amylose can influence aroma release and thus aroma perception (Heinemann et al. 2005, Pozo-Bayon et al. 2008). However, aroma retention by starch is not always explained by complexation with amylose and also often due to interaction with amylopectin (Arvisenet et al. 2002, Langourieux and Crouzet 1994) resulting from adsorption involving hydrogen bonds.

Maltodextrins and cyclodextrins (CDs) are well-known agents obtained from the degradation of starch. The interactions between these polysaccharides and aroma compounds have been extensively studied (Jouquand et al. 2008, Kant et al. 2004). The most common CDs are: α-cyclodextrin (α-CD), β-cyclodextrin (β-CD) and γ-cyclodextrin (γ-CD) composed of six, seven and eight glucosyl units, respectively (Ciobanu et al. 2013). CDs have a hydrophilic outer surface and a hydrophobic inner cavity able to trap the lipophilic compounds and to generate inclusion complex. A study of the interactions between 13 volatile aroma compounds and six CDs showed that stable 1:1 inclusion complexes are formed in all cases. However, it appears that β-CDs are the most versatile CDs for the studied compounds and showed the best complexation efficiency (Ciobanu et al. 2013). Indeed, complexation is dependent both on the hydrophobicity of the aroma compounds and the molecular size and geometry of the cyclodextrin which will drive the strength of the inclusion complex (Astray et al. 2010, Goubet et al. 1998). Extrinsic parameters such as temperature can also impact the retention/release behaviour in carbohydrate systems differently according to the polysaccharide studied. In a maltodextrin solution for example, retention is favoured by an increase of temperature from 60 to 80°C which suggests an hydrophobic effect, whereas no such temperature effect influences the retention by β-cyclodextrin due to energetically favourable nonpolar–nonpolar interaction of hydrophobic compounds with the inner cavity of β-cyclodextrin (Jouquand et al. 2004).

In sugar confectionery, glucose syrup and corn syrup are commonly used as sweeteners. They are derived from the hydrolysis of starch and contain a complex mixture of mono- and polysaccharides such as polydextrose. The Dextrose Equivalent (DE) is usually measured to evaluate the amount of reducing sugars and thus the degree of hydrolysis. In model fruit pastilles, the addition of corn syrups DE40 and DE60 tend to reduce aroma release in comparison to simple glucose-sucrose mixtures (Lubbers and Guichard 2003). In accordance, in low concentration solutions, the more volatile compounds (benzaldehyde, ethyl butanoate and butyl isovalerate) are retained more extensively in the presence of polydextrose compared to a pure water solution. However, an increase in the release of the less volatile compounds *cis*-3-hexen-1-ol is observed (Siefarth et al. 2011). Authors assumed that the diffusion of the higher molecular weight volatile compounds can be affected by the meshes of the carbohydrate network. In contrast, compounds with lower molecular weight have a greater ability to diffuse through the matrix. More recently, in solid systems such as cereal bars, the addition of polydextrose has been demonstrated to favour the release of acetaldehyde, esters, menthol and menthone, which was attributed to a salting-out effect due to the reduction of free water (Heenan et al. 2012).

Other polysaccharides are often added in foods as thickening agents. Even if the main effect of thickener agents is usually attributed to a modification of the viscosity, they also often induce a significant decrease in perceived flavour above the critical concentration (c*) at which the carbohydrate network begins (Godshall 1997).

Pectin belongs to the family of complex polysaccharides composed of 1,4-linked α-D-galactosyluronic acid (Ridley et al. 2001). In nature, pectins occur in plant cells and are commonly used as thickeners in food products. In gel model systems, both the rigidity of the gel and the nature of hydrocolloid have a considerable effect on aroma release and perception (Boland et al. 2006). These authors showed that when the rigidity of the gel increased (from 0.75 to 1% w/w of pectin), the air/gel partition coefficient decreased and the flavour perception decreased. However, in soft drink model systems, other authors showed that the addition of pectin below than 2.5% did not impact aroma release suggesting that viscosity had no effect on aroma release (Hansson et al. 2001). In yoghurts, addition of low methoxyl pectin reduced aroma concentration in headspace, but this effect was not significant (Decourcelle et al. 2004, Lubbers and Guichard 2003). On the contrary, in fat-free dairy models, it has been shown that an increase in pectin concentration in gel from 0.04 to 0.15% caused a significant increase in aroma release (Lubbers et al. 2007). These discrepant results emphasize once again the importance of the matrix composition and the properties of aroma compounds on aroma release. In order to understand the retention/release behaviour in the pectin gels and to investigate the nature of the interactions involved, Ayed et al. (2014) performed a statistical analysis of liquid/vapour partition coefficients of 51 aroma compounds. They explained the retention/release balance of solvated hydrophobic molecules by the presence of electrostatic charges on the pectin altering the network of water molecules. They also identified different release behaviours as a function of the chemical functionalities of aroma compounds and the structure of the carbon chains.

Among the thickening agents, carrageenans, principally composed of highly sulphated, alternating $\alpha(1\rightarrow3)$-and $\beta(1\rightarrow4)$-linked galactase residues, are widely used in the food industry, κ and ι–carrageenan as gelling agents and λ–carrageenan as thickening agent. The impact of addition of ι–carrageenan to water on aroma release is rather low (Chana et al. 2006), whereas addition of λ–carrageenan induced a significant decrease in air-liquid partition coefficient of the most hydrophobic compounds within a same chemical class (Bylaite et al. 2004). However in presence of sodium chloride, a three-dimensional network is formed between ι–carrageenan molecules leading to an increase in gel hardness and a decrease in air/matrix partition coefficients (Juteau et al. 2004).

9.3.4 Polyphenols

Other macromolecules which are able to bind aroma compounds are polyphenols, which are important compounds in fruits and vegetables. Polyphenols contribute to colour, astringency and taste of such products. Even if the interactions between wine polyphenols and proteins have been the subject of many studies (Siebert et al. 1996), their interactions with aroma compounds are less understood. The elimination of polyphenols through filtration or fining treatment and precipitation by polymerization during wine aging has been suspected to produce flavour balance modifications. However, only few studies demonstrated the effect of monomeric phenols on partitioning of odorants. The aqueous solubility of anisole, 2,3-diethylpyrazine, and ethyl benzoate increased in solutions containing propyl gallate, chlorogenic acid, caffeine or naringin (King and Solms 1982). A weak retention of different aroma compounds by catechin has also been observed (Dufour and Bayonove 1999), with the highest affinity for benzaldehyde ($K_d = 0.187$), the lowest for 3,5-dimethoxyphenol ($K_d = 0.427$), and intermediate for the two esters. The dissociation constants (K_d) were measured by ^1H NMR. Jung and Ebeler (2003) also found a small retention of ethyl hexanoate and hexanal by catechin, similar to that observed with β-lactoglobulin. Gallic acid significantly decreased the volatility of 2-methylpyrazine, while naringin had less effect on the headspace volatility of this aroma (Aronson and Ebeler 2004).

9.3.5 Maillard melanoidins

Melanoidins are the non-volatile compounds resulting from the non-enzymatic browning Maillard reaction and also from the reaction between ascorbic acid and amino acid. They occur in most of thermally treated food products. In coffee beverage for example, they compose up to 30% of the non-volatile fraction. Melanoidins are nitrogen polymers but their chemical structure is not fully elucidated yet. However, their interactions with aroma compounds have been extensively studied (Ames 1997, Charles-Bernard et al. 2005, Hofmann and Schieberle 2002, Lopez-Galilea et al. 2008, Obretenov et al. 2002). Particularly the interactions between melanoidins and thiols have been deeply studied due to the typical sulfuric and roasted odour notes of thiols responsible for the typical aroma of coffee. In a model system, melanoidins induced a significant decrease of the headspace concentration of three important coffee odorants, 2-furfurylthiol (FFT), 3-methyl-2-butene-1-thiol and 3-mercapto-3-methylbutyl formate (Hofmann et al. 2001). However, aldehydes were not impacted by the addition of melanoidins in the system. Indeed, further studies highlighted the existence of covalent bond between melanoidins and thiols through a nitrogen-containing intermediate (Hofmann and Schieberle 2003). Another study showed that interactions between coffee melanoidins and aroma compounds highly depend on the method used for the extraction of melanoidins and on the properties of aroma compounds (Andriot et al. 2004). These authors showed that melanoidins induced the retention of methylketone and esters most likely through hydrophobic interactions. Moreover, it appeared that the lyophilisation process, commonly used to isolate melanoidins, modifies the retention effect suggesting a denaturation of the polymers.

9.3.6 Artificial sweeteners

The interest for artificial sweeteners has been growing for years which led food industry to reformulate their products and thus scientists to study their impact on aroma release in different media. However, as seen previously for carbohydrates, the impact of artificial sweeteners is not predictable yet. In yoghurts for example, the findings are quite contradictory. An *in vitro* study in fat-free yoghurts has shown that the addition of aspartame and potassium acesulfame did not change the aroma release behaviour (Decourcelle et al. 2004). However, other authors showed that during yoghurt consumption, the release of ethyl butanoate was decreased by about 20% in presence of 400 ppm of aspartame (Mei et al. 2004). The effects are also different depending on the food product. In beverages, a greater volatility of aroma compounds was found in products containing aspartame or acesulfame K compared to sucrose (Deibler and Acree 2000, King et al. 2006). However, Deibler and Acree (2000) also observed selective effects of sucralose, able to either increase or decrease the volatility of some aroma compounds. For example, sucralose boosted the concentration of citronellyl acetate in the gas phase whereas it suppressed neral release. In alcoholic beverages, the addition of acesulfame K or sorbitol did not change the volatility of different aroma compounds such as esters and terpenes, whereas the addition of aspartame decreased the volatility of linalool and benzaldehyde (Da Porto et al. 2006). Authors attributed the increase in gas phase concentrations to a salting-out effect. However, this does not explain the selectivity of the sweeteners effects on some aroma compounds.

9.3.7 Other effects

9.3.7.1 Alcohol

Ethanol has been shown to decrease the partition coefficient of various classes of volatile compounds by increasing their solubility (Voilley et al. 1991). Thus, in systems which contain ethanol, aroma release is changed partly due, to a change in air/liquid partition as a function of aroma compound log *P* value (Aznar et al. 2004), but also due to other physico-chemical effects such as micelle formation (Conner et al. 1998, Escalona et al. 1999) and surface tension effects.

Glycerol, present in relatively high amount in wines, seems to have no additional effect to ethanol on volatile partitioning (Robinson et al. 2009), which is consistent with the report that an increase in glycerol concentration in Chardonnay wine does not change the overall flavour perception (Lubbers et al. 2001).

9.3.7.2 Salt

An increased amount of salt induces an increase in the air/water partitioning of most aroma compounds, due to a salting-out effect, that is, a reduction of the available solvent in the liquid phase (Deak et al. 2006, Endo et al. 2012, Zhang et al. 2011). This effect is more pronounced for hydrophobic compounds, less soluble in water (Lauverjat et al. 2009). Indeed, Na^+ and Cl^- ions are able to mobilise water molecules for their hydration so less water is available for the solubilisation of aroma compounds, which are therefore more released into the vapour phase with higher air-matrix partition coefficients (Rabe et al. 2003).

9.4 Simultaneous effects of different components in complex matrices

9.4.1 Emulsions

The volatility of aroma compounds in oil/water emulsions depends on the amount of fat but also on the type of fat (Relkin et al. 2004). An increase in fat content decreases the volatility of the more hydrophobic compounds (Jo and Ahn 1999). As an example, Table 9.2 shows that the air/matrix partition coefficients increase from water to emulsion for diacetyl, 1-hexanol and 3-methyl-1-butanol and decrease for aldehydes, ketones and esters. An increase in oil fraction in emulsion leads to a higher retention of hydrophobic compounds (Van Ruth et al. 2002a). In emulsions, the influence of the droplet size has been reported by some authors to be not significant or rather limited (Carey et al. 2002, Landy et al. 1996). Others found that an increase in droplet size decreases the volatility of the more hydrophobic aroma compounds (Charles et al. 2000, Miettinen et al. 2002, Van Ruth et al. 2002a).

The presence of proteins (β-lactoglobulin or sodium caseinate) as emulsifiers at the oil/water interface increases the resistance to transfer of hydrophobic aroma compounds from oil to water and thus induces a decrease in their release in the air phase (Harvey et al. 1995, Rogacheva et al. 1999).

Moreover NaCl modifies the structure of proteins, as has been observed with β-lactoglobulin which is mainly present in its dimeric form in the presence of NaCl because of a modification of electrostatic forces (Sakurai et al. 2001). This increases the binding of 2-octanone by β-lactoglobulin (Jouenne and Crouzet 1997). NaCl also modifies the structure of the food matrix; for example, a higher NaCl content increases the strength of ι-carrageenan gels and the self-diffusion of ethyl butanoate (Gostan et al. 2004), which can be explained by the greater size of the open space located between the gel chains, resulting in fewer obstacles to the free diffusion of solutes. In model cheeses, the addition of salt may modify the texture by decreasing water activity, increasing firmness and thereby increasing the release of aroma compounds (Saint-Eve et al. 2009). Using a model mouth system, the effect of emulsifier concentration on air/emulsion partition coefficients of aroma compounds was reduced by addition of artificial saliva and aroma release under mouth conditions generally decreased with the oil fraction/particle diameter (Van Ruth et al. 2002b). This observation is in agreement with predicted models of flavour release (Harrison et al. 1997).

9.4.2 Hydroalcoholic systems

Wine is a complex matrix in which ethanol, polyphenols and polysaccharides are able to interact individually with aroma compounds. When present in mixtures, tannins generally enhance the release of aroma compounds while fructose induces a retention effect, these effects being modulated by ethanol (Villamor et al. 2013). The retention of hydrophobic aroma compounds in the wine matrix was higher in formulations with high ethanol, tannins and fructose concentration (respectively 14% ethanol, 1.5 g/L tannins and 2 g/L fructose). The salting out effect of tannins is more pronounced at low ethanol concentrations (8–10%), which can be explained by tannin self-aggregation (Poncet-Legrand et al. 2003). In model wine the effect of addition of tannins on aroma release varied in function of the nature of tannin extract: the extracts derived from grape skin induced a higher retention of esters than extracts derived from grape seeds (Mitropoulou et al. 2011). As polysaccharides are usually present in wines along with tannins, their combined effects were studied and showed for example an increased retention for linalool and long chain ethyl esters and a decreased retention for alcohols. The authors suggested the formation of complex polysaccharide-tannins entities which should modify the nature of the interactions with aroma compounds. The effects of polyphenols are also modulated by the concentration in glycerol and for example the retention effect of catechin on aroma compounds (Dufour and Bayonove 1999) seems to be reduced in the presence of glycerol (Robinson et al. 2009). However, catechin and glycerol do not affect the dynamic headspace concentration of aroma at acidic pH, whereas the addition of casein or β-lactoglobulin tends to decrease significantly the amount of aroma compounds in the headspace (Tsachaki et al. 2009). This observation suggested perturbation of the ethanol monolayer at the air/liquid interface and disruption of the Marangoni effect, which causes bulk convection within ethanolic solutions. Convection carries volatile compounds and warm liquid from the bulk phase to the air/liquid interface, thus replenishing the interfacial concentration and maintaining the gas phase concentration and interfacial surface temperature during headspace dilution. It is postulated that some proteins may exert a similar effect in wine.

The effect of increasing levels of polysaccharides in a model wine containing tannins differs according to the type of polysaccharide (Mitropoulou et al. 2011). At low concentration, arabinolgalactan induced an increase in aroma release regardless the aroma compound (salting out effect), whereas at concentration higher than 1g/L the retention of the more hydrophobic aroma compounds is observed, suggesting intermolecular binding, not observed at a concentration of 5g/L, due to saturation of the binding sites. A similar salting out effect was observed with addition of pectins up to 2g/L, whereas at higher pectin concentration, a continuous increase in the release of the less hydrophobic esters and a decrease in the release of the more hydrophobic esters and of alcohols were observed. The authors concluded that the different behaviours could be related to the structure of both aroma and polysaccharide. Pectins contains a great number of sugar units and are thus able to trap a significant amount of water, which can explain the high salting out effect. The retention of the more hydrophobic esters can be explained by binding in the esterified regions (Kim et al. 2010).

The mannoproteins and glucans released from yeast during alcoholic fermentation are able to decrease the volatility of aroma compounds, depending on the nature of aroma compounds (Lubbers et al. 1994). The retention of β-ionone, ethyl hexanoate and octanal can be attributed to mannoproteins containing a high proportion of proteins.

9.5 Modelling binding and release of flavour compounds

Modelling of phase partitioning and aroma transfer has been extensively studied (De Roos 2006). Simple models were first developed to predict oil/water partition coefficients from their oil/air and water/air coefficients (Buttery et al. 1973) and then validated using a homologous series of alcohols (Mc Nulty and Karel 1973). A number of useful mathematical relationships that relate release to both partition and diffusion coefficients were also developed (Overbosch et al. 1991). These different models were successively applied to emulsions made with different types of fat and emulsifiers (Carey et al. 2002, Guth and Rusu 2008, Landy et al. 1996, Roberts et al. 2003), showing that fat is the prominent factor affecting aroma release compared to proteins and viscosity.

A rigorous physico-chemical model taking into account the composition of food and the resistance to mass transfer resulted in a good agreement between observed and calculated data on flavour retention (de Roos and Wolswinkel 1994). In general, the food constituents having the most pronounced effect on the equilibrium headspace concentration of flavour compounds are water, lipids and alcohol.

Different approches were used to better understand the interactions between food ingredients and aroma compounds with the aim to optimize the formulation of food products. In fat-free stirred yoghurts, an optimal experimental design was used to determine which combination of pectin, starch, locust bean gum, guar, fructose, fructo-oligosaccharide syrup and aspartame would deliver the best aroma (Decourcelle et al. 2004). The results show that only starch and locust bean gum had a significant effect on aroma release, with lower aroma release for high starch concentration and low locus bean gum concentration. The authors concluded that aroma release under shear conditions was more influenced by composition than by rheology. Response surface methodology was used to determine the optimum formulation that gives maximum retention and partition coefficient of ethyl acetate from food emulsion model systems (Samavati and D-jomeh 2013). The minimum partition coefficient for ethyl acetate was reached with 4% protein and 10% oleic acid. To study the effect of emulsion components on equilibrium headspace concentration of soursop aroma compounds, a four-component (modifed starch, whey protein, flavour oil and water) design was used (Cheong et al. 2014). The individual effect and the interactions were tested showing that a higher decrease in the release of hydrophobic aroma compounds was observed for systems with single emulsifiers than those containing a mixture of modified starch and whey protein.

Aroma release is time dependent and partitioning is not a sufficient key for an overall understanding of the behaviour of volatiles in food matrices. Marin et al. (1999) proposed and validated a mechanistic model which mimics the dilution of volatiles in the surrounding air. They showed that the hydrodynamic regime in the gas phase had a significant effect on the release of aroma compounds. A modelling of theoretical physicochemical data of aroma compounds in mygliol/water model emulsions showed that the rate-limiting factor determining initial flavour release was the dynamic partitioning from the aqueous phase into the gas phase (Rabe et al. 2004). Another mechanistic mathematical model was developed for the prediction of aroma release from dairy emulsions and for the calculation of apparent diffusion properties (Deleris et al. 2008, Deleris et al. 2009). The most influent parameters of the apparent diffusivity in dairy fat emulsions were the diffusivity in the aqueous phase, the liquid fat fraction and the partition between the fat and the aqueous phase.

Molecular modelling can be useful to better understand the interactions between small molecules and macromolecules. In the field of ligand-protein interactions, structural data on target proteins are used in molecular modelling and binding simulation studies (Colmenarejo et al. 2001, Mason et al. 2004, Perola and Charifson 2004). However, the three-dimensional structure must be determined either by X-ray crystallography or NMR, and the localization of the binding site should be known. Molecular modelling can also be used to propose a structural conformation for a protein lacking direct structural data (Kumosinski et al. 1991a, Kumosinski et al. 1991b, Kumosinski et al. 1993a, Kumosinski et al. 1993b). In the case of unavailability of the three-dimensional structures of a target protein, homology models can be built if suitable target proteins with high sequence identity are available (Kister et al. 2002, Schmitt et al. 2002).

If the structure of macromolecule is unknown, QSAR (Quantitative Structure Activity Relationships) or QSPR (Quantitative Structure Property Relationships) ligand approaches are suitable ways to improve the comprehension of interactions between aroma-ligands and food proteins (Guth and Fritzler 2004, Marabotti et al. 2000, Tromelin and Guichard 2003, Tromelin and Guichard 2004). Such approaches showed for example that not only hydrophobic effect but also hydrogen bonding were involved in the interactions between β-lactoglobulin and aroma compounds (Tromelin and Guichard 2004). Similar approaches conducted on the release of aroma compounds from polysaccharide gels highlighted the role of polar effects on the retention of odorant molecules, and suggested that retention/release properties are not determined by the chemical classes but by an assembly of chemical properties (Chana et al. 2006,

Tromelin et al. 2007, Tromelin et al. 2010). In this way, the QSPR approach represents a tool for the identification of interactions involved in complex food matrices.

9.6 Conclusion

In conclusion, this chapter showed that despite the profusion of studies dealing with aroma and food matrix interactions, the physico-chemical nature of interactions involved is not fully characterized, mainly due to a lack of information on the structure of the different macromolecules used in food formulations. The studies conducted in complex systems show that it is still difficult to predict aroma release in real food based on the results obtained with single ingredients, due to other interactions between macromolecules themselves. Nevertheless some models have been developed to help our understanding in the retention/release behaviour of aroma compounds from different systems of increasing complexity. The next step is to take into account in-mouth mechanisms which occur during real food consumption, such as the interactions between aroma compounds and salivary proteins which are developed in Chapter 12.

References

Ames, J. (1997) Characterization of melanoidins and their role in flavour binding. in Schieberle, P., (ed.) *Interaction of food matrix with small ligands influencing flavour and texture*, Garching (D): European Communities. pp. 161–165.

Andriot, I., Marin, I., Feron, G., Relkin, P. and Guichard, E. (1999) Binding of benzaldehyde by β-lactoglobulin, by static headspace and high performance liquid chromatography in different physico-chemical conditions. *Le Lait, 79*(6), pp. 577–586.

Andriot, I., Le Quere, J.-L. and Guichard, E. (2004) Interactions between coffee melanoidins and flavour compounds: impact of freeze-drying (method and time) and roasting degree of coffee on melanoidins retention capacity. *Food Chemistry, 85*(2), pp. 289–294.

Aronson, J. and Ebeler, S. E. (2004) Effect of Polyphenol compounds on the headspace volatility of flavors. *American Journal of Enology and Viticulture, 55*(1), pp. 13–21.

Arvisenet, G., Voilley, A. and Cayot, N. (2002) Retention of aroma compounds in starch matrices: Competitions between aroma compounds toward amylose and amylopectin. *Journal of Agricultural and Food Chemistry, 50*(25), pp. 7345–7349.

Astray, G., Mejuto, J. C., Morales, J., Rial-Otero, R. and Simal-Gandara, J. (2010) Factors controlling flavors binding constants to cyclodextrins and their applications in foods. *Food Research International, 43*(4), pp. 1212–1218.

Ayed, C., Lubbers, S., Andriot, I., Merabtine, Y., Guichard, E. and Tromelin, A. (2014) Impact of structural features of odorant molecules on their retention/release behaviours in dairy and pectin gels. *Food Research International, 62*, pp. 846–859.

Aznar, M., Tsachaki, M., Linforth, R. S. T., Ferreira, V. and Taylor, A. J. (2004) Headspace analysis of volatile organic compounds from ethanolic systems by direct APCI-MS. *International Journal of Mass Spectrometry, 239*(1), pp. 17–25.

Boland, A. B., Delahunty, C. M. and van Ruth, S. M. (2006) Influence of the texture of gelatin gels and pectin gels on strawberry flavour release and perception. *Food Chemistry, 96*(3), pp. 452–460.

Buttery, R., Guadagni, D. G. and Ling, L. C. (1973) Flavor compounds: volatilitites in vegetable oil and oil-water mixtures. Estimation of odor thresholds. *Journal of Agricultural and Food Chemistry, 21*(2), pp. 198–201.

Bylaite, E., Ilgunaite, Z., Meyer, A. S. and Adler-Nissen, J. (2004) Influence of lambda-carrageenan on the release of systematic series of volatile flavor compounds from viscous food model systems. *Journal of Agricultural and Food Chemistry, 52*(11), pp. 3542–3549.

Carey, M. E., Asquith, T., Linforth, R. S. T. and Taylor, A. J. (2002) Modeling the partition of volatile aroma compounds from a cloud emulsion. *Journal of Agricultural and Food Chemistry, 50*(7), pp. 1985–1990.

Chana, A., Tromelin, A., Andriot, I. and Guichard, E. (2006) Flavor release from iota-carrageenan matrix: a Quantitative Structure-Property Relationships approach. *Journal of Agricultural and Food Chemistry, 54*(10), pp. 3679–3685.

Charles-Bernard, M., Kraehenbuehl, K., Rytz, A. and Roberts, D. D. (2005) Interactions between volatile and nonvolatile coffee components. 1. Screening of nonvolatile components. *Journal of Agricultural and Food Chemistry, 53*(11), pp. 4417–4425.

Charles, M., Rosselin, V., Beck, L., Sauvageot, F. and Guichard, E. (2000) Flavor release from salad dressings: Sensory and physicochemical approaches in relation with the structure. *Journal of Agricultural and Food Chemistry, 48*(5), pp. 1810–1816.

Cheong, K. W., Tan, C. P., Mirhosseini, H., Joanne-Kam, W. Y., Hamid, N. S. A., Osman, A. and Basri, M. (2014) The effect of prime emulsion components as a function of equilibrium headspace concentration of soursop flavor compounds. *Chemistry Central Journal, 8*: 23.

Ciobanu, A., Mallard, I., Landy, D., Brabie, G., Nistor, D. and Fourmentin, S. (2013) Retention of aroma compounds from Mentha piperita essential oil by cyclodextrins and crosslinked cyclodextrin polymers. *Food Chemistry, 138*(1), pp. 291–297.

Colmenarejo, G., Alvarez Pedraglio, A. and Lavandera, J.L. (2001) Cheminformatic models to predict binding affinities to human serum albumin. *Journal of Medicinal Chemistry, 44*(25), pp. 4370–4378.

Conner, J. M., Birkmyre, L., Paterson, A. and Piggott, J. R. (1998) Headspace concentrations of ethyl esters at different alcoholic strengths. *Journal of the Science of Food and Agriculture, 77*(1), pp. 121–126.

Da Porto, C., Cordaro, F. and Marcassa, N. (2006) Effects of carbohydrate and noncarbohydrate sweeteners on the orange spirit volatile compounds. *Lwt-Food Science and Technology, 39*(2), pp. 159–165.

Decourcelle, N., Lubbers, S., Vallet, N., Rondeau, P. and Guichard, E. (2004) Effect of thickeners and sweeteners on the release of blended aroma compounds in fat-free stirred yoghurt during shear conditions. *International Dairy Journal, 14*(9), pp. 783–789.

Deibler, K. and Acree, T. (2000) The effect of soft drink base composition on flavour release. In Schieberle, P. and Engel, K.H., (eds). *Frontiers of Flavour Science*, Garching (Germany): Deutsche Forshunganstalt für Lebensmittelchemie. pp. 283–286.

Delarue, J. and Giampaoli, P. (2006) Carbohydrate-flavour interactions. in Voilley, A. and Etievant, P., (eds.) *Flavour in food*, Cambridge (GBR): Woodhead Publishing Limited and CRC Press LLC. pp. 208–228.

De Roos, K. B. (2006) Modeling aroma interactions in food matrices. in Voilley, A. and Etievant, P., (eds.) *Flavour in food*, Cambridge (GBR): Woodhead publishing limited. pp. 229–259.

de Roos, K. B. and Wolswinkel, K. (1994) Non-equilibrium partition model for predicting flavour release in the mouth. in Maarse, H. and van der Heij, D. G., (eds.) *Trends in Flavour Research*, Amsterdam (NLD): Elsevier Science. pp. 15–32.

Deak, N. A., Murphy, P. A. and Johnson, L. A. (2006) Effects of NaCl concentration on salting-in and dilution during salting-out on soy protein Fractionation. *Journal of Food Science, 71*(4), pp. C247–C254.

Decourcelle, N., Lubbers, S., Vallet, N., Rondeau, P. and Guichard, E. (2004) Effect of thickeners and sweeteners on the release of blended aroma compounds in fat-free stirred yoghurt during shear conditions. *International Dairy Journal, 14*(9), pp. 783–789.

Deleris, I., Atlan, S., Souchon, I., Marin, M. and Trelea, L. C. (2008) An experimental device to determine the apparent diffusivities of aroma compounds. *Journal of Food Engineering, 85*(2), pp. 232–242.

Deleris, I., Zouid, I., Souchon, I. and Trelea, I. C. (2009) Calculation of apparent diffusion coefficients of aroma compounds in dairy emulsions based on fat content and physicochemical properties in each phase. *Journal of Food Engineering, 94*(3-4), pp. 205–214.

Dufour, C. and Bayonove, C. L. (1999) Interactions between wine polyphenols and aroma substances. An insight at the molecular level. *Journal of Agricultural and Food Chemistry, 47*(2), pp. 678–684.

Ebeler, S. E., Pangborn, R. M. and Jennings, W. G. (1988) Influence of dispersion medium on aroma intensity and headspace concentration of menthone and isoamyl acetate. *Journal of Agricultural and Food Chemistry, 36*(4), pp. 791–796.

Endo, S., Pfennigsdorff, A. and Goss, K.-U. (2012) Salting-Out Effect in Aqueous NaCl Solutions: Trends with Size and Polarity of Solute Molecules. *Environmental Science & Technology, 46*(3), pp. 1496–1503.

Escalona, H., Piggott, J. R., Conner, J. M. and Paterson, A. (1999) Effect of ethanol strength on the volatility of higher alcohols and aldehydes. *Italian Journal of Food Science, 11*(3), pp. 241–248.

Famelart, M. H., Tomazewski, J., Piot, M. and Pezennec, S. (2004) Comprehensive study of acid gelation of heated milk with model protein systems. *International Dairy Journal, 14*, pp. 313–321.

Farrell, H. M., Qi, P. X., Wickham, E. D. and Unruh, J. J. (2002) Secondary structural studies of bovine caseins: Structure and temperature dependence of beta-casein phosphopeptide (1-25) as analyzed by circular dichroism, FTIR spectroscopy, and analytical ultracentrifugation. *Protein Chemistry, 21*(5), pp. 307–321.

Gostan, T., Moreau, C., Juteau, A., Guichard, E. and Delsuc, M.-A. (2004) Measurement of aroma compound self-diffusion in food models by DOSY. *Magnetic Resonance in Chemistry, 42*(6), pp. 496–499.

Goubet, I., Le Quere, J.-L. and Voilley, A. J. (1998) Retention of aroma compounds by carbohydrates; influence of their physicochemical characteristics and of their physical state. *A review. Journal of Agricultural and Food Chemistry, 46*(5), pp. 1981–1990.

Guichard, E. (2002) Interactions between flavor compounds and food ingredients and their influence on flavor perception. *Food Reviews International, 18*(1), pp. 49–70.

Guth, H. and Fritzler, R. (2004) Binding studies and computer-aided modelling of macromolecule/odorant interactions. *Chemistry & Biodiversity, 1*(12), pp. 2001–2023.

Guth, H. and Rusu, M. (2008) Food matrices - determination of odorant partition coefficients and application of models for their prediction. *Food Chemistry, 108*(4), pp. 1208–1216.

Godshall, M. A. (1997) How carbohydrates influence food flavor. *Food Technology, 51*(1), pp. 63–67.

Hansen, A. (1997) A review of the interactions between milk proteins and dairy flavor compounds. in Damodaran , (ed.) *Food proteins and lipids*, New York (USA): Plenum Press. pp. 67–75.

Hansson, A., Andersson, J. and Leufven, A. (2001) The effect of sugars and pectin on flavour release from a soft drink-related model system. *Food Chemistry, 72*(3), pp. 363–368.

Harrison, M., Hills, B., Bakker, J. and Clothier, T. (1997) Mathematical models of flavor release from liquid emulsions. *Journal of Food Science, 62*(4), pp. 653–658.

Harvey, B. A., Druaux, C. and Voilley, A. (1995) Effect of protein on the retention and transfer of aroma compounds at the lipd-water interface. in Lorient, E. D. a. D., (ed.) *Food Macromolecule and Colloids,* Cambridge (UK): The Royal Society of Chemistry. pp. 154–163.

Heenan, S., Soukoulis, C., Silcock, P., Fabris, A., Aprea, E., Cappellin, L., Mark, T. D., Gasperi, F. and Biasioli, F. (2012) PTR-TOF-MS monitoring of in vitro and in vivo flavour release in cereal bars with varying sugar composition. *Food Chemistry, 131*(2), pp. 477–484.

Heilig, A., Cetin, S., Erpenbach, K., Hohn, J. and Hinrichs, J. (2011) Inherent and process-induced hydrophobicity influences aroma retention in fat-free dairy matrices. *International Dairy Journal, 21*(9), pp. 696–702.

Heinemann, C., Zinsli, M., Renggli, A., Escher, F. and Conde-Petit, B. (2005) Influence of amylose-flavor complexation on build-up and breakdown of starch structures in aqueous food model systems. *Lebensmittel-Wissenschaft und -Technologie, 38*(8), pp. 885–894.

Heng, L., van Koningsveld, G. A., Gruppen, H., van Boekel, M., Vincken, J. P., Roozen, J. P. and Voragen, A. G. J. (2004) Protein-flavour interactions in relation to development of novel protein foods. *Trends in Food Science & Technology, 15*(3-4), pp. 217–224.

Hofmann, T., Czerny, M., Calligaris, S. and Schieberle, P. (2001) Model studies on the influence of coffee melanoidins on flavor volatiles of coffee beverages. *Journal of Agricultural and Food Chemistry, 49,* pp. 2382–2386.

Hofmann, T. and Schieberle, P. (2002) Chemical interactions between odor-active thiols and melanoidins involved in the aroma staling of coffee beverages. *Journal of Agricultural and Food Chemistry, 50*(2), pp. 319–326.

Hofmann, T. and Schieberle, P. (2003) Influence of Melanoidins on the Aroma Staling of Coffee Beverage. in *Nutraceutical Beverages*: American Chemical Society. pp. 200–215.

Holt, C., Carver, J. A., Ecroyd, H. and Thorn, D. C. (2013) Invited review: Caseins and the casein micelle: Their biological functions, structures, and behavior in foods. *Journal of Dairy Science, 96*(10), pp. 6127–6146.

Jo, C. and Ahn, D. U. (1999) Fat reduces volatiles production in oil emulsion system analysed by purge and trap dynamic headspace/gas chromatography. *Journal of Food Science, 64*(4), pp. 641–643.

Jorgensen, A. D., Jensen, S. L., Ziegler, G., Pandeya, A., Buleon, A., Svensson, B. and Blennow, A. (2012) Structural and physical effects of aroma compound binding to native starch granules. *Starch-Starke, 64*(6), pp. 461–469.

Jouenne, E. and Crouzet, J. (1997) Influence of sodium chloride and urea on volatile compounds-β-lactoglobulin interactions. in Zacharoff, A. B., (ed.) *Interaction of food matrix with small ligands influencing flavour and texture,* Gothenburg (Sweden): European Communities. pp. 37–43.

Jouenne, E. and Crouzet, J. (2000) Effect of pH on retention of aroma compounds by β-lactoglobulin. *Journal of Agricultural and Food Chemistry, 48*(4), pp. 1273–1277.

Jouquand, C., Aguni, Y., Malhiac, C. and Grisel, M. (2008) Influence of chemical composition of polysaccharides on aroma retention. *Food Hydrocolloids, 22*(6), pp. 1097–1104.

Jouquand, C., Ducruet, V. and Giampaoli, P. (2004) Partition coefficients of aroma compounds in polysaccharide solutions by the phase ratio variation method. *Food Chemistry, 85*(3), pp. 467–474.

Jouquand, C., Ducruet, V. and Le Bail, P. (2006) Formation of amylose complexes with C6-aroma compounds in starch dispersions and its impact on retention. *Food Chemistry, 96*(3), pp. 461–470.

Jung, D. M. and Ebeler, S. E. (2003) Headspace solid-phase microextraction method for the study of the volatility of selected flavor compounds. *Journal of Agricultural and Food Chemistry, 51*(1), pp. 200–205.

Juteau, A., Doublier, J. L. and Guichard, E. (2004) Flavor release from iota-carrageenan matrices: A kinetic approach. *Journal of Agricultural and Food Chemistry, 52*(6), pp. 1621–1629.

Kant, A., Linforth, R. S. T., Hort, J. and Taylor, A. J. (2004) Effect of beta-cyclodextrin on aroma release and flavor perception. *Journal of Agricultural and Food Chemistry, 52*(7), pp. 2028–2035.

Kim, Y., Kim, Y. S., Yoo, S. H. and Kim, K. O. (2010) Molecular differences of low methoxy pectins induced by pectin methyl esterase I: Effects on texture, release and perception of aroma in gel systems. *Food Chemistry, 123*(2), pp. 451–455.

King, B. M. and Solms, J. (1982) Interactions of volatile flavor compounds with propyl gallate and other phenols as compared with caffeine. *Journal of Agricultural and Food Chemistry, 30*(5), pp. 838–840.

King, B. M., Arents, P., Bouter, N., Duineveld, C. A. A., Meyners, M., Schroff, S. I. and Soekhai, S. T. (2006) Sweetener/sweetness-induced changes in flavor perception and flavor release of fruity and green character in beverages. *Journal of Agricultural and Food Chemistry, 54*(7), pp. 2671–2677.

Kister, A. E., Finkelstein, A. V. and Gelfand, I. M. (2002) Common features in structures and sequences of sandwich-like proteins. *Proceedings of the National Academy of Sciences of the United States of America, 99*(22), pp. 14137–14141.

Kopjar, M., I., A., Saint-Eve, A., Souchon, I. and Guichard, E. (2010) Retention of aroma compounds: an interlaboratory study on the effect of the composition of food matrices on thermodynamic parameters in comparison with water. *Journal of the Science of Food and Agriculture, 90*(8), pp. 1285–1292.

Kumosinski, T. F., Brown, E. M. and Farrell, H. M., Jr. (1991a) Three-dimensional molecular modeling of bovine caseins: alpha s1-casein. *Journal of Dairy Science, 74*(9), pp. 2889–2895.

Kumosinski, T. F., Brown, E. M. and Farrell, H. M., Jr. (1991b) Three-dimensional molecular modeling of bovine caseins: kappa-casein. *Journal of Dairy Science, 74*(9), pp. 2879–2887.

Kumosinski, T. F., Brown, E. M. and Farrell, H. M., Jr. (1993a) Three-dimensional molecular modeling of bovine caseins: a refined, energy-minimized kappa-casein structure. *Journal of Dairy Science, 76*(9), pp. 2507–2520.

Kumosinski, T. F., Brown, E. M. and Farrell, H. M., Jr. (1993b) Three-dimensional molecular modeling of bovine caseins: an energy-minimized beta-casein structure. *Journal of Dairy Science, 76*(4), pp. 931–45.

Landy, P., Courthaudon, J. L., Dubois, C. and Voilley, A. (1996) Effect of Interface in Model Food Emulsions on the Volatility of Aroma Compounds. *Journal of Agricultural and Food Chemistry, 44*, pp. 526–530.

Langourieux, S. and Crouzet, J. (1994) Study of aroma compounds-polysaccharides interactions by dynamic exponential dilution. *Lebensmittel-Wissenschaft und -Technologie, 27*(6), pp. 544–549.

Lauverjat, C., Deleris, I., Trelea, C. I., Salles, C. and Souchon, I. (2009) Salt and aroma compound release in model cheeses in relation to their mobility. *Journal of Agricultural and Food Chemistry, 57*(21), pp. 9878–9887.

Lopez-Galilea, I., Andriot, I., de Pena, M. P., Cid, C. and Guichard, E. (2008) How Does Roasting Process Influence the Retention of Coffee Aroma Compounds by Lyophilized Coffee Extract? *Journal of Food Science, 73*(3), pp. S165–S171.

Lubbers, S., Decourcelle, N., Martinez, D., Guichard, E. and Tromelin, A. (2007) Effect of thickeners on aroma compound behavior in a model dairy gel. *Journal of Agricultural and Food Chemistry, 55*(12), pp. 4835–4841.

Lubbers, S. and Guichard, E. (2003) The effects of sugars and pectin on flavour release from a fruit pastille model system. *Food Chemistry, 81*(2), pp. 269–273.

Lubbers, S., Landy, P. and Voilley, A. (1998) Retention and release of aroma compounds. *Food Technology, 52*(5), pp. 68, 70, 72, 74, 208, 210, 212, 214.

Lubbers, S., Verret, C. and Voilley, A. (2001) The effect of glycerol on the perceived aroma of a model wine and a white wine. *Lebensmittel-Wissenschaft und -Technologie, 34*(4), pp. 262–265.

Lubbers, S., Voilley, A., Feuillat, M. and Charpentier, C. (1994) Influence of mannoproteins from yeast on the aroma intensity of a model wine. *Food Science and Technology-Lebensmittel-Wissenschaft & Technologie, 27*(2), pp. 108–114.

Maier, H. G. (1975) Binding of volatile aroma substances to nutrients and foodstuffs. in Maarse, H. and Groenen, P. J., (eds.) *International symposium on aroma research*, Zeist (the Netherlands): Pudoc, Wageningen. pp. 143–157.

Marabotti, A., Balestreri, L., Cozzini, P., Mozzarelli, A., Kellogg, G. E. and Abraham, D. J. (2000) HINT predictive analysis of binding between retinol binding protein and hydrophobic ligands. *Bioorganic & Medicinal Chemistry Letters, 10*(18), pp. 2129–2132.

Marin, M., Baek, I. and Taylor, A. J. (1999) Volatile release from aqueous solutions under dynamic headspace dilution conditions. *Journal of Agricultural and Food Chemistry, 47*(11), pp. 4750–4755.

Mason, K., Patel, N. M., Ledel, A., Moallemi, C. C. and Wintner, E. A. (2004) Mapping protein pockets through their potential small-molecule binding volumes: QSCD applied to biological protein structures. *Journal of Computer-Aided Molecular Design, 18*(1), pp. 55–70.

Mc Nulty, P. B. and Karel, M. (1973) Factors affecting flavour release and uptake in O/W emulsion. III. Scale-up model and emulsion studies. *Journal of Food Technology, 8*, pp. 415–427.

Mei, J. B., Reineccius, G. A., Knighton, W. B. and Grimsrud, E. P. (2004) Influence of strawberry yogurt composition on aroma release. *Journal of Agricultural and Food Chemistry, 52*(20), pp. 6267–6270.

Meynier, A., Rampon, V., Dalgalarrondo, M. M. and Genot, C. (2004) Hexanal and t-2-hexenal form covalent bonds with whey proteins and sodium caseinate in aqueous solution. *International Dairy Journal, 14*(8), pp. 681–690.

Miettinen, S. M., Tuorila, H., Piironen, V., Vehkalahti, K. and Hyvönen, L. (2002) Effect of emulsion characteristics on the release of aroma as detected by sensory evaluation, static headspace gas chromatography, and electronic nose. *Journal of Agricultural and Food Chemistry, 50*(15), pp. 4232–4239.

Mitropoulou, A., Hatzidimitriou, E. and Paraskevopoulou, A. (2011) Aroma release of a model wine solution as influenced by the presence of non-volatile components. Effect of commercial tannin extracts, polysaccharides and artificial saliva. *Food Research International, 44*(5), pp. 1561–1570.

Nahon, D. F., Harrison, M. and Roozen, J. P. (2000) Modeling flavor release from aqueous sucrose solutions, using mass transfer and partition coefficients. *Journal of Agricultural and Food Chemistry, 48*(4), pp. 1278–1284.

Nawar, W. W. (1971) Some variables affecting composition of headspace aroma. *Journal of Agricultural and Food Chemistry, 19*(6), pp. 1057–1059.

O'Neill, T. E. (1996) Flavor binding by food proteins: An overview. in Robert J. McGorrin, J. V. L., (ed.) *Flavor-Food Interactions*: Washington, DC : American Chemical Society. pp. 59–74.

O'Neill, T. E. and Kinsella, J. E. (1987) Binding of alkanone flavors to beta-lactoglobulin: effects of conformational and chemical modification. *Journal of Agricultural and Food Chemistry, 35*(5), pp. 770–774.

Obretenov, C., Demyttenaere, J., Tehrani, K. A., Adams, A., Kersiene, M. and De Kimpe, N. (2002) Flavor release in the presence of melanoidins prepared from L-(+)-ascorbic acid and amino acids. *Journal of Agricultural and Food Chemistry, 50*(15), pp. 4244–4250.

Overbosch, P., Afterof, W. G. M. and Haring, P. G. M. (1991) Flavor release in the mouth. *Food Reviews International, 7*(2), pp. 137–184.

Perola, E. and Charifson, P. S. (2004) Conformational analysis of drug-like molecules bound to proteins: An extensive study of ligand reorganization upon binding. *Journal of Medicinal Chemistry*, *47*(10), pp. 2499–2510.

Piccone, P., Lonzarich, V., Navarini, L., Fusella, G. and Pittia, P. (2012) Effect of sugars on liquid-vapour partition of volatile compounds in ready-to-drink coffee beverages. *Journal of Mass Spectrometry*, *47*(9), pp. 1120–1131.

Poncet-Legrand, C., Cartalade, D., Putaux, J. L., Cheynier, W. and Vernhet, A. (2003) Flavan-3-ol aggregation in model ethanolic solutions: Incidence of polyphenol structure, concentration, ethanol content, and ionic strength. *Langmuir*, *19*(25), pp. 10563–10572.

Pozo-Bayon, M.-A., Biais, B., Rampon, V., Cayot, N. and Le Bail, P. (2008) Influence of complexation between amylose and a flavored model sponge cake on the degree of aroma compound release. *Journal of Agricultural and Food Chemistry*, *56*(15), pp. 6640–6647.

Rabe, S., Krings, U. and Berger, R. G. (2003) Initial dynamic flavour release from sodium chloride solutions. *European Food Research and Technology*, *218*(1), pp. 32–39.

Rabe, S., Krings, U. and Berger, R. G. (2004) Dynamic flavour release from miglyol/water emulsions: modelling and validation. *Food Chemistry*, *84*(1), pp. 117–125.

Reiners, J., Nicklaus, S. and Guichard, E. (2000) Interactions between beta-lactoglobulin and flavour compounds of different chemical classes. Impact of the protein on the odour perception of vanillin and eugenol. *Le Lait*, *80*, pp. 347–360.

Relkin, P., Fabre, M. and Guichard, E. (2004) Effect of fat nature and aroma compound hydrophobicity on flavor release from complex food emulsions. *Journal of Agricultural and Food Chemistry*, *52*(20), pp. 6257–6263.

Ridley, B. L., O'Neill, M. A. and Mohnen, D. A. (2001) Pectins: structure, biosynthesis, and oligogalacturonide-related signaling. *Phytochemistry*, *57*(6), pp. 929–967.

Roberts, D. D., Elmore, J. S., Langley, K. R. and Bakker, J. (1996) Effects of sucrose, guar gum, and carboxymethycellulose on the release of volatile flavor compounds under dynamic conditions. *Journal of Agricultural and Food Chemistry*, *44*, pp. 1321–1326.

Roberts, D. D., Pollien, P. and Watzke, B. (2003) Experimental and Modeling Studies Showing the Effect of Lipid Type and Level on Flavor Release from Milk-Based Liquid Emulsions. *Journal of Agricultural and Food Chemistry*, *51*(1), pp. 189–195.

Robinson, A. L., Ebeler, S. E., Heymann, H., Boss, P. K., Solomon, P. S. and Trengove, R. D. (2009) Interactions between wine volatile compounds and grape and wine matrix components influence aroma compound headspace partitioning. *Journal of Agricultural and Food Chemistry*, *57*(21), pp. 10313–22.

Rogacheva, S., Espinosa-Diaz, M. A. and Voilley, A. (1999) Transfer of aroma compounds in water-lipid systems : binding tendency of β–lactoglobulin. *Journal of Agricultural and Food Chemistry*, *47*(1), pp. 259–263.

Roudnitzky, N., Irl, H., Roudaut, G. and Guichard, E. (2003) Influence of fat nature on flavour release. in Le Quere, J. L. and Etievant, P. X., (eds.) *Flavour research at the dawn of the twenty-first century*, Paris (France): Lavoisier Tec & Doc. pp. 136–139.

Rutschmann, M. A. and Solms, J. (1990a) Formation of inclusion complexes of starch with different organic compounds. II. Study of ligand binding in binary model systems with decanal, 1-naphthol, monosteatate and monopalmitate. *Lebensmittel-Wissenschaft und -Technologie*, *23*, pp. 70–79.

Rutschmann, M. A. and Solms, J. (1990b) Formation of inclusion complexes of starch with different organic compounds. III. Study of ligand binding in binary model systems with (−)limonene. *Lebensmittel-Wissenschaft und -Technologie*, *23*(1), pp. 80–83.

Rutschmann, M. A. and Solms, J. (1990c) Formation of inclusion complexes of starch with different organic compounds. IV. Ligand binding and variability in helical conformations of V amylose complexes. *Lebensmittel-Wissenschaft und -Technologie*, *23*(1), pp. 84–87.

Saint-Eve, A., Lauverjat, C., Magnan, C., Deleris, I. and Souchon, I. (2009) Reducing salt and fat content: Impact of composition, texture and cognitive interactions on the perception of flavoured model cheeses. *Food Chemistry, 116*(1), pp. 167–175.

Sakurai, K., Oobatake, M. and Goto, Y. (2001) Salt-dependent monomer-dimer equilibrium of bovine beta-lactoglobulin at pH 3. *Protein Science, 10*(11), pp. 2325–35.

Samavati, V. and D-jomeh, Z. E. (2013) Multivariate-parameter optimization of aroma compound release from carbohydrate-oil-protein model emulsions. *Carbohydrate Polymers, 98*(2), pp. 1667–1676.

Savary, G., Guichard, E., Doublier, J.-L., Cayot, N. and Moreau, C. (2006) Influence of ingredients on the self-diffusion of aroma compounds in a model fruit preparation: An nuclear magnetic resonance-diffusion-ordered spectroscopy investigation. *Journal of Agricultural and Food Chemistry, 54*(3), pp. 665–671.

Schmitt, S., Kuhn, D. and Klebe, G. (2002) A new method to detect related function among proteins independent of sequence and fold homology. *Journal of Molecular Biology, 323*(2), pp. 387–406.

Semenova, M. G., Antipova, A. S., Belyakova, L. E., Polikarpov, Y. N., Wasserman, L. A., Misharina, T. A., Terenina, M. B. and Golovnya, R. V. (2002a) Binding of aroma compounds with legumin. III. Thermodynamics of competitive binding of aroma compounds with 11S globulin depending on the structure of aroma compounds. *Food Hydrocolloids, 16*(6), pp. 573–584.

Semenova, M. G., Antipova, A. S., Misharina, T. A. and Golovnya, R. V. (2002b) Binding of aroma compounds with legumin. I. Binding of hexyl acetate with 11S globulin depending on the protein molecular state in aqueous medium. *Food Hydrocolloids, 16*(6), pp. 557–564.

Siebert, K. J., Troukhanova, N. V. and Lynn, P. Y. (1996) Nature of polyphenol-protein interactions. *Journal of Agricultural and Food Chemistry, 44*(1), pp. 80–85.

Siefarth, C., Tyapkova, O., Beauchamp, J., Schweiggert, U., Buettner, A. and Bader, S. (2011) Influence of polyols and bulking agents on flavour release from low-viscosity solutions. *Food Chemistry, 129*(4), pp. 1462–1468.

Sostmann, K., Bernal, B., Andriot, I. and Guichard, E. (1997) Flavour binding by β-lactoglobulin: different approaches. in Kruse, H.-P. and Rothe, M., (eds.) *Flavor Perception*. Aroma Evaluation, Eisenach (Germany): Eigenverlag Universität Postdam. pp. 425–434.

Swaisgood, H. E. (1993) Review and update of casein chemistry. *Journal of Dairy Science, 76*(10), pp. 3054–3061.

Tavel, L., Andriot, I., Moreau, C. and Guichard, E. (2008) Interactions between beta-Lactoglobulin and Aroma Compounds: Different Binding Behaviors as a Function of Ligand Structure. *Journal of Agricultural and Food Chemistry, 56*(21), pp. 10208–10217.

Tavel, L., Moreau, C., Bouhallab, S., Li-Chan, E. C. Y. and Guichard, E. (2010) Interactions between aroma compounds and beta-lactoglobulin in the heat-induced molten globule state. *Food Chemistry, 119*(4), pp. 1550–1556.

Tromelin, A., Andriot, I. and Guichard, E. (2006) Protein-flavour interactions. in Voilley, A. and Etievant, P., (eds.) *Flavour in food*, Cambridge (GBR): Woodhead Publishing Limited. pp. 172–207.

Tromelin, A. and Guichard, E. (2003) Use of Catalyst in a 3D-QSAR study of the interactions between flavor compounds and β-lactoglobulin. *Journal of Agricultural and Food Chemistry, 51*(7), pp. 1977–1983.

Tromelin, A. and Guichard, E. (2004) 2D-and 3D-QSAR models of interaction between flavor compounds and beta-lactoglobulin using catalyst and Cerius2. *Qsar & Combinatorial Science, 23*(4), pp. 214–233.

Tromelin, A., Lubbers, S., Martinez, D. and Guichard, E. (2007) QSAR/QSPR approach of aroma compounds behaviour in food products. in This, H. , (ed.) *Euro Food Chem XIV*, Paris (France): Société Française de Chime, Paris. pp. 474–478.

Tromelin, A., Merabtine, Y., Andriot, I., Lubbers, S. and Guichard, E. (2010) Retention-release equilibrium of aroma compounds in polysaccharide gels: study by Quantitative Structure-Activity/Property Relationships approach. *Flavour and Fragrance Journal, 25*(6), pp. 431–442.

Tsachaki, M., Linforth, R. S. T. and Taylor, A. J. (2009) Aroma Release from Wines under Dynamic Conditions. *Journal of Agricultural and Food Chemistry, 57*(15), pp. 6976–6981.

Van Ruth, S. M., de Vries, G., Geary, M. and Giannouli, P. (2002a) Influence of composition and structure of oil-in-water emulsions on retention of aroma compounds. *Journal of the Science of Food and Agriculture, 82*, pp. 1028–1035.

Van Ruth, S. M., King, C. and Giannouli, P. (2002b) Influence of lipid fraction, emulsifier fraction, and mean particle diameter of oil-in-water emulsions on the release of 20 aroma compounds. *Journal of Agricultural and Food Chemistry, 50*(8), pp. 2365–2371.

Villamor, R. R., Evans, M. A., Mattinson, D. S. and Ross, C. F. (2013) Effects of ethanol, tannin and fructose on the headspace concentration and potential sensory significance of odorants in a model wine. *Food Research International, 50*(1), pp. 38–45.

Voilley, A., Beghin, V., Charpentier, C. and Peyron, D. (1991) Interactions between aroma substances and macromolecules in a model wine. *Lebensmittel Wissenschaft und Technologie, 24*(5), pp. 469–472.

Voilley, A., Simatos, D. and Loncin, M. (1977) Gas Phase Concentration of Volatiles in Equilibrium with a Liquid Aqueous Phase. *Lebensmittel-Wissenschaft und -Technologie, 10*, pp. 45–49.

Wong, D. W. S., Camirand, W. M. and Pavlath, A. E. (1996) Structures and functionalities of milk proteins. *Critical Reviews in Food Science and Nutrition, 36*(8), pp. 807–844.

Zhang, C., Huang, K., Yu, P. and Liu, H. (2011) Salting-out induced three-liquid-phase separation of Pt(IV), Pd(II) and Rh(III) in system of S201-acetonitrile-NaCl-water. *Separation and Purification Technology, 80*(1), pp. 81–89.

CHAPTER 10

Aroma release during in-mouth process

Andrea Romano[1,2]

[1] *Department of Food Quality and Nutrition, Research and Innovation Centre, Fondazione Edmund Mach (FEM), San Michele all'Adige, Italy*

[2] *Faculty of Science and Technology, Freie Universität Bozen – Libera Università di Bolzano, Bozen-Bolzano, Italy*

10.1 Instrumental techniques for monitoring in-mouth aroma release

10.1.1 In-mouth aroma release: a complex and dynamic process

The release of aroma compounds within the nasal and oral cavities is the result of a complex series of actions and phenomena. Imaging techniques such as Magnetic Resonance Imaging (MRI) and videofluoroscopy can be employed in the visualization of physiological processes in real time. The application of such techniques to the online monitoring of eating or drinking has allowed visualizing the main events linked to oral food processing (Buettner et al. 2002). In an early phase, when liquid and solid foods are first introduced into the mouth, the velum-tongue border is closed, and theoretically constitutes a barrier for aroma transfer to the nose. Nevertheless, due to factors that will be further discussed in this chapter, the occasional opening of the velum-tongue border during chewing allows the transfer of aroma compounds from the mouth to the nose. Upon swallowing, the epiglottis and the velum-pharynx border are open and volatile compounds are conveyed by exhalation to the nose. In a further stage, the remnants of liquid or solid food adhere to the oral and pharyngeal mucosae, originating a late release phase. All the events related to oral processing are relatively fast: time scales range from a few seconds to a few minutes, also depending upon the type of food. Moreover, due to the tidal airflow generated by breathing, the transfer of aroma compounds to the nose is not steady but it proceeds following rapid "aroma pulses". Considering these factors, it becomes obvious that, in order to capture the aroma pulses transiently generated during mastication, swallowing and breathing, an analytical technique should ideally have high time resolution. In addition to that, it must be noted that most food matrices are

Flavour: From food to perception, First Edition.
Edited by Elisabeth Guichard, Christian Salles, Martine Morzel, and Anne-Marie Le Bon.

characterized by a remarkable diversity in terms of aroma compound composition, as discussed in chapter 6. For instance, a compendium on coffee flavour chemistry (Flament and Bessière-Thomas 2002) mentions nearly 850 distinct volatile compounds, with very diverse chemical structures and properties, and concentrations ranging over seven orders of magnitude. This means that another important requisite will be analytical versatility, and therefore the possibility to measure many different and very chemically diverse compounds, with a linear and dynamic range as wide as possible.

Several different experimental approaches have been adopted in the last two decades, in the attempt to meet the aforementioned analytical challenges. They are generally based upon the employment of direct-injection-mass-spectrometry (DIMS). Several DIMS-based techniques have been developed, but as far as *in vivo* aroma release is concerned, most applications are based upon Atmospheric Pressure Chemical Ionization-Mass Spectrometry (APCI-MS) or Proton Transfer Reaction-Mass Spectrometry (PTR-MS). The main characteristics of these techniques will be reviewed in sections 1.2 and 1.3 of this chapter but for more theoretical and technical details, the reader is advised to refer to some recent and very exhaustive works (Biasioli et al. 2011, Ellis and Mayhew 2014).

10.1.2 Atmospheric pressure chemical ionization-mass spectrometry

Atmospheric Pressure Chemical Ionization-Mass Spectrometry (APCI-MS) is an analytical technique that has known a considerable development in the 1980s for the analysis of air and airborne substances. The possibility to employ APCI-MS to analyse *in vivo* aroma release was explored for the first time by Taylor and Linforth (1996), with the development of a dedicated source.

A schematic representation of a more recent version of APCI source (Le Quéré et al. 2014) is given in Figure 10.1. The sample (i.e., the subject's breath, exiting the mouth or the nostrils) enters the source through a venturi inlet, made with a capillary, sheathed with a second tube, which is flushed with a constant flow of nitrogen. Ionization is achieved by means of a corona discharge; typically, an excess of reagent ion is generated, which then reacts with the other compounds present in the mixture, which are finally detected by the mass analyser (typically a quadrupole mass spectrometer). When sampling room air, the main reagent ion thus generated is the hydronium ion (i.e., protonated water, H_3O^+). The generation of H_3O^+ is not straightforward but its production is the result of cascade reactions triggered by other chemical species, such as H_2O^+ and $N_2{}^+$.

The hydronium ion protonates neutral compounds (M) according to the reaction:

$$H_3O^+ + M \rightarrow H_2O + MH^+$$

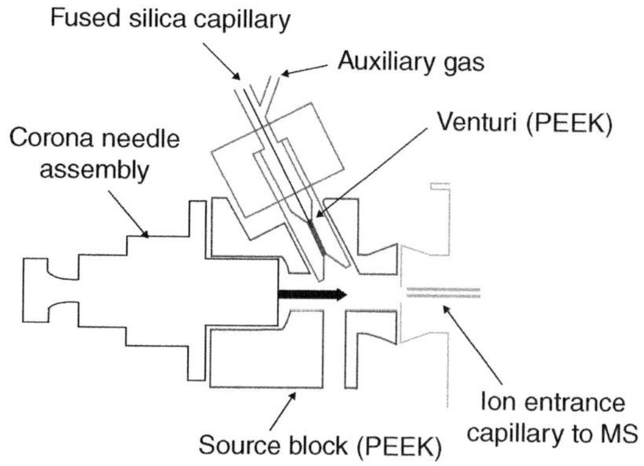

Figure 10.1 Schematic representation of an APCI source. Source: Quéré et al. (2014). Reproduced with permission of John Wiley & Sons.

The reactivity of a compound M as proton acceptor is expressed by its proton affinity. Only compounds with a proton affinity higher than that of water will react.

Table 10.1 reports a list of proton affinities for some representative compounds. It must be noted that, whereas the most abundant constituents of atmospheric air will not be ionized, most volatile organic compounds (with the exclusion of some aliphatic hydrocarbons), will be protonated and eventually detected. This is of pivotal importance to aroma analysis, as it means that most odorant molecules can be detected whereas air can be directly employed as buffer. Another advantage lies in the fact that, among the different types of chemical ionization, proton transfer from H_3O^+ is known to be particularly "soft". This means that upon ionization a parent ion, with general formula MH+, is formed and, ideally, does not undergo further fragmentation. The possibility of soft ionization significantly simplifies the interpretation of mass spectra, increasing the informational content of the analytical data. Nevertheless, some classes of molecules are more prone to undergo fragmentation than others; in APCI-MS, the fragmentation rate can be controlled to some extent by acting on the voltage applied to the cone element, typically placed right before the aperture leading to the mass analyser.

In comparison with other DIMS techniques, APCI-MS is characterized by great simplicity, as ionization is performed directly in ambient air. Instruments can be extremely lightweight and even portable (Huang et al. 2010). Nevertheless, due to the fact that the reaction conditions are not controlled, reagent ions also include charged water clusters with general formula $(H_2O)_nH^+$, with n being 2 or 3. The presence of water clusters is detrimental to standardization, as these species can also undergo proton transfer with other compounds, but they have different proton affinities and reaction constants. Moreover, water cluster concentration varies according to sample humidity: this is of particular relevance to the analysis of *in vivo* aroma release, as the sample undergoes continuous changes

Table 10.1 Proton affinities of the main air constituents and some volatile organic compounds representative of different chemical classes.

Compound	Proton affinity (kJ mol^{-1})
Argon	369
Oxygen	421
Nitrogen	494
Isobutane	678
Ethene	681
Water	**691**
Benzene	750
Acetaldehyde	769
Methyl sulfide	773
Ethanol	776
Acetonitrile	779
Acetic acid	784
Dimethyl ether	792
Propionic acid	797
Furan	803
Acetone	812
Thiophene	815
Dimethyl sulfide	831
Ethyl acetate	836
Acetamide	864
Pyrrole	875
Pyrazine	877
Methylamine	899
Thiazole	904

Source: Adapted from Hunter and Lias (1998).

in humidity throughout inhalation and exhalation cycles. Cluster formation can be reduced by increasing the temperature of the source and using a flow of dry nitrogen, acting as curtain gas. Introducing a dry inert gas between the sample and the source is beneficial, due to the low water content of the curtain gas itself, but also thanks to the de-clustering induced by collisions between charged clusters and inert nitrogen molecules.

When the performance of APCI-MS in the quantification of aroma compounds was evaluated (Taylor 2000), the results showed linear ranges of two orders of magnitude, even if deviations from linearity were observed for some compounds within complex mixtures, due to saturation (Buffo and Reineccius 2008). Detection limits are usually in the order of one ppb (part per billion, in volume), even though some researchers (Jost et al. 2003) have reported limits of detection in the range of a few tens of ppt (parts per trillion) in the quantification of airborne compounds, with considerable variation due to sample humidity.

One last caveat, which is in fact common to all techniques for volatile compound analysis, concerns the risk of absorption of odorants to all solid surfaces of the source and sample inlet. This is generally reduced by avoiding "cold spots" that might cause condensation, and by employing inert materials, such as Poly-Ether-Ether-Ketone (PEEK) or deactivated silica.

10.1.3 Proton transfer reaction-mass spectrometry (PTR-MS)

Proton Transfer Reaction-Mass Spectrometry was developed in the 1990s by Werner Lindinger and co-workers (Hansel et al. 1995). Interestingly, one of the first applications of the technique to be published concerned the measurement of sulfides in breath upon garlic consumption (Taucher et al. 1996). Like APCI-MS, PTR-MS is a direct injection mass spectrometry technique, where ionization is achieved by using protonated water as reagent ion. Therefore, most of the considerations regarding reaction mechanisms in APCI-MS hold valid for PTR-MS too.

Unlike APCI-MS, in PTR-MS a different technological philosophy is applied and greater attention is devoted to the control of the ionization conditions.

The instrumental apparatus is more complex (Figure 10.2). One key difference lies in the fact that primary ion generation and sample ionization are confined to two separate compartments of the instrument. The ion source (hollow cathode) generates a beam of hydronium ions; these enter a separated chamber (drift tube) where ionization takes place. Charged molecules are finally addressed to

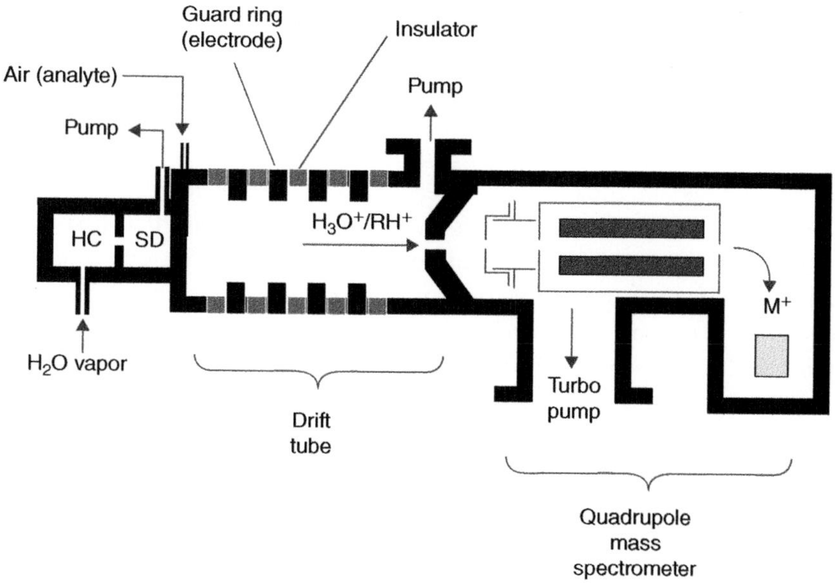

Figure 10.2 Schematic representation of a PTR-MS instrument. Source: Blake et al. (2009). Reproduced with permission of American Chemical Society.

the detector (a quadrupole mass analyser in Figure 10.2) by means of an electrical field applied to the drift tube. The drift tube is key for the functioning of the technique as it allows for the ionization to take place in controlled conditions of temperature, pressure and most importantly voltage. An increase in voltage results in an increased energy of the collisions between ions and uncharged molecules, this means that fragmentation is favoured while water clusters decrease. Conversely, when drift voltage is low, clusters increase but the ratio between parent ions and fragments is modified in favour of the former; moreover, a reduced drift voltage results in increased ionization efficiency and therefore higher sensitivity. In general, voltage and other drift tube parameters are optimized with the aim to find the best compromise between sensitivity, standardization and informational content of mass spectra.

The most widespread type of mass spectrometer in PTR-MS and APCI-MS instruments is the quadrupole. A quadrupole mass analyser acts as a mass filter: only ions having a given mass are allowed to reach the detector; by scanning a given range of masses, a mass spectrum is generally produced. This type of detector is robust, relatively inexpensive and the data it generates are easy to process, hence its widespread diffusion. The main drawback is represented by the low mass resolution: using a quadrupole it is only possible to measure the nominal mass of an ion, that is, the rounded up mass with no decimal digits. This means for instance that, in a hypothetical measurement of in-mouth aroma release, carbon dioxide (a typical constituent of breath gas) and acetaldehyde (a volatile compound released by food) would result in the same mass peak, having mass-to-charge ratio (m/z) equal to 45 (since z is always equal to one, this value always corresponds to the protonated mass). Another drawback results from the fact that, since the quadrupole acts as a mass filter, of all ions generated by a single ionization event only those having a given nominal mass are selected. A single mass spectrum results from the repetition of several analytical cycles and a full spectral scan (e.g., from m/z 18 to m/z 200) typically requires a few seconds.

In alternative to quadrupole-based PTR-MS, instruments coupled to time-of-flight (ToF) mass analysers were also developed: the first report on a PTR-ToF-MS was by Blake and co-authors (Blake et al. 2004); another better performing version of this type of instrument was commercialized more recently (Jordan et al. 2009). All ions generated by a single ionization event migrate within the ToF with a speed that only depends upon their mass, eventually reaching the detector. Ion mass is therefore estimated based upon the time taken to reach the detector. A very important consequence of this mode of functioning is the very high time resolution: a complete spectrum is generated for each ionization event and acquisition can typically be performed at frequencies of 0.1 Hz or lower. These values perfectly comply with the high time resolution required by *in vivo* monitoring of aroma release during mastication, breathing or swallowing events. The ToF mass analyser also guarantees high mass resolution: if mass calibration is performed correctly, ion masses can be determined up to the third decimal digit, allowing the assignment of a sum formula to most mass peaks. Like in other direct injection mass spectrometry techniques, the separation of isomers (i.e., molecules having the same sum formula but different chemical structures) is not possible.

In summary, the introduction of PTR-MS has granted an improved overall performance in volatile compound detection, obtained through an improved controllability of the ionization reaction. Limits of detection are in the low pptv range (Lindinger and Jordan 1998) and, if data are processed correctly, a dynamic range spanning over six orders of magnitude is obtained (Cappellin et al. 2011a). The recent implementation of the time-of-flight detector has opened new potentialities for online process monitoring, including the study of *in vivo* aroma release. One important consequence of this increased time and mass resolution is the large amount of data thus generated: PTR-ToF-MS typically generates one gigabyte of data per hour of analysis, also varying upon the rate of acquisition; this adds a further challenge to the analysis of in-mouth aroma release.

10.1.4 Gas chromatography-mass spectrometry (GC-MS)

Gas chromatography, often coupled with the use of mass analysers, represents the reference technique for the analysis of volatiles in food. With respect to DIMS techniques, GC-MS offers the possibility to separate and identify the single constituents of a gaseous mix, including isomers. On the other hand, GC-MS suffers from some inherent limitations:

(i) a single chromatographic run rarely allows a perfect separation of all the compounds present in the sample and some molecules can be co-eluted or not retained by the stationary phase, requiring the use of different types of columns, sometimes mounted in sequence;
(ii) injection is generally performed at high temperatures (200–250 °C). This reduces the adhesion of analytes to the inner surfaces of the instrument but also limits the applicability of the technique to some heat-unstable compounds;
(iii) in order to attain a sensitivity comparable to that of direct-injection-based techniques a step of pre-concentration is generally required.

The length of analysis represents the greatest limitation to the application of gas chromatography to the study of in-mouth aroma release; a typical analytical cycle lasts from a few minutes (in the so-called "FastGC") to more than an hour. Nevertheless, since GC-MS has been a widespread and consolidated technique for several decades, some of the first attempts at performing *in vivo* aroma release analysis were carried out using this instrumentation (Linforth and Taylor 1993). Some solutions have been devised with the aim to adapt GC-MS to the monitoring of eating and drinking. The compounds released during the in-mouth processing of food can be trapped using a glass capsule containing a stirring bar coated with a solid phase, acting as sorbent, which is held in the mouth during drinking (Buettner and Welle 2004, Buettner 2004). Odour-active compounds exiting the nostrils can also be captured using a trapping material (Buettner and Schieberle 2000, Itobe et al. 2009). Following desorption, GC-MS analysis is finally performed.

This type of approach has indeed been useful in identifying compounds released during eating or drinking. Nevertheless, its limitations, especially with respect to DIMS-based techniques, are evident. It must in fact be taken into account that the construction of a single release curve requires a several cycles of absorption, desorption and analysis. Therefore, due to practical reasons, the employment of GC-MS in the monitoring of *in vivo* aroma release has always been limited. DIMS and GC-MS are complementary techniques and the latter remains extremely valuable whenever exact compound identification is demanded.

10.2 Model and real systems of aroma release

10.2.1 *In silico* models of eating and drinking

In the last two decades, several mechanistic and empirical models were devised with the aim to simulate and predict the release of selected aroma compounds from liquid and solid foods during drinking or eating. Given the complexity of the processes involved, the development of such models can be challenging. Since the development of instrumental techniques for the online monitoring of *in vivo* aroma release (e.g., APCI-MS and PTR-MS), the *in silico* studies have acquired a precious tool for validation.

In the model developed by Wright et al. (2003), the amount of a given aroma compound released in the nose is predicted taking into account mass transfer from saliva to the gas phase along with physiological factors, such as breath period and air flow rate. The model is successful in predicting breath-by-breath variations in concentration for several different odorants in aqueous solution. Further studies (Doyennette et al. 2011, Normand 2004) presenting models of flavour release from liquid or semi-liquid foods, also take into account the formation of a pharyngeal deposit containing a bolus residue that is responsible for persistence effects in release. The models were also effective in predicting aroma release, as measured by APCI-MS or PTR-MS. Only recently (Doyennette et al. 2014), a mechanistic model was developed for the release from solid food; the study was successful in simulating aroma release curves from flavoured cheese products in different tasters. The increased complexity of the described system required the inclusion of additional factors related to product dissolution and chewing. Moreover, an accurate prediction of release curves required the measurement of individual physiological parameters, such as salivary flow, mouth coating, or the volumes of the nasal, oral and pharyngeal cavities.

The main outcomes of *in silico* studies include the possibility to develop a rationale for the formulation of foods and drinks with the aim to optimize volatile release kinetics. For example, it should be possible to predict the impact of viscosity changes on the release pattern for a given aroma compound. In addition, model sensitivity analysis helps to highlight the parameters having a major impact on aroma release.

10.2.2 Artificial mouths and model systems

One of the main drawbacks of *in vivo* studies resides in the high degree of experimental variability, mostly due to inter-individual differences among subjects. Before a trial, tasters are requested to refrain from eating or drinking strong-flavoured beverages for a certain time and, due to carry-over effects and acceptability issues, only a limited number of replicates is possible. All the aforementioned factors pose obvious practical constraints to aroma release studies employing human subjects: the development of model mouth systems has granted the possibility to simulate *in vitro* some aspects of in-mouth food processing, bypassing at least in part the need to employ human subjects. Artificial devices, aiming at reproducing one or more functions related to in-mouth processing, have been developed for dental, pharmaceutical and food research applications.

Model systems employed in food research vary upon the degree of complexity and standardization they provide. One of the simplest artificial mouths is the one developed by van Ruth (2000); this consists of a jacketed reactor (Figure 10.3a) with a volume (70 mL) comparable to that of the human mouth, and equipped with a plunger that performs simultaneous vertical and rotational movements, with the aim to simulate compression and shearing typical of chewing. Typically, the reactor is filled with a fixed volume of artificial saliva. The reactor can be interfaced with systems for volatile absorption (e.g., purge-and-trap systems equipped with Tenax® traps) for subsequent off-line volatile analysis (i.e., by GC-MS) or it can be coupled with instruments for on-line volatile monitoring (i.e., PTR-MS or APCI-MS). This artificial mouth was tested on several model and real foods. When the apparatus was tested on French fries (van Loon et al. 2005), it was possible to mimic the release of several volatile compounds, which were known to be key aroma constituents for this particular food product. The comparison between *in vitro* and *in vivo* analysis showed reasonable similarities between the corresponding release curves. A higher number of volatiles was detected in the artificial mouth experiment than during *in vivo* release. The effect of several factors (e.g., salt content, type of cooking process) was assessed. In comparison to the in-mouth experiment, some factors were indeed missing in the simulation, such as the possible change in salivary volume induced by differences in salt content.

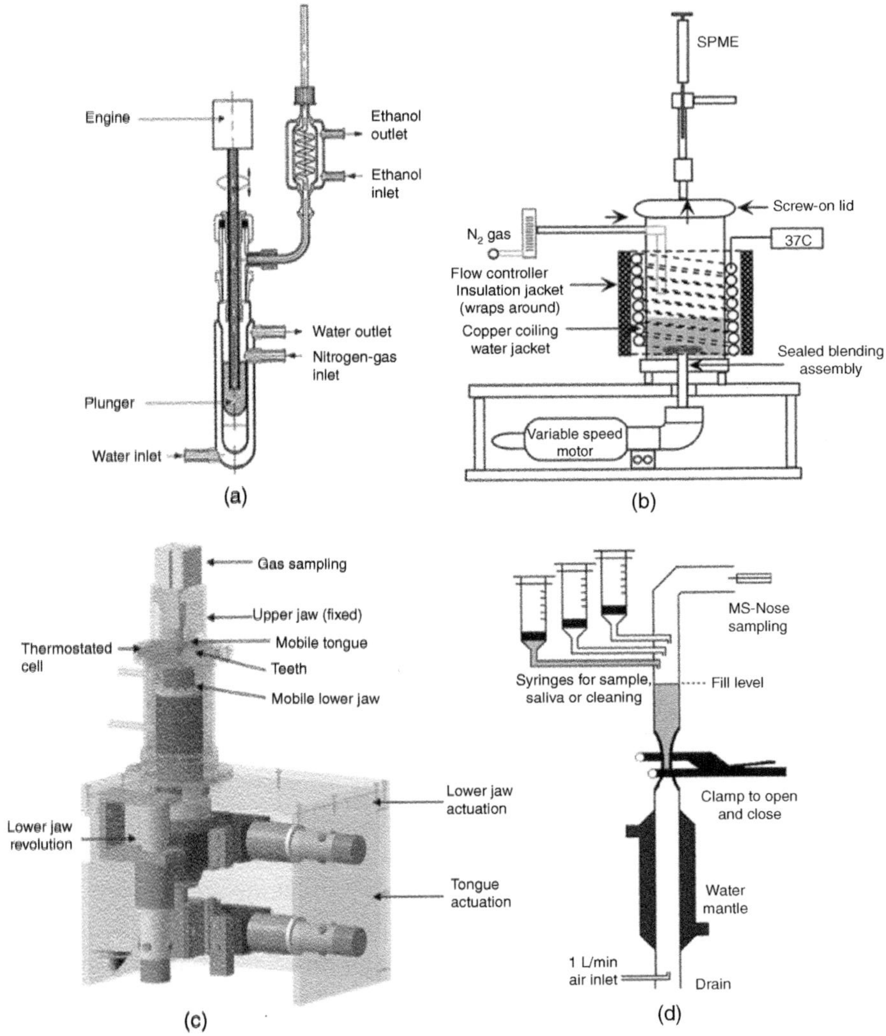

Figure 10.3 Schematic representations of three model mouths (a–c) and an artificial throat (d). Source: (a) Ruth (2000). Reproduced with permission of Elsevier; (b) Deibler et al. (2001). Reproduced with permission of American Chemical Society; (c) Salles et al. (2007). Reproduced with permission of Elsevier; (d) Weel et al. (2004). Reproduced with permission of American Chemical Society.

Other types of reactor were developed with overall volumes much higher than that of the human mouth (1-5 L), also with the aim to increase sensitivity. In spite of this higher volume, in the experiments performed using these reactors, a realistic ratio between liquid and gaseous phase was generally respected. In one of these model systems (Rabe et al. 2002), a great attention was devoted to the control of some physiological parameters (saliva and airflow) and to the flow conditions existing within the reactor, even though chewing and shear actions were not simulated. This model mouth was successful in reproducing in-mouth release for liquid model systems and real beverages, as measured by APCI-MS (Rabe 2004). The Retronasal Aroma Simulator (RAS) developed by Roberts and Acree (1995, Figure 10.3b) has a relatively high volume (1L); this reactor allows for a precise control of shear rate and, unlike the previously cited system, is also suited to solid food processing. In a comparison between RAS and real tasters (Deibler et al. 2001), relatively good correlations ($r^2 = 0.8$ or higher) were obtained for volatile release parameters measured by APCI-MS on very diverse food products (e.g., orange juice, chocolate, ice cream, and cookies). With respect to *in vivo* conditions, a much higher sensitivity and better reproducibility were granted.

Among artificial mouths, the device developed by Salles et al. (2007) is probably the most complex, granting precise control and mimicking of several eating-related functions. The device (Figure 10.3c) is equipped with a fixed upper jaw and a mobile lower jaw, both presenting molar-like indentations. A cylinder made of inert material is employed as "artificial tongue". The movement of these elements can be regulated in order to reproduce compression and shearing actions. The flows of air and artificial saliva can also be controlled. Similarly to other model devices, the system was connected to an APCI-MS in order to obtain profiles of volatile release during the processing of model and real foods (Mielle et al. 2010).

In spite of a certain level of sophistication, to date none of the aforementioned artificial mouth systems allows the full reproduction of *in vivo* processing of food from the point of view of aroma release. In all aforementioned systems, the swallowing event is absent. Moreover, as suggested by De Roos and Wolswinkel (1994), in the case of liquid and semi-liquid foods, release might be influenced by the formation of a liquid layer on the pharyngeal mucosa. This hypothesis was further corroborated by observations performed by videofluoroscopy (Buettner et al. 2002), and *in silico* simulations (Doyennette et al. 2011; Normand 2004). Another type of device, called "artificial throat", aims at reproducing this phenomenon. The artificial throat (Weel et al. 2004a, and Figure 10.3d) consists of a jacketed glass tube with improved wetting properties. A rubber tube equipped with a valve simulates swallowing. Sampling is performed from the upper end (by APCI-MS or PTR-MS) while an airflow can be applied from the lower end in order to reproduce exhalation. This relatively simple system aims at mimicking aroma release phenomena occurring in the throat, where volatile compounds contained within a thin layer are released through the action of a relatively high airflow. This system was employed to study the effect of oil content (Weel et al. 2004b) or carbonation (Pozo-Bayón et al. 2009) in the release of selected aroma compounds from liquid systems.

In conclusion, some mouth model systems focus on specific functions related to in-mouth processing (e.g., tongue action or chewing), while others aim at providing an all-round mimicking of the process. Even though none of these systems allows to completely reproduce in-mouth processing, their employment has provided precious information, also granting higher reproducibility and sensitivity than those obtained during *in vivo* experimentation. The development of mouth or throat model devices represents an inspiring example of the successful integration of scientific knowledge deriving from diverse fields of study, such as mathematical modelling, physiology, physical chemistry, and engineering.

10.2.3 Inter-individual differences in aroma release

The description of phenomena concerning in-mouth flavour release is sometimes influenced by the great inter-individual variability that characterizes all experimentation involving human subjects. Some studies have addressed this issue, attempting to describe and explain inter-subject variation. Typically, researchers look for correlations between some parameters characterizing the tasters' oral physiology and a limited number of target aroma compounds. These are supplemented to the model food, or known to be key compounds for a particular food product.

In a study performed on flavoured candies with a panel of 35 subjects (Blissett et al. 2006), a negative correlation was detected between mastication and aroma (limonene) release, as measured by APCI-MS. It was found that "fast" and "energetic" chewers tended to release lower amounts of flavouring agent from the confectionery system during chewing, probably because of reduced in-mouth processing time. Due to the great number of factors involved and to the interaction between them, these conclusions cannot always be extrapolated to other model foods: in fact, in a similar research performed on flavoured model cheese (Pionnier et al. 2004), a positive correlation was observed between respiratory rate, chewing activity and aroma release.

For semi-liquid food, swallowing time seems to play a major role. In a study carried out on strawberry-flavoured model custards supplemented with different amounts of thickening agent (Aprea et al. 2006), the panel could be divided into two subgroups, displaying either a late or an early swallow phase and also having different patterns of flavour release. Interestingly, inter-subject variability was more pronounced than the differences observed among the custard samples. When a tasting protocol involving the introduction of fixed swallow time was imposed to the panellists, the range of inter-individual variability was reduced.

In an investigation performed on flavoured milk gels of variable hardness (Gierczynski et al. 2008), the measurement of aroma release parameters determined by APCI-MS permitted the division of the panel into two subgroups (Figure 10.4). For the subjects belonging to group A, aroma compounds were detected throughout the chewing phase; this was explained with the possible occurrence of repeated events of velum opening during chewing. Subjects from group B displayed an onset of flavour release immediately after swallowing; this was probably because the velum-tongue closure formed a permanent barrier

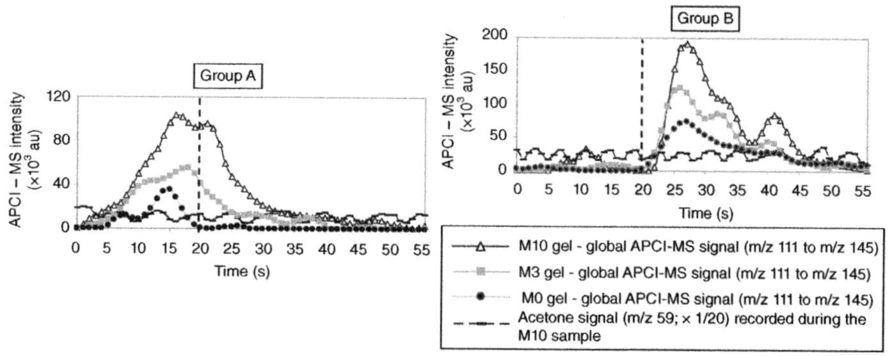

Figure 10.4 Typical aroma-release profile presented by two groups of panellists for three milk gels of different hardness (M0, M3, and M10). The dashed line represents the main swallow event. Source: Gierczynski et al. (2008). Reproduced with permission of American Chemical Society.

during chewing, while aroma compounds were transferred to the nose only in correspondence of the so-called "breath-swallow". Similar release patterns were also reported by other researchers (Mestres et al. 2005; 2006) in the chewing of flavoured protein gels of different hardness.

The aforementioned works display interesting analogies with a latter study performed on candies (Mishellany-Dutour et al. 2012), and involving the online monitoring of menthone release by APCI-MS, and the measurement of in-mouth air cavity (IMAC) by means of an acoustic pharyngometer device. In this case, a group of panellists was characterized by a constant IMAC value throughout tasting; this was accompanied by an equally constant aroma release pattern. A second group showed instead intermittent release, occurring in correspondence of the swallowing event and accompanied by an increase in IMAC. These observations underline the important role played by involuntary motor patterns in influencing aroma release. It must also be noted that some movements, such as the opening of the velum, are involuntary in most subjects but can also be controlled by trained individuals (e.g., professional wine tasters).

In a more recent research, Feron et al. (2014) have studied aroma release from flavoured cheese on a panel of 48 subjects. A wide series of measurements of oral parameters was carried out, including the determination of bolus properties, chewing activity, mouth coating and saliva composition. Aroma release and oral parameters were correlated by means of multivariate regression. This approach, involving the simultaneous determination of diverse aspects related to oral physiology and their inclusion within a single predictive model, allowed to highlight the most influential variables. Among all the measured variables, masticatory behaviour appeared to be most influential. Bolus rheology was found to be particularly significant for aroma release in hard cheeses: it is indeed possible that some properties, such as bolus spreadability, can be considered an indirect way to measure the formation of a pharyngeal residue, which is allegedly responsible of persistence phenomena in aroma release.

In aroma release studies, the attention is generally focused on monitoring a limited number of volatiles; this is partly because most instruments used to measure *in vivo* aroma release (i.e., APCI-MS and PTR-MS) are equipped with a quadrupole mass analyser. This type of detector allows the monitoring of a limited number of nominal masses. Increasing the number of monitored masses implies a reduction in sensitivity and/or time resolution. More recently, the introduction of PTR-MS instruments equipped with time-of-flight mass analysers (see section 1.3 of this chapter) has allowed for a drastic improvement in performance, including the possibility to perform untargeted studies on real and unmodified food products. In a recent work (Romano et al. 2014), three different coffees were consumed by five panellists and release profiles were determined by PTR-ToF-MS. As observed in a previous study (Aprea et al. 2006), the panellist effect appeared to be more important than the sample effect. The generation of individualized release profiles, based on 40 distinct ion peaks, has revealed a surprising variability in the interaction between tasters and products (Figure 10.5): for 4 out of 5 panellists aroma release showed positive correlation with the degree of roasting; this result is relatively unsurprising as volatiles are generally more concentrated in dark roasted coffee.

Unexpectedly, one panellist presented exactly opposite results, thus emphasizing the significance that the measurement of these inter-individual differences could have for consumer profiling purposes.

In 2004 Yeretzian et al. (2004), when presenting some recent advancements in the field of real-time measurement of aroma release, proposed the use of instrumental techniques in elucidating the mechanisms behind individual tastes and preferences. The examples presented in this section show how, in the last ten years, research has allowed food scientists advancing in this direction. Fundamental links between oral physiology and in-mouth aroma release have been disclosed. Moreover, the introduction of PTR-MS coupled to time-of-flight mass analysers has greatly improved the applicability of instrumental techniques in the characterization of panellists and consumers. Further achievements, obtained in the characterization of products and in the comparison between instrumental and sensory data, will be discussed in the forthcoming sections of this chapter.

10.3 Aroma release and matrix interactions

10.3.1 The influence of non-volatile compounds on aroma release

Non-volatile compounds have a key role in defining the sensory properties of food, including appearance, mouthfeel, taste and texture. The release of volatile compounds is also influenced by the overall composition of the food matrix itself:

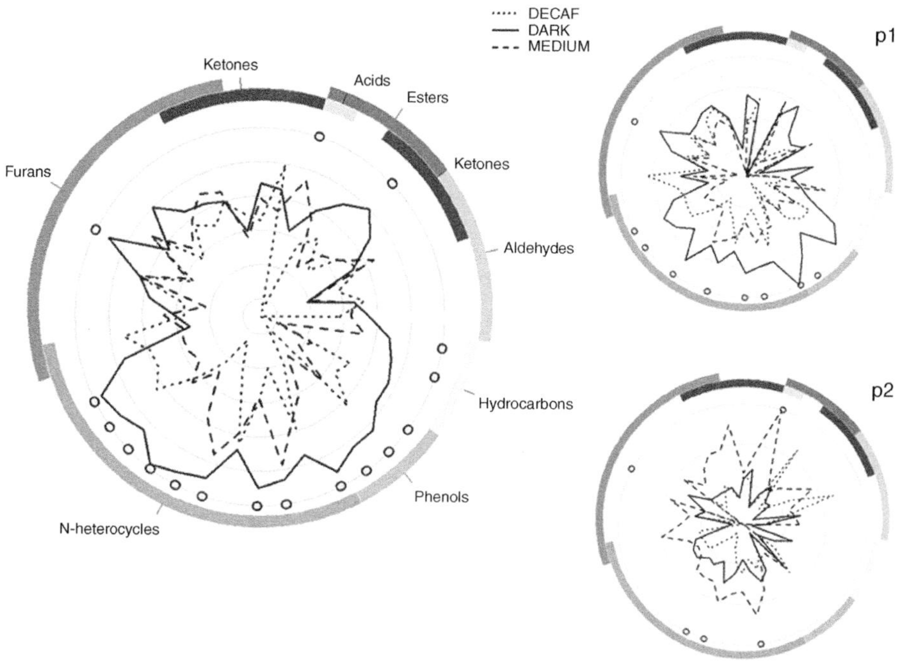

Figure 10.5 Radial plots representing aroma release profiles for three different coffees (DECAF, DARK, and MEDIUM). Profiles are portrayed in aggregate (left) or individual (right) form for two selected panellists (p1 and p2). The semi-circular bands on the outer margins represent chemical classes, based upon tentative identification of the peaks. Circles mark significant differences among coffee types (ANOVA and Tukey's test, $p < 0.05$). Source: Romano et al. (2014). Reproduced with permission of Elsevier.

fat, sugar, salt, proteins and additives do affect the dynamics of aroma release by acting on texture, hydration, lipophilicity, due to "salting out" effects, or by binding to volatile compounds, as detailed in Chapter 9.

Non-volatile compounds also define the nutritional properties of food. The consumption of food rich in salt, sugar and fat represents a risk factor for the development of obesity, heart conditions and metabolic diseases. In the development of healthier foods, the attention is often directed towards the reformulation of food products with the aim to reduce one of more of the aforementioned components, or to substitute sugars and saturated fats with ingredients providing a lower caloric intake. Unfortunately, this often results in products with altered sensory properties and lower acceptability. The measurement of aroma release as a function of food composition is therefore of major interest in the field.

When aroma release is monitored under *in vivo* conditions, the results are the product of a complex interaction of factors, including mastication, swallowing and interaction with saliva and mucosal surfaces. When volatile compound release is studied on the headspace of the product alone or by employing model mouth devices, the information related to oral processing is partially lost. Therefore, *in vivo* studies provide valuable indications on how to optimize aroma release and, when needed, find the best compromise between healthiness and acceptability. This concept will be further demonstrated in the following sections through a series of case studies, focusing on the influence of fat (section 3.2) and texture (section 3.3) on *in vivo* aroma release.

10.3.2 Fat and *in vivo* aroma release

With regard to the overall properties of a food product, the role of fat is manifold. Fat influences appearance, structure and mouthfeel of food, and, by affecting partition and mass transfer phenomena, it acts on aroma release. Fat is known to act as reservoir for aroma compounds, especially for lipophilic substances, promoting a more persistent release of the olfactory stimulus. In low-fat foods, the action of odorants, especially lipophilic ones, exhausts itself in rapid "aroma bursts", resulting in a product perceived as "unbalanced" and having lower acceptability (Plug and Haring, 1993).

The impact of fat on aroma release has been studied employing model systems, composed by oil/water emulsions, spiked with volatile compounds of variable hydrophobicity, as detailed in chapter 9. Such a system has the advantage of being simple and easily modifiable by changing the amount of added fat and the hydrophobicity of the target volatile compound.

Headspace and *in vivo* aroma release experiments have been performed using APCI-MS (Doyen et al. 2001; Malone et al. 2003), or PTR-MS (Weel et al. 2004b). The release curves obtained in the two experimental modes, and under different conditions, confirm the role of "volatile sink" of the lipid fraction. With lipophilic compounds, an increase in fat content results in a reduction in maximum intensity and overall release, accompanied by an increased persistence. Fat addition becomes less and less influential with a decrease in hydrophobicity of the target compound. This behaviour is apparent when comparing release kinetics of two volatile compounds of different lipophilicity, such as ethyl-hexanoate and acetone: fat addition reduces maximum intensity by one order of magnitude for the former, whereas no apparent reduction is observed for the latter. Interestingly, if headspace experiments are used to predict the effect of oil addition on *in vivo* time-dependent profiles, release values are systematically underestimated; that is, when performing in-mouth aroma release studies, volatile release is less influenced by an increase in fat content of the sample.

In headspace measurements, aroma release from oil/water emulsions is mostly influenced by the partitioning behaviour of the target compound. Expectedly, an increase in oil content reduces the release of hydrophobic compounds from emulsions. The reduced impact of fat additions observed during *in vivo* experiments is likely due to the peculiar conditions that characterize in-mouth processing. When the sample is first introduced into the mouth, it is mixed with saliva.

After swallowing, a thin layer, containing oil, water and aroma compounds is formed on the pharyngeal mucosa. Mechanistic models and experimental evidence (see sections 2.1 and 2.2 of this chapter) consider the formation of a bolus residue within the pharynx to be very important for persistence phenomena in aroma release. Throughout this process, an alteration in oil/water and oil/air partitioning phenomena is likely to take place, thus explaining the poor agreement between headspace and *in vivo* release data.

In studies carried out on water/oil systems, Frank et al. (2011) have measured *in vivo* release parameters, distinguishing between a pre-swallow and a post-swallow phase. As discussed in section 2.3, due to the opening of the velum-tongue barrier, aroma compounds can be transferred to the nasal cavity also before swallowing. Interestingly, differences in the effect of fat addition are observed between the pre-swallow and post-swallow phase and the pre-swallow release data show better agreement with dynamic headspace experiments. This corroborates the hypothesis that the peculiar conditions that characterize aroma release after the swallowing event cannot be reproduced by dynamic headspace experiments alone. On the contrary, when these conditions are simulated within an artificial throat device (Weel et al. 2004b), a good agreement is found between *in vitro* and *in vivo* results.

Overall, experiments have shown that, upon reduction of fat content, persistence effects in aroma release are reduced. This generally implies lower product acceptability. This justifies the interest of the food industry in smart food formulations that allow reduction in fat content of food without compromising persistence effects in aroma release. An example is represented by the system developed by Malone et al. (2003), called microstructured emulsions. This is based upon the incorporation of oil droplets within gel particles in the attempt to slow down the diffusion of hydrophobic compounds from the oil phase. When using ethyl-hexanoate as model compound, a reduction in fat content from 40% to 3% (expressed as volume fraction of the emulsion) is possible with no apparent reduction of persistence effects in aroma release.

Another aspect of practical interest is represented by the possibility to employ experimental data to develop predictive models with the aim to provide indications as to the formulation of low-fat foods with improved organoleptic properties. In order to predict the effect of a change in fat content on aroma release, Linforth et al. (2010) have developed an empirical model based on a total of 345 experiments derived from literature data and unpublished results. The model allowed expressing the expected reduction in aroma release as a function of the initial and final fat contents of the food product and the log$P$ of the target compound. The log$P$ value, representing the logarithm of the partition coefficient between octanol and water, was employed as index of hydrophobicity of a compound. Subdividing the dataset into a training and test set, a relatively good (R^2=0.71) correlation was obtained between predicted and observed responses. One limitation resided in the non-mechanistic nature of the model, which did not provide information as to the mechanisms governing aroma release. On the other hand, several mechanistic models were successful in predicting aroma release from liquid (Normand 2004), semiliquid (Trelea et al. 2007) or solid foods (Doyennette et al. 2014), and it is conceivable that these could be adapted in order to include parameters relative to bolus fat content.

The in-mouth release of aroma compounds of variable lipophilicity was also studied in systems that are more complex. Low-fat and high-fat milk showed behaviour similar to that observed for emulsions (Roberts et al. 2003; Shojaei et al. 2006). Analogous effects on intensity and persistence of aroma release were obtained in biscuits (Brauss et al. 1999a) and yogurt (Brauss et al. 1999b). Yogurt represents an interesting example of the multimodal impact of fat on food properties. Indeed, it was observed that, apart from the aforementioned impact on partitioning phenomena, fat addition increased yogurt viscosity, possibly reducing bolus spreadability and aroma release. Cheese constitutes another model of interest for the study of the action of fat on flavour release within complex systems. Boisard et al. (2014) have measured aroma release on cheeses of different composition, studying, among others, the effect of varying the lipid/protein (L/P) ratio. Again, persistence effects in flavour release were observed in samples with higher L/P ratio, and especially for hydrophobic compounds. Overall, the study of the effect of fat content on volatile release reveals its role of "flavour sink"; *in vitro* and *in vivo* models are not always in complete agreement and the former tend to overestimate the effect of fat addition. These discrepancies can be explained if we take into account the presence of some factors related to oral processing.

10.3.3 Texture and *in vivo* aroma release

Textural parameters of food, such as viscosity or hardness, can have an influence on the release of aroma compounds. Changes in texture affect diffusion of volatile compounds through the food matrix, by influencing its resistance to mass transfer. Under *in vivo* conditions, textural parameters also have an impact on the rate of food breakdown as well as on oral behaviour, ultimately influencing aroma release.

The viscosity of model solutions can be varied by adding different amount of thickening agents. Solutions of aroma compounds were supplemented with different amounts of guar gum, xanthan, or alginate, and submitted to APCI-MS analysis under *in vivo* conditions (Malone et al. 2003). An increase in viscosity caused a decrease in maximum intensity of the release profiles (*Imax*). Compared to the effect of fat addition under analogous conditions, the impact of viscosity is limited, amounting to a 50% decrease in *Imax*. This reduction in *Imax* is observed in correspondence of an increase in viscosity by six orders of magnitude. This goes well beyond realistic conditions. Indeed, for real food products, the range of acceptability in terms of viscosity usually spans over roughly one order of magnitude. In a study performed on flavoured yogurt (Saint-Eve et al. 2006), mechanical treatments of different intensities were used to obtain products having the same composition but differing in overall viscosity. The low viscosity yogurts displayed lower *Imax* in aroma release. Based upon mass transfer analysis, the authors attributed this effect to a higher spreadability of low viscosity products, resulting in a higher exchange surface between product and air during in-mouth processing.

The impact of texture on volatile release was also studied in solid model systems, composed by protein or pectin gels of different hardness, supplemented with different aroma blends or single compounds (Boland et al. 2006; Mestres et al. 2006, 2005). In summary, gels having higher hardness generate aroma release profiles where *Imax*, area under the curve (*AUC*) and time to reach maximum intensity (*tmax*) are higher, and initial release rates (*RR*) are lower. These differences were attributed to the interaction between texture and oral behaviour. Soft gels were processed by panellists using the incisors, or pressed between the tongue and the hard palate, whereas hard gels were positioned in the back part of the mouth and chewed with the molars. Chewing of harder gels was associated with higher salivation, possibly because salivary glands, also located in the back of the mouth, were subjected to a higher degree of stimulation. The inspection of the aroma release profiles also suggested that, during chewing of softer gels, no opening of the velum-tongue barrier took place and the biggest pulse of volatile release was observed in correspondence of the "breath-swallow". On the other hand, during in-mouth processing of softer gels, intermittent opening of the velum-tongue barrier took place, generating minor pulses of aroma release. An explanation for these differences in oral behaviour could depend on an unconscious adaptation to texture perception: upon perceiving liquid or semi-liquid foods, the velum-tongue barrier closure becomes tighter in order to prevent accidental transfer of liquid bolus into the airways. Analogies are encountered between the aforementioned work and the results of Gierczynski et al. (2008), also showing two groups of panellists with different oral behaviours in the oral processing of milk gels (see also Figure 10.4 and section 2.3 of this chapter). Interestingly, the results of Gierczynski et al. (2008) showed that the two different behaviours were invariable for most panellists, but they could occasionally be altered by a change in gel hardness.

Other studies performed on flavoured gel systems of different hardness, although confirming this general trend, show occasional discrepancies with the aforementioned results. In the works of Gierczynski et al. (2007) on cheese-like gels and Repoux et al. (2012) on flavoured model cheeses, higher rates of release are obtained for products having harder texture. Other researchers (Baek et al. 1999; Weel et al. 2002) report no difference in aroma release from gels having different textures. These contradictions can be explained considering that all the tested model foods display differences in preparation and composition. This could result in products with different brittleness, which undergo oral breakdown according to different modalities. Moreover, it must be taken into account that researchers impose different protocols for product consumption. This factor was examined in a study carried out on candies prepared using different gelatin contents (Déléris et al. 2011). Samples were either chewed or melted in the mouth by the panellists, showing that with the former protocol much faster aroma release kinetics and shorter times of residence in mouth were obtained.

Most of the examples presented so far are relative to model systems or processed foods. These products present the advantage that it is relatively simple to modify one property (e.g., texture) without changing the aroma compound content. This is instead not possible in the case of non-processed foods such as fruits. When studying the impact of fruit piece hardness on volatile release from low-fat pear yogurts (Mesurolle et al. 2013), it was hard to establish an unambiguous relationship between fruit texture and aroma release. This is probably because, in order to obtain fruit pieces with different hardness, different cultivars must be employed. The analysis of textural parameters, headspace and *in vivo* volatile release in apples (Ting et al. 2012) reveals three types of behaviour: (i) firm (hard) cultivars with low *AUC*, (ii) firm cultivars with high *AUC* and *Imax*, and (iii) mealy (soft) cultivars with high *AUC* and *Imax*. It is not possible to establish a clear connection between texture and release but the classification is rather the result of the interaction between inter-varietal differences in volatile compound contents and textural parameters.

In the examples mentioned so far, the impact of textural parameters (mainly viscosity and hardness) on in-mouth aroma release was examined. It must also be considered that the tactile sensations associated to texture perceptions also modify the way aroma is perceived. The measurement of *in vivo* aroma release can help to enlighten the perceptual interaction between texture and aroma and this aspect will be further addressed in section 4.3 of this chapter.

10.4 Sensory analysis and *in vivo* aroma release

10.4.1 Static and dynamic methods in sensory analysis

The interest in methods for *in vivo* analysis of aroma release arises, among others, from the idea that time evolution of the volatile profile of a food product during its consumption contains precious information, which would not be obtained by headspace measurements alone. It is also reasonable to expect that this approach would help explaining the sensory properties of a product acquired during consumption.

In the generation of a sensory profile, "static" methods are most often used, such as Quantitative Descriptive Analysis (Murray et al. 2001). In such approaches, judges are requested to rate a product according to one or more attributes, using intensity scores. When doing so, the overall intensity of a sensory attribute is extracted across a given time period, while no information about its time evolution is collected.

Since the 1930s, researchers have been aware of the temporal character of perception (Holway and Hurvich 1937). During the following decades, a dynamic methodology of sensory analysis, known as Time-Intensity (TI), was developed and consolidated (see Cliff and Heymann 1993 for review of the technique). In TI, the approach to attribute rating is similar to that adopted in static descriptive analysis but every attribute is separately rated over time, generating a bell-shaped profile that shows an undeniable similarity to aroma release curves obtained by APCI-MS or PTR-MS. In comparison to Quantitative Descriptive Analysis, where several attributes are evaluated simultaneously, the technique is much more time-consuming, as the panellist rates only one attribute at a time. Moreover, when evaluating only one attribute, dumping effect may occur (Clark and Lawless 1994). For example, when evaluating of bitterness in wine, ratings may be exaggerated due to the interaction with sour taste or to the chemestetic perception of astringency (Sokolowsky and Fischer 2012).

More recently, a novel procedure, called Temporal Dominance of Sensations (TDS), was introduced (Pineau et al. 2003) for the dynamic sensory analysis of food. In TDS, a list of attributes is generated for the product of interest; during the evaluation, the judge is requested to choose from the list which attribute is dominant at a given time. The dominant attribute is defined as the one that most captures the panellist's attention, which is not necessarily the most intense. At a given moment of the analysis, each attribute will be considered as dominant by a fraction of the panel: this corresponds to the dominance rate of the given attribute. Dominance rates are submitted to a statistical test to establish whether they are significantly higher than the chance level. For a recent review on TDS, the reader is advised to refer to Di Monaco et al. (2014).

The time-dependent profiles obtained using either TI or TDS were compared on the analysis of dairy products (Pineau et al. 2009). Comparing the profiles obtained on the same product some similarities are evident. The attributes that are rated as most intense by TI also have the highest dominance rates in TDS. On the other hand, since in TDS judges are forced to choose only one attribute at a time, this technique inevitably generates a temporal sequence of sensory notes, thus enhancing the sequence of sensations over time. Moreover, TDS is more high-throughput than TI, as in this methodology multiple characteristics can be evaluated simultaneously. However, it must also be noted that the two methodologies are based on different approaches and the choice of the appropriate technique will be based on the aim of the research. TI is more suited to the analysis of a single sensory attribute, while TDS focuses on the temporal sequence of sensations elicited during drinking or eating. Each of the two techniques requires a specific training and judges must develop a precise mindset, centred on evaluating either intensity or dominance. Therefore, when selecting panellists for TDS, it is advisable to choose subjects that have not been previously trained to classic descriptive analysis (Meyners, 2011).

10.4.2 *In vivo* aroma release and sensory analysis

Among all sensory analysis techniques, the typical output of Time-Intensity (TI) is the one that most closely resembles the curve obtained in an *in vivo* aroma release experiment. Typically, when studying a model food supplemented with a target compound having an easily recognizable sensory note, the TI curve and the aroma release curve obtained by APCI-MS or PTR-MS show closely related profiles. In the work of Linforth et al. (1999), aroma release and TI curves were produced simultaneously during eating of flavoured gels. Interestingly, in products having relatively fast oral breakdown rates, the time to reach the maximum (*tmax*) was lower in the TI curve than in the aroma release curve. Following the generally accepted theory of perception, this could be because of adaptation (Overbosch et al.1986): due to this mechanism, after the first aroma burst, the response to the sensory stimulus decreases, explaining the time discrepancy between perceived maximum and aroma release maximum. On the other hand, a delay in the *tmax* of the TI curve was highlighted in gels that underwent rapid in-mouth processing. This delay can be explained considering the temporal integration of the sensory stimulus (Overbosch et al. 1989). This can be directly verified if an olfactory stimulus is directly transferred to the nasal mucosa by diluting an aroma compound in a gas stream, typically by means of an olfactometer (Leclercq and Blancher 2012). At the same time, a TI curve is acquired, showing an average 9-10s delay between stimulus and response. It is interesting to note that, when the same aroma is supplied in the form of a flavoured gelled candy, this delay is reduced to 2-5 s. This suggests that the presence of texture and taste cues reduces temporal integration of olfactory stimuli.

Nowadays, a great deal of experimental evidence shows that flavour perception results from the integration of cues originating from different sensory domains (Auvray and Spence 2008), as reviewed in detail in Chapter 16. In this section, some examples will be provided, showing the contribution of *in vivo* aroma release studies to the discovery of perceptual interactions during eating or drinking. In a study on whey protein gels of different texture supplemented with either ethyl-butyrate or diacetyl, Weel et al. (2002) have observed that, upon increasing gel hardness, the TI curves indicated a reduction in the maximum and overall sensory stimulus perceived. The *in vivo* release curves generated by APCI-MS did not show significant changes in aroma release kinetics, thus suggesting that in this particular case aroma perception is driven by texture perception rather than by the amount of aroma present in the nose. The use of APCI-MS or PTR-MS has also allowed highlighting perceptual interaction between taste and aroma. The analysis of model dairy desserts having different texture and supplemented with a fruit aroma blend (Lethuaut et al. 2004) has revealed that perception of fruity aroma was mostly influenced by sweetness: products with a higher amount of sucrose were perceived with a more intense aroma according to TI analysis, whereas *in vivo* aroma release was not influenced by sweetness or texture. Analogously, the combined study of aroma release and perception on different chewing-gum formulations has shown that aroma perception was correlated with sweetener release, rather than with the in-nose concentration of aroma compounds (Potineni and Peterson 2008).

A similar interaction between saltiness and savoury aroma (i.e., garlic or mushroom aroma) was demonstrated by (Cook et al. 2003). It is reasonable to imagine that all these cognitive interactions (e.g., apple-sweet, mushroom-salty) derive from the habit of perceiving certain smell and taste stimuli in association with one another; the prevalence of such associations over aroma release kinetics remains indeed noteworthy.

Davidson et al. (1999) have measured menthone and sucrose release from chewing-gums during in-mouth processing employing APCI-MS. The online monitoring of sucrose release was performed using cotton swabs, also thanks to an instrument equipped with a liquid sampling interface. The evaluation of mint flavour perception by TI showed that perception was related to sucrose, rather than to menthone release. Interestingly, mint flavour is in itself an example of multi-modal perception, involving both a distinctive aroma and a cooling (trigeminal) effect. This case study also demonstrates the suitability of Time-Intensity to products requiring long periods of oral processing, and designed to elicit a persistent perception such as chewing gum.

Temporal Dominance of Sensations (TDS) has existed for only a decade, and applications combining this dynamic sensory analysis with *in vivo* aroma release are relatively uncommon. The works reported in the literature apply PTR-MS with TDS to complex unmodified foods (e.g., coffee), or model foods supplemented with aroma blends, thus exploiting the possibility to follow several sensory notes and aroma compounds simultaneously. This approach was applied to candies having different texture, also examining the impact of the consumption protocol on perception and *in vivo* aroma release (Déléris et al. 2011). Similar researches were conducted on flavoured vodka (Deleris et al. 2011), fruit yogurt (Mesurolle et al. 2013), and espresso coffee (Barron et al. 2012; Charles et al. 2015). In some instances, it was possible to establish a correlation between the dominance of some sensory attributes and specific mass peaks. In flavoured candies, the dominance of a strawberry note could be linked to peak *m/z* 145, tentatively assigned to ethyl-hexanoate, also associated to fruity aroma (Déléris et al. 2011). In espresso coffees, the sequential release of mass peaks attributed to different heterocyclic compounds could be related to distinct temporal phases of perception of the "burnt" note (Charles et al. 2015). Nevertheless, in spite of a few successful examples, TDS and PTR-MS remain difficult to compare, also due to intrinsic differences between the two techniques. In fact, instrumental data is based upon measuring aroma concentration, whereas the concept of dominance implied by TDS does not necessarily relate to stimulus intensity. Moreover, perceptual interactions between texture and aroma, and taste and aroma are possible. Finally, the list of attributes employed in TDS inevitably includes both taste and aroma attributes while PTR-MS only detects volatile compounds; hence, the overt dominance of a taste-related attribute could mask information about olfactory notes possibly perceived by the judges.

Overall, the association of dynamic sensory and instrumental methods represents a valid example of complementarity between analytical techniques, the comparison of data originating from different domains is sometimes challenging but the combination of the two approaches brings original evidence regarding the mechanisms of perception.

10.5 Conclusion and perspectives

The hypothesis that justifies the need to perform *in vivo* aroma release studies is relatively simple: measuring the volatile compounds released by a given food product during drinking or eating (the so-called "nosespace") is likely to provide more useful information than those obtained from headspace alone. According to this assumption, the nosespace composition of a given food should be more easily linkable to sensory perception. Indeed, over the last two decades such a straightforward idea has generated a surprising amount of useful information. *In vivo* aroma release studies could help researchers to reduce fat and/or salt contents in food without compromising its palatability. Moreover, we have a better understanding as to why we perceive food in a certain way and what are the reasons that make all tasters and consumers so remarkably different from each other.

Instrumentation based upon direct injection-mass spectrometry has played a key role in the advancement of *in vivo* aroma release studies. The coupling of PTR-MS to time-of-flight mass analysers represents the latest technical advancement. As discussed in section 1.3 of this chapter, the introduction of PTR-ToF-MS guarantees, among others, the possibility to perform untargeted analysis, monitoring the evolution of hundreds of volatile markers of potential interest in real time. It appears evident from the various examples presented throughout this chapter that most *in vivo* aroma release studies rely on the measurement of one or a few target compounds added to a model food. These conditions comply with the need of standardization required by experimentation, and provide indications that are indeed applicable to contexts that are more complex. This choice is also dictated by the limitations of instruments equipped with quadrupole mass analysers, and namely (i) it is not possible to scan more than a few masses per second of analysis, and (ii) only nominal masses of ions can be determined. On the other hand, this approach is less successful on food products that are not easy to modify (e.g., fruits) or whose chemical composition is particularly complex and where by markers of a given sensory notes are not entirely known (e.g., coffee or wine). The use of ToF mass analysers also allows relating unambiguously ion peaks to chemical formulae, providing precious hints as to the possible identities of new chemical markers of sensory notes of interest.

In spite of this analytical power and of the availability of a commercial version of the instrument since 2009, the application of PTR-ToF-MS to *in vivo* aroma release monitoring remains limited. To date, and to the best of our knowledge, only three original papers were published (Heenan et al. 2012, Romano et al. 2014, Charles et al. 2015), employing the technique on flavoured cereal bars and coffee. Apart from considerations regarding obvious economic constraints (the ToF-based instrument is more expensive than the quadrupole-based version), another serious issue is represented by data treatment. Unlike quadrupole based PTR-MS, where a classic bi-dimensional spreadsheet (masses x time) constitutes the typical throughput, PTR-ToF-MS generates a file in the form of a non-user-friendly multi-dimensional array. Moreover, in order to exploit its potentialities, data must be properly processed. The company commercializing the PTR-ToF-MS (Ionicon Analytik GmbH, Innsbruck, Austria) supplies its own proprietary software for data processing and in addition to that, some research groups have developed in-house software packages (Cappellin et al. 2011b, Müller et al. 2013).

In summary, even in the absence of a unified approach to data analysis, PTR-ToF-MS data processing is indeed possible, even though it can be lengthy. Moreover, even after post-processing, the dataset resulting from an *in vivo* aroma release experiment can be massive. For example, in an experiment performed in the author's lab (Charles et al. 2015), 216 coffee tasting sessions were performed, each lasting about two minutes. Data processing alone required approximately one month of computing time, and the final dataset consisted of a spreadsheet containing nearly 10^7 numeric values (465 mass peaks x 19,440 time points). These figures can indeed be discouraging, but this example also highlights the importance of a multi-disciplinary approach in food research, also involving computer science and chemometrics. As it often happens in experimental science, progress was driven by the development and application of new instrumental techniques. Similarly, it is likely that further technical developments will again pave the way to future discoveries.

References

Aprea, E., Biasioli, F., Gasperi, F., Märk, T. D. and van Ruth, S. (2006) In vivo monitoring of strawberry flavour release from model custards: effect of texture and oral processing. *Flavour and Fragrance Journal, 21*(1), pp. 53–58.

Auvray, M. and Spence, C. (2008) The multisensory perception of flavor. *Consciousness and Cognition, 17*(3), pp. 1016–1031.

Baek, I., Linforth, R. S., Blake, A. and Taylor, A. J. (1999) Sensory perception is related to the rate of change of volatile concentration in-nose during eating of model gels. *Chemical Senses, 24*(2), pp. 155–160.

Barron, D., Pineau, N., Matthey-Doret, W., Ali, S., Sudre, J., Germain, J. C., Kolodziejcyk, E., Pollien, P., Labbe. D., Jarisch, C., Dugas, V., Hartmann, C. and Folmer, B. (2012) Impact of crema on the aroma release and the in-mouth sensory perception of espresso coffee. *Food & Function, 3*(9), pp. 923–930.

Biasioli, F., Yeretzian, C., Märk, T. D., Dewulf, J. and Van Langenhove, H. (2011) Direct-injection mass spectrometry adds the time dimension to (B)VOC analysis. *TrAC Trends in Analytical Chemistry, 30*(7), pp. 1003–1017.

Blake, R. S., Monks, P. S. and Ellis, A. M. (2009) Proton-Transfer Reaction Mass Spectrometry. *Chemical Reviews, 109*(3), pp. 861–896.

Blake, R. S., Whyte, C., Hughes, C. O., Ellis, A. M. and Monks, P. S. (2004) Demonstration of Proton-Transfer Reaction Time-of-Flight Mass Spectrometry for Real-Time Analysis of Trace Volatile Organic Compounds. *Analytical Chemistry, 76*(13), pp. 3841–3845.

Blissett, A., Hort, J. and Taylor, A. J. (2006) Influence of chewing and swallowing behavior on volatile release in two confectionery systems. *Journal of Texture Studies, 37*(5), pp. 476–496.

Boisard, L., Andriot, I., Martin, C., Septier, C., Boissard, V., Salles, C. and Guichard, E. (2014) The salt and lipid composition of model cheeses modifies in-mouth flavour release and perception related to the free sodium ion content. *Food Chemistry, 145*, pp. 437–444.

Boland, A., Delahunty, C. and Vanruth, S. (2006) Influence of the texture of gelatin gels and pectin gels on strawberry flavour release and perception. *Food Chemistry, 96*(3), pp. 452–460.

Brauss, M. S., Balders, B., Linforth, R. S. T., Avison, S. and Taylor, A. J. (1999a) Fat content, baking time, hydration and temperature affect flavour release from biscuits in model-mouth and real systems. *Flavour and Fragrance Journal, 14*(6), pp. 351–357.

Brauss, M. S., Linforth, R. S. T., Cayeux, I., Harvey, B. and Taylor, A. J. (1999b) Altering the Fat Content Affects Flavor Release in a Model Yogurt System. *Journal of Agricultural and Food Chemistry, 47*(5), pp. 2055–2059.

Buettner, A. (2004) Investigation of potent odorants and afterodor development in two Chardonnay wines using the buccal odor screening system (BOSS). *Journal of Agricultural and Food Chemistry, 52*(8), pp. 2339–2346.

Buettner, A., Beer, A., Hannig, C., Settles, M. and Schieberle, P. (2002) Physiological and analytical studies on flavor perception dynamics as induced by the eating and swallowing process. *Food Quality and Preference, 13*(7-8), pp. 497–504.

Buettner, A. and Schieberle, P. (2000). Exhaled Odorant Measurement (EXOM) — A New Approach to Quantify the Degree of In-mouth Release of Food Aroma Compounds. *LWT - Food Science and Technology, 33*(8), pp. 553–559.

Buettner, A. and Welle, F. (2004) Intra-oral detection of potent odorants using a modified stir-bar sorptive extraction system in combination with HRGC-O, known as the buccal odour screening system (BOSS). *Flavour and Fragrance Journal, 19*(6), pp. 505–514.

Buffo, R. A. and Reineccius, G. A. (2008) Determination of linear response in the detection of mixtures of aroma compounds by atmospheric pressure ionization–mass spectrometry (API–MS). *Flavour and Fragrance Journal, 23*(1), 16–22.

Cappellin, L., Biasioli, F., Granitto, P. M., Schuhfried, E., Soukoulis, C., Costa, F., Märk, T. D. and Gasperi, F. (2011a) On data analysis in PTR-TOF-MS: From raw spectra to data mining. *Sensors and Actuators B: Chemical, 155*(1), pp. 183–190.

Cappellin, L., Biasioli, F., Schuhfried, E., Soukoulis, C., Märk, T. D. and Gasperi, F. (2011b) Extending the dynamic range of proton transfer reaction time-of-flight mass spectrometers by a novel dead time correction. *Rapid Communications in Mass Spectrometry, 25*(1), pp. 179–183.

Charles, M., Romano, A., Yener, S., Barnabà, M., Navarini, L., Märk, T. D., Biasioli, F. and Gasperi, F. (2015) Understanding flavour perception of espresso coffee by the combination of a dynamic sensory method and in-vivo nosespace analysis. *Food Research International, 69*, pp. 9–20.

Clark, C. C. and Lawless, H. T. (1994) Limiting response alternatives in time-intensity scaling: an examination of the halo-dumping effect. *Chemical Senses, 19*(6), pp. 583–594.

Cliff, M., and Heymann, H. (1993) Development and use of time-intensity methodology for sensory evaluation: A review. *Food Research International, 26*(5), pp. 375–385.

Cook, D. J., Linforth, R. S. T. and Taylor, A. J. (2003) Effects of Hydrocolloid Thickeners on the Perception of Savory Flavors. *Journal of Agricultural and Food Chemistry, 51*(10), pp. 3067–3072.

Davidson, J. M., Linforth, R. S. T., Hollowood, T. A. and Taylor, A. J. (1999) Effect of Sucrose on the Perceived Flavor Intensity of Chewing Gum. *Journal of Agricultural and Food Chemistry, 47*(10), pp. 4336–4340.

Deibler, K. D., Lavin, E. H., Linforth, R. S., Taylor, A. J. and Acree, T. E. (2001) Verification of a mouth simulator by in vivo measurements. *Journal of Agricultural and Food Chemistry, 49*(3), pp. 1388–1393.

Déléris, I., Saint-Eve, A., Dakowski, F., Sémon, E., Le Quéré, J.-L., Guillemin, H. and Souchon, I. (2011) The dynamics of aroma release during consumption of candies of different structures, and relationship with temporal perception. *Food Chemistry, 127*(4), pp. 1615–1624.

Deleris, I., Saint-Eve, A., Guo, Y., Lieben, P., Cypriani, M.-L., Jacquet, N., Brunerie, P. and Souchon, I. (2011) Impact of Swallowing on the Dynamics of Aroma Release and Perception during the Consumption of Alcoholic Beverages. *Chemical Senses, 36*(8), pp. 701–713.

De Roos, K. B. and Wolswinkel, K. (1994) Non-equilibrium partition model for predicting flavour release in the mouth. In Maarse, C. and van der Heij, D. G. (eds.) Trends in Flavour Research, Amsterdam (the Netherlands) : Elsevier Science B.V. pp. 15–32.

Di Monaco, R., Su, C., Masi, P. and Cavella, S. (2014) Temporal Dominance of Sensations: A review. *Trends in Food Science & Technology, 38*(2), pp. 104–112.

Doyen, K., Carey, M., Linforth, R. S. T., Marin, M. and Taylor, A. J. (2001) Volatile Release from an Emulsion: Headspace and In-Mouth Studies. *Journal of Agricultural and Food Chemistry, 49*(2), pp. 804–810.

Doyennette, M., Déléris, I., Féron, G., Guichard, E., Souchon, I. and Trelea, I. C. (2014) Main individual and product characteristics influencing in-mouth flavour release during eating masticated food products with different textures: Mechanistic modelling and experimental validation. *Journal of Theoretical Biology, 340*, pp. 209–221.

Doyennette, M., de Loubens, C., Déléris, I., Souchon, I. and Trelea, I. C. (2011) Mechanisms explaining the role of viscosity and post-deglutitive pharyngeal residue on in vivo aroma release: A combined experimental and modeling study. *Food Chemistry, 128*(2), pp. 380–390.

Ellis, A. M. and Mayhew, C. M. (2014) *Proton transfer reaction mass spectrometry: principles and applications*. Chichester (Great Britain): John Wiley & Sons, Inc.

Feron, G., Ayed, C., Qannari, E. M., Courcoux, P., Laboure, H. and Guichard, E. (2014) Understanding Aroma Release from Model Cheeses by a Statistical Multiblock Approach on Oral Processing. *PLoS ONE, 9*(4), e93113.

Flament, I. and Bessière-Thomas, Y. (2002) *Coffee flavor chemistry*. Chichester (Great Britain): John Wiley & Sons, Inc.

Frank, D., Appelqvist, I., Piyasiri, U., Wooster, T. J. and Delahunty, C. (2011) Proton Transfer Reaction Mass Spectrometry and Time Intensity Perceptual Measurement of Flavor Release from Lipid Emulsions Using Trained Human Subjects. *Journal of Agricultural and Food Chemistry, 59*(9), pp. 4891–4903.

Gierczynski, I., Laboure, H. and Guichard, E. (2008) In Vivo Aroma Release of Milk Gels of Different Hardnesses: Inter-individual Differences and Their Consequences on Aroma Perception. *Journal of Agricultural and Food Chemistry, 56*(5), pp. 1697–1703.

Gierczynski, I., Labouré, H., Sémon, E. and Guichard, E. (2007) Impact of hardness of model fresh cheese on aroma release: in vivo and in vitro study. *Journal of Agricultural and Food Chemistry, 55*(8), pp. 3066–3073.

Hansel, A., Jordan, A., Holzinger, R., Prazeller, P., Vogel, W. and Lindinger, W. (1995) Proton transfer reaction mass spectrometry: on-line trace gas analysis at the ppb level. *International Journal of Mass Spectrometry and Ion Processes, 149-150*, pp. 609–619.

Heenan, S., Soukoulis, C., Silcock, P., Fabris, A., Aprea, E., Cappellin, L., Märk, T.D., Gasperi, F. and Biasioli, F. (2012) PTR-TOF-MS monitoring of in vitro and in vivo flavour release in cereal bars with varying sugar composition. *Food Chemistry, 131*(2), pp. 477–484.

Holway, A. M. and Hurvich, D. M. (1937) Differential gustatory sensitivity to salt. *American Journal of Psychology, 49*, pp. 37–38.

Huang, G., Gao, L., Duncan, J., Harper, J. D., Sanders, N. L., Ouyang, Z. and Cooks, R. G. (2010) Direct detection of benzene, toluene, and ethylbenzene at trace levels in ambient air by atmospheric pressure chemical ionization using a handheld mass spectrometer. *Journal of the American Society for Mass Spectrometry, 21*(1), pp. 132–135.

Hunter, E. P. L. and Lias, S. G. (1998) Evaluated Gas Phase Basicities and Proton Affinities of Molecules: An Update. *Journal of Physical and Chemical Reference Data, 27*(3), p. 413.

Itobe, T., Kumazawa, K. and Nishimura, O. (2009) New Factor Characterizing the In-Mouth Release of Odorants (Volatile Thiols): Compositional Changes in Odorants Exhaled from the Human Nose during Drinking. *Journal of Agricultural and Food Chemistry, 57*(23), pp. 11297–11301.

Jordan, A., Haidacher, S., Hanel, G., Hartungen, E., Märk, L., Seehauser, H., Schottkowsky, R., Sulzer, P. and Märk, T. D. (2009) A high resolution and high sensitivity proton-transfer-reaction time-of-flight mass spectrometer (PTR-TOF-MS). *International Journal of Mass Spectrometry, 286*(2-3), pp. 122–128.

Jost, C., Sprung, D., Kenntner, T. and Reiner, T. (2003) Atmospheric pressure chemical ionization mass spectrometry for the detection of tropospheric trace gases: the influence of clustering on sensitivity and precision. *International Journal of Mass Spectrometry, 223-224*, pp. 771–782.

Leclercq, S. and Blancher, G. (2012) Multimodal Sensory Integration during Sequential Eating--Linking Chewing Activity, Aroma Release, and Aroma Perception over Time. *Chemical Senses, 37*(8), pp. 689–700.

Le Quéré, J.-L., Gierczynski, I. and Sémon, E. (2014) An atmospheric pressure chemical ionization-ion-trap mass spectrometer for the on-line analysis of volatile compounds in foods: a tool for linking aroma release to aroma perception: APCI-ion-trap MS for *in vitro* and *in vivo* analysis of odorants in foods. *Journal of Mass Spectrometry, 49*(9), pp. 918–928.

Lethuaut, L., Weel, K. G. C., Boelrijk, A. E. M. and Brossard, C. D. (2004) Flavor Perception and Aroma Release from Model Dairy Desserts. *Journal of Agricultural and Food Chemistry, 52*(11), pp. 3478–3485.

Lindinger, W. and Jordan, A. (1998) Proton-transfer-reaction mass spectrometry (PTR–MS): on-line monitoring of volatile organic compounds at pptv levels. *Chemical Society Reviews, 27*(5), pp. 347–375.

Linforth, R., Cabannes, M., Hewson, L., Yang, N. and Taylor, A. (2010). Effect of Fat Content on Flavor Delivery during Consumption: an in Vivo Model. *Journal of Agricultural and Food Chemistry, 58*(11), pp. 6905–6911.

Linforth, R. S., Baek, I. and Taylor, A. (1999) Simultaneous instrumental and sensory analysis of volatile release from gelatine and pectin/gelatine gels. *Food Chemistry, 65*(1), pp. 77–83.

Linforth, R. S. T. and Taylor, A. J. (1993) Measurement of volatile release in the mouth. *Food Chemistry, 48*(2), pp. 115–120.

Malone, M. E., Appelqvist, I. A. M. and Norton, I. T. (2003) Oral behaviour of food hydrocolloids and emulsions. Part 2. Taste and aroma release. *Food Hydrocolloids, 17*(6), pp. 775–784.

Mestres, M., Kieffer, R. and Buettner, A. (2006) Release and Perception of Ethyl Butanoate during and after Consumption of Whey Protein Gels: Relation between Textural and Physiological Parameters. *Journal of Agricultural and Food Chemistry, 54*(5), pp. 1814–1821.

Mestres, M., Moran, N., Jordan, A. and Buettner, A. (2005) Aroma Release and Retronasal Perception during and after Consumption of Flavored Whey Protein Gels with Different Textures. 1. in Vivo Release Analysis. *Journal of Agricultural and Food Chemistry, 53*(2), pp. 403–409.

Mesurolle, J., Saint-Eve, A., Déléris, I. and Souchon, I. (2013) Impact of Fruit Piece Structure in Yogurts on the Dynamics of Aroma Release and Sensory Perception. *Molecules, 18*(5), pp. 6035–6056.

Meyners, M. (2011) Panel and panelist agreement for product comparisons in studies of Temporal Dominance of Sensations. *Food Quality and Preference, 22*(4), pp. 365–370.

Mielle, P., Tarrega, A., Sémon, E., Maratray, J., Gorria, P., Liodenot, J. J., Liaboeuf, J., Andrejewski, J. L. and Salles, C. (2010) From human to artificial mouth, from basics to results. *Sensors and Actuators B: Chemical, 146*(2), pp. 440–445.

Mishellany-Dutour, A., Woda, A., Labouré, H., Bourdiol, P., Lachaze, P., Guichard, E. and Feron, G. (2012). Retro-Nasal Aroma Release Is Correlated with Variations in the In-Mouth Air Cavity Volume after Empty Deglutition. *PLoS ONE, 7*(7), e41276.

Müller, M., Mikoviny, T., Jud, W., D'Anna, B. and Wisthaler, A. (2013) A new software tool for the analysis of high resolution PTR-TOF mass spectra. *Chemometrics and Intelligent Laboratory Systems, 127*, pp. 158–165.

Murray, J., Delahunty, C. and Baxter, I. (2001). Descriptive sensory analysis: past, present and future. *Food Research International, 34*(6), pp. 461–471.

Normand, V. (2004) Modeling the Kinetics of Flavour Release during Drinking. *Chemical Senses, 29*(3), pp. 235–245.

Overbosch, P., De Wijk, R., De Jonge, T. J. R. and Köster, E. P. (1989) Temporal integration and reaction times in human smell. *Physiology & Behavior, 45*(3), pp. 615–626.

Overbosch, P., Enden, J. C. van den and Keur, B. M. (1986) An improved method for measuring perceived intensity/time relationships in human taste and smell. *Chemical Senses, 11*(3), pp. 331–338.

Pineau, N., Cordelle, S. and Schlich, P. (2003) Temporal dominance of sensations: a new technique to record several sensory attributes simultaneously over time. In *Abstract book of 5th Pangborn Sensory Science Symposium*, Boston, *MA, USA, July 20-24, 2003* (p. 121).

Pineau, N., Schlich, P., Cordelle, S., Mathonnière, C., Issanchou, S., Imbert, A., Rogeaux, M., Etiévant, P. and Köster, E. (2009) Temporal Dominance of Sensations: Construction of the TDS curves and comparison with time–intensity. *Food Quality and Preference, 20*(6), pp. 450–455.

Pionnier, E., Chabanet, C., Mioche, L., Le Quéré, J.-L. and Salles, C. (2004) 1. In Vivo Aroma Release during Eating of a Model Cheese: Relationships with Oral Parameters. *Journal of Agricultural and Food Chemistry, 52*(3), pp. 557–564.

Plug, H. and Haring, P. (1993) The role of ingredient-flavour interactions in the development of fat-free foods. *Trends in Food Science & Technology, 4*(5), pp. 150–152.

Potineni, R. V. and Peterson, D. G. (2008) Influence of Flavor Solvent on Flavor Release and Perception in Sugar-Free Chewing Gum. *Journal of Agricultural and Food Chemistry, 56*(9), pp. 3254–3259.

Pozo-Bayón, M. Á., Santos, M., Martín-Álvarez, P. J. and Reineccius, G. (2009) Influence of carbonation on aroma release from liquid systems using an artificial throat and a proton transfer reaction-mass spectrometric technique (PTR-MS). *Flavour and Fragrance Journal, 24*(5), pp. 226–233.

Rabe, S. (2004) Volatile Release from Liquids: A Comparison of In Vivo APCI-MS, In-mouth Headspace Trapping and In vitro Mouth Model Data. *Chemical Senses, 29*(2), pp. 163–173.

Rabe, S., Krings, U., Banavara, D. S. and Berger, R. G. (2002) Computerized Apparatus for Measuring Dynamic Flavor Release from Liquid Food Matrices. *Journal of Agricultural and Food Chemistry, 50*(22), pp. 6440–6447.

Repoux, M., Labouré, H., Courcoux, P., Andriot, I., Sémon, É., Yven, C., Feron, G. and Guichard, E. (2012) Combined effect of cheese characteristics and food oral processing on *in vivo* aroma release: *In vivo* aroma release during cheeses oral processing. *Flavour and Fragrance Journal, 27*(6), pp. 414–423.

Roberts, D. D. and Acree, T. E. (1995) Simulation of Retronasal Aroma Using a Modified Headspace Technique: Investigating the Effects of Saliva, Temperature, Shearing, and Oil on Flavor Release. *Journal of Agricultural and Food Chemistry, 43*(8), pp. 2179–2186.

Roberts, D. D., Pollien, P., Antille, N., Lindinger, C. and Yeretzian, C. (2003). Comparison of Nosespace, Headspace, and Sensory Intensity Ratings for the Evaluation of Flavor Absorption by Fat. *Journal of Agricultural and Food Chemistry, 51*(12), pp. 3636–3642.

Romano, A., Cappellin, L., Ting, V., Aprea, E., Navarini, L., Gasperi, F. and Biasioli, F. (2014) Nosespace analysis by PTR-ToF-MS for the characterization of food and tasters: The case study of coffee. *International Journal of Mass Spectrometry,*pp. 365–366, 20–27.

Saint-Eve, A., Martin, N., Guillemin, H., Sémon, E., Guichard, E. and Souchon, I. (2006) Flavored yogurt complex viscosity influences real-time aroma release in the mouth and sensory properties. *Journal of Agricultural and Food Chemistry, 54*(20), pp. 7794–7803.

Salles, C., Tarrega, A., Mielle, P., Maratray, J., Gorria, P., Liaboeuf, J. and Liodenot, J.-J. (2007) Development of a chewing simulator for food breakdown and the analysis of in vitro flavor compound release in a mouth environment. *Journal of Food Engineering, 82*(2), pp. 189–198.

Shojaei, Z. A., Linforth, R. S. T., Hort, J., Hollowood, T. and Taylor, A. J. (2006) Measurement and manipulation of aroma delivery allows control of perceived fruit flavour in low- and regular-fat milks. *International Journal of Food Science and Technology, 41*(10), pp. 1192–1196.

Sokolowsky, M. and Fischer, U. (2012) Evaluation of bitterness in white wine applying descriptive analysis, time-intensity analysis, and temporal dominance of sensations analysis. *Analytica Chimica Acta, 732*, pp. 46–52.

Taucher, J., Hansel, A., Jordan, A. and Lindinger, W. (1996) Analysis of Compounds in Human Breath after Ingestion of Garlic Using Proton-Transfer-Reaction Mass Spectrometry. *Journal of Agricultural and Food Chemistry, 44*(12), pp. 3778–3782.

Taylor, A. (2000) Atmospheric pressure chemical ionisation mass spectrometry for in vivo analysis of volatile flavour release. *Food Chemistry, 71*(3), pp. 327–338.

Taylor, A. J. and Linforth, R. S. T. (1996) Flavour release in the mouth. *Trends in Food Science & Technology, 7*(12), pp. 444–448.

Ting J.L., V., Soukoulis, C., Silcock, P., Cappellin, L., Romano, A., Aprea, E., Bremer, P. J., Märk, T. D., Gasperi, F. and Biasioli, F. (2012) *In Vitro* and *In Vivo* Flavor Release from Intact and Fresh-Cut Apple in Relation with Genetic, Textural, and Physicochemical Parameters. *Journal of Food Science, 77*(11), pp. C1226–C1233.

Trelea, I. C., Atlan, S., Deleris, I., Saint-Eve, A., Marin, M. and Souchon, I. (2007) Mechanistic Mathematical Model for In Vivo Aroma Release during Eating of Semiliquid Foods. *Chemical Senses, 33*(2), pp. 181–192.

Van Loon, W. A. M., Linssen, J. P. H., Boelrijk, A. E. M., Burgering, M. J. M. and Voragen, A. G. J. (2005) Real-Time Flavor Release from French Fries Using Atmospheric Pressure Chemical Ionization–Mass Spectrometry. *Journal of Agricultural and Food Chemistry, 53*(16), pp. 6438–6442.

Van Ruth, S. (2000) Influence of mastication and saliva on aroma release in a model mouth system. *Food Chemistry, 71*(3), pp. 339–345.

Weel, K. G. C., Boelrijk, A. E. M., Alting, A. C., van Mil, P. J. J. M., Burger, J. J., Gruppen, H., Voragen, A. G. J. and Smit, G. (2002) Flavor Release and Perception of Flavored Whey Protein Gels: Perception Is Determined by Texture Rather than by Release. *Journal of Agricultural and Food Chemistry, 50*(18), pp. 5149–5155.

Weel, K. G. C., Boelrijk, A. E. M., Burger, J. J., Jacobs, M. A., Gruppen, H., Voragen, A. G. J. and Smit, G. (2004a) Effect of Emulsion Properties on Release of Esters under Static Headspace, in Vivo, and Artificial Throat Conditions in Relation to Sensory Intensity. *Journal of Agricultural and Food Chemistry, 52*(21), pp. 6572–6577.

Weel, K. G. C., Boelrijk, A. E. M., Burger, J. J., Verschueren, M., Gruppen, H., Voragen, A. G. J. and Smit, G. (2004b) New Device To Simulate Swallowing and in Vivo Aroma Release in the Throat from Liquid and Semiliquid Food Systems. *Journal of Agricultural and Food Chemistry, 52*(21), pp. 6564–6571.

Wright, K. M., Hills, B. P., Hollowood, T. A., Linforth, R. S. T. and Taylor, A. J. (2003) Persistence effects in flavour release from liquids in the mouth. *International Journal of Food Science and Technology, 38*(3), pp. 343–350.

Yeretzian, C., Pollien, P., Lindinger, C. and Ali, S. (2004) Individualization of Flavor Preferences: Toward a Consumer-centric and Individualized Aroma Science. *Comprehensive Reviews in Food Science and Food Safety, 3*(4), pp. 152–159.

CHAPTER 11

Release of tastants during in-mouth processing

Christian Salles
Center for Taste and Feeding Behaviour, CNRS, INRA, University of Bourgogne Franche-Comté, Dijon, France

11.1 Introduction

Taste perception is one of the components of flavour perception, which is defined as the integrated perception of the olfactive, taste and trigeminal cues caused by different chemical compounds contained in a food product (Salles et al. 2012). Humans are able to discriminate at least five taste qualities: bitterness, sourness, sweetness, saltiness and umami. The nature and structure of the tastants are varied, as discussed in detail in chapter 7. During the in-mouth process, tastants are dissolved in saliva, either directly by dilution in the case of liquid or semi-solid food, or progressively during mastication for solid food (Salles et al. 2011). The delivery of these tastants to the receptors located on the tongue is one of the key factors determining the perceived taste of foods. However, the composition of the food matrix highly influences the release kinetics of taste compounds because of interactions between the different components. The oral physiology characteristics also highly modulate taste compounds release and are responsible for inter-individual variability.

Few studies have been dedicated to in-mouth taste compound release compared to the numerous studies on volatile compound release (see chapter 10), probably because no direct and rapid technique is able to analyse on line and in real time the continuous release of several taste compounds (which is the case for example of PTR-MS for the volatile fraction). Most of the in-mouth taste compound release studies have been performed on salt (sodium), in relation to temporal saltiness perception. This can be easily explained by the recommendations of the public health authorities to decrease significantly the total amount of dietary sodium chloride (WHO 2007) which motivated studies on salt perception in relation with the food matrix structure and composition.

The aim of the chapter is to present the relationships between food structure and composition, release of tastants in mouth conditions and overall perception,

Flavour: From food to perception, First Edition.
Edited by Elisabeth Guichard, Christian Salles, Martine Morzel, and Anne-Marie Le Bon.

taking into account the oral physiology of individuals, in particular the mastication and salivation characteristics. Concerning salivation, our discussion will be limited to the salivary flow rate as the other aspects of saliva (composition, regulation ...) are discussed in chapters 4 and 12 of this book. We here mainly focus on studies which directly related in-mouth taste compounds release with temporal taste perception.

11.2 Instrumental techniques to measure in-mouth release of tastants

11.2.1 Saliva sampling

The measurement of *in vivo* non-volatile compounds release is most often based on discontinuous saliva sampling using cotton buds (Phan et al. 2008) or pieces of adsorbent paper (de Loubens et al. 2011b). The use of adsorbents allows saliva sampling to be performed on the surface of the tongue, as close as possible to the taste receptor cells. The sample is then extracted with an aqueous polar solvent prior to analysis. The main drawback of this method is the risk to contaminate the low amount of saliva sampled with small food particles. An alternative method is the direct sampling of saliva. This consists in asking the subjects to spit out a small aliquot of saliva in tubes at different times during food consumption (Lawrence et al. 2012a). Each saliva sample should be quickly chilled and centrifuged to avoid alteration before analysis. An original device was proposed to attempt an *in vivo* continuous sampling of saliva during eating (Davidson et al. 2000). It consists in a motor with a barrel attachment which reels a length of ribbon at a constant rate through the mouth during eating, thus absorbing saliva. Different portions of ribbons corresponding to different ranges of time are then extracted and analysed.

11.2.2 Analysis of tastants

Analyses of saliva samples require sensitive methods to identify and quantify the tastants because in most cases very small volumes of liquid are sampled from the mouth during eating.

The analysis of the components released in saliva when eating model cheese by HPLC coupling ion exchange chromatography and conductimetric or pulsed-amperometric detection allowed released cations and anions to be analysed in the saliva samples collected during eating a solid food with a good reproducibility (Pionnier et al. 2004a). This reliable and sensitive method has been successfully used to analyse in mouth sodium release from solid foods in several studies (Chabanet et al. 2013, Lawrence et al. 2012a, Phan et al. 2008). A quicker alternative method to analyse saliva samples consisted in using a direct liquid mass spectrometric procedure (Davidson et al. 1999).

Following simultaneously in-nose menthone release using direct gas phase atmospheric pressure mass spectrometry (APCI-MS) and temporal perception of mint flavour, saliva was sampled at different times during chewing a mint flavoured chewing-gum and sucrose release was monitored by direct injection of diluted saliva samples in a Liquid-APCI-MS. The authors were thus able to show perceptual interactions between sweetness and mint perceptions.

Different systems for continuously measuring in-mouth changes in conductivity in real time during chewing food were proposed (Davidson et al. 1998, Jack et al. 1995, Neyraud et al. 2003). Davidson et al. (1998) proposed a combined system of sensors to successfully measure both variations in pH and conductivity. The main drawback of these systems is the important lag between the introduction of the food in the mouth and the significant increase of conductivity, due to the time for saliva (containing released salt) to reach the sensors. Later, a mini conductimeter stuck on the lower part of the mouth with a very short response time and lag time was proposed (Emorine et al. 2012). The authors were able to follow the in-mouth release of salt in real time during chewing solid foods of different textures. However, these measurements in real time were nonspecific of sodium ion but gave only an idea on the overall ionic content of saliva.

The electronic tongues (Tahara and Toko 2013) can be considered as an artificial gustatory sense and consist of sensor arrays and pattern recognition systems. They aim at discriminating and analysing food and beverages (Beullens et al. 2008, Tian et al. 2013) but generally, these systems need a large volume of liquid sample and are limited in sensitivity. Moreover, consisting only of the simultaneous measure of chemical components by sensor arrays, they are poorly correlated to taste attributes described and rated by a sensory panel. Finally, the large size of these systems and their limits in terms of biocompatibility do not allow direct measurements in the mouth. More recently, sensory cell-based sensors were developed and show some advantages such as fast response, excellent selectivity and high sensitivity (Hui et al. 2012). Such sensors are able to respond specifically to compounds of a given basic taste. This seems to be a promising way to mimic taste perception or follow the release of tastants in mastication condition. However, these sensory cell-based sensors have been tested only in model single or mixed solutions and they do not seem adapted yet for in-mouth conditions.

11.3 Influence of food composition on the release of tastants at the macromolecular level

The tastants can be retained in the food matrix through physicochemical interactions. In general, the quantity of non-volatile compounds released in the saliva increase at the beginning of the chewing process, reach a maximum and decrease more or less rapidly at the end of the mastication. Most of the studies on in-mouth taste compounds release were conducted on dairy products and breads and focused mainly on salt release. They reported a large influence of the food composition on temporal release and perception but with large differences according to the nature of the macromolecules present in the food matrix and their structure.

11.3.1 Dairy products

A primary study reported that in-mouth salt release during the mastication process measured by conductimetry was related to Cheddar cheese breakdown and thus to its texture (Jack et al. 1995). In the same way, Davidson et al. (2000, 1998) reported that in-mouth salt release depended on the texture and water content of food, in terms of maximal concentration of stimuli and its corresponding time. For example, the release of salt from crisps was quicker than from Cheddar cheese as salt was mainly located on the surface of crisps. In addition, for a similar initial salt concentration, the quantity of salt released from a soft cheese was higher than that released from a hard cheese. The study of in-mouth non-volatile compounds release was extended to tastants other than sodium chloride by following their concentration in saliva during chewing a model cheese and in the same time the temporal taste perception (Pionnier et al. 2004a).

Considering the time at which the maximum concentration of tastant is released in saliva, no significant effect of the nature of non-volatile compound was observed during the chewing of either Cheddar cheese or fresh orange segments (Davidson et al. 2000). In contrast, the nature of the released compounds had an impact on the maximum concentration in saliva, the total released quantity and the release rates. The most highly concentrated compound in the cheese was not necessarily the most concentrated in the saliva. Though some non-volatile compounds are permanently present in saliva, this observation could be also explained by some specific distributions of each non-volatile compound in the different phases of the bolus (water, proteins, and fat). For example, concerning the decreasing phase of the temporal release curve, Pionnier et al. (2004a) reported that the quantity of each non-volatile compound was more or less persistent in saliva after the end of chewing process in function of the nature of the compounds. Although different patterns of release were observed according to the subject, for a given subject, the shapes of the curves were similar for most of the twelve non-volatile compounds studied. This was explained by their rather good solubility in aqueous medium. These temporal releases of specific cheese components were related to temporal taste perception evaluated by time-intensity experiments (Pionnier et al. 2004b). The authors observed in particular that perceived tastes persisted after the swallowing of the product but they reported very few correlations between taste compound release and temporal taste perception. Only the time to reach the maximum intensity for sourness and saltiness perception were correlated to citric acid and sodium ion release, respectively.

The composition of such model cheeses influenced both temporal sodium release and saltiness perception (Phan et al. 2008, Lawrence et al. 2012a).

11.3.1.1 Effect of fat content

The increase of fat content was related to a global decrease of sodium release and an increase of saltiness perception (Phan et al. 2008). Several hypotheses concerning the relationships between saltiness perception and matrix composition were proposed. Fat could act as a sodium barrier in the matrix to slow down its diffusion into the saliva phase, or it could favour the formation of fat/water emulsion retaining salt inside the water phase. Concerning saltiness perception, a highly significant negative effect of lipid content could be related to a food matrix effect masking saltiness perception (Engel et al. 2001).

11.3.1.2 Effect of water content

The water content of cheeses directly influences in-mouth salt concentration. A decrease of sodium release when water content is low could be explained by a lower solvating capacity, and the higher saltiness intensity could be due to a higher concentration of salt in the smaller water phase of Cheddar cheeses (Jack et al. 1995). Phan et al. (2008) observed a positive effect of the water content of model cheeses on sodium release at the beginning of chewing, when the ratio between the quantity of saliva and the quantity of food was low and when only the water contained in the product intervened in the release of sodium. A lower mixing efficiency of food with water and consequently a reduced transport of sodium to the saliva phase and then to taste receptors can affect saltiness perception. Phan et al. (2008) reported that only when the water content of model cheeses was high, sodium release seemed to be favoured by a high fat content, thus suggesting interactions between fat and water content for in-mouth sodium release.

11.3.1.3 Effect of salt content

Variation of dairy lipoprotein matrices composition factors: fat/dry matter ratio, dry matter, salt contents and pH at renneting influence sodium release and saltiness perception (Lawrence et al. 2012a). Composition and textural changes modified temporal sodium release and saltiness perception but the effects varied with the salt content. At low salt contents (0.5 and 1%, respectively), no effect of composition on the maximum concentration of sodium released was found, suggesting that the variation in texture did not induce a variation in sodium release. At higher salt content (1.5%), the dry matter level, fat/dry matter ratio and pH at renneting all exerted a significant effect on the maximum concentration of sodium released (Figure 11.1). Moreover, the maximum concentration of sodium released was higher for products with a lower dry matter content, lower fat/dry

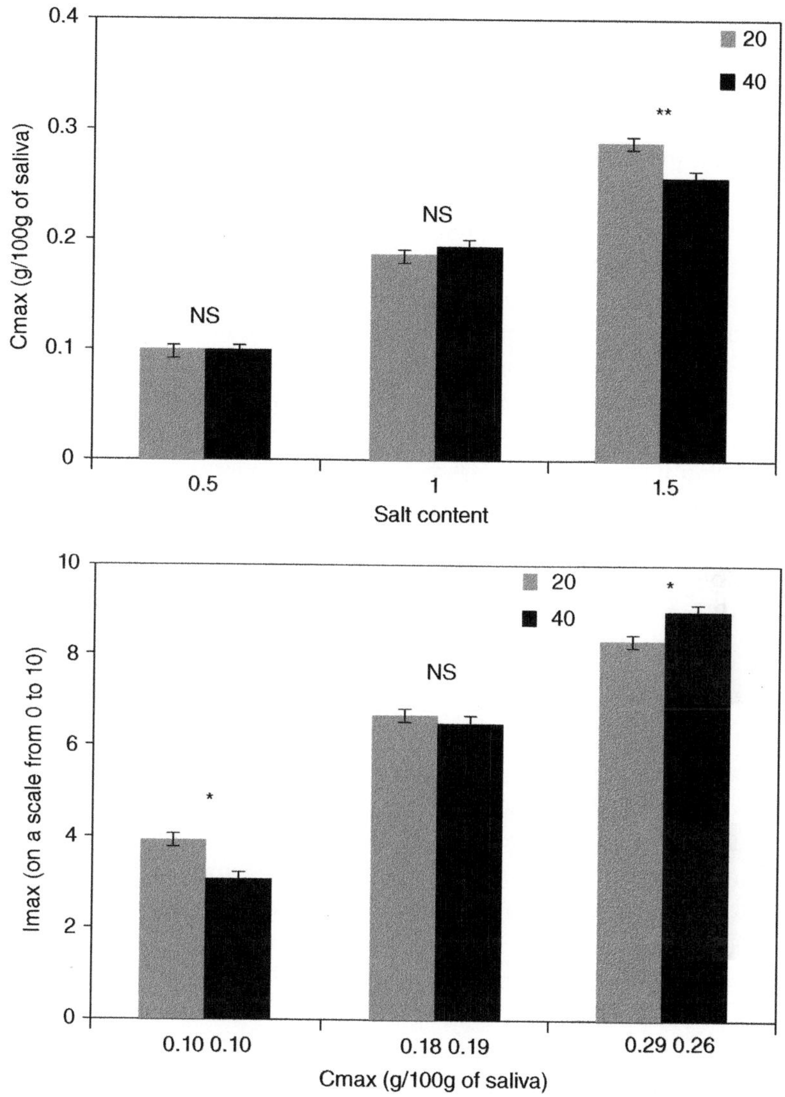

Figure 11.1 Influence of Fat/Dry matter ratio of a model cheese on in-mouth sodium release (C max, top panel) and saltiness perception (I max, lower panel). Source: Lawrence et al. (2012a). Reproduced with permission of American Chemical Society.

matter ratio and higher pH at renneting level. Concerning saltiness perception, the maximum saltiness perceived increased in line with the salt content. At the lowest salt content, an effect of the dry matter level and fat/dry matter ratio was observed on the maximum saltiness perception. Saltiness perception was more intense when the dry matter level was high and when the fat/dry matter ratio was low. When the salt level was intermediate, no significant effect of composition factors was observed. When the salt content was the highest, only a significant effect of the fat/dry matter ratio was observed on the maximum saltiness perceived. It is noteworthy that in this case, the saltiness perception was less intense when the fat/dry matter content was lower, which represented a behaviour that contrasted with what was observed at the lowest salt level.

Overall, in most cases, temporal sodium release parameters were well correlated to their equivalent for temporal saltiness perception. However, the effect of fat content on in-mouth sodium release and saltiness perception seems rather complex leading to contradicting results published in the literature. The in-mouth process of mixing food with saliva and its evolution may also affect the release and perception phenomena. Thus, the effect of fat and water on sodium release and saltiness perception may also vary according to the chewing time due to the variation of the relative composition, fat/saliva ratio in mouth, and changes in the structure of the bolus during oral processing. Probably different physico-chemical mechanisms such as transport kinetics or lipid coating on the tongue surface are involved, thus limiting the molecular diffusion of the sodium ions to their respective channels, causing a decrease in the perception of salty taste. Perceptual interactions with fat perception probably also play a role in salt perception.

11.3.2 Bakery products

Bread is a heterogeneous airy food containing starch and composed of a rigid material on surface (Chaunier et al. 2008). It is also the food responsible for the highest sodium intake in daily diets (WHO 2007). Texture and bread composition were found to highly influence salt release and saltiness perception.

Sodium released in the mouth into saliva was quantified during oral processing. Tournier et al. (2014) reported that the sodium content in saliva varied between breads. A higher sodium concentration in saliva was observed for toast bread and rye bread compared to baguettes, and the highest concentrations were also reached faster in those products, probably due to their differences in structure and composition. These effects were not subject-dependent but large interindividual differences were observed, which is developed in section 5 of this chapter. The influence of texture and initial water content on salt release and saltiness perception was particularly pointed out by a comparison between salt release in dough and the corresponding bread. Pflaum et al. (2013) showed that significant lower amounts of sodium were extracted from bread crumb than from dough for short durations of mastication while they were comparable for longer durations of mastication. This was explained first by a more rapid transfer of the sodium ions to the saliva phase from the dough which contains more free water than bread. In addition, the release of sodium from dough was slower than from bread crumb because of the higher viscoelastic texture of the dough leading to a more difficult homogenization with saliva. These results were coherent with the dough found significantly saltier at the beginning of the mastication. In addition, sodium was faster released from the more coarse-pored crumb compared to fine-pored crumbs. This was particularly observed at the first steps of chewing, with a high impact on saltiness perception.

Le Bleis et al. (2013) suggested that plasticization of bread by water from saliva had a larger effect on the rheological behaviour of the bolus than its fragmentation. Comparing three different breads, Panouillé et al. (2014) reported that denser bread was perceived as being less salty and displayed a less complex texture perception pattern but hypothesised perceptual interactions between saltiness and texture perception rather than physicochemical phenomena such as bolus rheological behaviour influencing in-mouth salt release during chewing bread.

11.3.3 Other food systems

Few other foods have been investigated for in-mouth non-volatile compound release. Interesting results were reported concerning chicken sausages varying in fat, salt and water contents (Chabanet et al. 2013). Overall, they observed as in all studies inter-individual variations, and as expected a positive effect of salt content and a negative effect of fat content on salt release. However, the importance of these effects was different according to the chewing time. The fat content effect was clear: the lower the fat content, the higher the level of sodium release, probably due to differences of mass transfer of sodium in the matrix. A negative effect of water content on sodium release was also found though it was minor compared to the others. Concerning saltiness perception, a highly significant effect of salt and fat contents on saltiness intensity was observed at each chewing time, while a water content effect contributed to increasing saltiness intensity only at the medium chewing time. This effect of fat content was explained by a lower availability of taste receptors for released sodium in saliva due to tongue coating by fat material. The modelling of saltiness intensities with the quantity of salt released in saliva showed that fat content had an additional negative effect on saltiness intensity, suggesting that fat content exerted a masking effect on saltiness intensity which was not due to sodium release in saliva and was therefore a perceptual effect due to fat perception (Chabanet et al. 2013). This could explain why the negative effect of fat content on sodium released in saliva remained significant, even though the food product was no longer present in the mouth of the subjects, probably due to the persistence of fat coating in the mouth.

Some other studies reported relationships between salt release and saltiness perception for more specific food systems with relevant results concerning the understanding of non-volatile compounds release mechanisms. Fried sliced potato crisps were flavoured with sodium chloride of varying size fractions to investigate whether salt crystal size could impact upon the delivery rate and perceived saltiness (Rama et al. 2013). In particular, the smallest crystal size fraction was found to dissolve in saliva and to reach the taste receptor on the tongue faster than the medium and large crystal size fraction. The direct consequences were a higher maximum concentration and a higher total amount of sodium released. With this smallest crystal size, it was also faster to reach the maximum saltiness intensity and maximum total saltiness perception. The authors explained these different delivery rates by differential dissolution kinetics and enhanced mass transfer of sodium in the saliva phase. Another example is the evaluation of the effect of air inclusion in hydrogels on delivery of sodium and saltiness perception (Chiu et al. 2015). When comparing a non-aerated and aerated system of equal volumes with the same amount of salt on a volume basis, the latter system uses air to replace the aqueous phase, making a higher salt concentration in the aqueous phase. Thus samples with more air inclusions were perceived as overall saltier which is supported by a higher salt release rate for the most aerated system.

11.4 Influence of composition on the release of tastants at the microscopic and molecular levels

Few studies were performed at the microstructure and molecular levels to explain non-volatile compound release and perception. Most of these studies deal with the distribution and the state of sodium ion which can interact with the proteins and water in the food matrix.

Using the non-invasive method ^{23}Na NMR, total sodium, free and bound sodium contents were determined in both salted salmon and pork lean meat (Foucat et al. 2003). It was in particular possible to differentiate different kinds of fillets and salting technological processes based on their different salt distribution in the tissues. This heterogeneity in salt distribution in fillets was visualised by NMR imaging, showing for example that salt did not penetrate into connective tissues (Foucat, unpublished results). In bread crumb, sodium chloride distribution was visualised by confocal laser scanning microscopy using a sodium-selective fluorescent dye (Konitzer et al. 2013). The authors suggested that an increase of the contrast in sodium chloride distribution can explain that, using coarse-grained sodium chloride instead of thin salt in bread accelerated sodium chloride release and enhanced saltiness perception.

The molecular behaviours of sodium in model cheeses varying in lipid-protein ratios and salt content studied by ^{23}Na NMR provided information on mobility, state and release of sodium (Boisard et al. 2013, Guichard et al. 2013). The mobility of sodium was found to be linked to the microstructure of the model cheeses. In particular, sodium chloride modified the microstructure by sodium ion binding to caseins, and fat droplets modified it by disrupting the protein network continuity, mainly when their size increased.

Relationships between the ^{23}Na NMR parameters and kinetics of sodium release measured *in vitro* from the model cheeses to a water phase were reported (Boisard et al. 2013). For an identical lipid-protein ratio, sodium chloride addition was responsible for the increase in the sodium release rate and a greater mobility of sodium ions. Overall, the sodium mobility was also correlated with saltiness perception. In particular, the highest correlation was found for the calculated amount of free sodium in the model cheese (Boisard et al. 2014). Moreover, the amount of salt released in saliva was rather well correlated with saltiness perception.

Changes observed at the microscopic and molecular level may have consequences on the macroscopic level. As example, sodium ions were less mobile when the protein content increased and when the lipid content decreased due to the higher hardness of the products with higher protein and lower lipid contents (Boisard et al. 2013). The sodium ions were more mobile in salted model cheeses. This could be explained by larger fat globules and a weaker protein network. The main consequence at the macroscopic level is the reduction in hardness of the model cheeses. A more generic study showed that protein type, concentration and pH affected the rheological properties of the protein matrices, the ionic binding of sodium ions and taste intensity. As the protein concentration increased, the mobility of sodium ions decreased (Mosca et al. 2015). Moreover, the mobility of sodium ion depended on the type of protein and seemed to be related to their negative sites content.

11.5 Influence of individual oral physiology on the release of tastants

Food oral processing is a dynamic and complex process which modifies the physical properties of food and affects flavour and texture perception. The major in-mouth phenomena after food ingestion are mastication and salivation resulting in the formation of a bolus and transport of particles in different locations in the mouth, then swallowing (Hiiemae 2004). Great interindividual variation is reported in most studies on temporal taste compound release and temporal taste perception (Figure 11.2). These variations are mainly due to differences in food oral processing during eating which are related to individual's anatomical and physiological characteristics (Salles et al. 2011).

11.5.1 Effects of mastication

Several studies reported the influence of mastication on taste compounds release and perception. Using continuous monitoring of released sodium by conductivity measurement, it was shown that the manner in which salt is released into

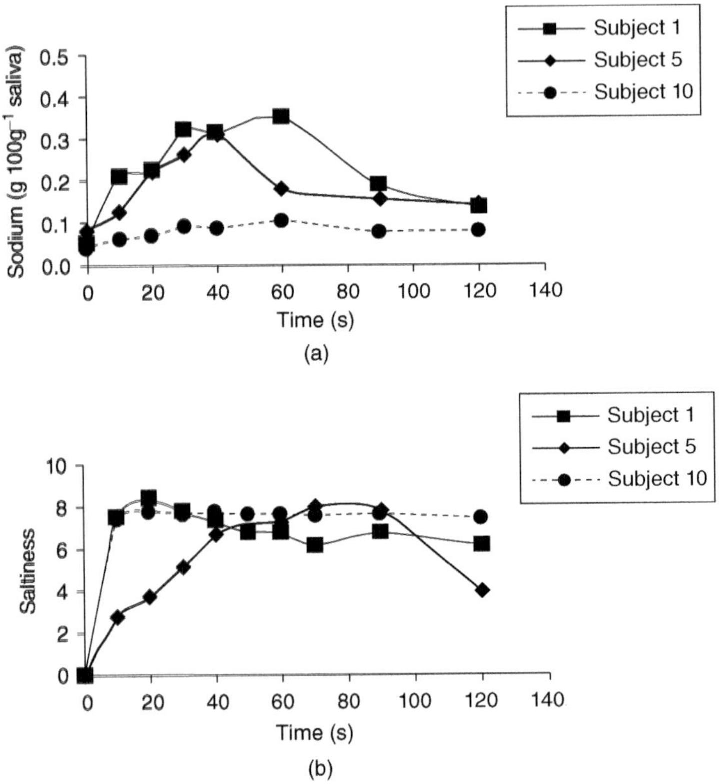

Figure 11.2 Concentrations of sodium in saliva (a) and saltiness perception (b) During the eating of model cheeses over time for three subjects. Source: Phan et al. (2008). Reproduced with permission of Elsevier.

the mouth during chewing different Cheddar cheeses is related to cheese breakdown (Jack et al. 1995). Dry and hard cheeses, such as low fat cheeses, required a higher chewing time and more salt was released. This may be also due to a higher salt concentration in the water phase of these cheeses. The mastication time required to reach maximum salt release was directly related to the smoothness and creaminess of the cheddars. Springier cheeses were characterized by lower salt release rates (Jack et al. 1995).

The kinetics of taste compounds release is subject-dependent. In model cheese, for calcium, potassium, sodium and leucine, the subjects with a high time to reach the maximum intensity were also the subjects for whom it was observed high values for chewing time and small values for chewing rate, swallowing frequency, and masticatory performances (Pionnier et al. 2004b). A trend was found where a long chewing time leads to a longer time to reach the maximum concentration associated with low chewing and low swallowing rate. High release rate was due to short chewing times and high chewing rates, high salivary flow rates, high swallowing rates, and high masticatory performances. However, no relationship between the maximum concentration of taste compounds released and oral parameters was reported.

Different salt release and saltiness perception profiles were observed (Figure 11.2), which are explained by the physiological differences between the subjects (Phan et al. 2008). Using model cheeses varying in texture and composition, the authors reported that a rapid sodium release in mouth was linked to a high chewing intensity and a slow sodium release was linked to a high chewing time. Chewing intensity and the total work of mastication were related to saltiness perception.

Working on another type of model cheese, Lawrence et al. (2012b) reported that the observed differences in chewing behaviour between subjects could explain most of the variability in in-mouth sodium release and saltiness perception. Subjects with a higher chewing force and lower salivary flow rate experienced higher level of sodium release and saltiness intensity. Texture, which varied as a function of composition, influenced both chewing behaviour and sodium release, whereas food composition influenced saltiness. This study pointed out that sodium release and saltiness perception were more influenced by chewing force and salivary flow rate, respectively.

Relevant observations were also reported from studies on bread. The heterogeneous characteristic of bread for both composition and colour contrast allowed different kinetics of bolus formation to be followed by measuring the bolus heterogeneity using image texture analysis between different breads and taking into account interindividual variability (Tournier et al. 2012). Mastication and salivation varied between products and between subjects (Tournier et al. 2014). The kinetics of bread degradation during food bolus formation and sodium released into the saliva were evaluated concomitantly. The initial release rate of sodium increased when the chewing muscles activity increased, and the maximum sodium concentration was reached later when more masticatory cycles were required to reach the swallowing point. It is noteworthy that *in vivo* salt release was mainly explained by mastication parameters. Though saliva uptake increased throughout the chewing process as the number of cycles increased, they found no correlation between saliva uptake at swallowing and the number of cycles required to reach swallowing.

11.5.2 Effects of saliva

Saliva is another factor that has a significant effect on taste compound release and perception. Saliva has numerous effects on mastication, food bolus formation, swallowing and digestion (Feron and Salles 2013, Pedersen et al. 2002). It is also a very complex biologic fluid with different roles in perception mechanisms (Neyraud 2014). The release of salt in mouth was shown to induce changes in chemical composition of saliva (Neyraud et al. 2003). The increase of pH in saliva was a consequence of the presence of salt, however, salt content and pH changed differently in saliva. The increase of pH is gradual with an increase of salt in saliva and the return time to the initial level of pH was longer than for salt concentration and saltiness perception. This effect could be related to a higher buffer effect of the stimulated saliva. During the chewing of salty chewing-gum, Neyraud et al. (2003) also showed that salt concentration in saliva varied in a cyclical manner. This was due to cyclical swallowing and saliva replacement by low-salt content saliva. Using a model cheese, Pionnier et al. (2004a) reported sodium release being negatively correlated with salivary flow rate. Moreover, when the salivary flow rate was low, more time was needed to extract the maximum concentration of taste compounds from the matrix during eating; therefore, the time to reach this concentration was longer.

Similar observation was done for cooked sausages: subjects with a low salivary flow rate displayed delayed effects on salt release and temporal saltiness perception (Chabanet et al. 2013). As observed in dairy products, the salivary sodium concentration measured during chewing bread clearly depended on the salivary flow rate of the subjects (Konitzer et al. 2013).

Thus the role of salivary flow rate seems relatively clear in taste compound release and taste perception. The influence of saliva composition on taste compound release is less clear and complex. It is known that ions such as HCO_3^- participate to the buffer effect and thus may affect sourness perception (Christensen et al. 1987). Other saliva compounds such as enzymes or other proteins can act directly on perception or interfere with the food matrix components (Feron and Salles 2013). As an example, a direct relationship was found between the alpha-amylase activity level in saliva and saltiness perception of starchy matrices by favouring the release of sodium in saliva after partial hydrolysis of the matrix and consequently altering the perception of saltiness (Ferry et al. 2006).

11.5.3 Combined effect of mastication and salivation

Generally, the mechanical action of chewing increases salivary flow rates. It was reported that both the mechanical action of chewing on a salty matrix and the release of sodium chloride increased salivary flow (Neyraud et al. 2003). However, several studies reported that salivation and mastication acted rather independently on taste compound release and taste perception. Salivary flow rate and mastication performance were found rather independent on saltiness intensity though high salivary flow rate and high masticatory performances were related to a high sodium release in model cheeses (Phan et al. 2008). In the same way, from model cheeses varying in compositions, it was reported that in-mouth sodium release and saltiness perception were more influenced, respectively, by the chewing force of the consumers and their salivary flow rate (Lawrence et al. 2012b).

11.6 Attempts to model the release of tastants

Mechanistic models were developed in order to better understand the release mechanisms of flavour compounds from food products in relation with composition and oral physiology. In food systems, the release phenomenon is limited by the resistance to the interfacial mass transfer (Druaux and Voilley 1997, Hills and Harrison 1995). Thus it is important to know the properties of the small molecules at the different transfer interfaces and the transfer kinetics between two phases or inside a phase (Déléris et al. 2012). The transfer of a component from one phase to another depends both on the thermodynamic equilibrium between the two phases and the simultaneous transfers inside each of the two adjacent phases. When food is eaten, the release of solutes can be described by mass transfer equations but it is necessary to know the partitioning properties at the interfaces level and the dynamic properties mainly characterised by the diffusion coefficients.

Concerning volatile compounds different models have been developed over these last decades (for review: De Roos 2006) and recent developments integrated the mastication process in modelling flavour release in the mouth (Harrison et al. 2014). In the case of non-volatile compounds, far less modelling approaches taking into account oral physiology in the release phenomena were proposed. Different methodologies have been developed to determine diffusion and partition coefficients of small solutes in foods including modern, complex and non-destructive methods for studies at the microscopic level (Floury et al. 2010). A simple method to measure the partition and diffusion coefficients of sodium chloride between a food product and a water phase consisted in the adaptation of the Phase Ratio Variation methodology (PRV) for the experimental determination of the water/product partition and diffusion coefficients of sodium chloride (Lauverjat et al. 2009a). These models highlighted in particular the influence of dry matter of cheese models on the diffusion of sodium chloride. A modelling approach of electrolytes diffusion based on gradient concentrations between different model matrices and artificial saliva was proposed by measuring their apparent diffusion coefficient by fitting the release of electrolytes data with the second Fick's law (Floury et al. 2009). The sodium chloride water/model cheeses partition coefficient highlighted interactions between proteins and sodium chloride but these interactions were not modified by the product composition and structure (Lauverjat et al. 2009b). On the contrary, model cheese composition (such as dry matter, fat and salt contents) and structure influenced sodium chloride diffusion from the model cheese to the saliva phase. Mechanistic models to better understand salt release and perception were developed in *in vivo* conditions for sodium chloride (Déléris et al. 2012). The release of sodium chloride during consumption of salted solutions was described as resulting from a mixing of the solutions with saliva and the diffusion of sodium chloride in the boundary layers above the tongue and palate mucous membranes

(Bakalis et al. 2009). Another model was developed using solid model cheeses varying in protein and fat contents to simulate *in vitro* salt release (de Loubens et al. 2011a). These authors proposed a mechanistic model for calculation of the contact area between the water and the product generated by compression. They reported in particular that fat had a major influence on breakdown behaviour and the calculated contact area that could be accounted for the microstructure of the products and its impact on salt release. A similar approach was carried out from *in vivo* data obtained with panellists characterized on a physiological basis (de Loubens et al. 2011b). A simple *in vivo* model was developed to study the specific contributions of the product and of the panellists on salt release. This model allowed a better understanding of the mechanisms of salt release and perception during food consumption with physiological parameters such as salivary flow rate and masticatory efficiency, and the rheological and breakdown parameters of food. Another model was developed for better understanding salt release from food matrices and saltiness perception considering oral processing as a physicochemical reactor (Le Reverend et al. 2013). The predictions of this model were compared favourably with temporal perception data, including response to pulsed salt release. Recently, a more complex dynamic 3D computational model was developed to predict the breakdown of food and the release of tastants (Harrison et al. 2014). This 3D model includes an accurate representation of: the anatomical movements of the oral cavity (including the teeth, tongue, cheeks and palate), the breakdown behaviour of the food, the interactions between comminuted food and saliva as the bolus is formed, the release and transport of taste and aromas and how these physical and chemical processes are perceived by a specific subject.

11.7 Conclusion

The composition of the products influences non-volatile compounds release and taste perception, which are subject-dependent because of their modulation by the oral physiology. In the case of salt, although it is clear that saltiness intensity is mainly due to sodium release, the kinetics of this release and temporal perception during chewing appear to be often disconnected. The effect of food composition on sodium release and saltiness perception is complex. It can vary according to the nature and the relative concentration of the food matrix constituents. Moreover, the importance of each composition factor can vary according to the chewing time, probably in relation to the water content in mouth due to the addition of saliva and also to the progressive change of the contact area and the texture of the food bolus during the eating process.

Thus, there is a great need to better understand the influence of food microstructure and physico-chemical and physiological factors on the mechanisms governing taste compound release in the mouth and taste perception during eating.

References

Bakalis, S., Le Révérend, B. J. D., Wolf, B. and Lian, G. P. (2009) Predicting saltiness: an in-mouth salt release model. in 5th *International Symposium on Food Rheology and Structure*, Zurich, Switzerland. pp. 626–627.

Beullens, K., Mészaros, P., Vermeir, S., Kirsanov, D., Legin, A., Buysens, S., Cap, N., Nicolaï, B. M. and Lammertyn, J. (2008) Analysis of tomato taste using two types of electronic tongues. *Sensors and Actuators B, 131*, pp. 10–17.

Boisard, L., Andriot, I., Arnould, C., Achilleos, C., Salles, C. and Guichard, E. (2013) Structure and composition of model cheeses influence sodium NMR mobility, kinetics of sodium release and sodium partition coefficients. *Food Chemistry, 136*(2), pp. 1070–1077.

Boisard, L., Andriot, I., Martin, C., Septier, C., Boissard, V., Salles, C. and Guichard, E. (2014) The salt and lipid composition of model cheeses modifies in-mouth flavour release and perception related to the free sodium ion content. *Food Chemistry, 145*, pp. 437–444.

Chabanet, C., Tarrega, A., Septier, C., Siret, F. and Salles, C. (2013) Fat and salt contents affect the in-mouth temporal sodium release and saltiness perception of chicken sausages. *Meat Science, 94*(2), pp. 253–261.

Chaunier, L., Chiron, H., Della-Valle, G. and Saulnier, L. (2008) Evaluation instrumentale de la texture contrastée croûte/mie de pains miniaturisés présentant des teneurs en fibres variées. *Industries des céréales, 160*, pp. 2–8.

Chiu, N., Hewson, L., Yang, N., Linforth, R. and Fisk, I. (2015) Controlling salt and aroma perception through the inclusion of air fillers. *LWT-Food Science and Technology, 63*(1), pp. 65–70.

Christensen, C. M., Brand, J. G. and Malanud, D. (1987) Salivary changes in solution pH - A source of individual-differences in sour taste perception. *Physiology and Behavior, 40*(2), pp. 221–227.

Davidson, J. M., Linforth, R. S. T., Hollowood, T. A. and Taylor, A. J. (1999) Effect of sucrose on the perceived flavor intensity of chewing gum. *Journal of Agricultural and Food Chemistry, 47*(10), pp. 4336–4340.

Davidson, J. M., Linforth, R. S. T., Hollowood, T. A. and Taylor, A. J. (2000) Release of non-volatile flavor compounds in-vivo. in Roberts, D. D. and Taylor, A. J., (eds.) *Flavor release*,Washington: American Chemical Society. pp. 99–111.

Davidson, J. M., Linforth, R. S. T. and Taylor, A. J. (1998) In-mouth measurement of pH and conductivity during eating. *Journal of Agricultural and Food Chemistry, 46*(12), pp. 5210–5214.

de Loubens, C., Panouille, M., Saint-Eve, A., Deleris, I., Trelea, I. C. and Souchon, I. (2011a) Mechanistic model of *in vitro* salt release from model dairy gels based on standardized breakdown test simulating mastication. *Journal of Food Engineering, 105*(1), pp. 161–168.

de Loubens, C., Saint-Eve, A., Deleris, I., Panouille, M., Doyennette, M., Trelea, I. C. and Souchon, I. (2011b) Mechanistic model to understand *in vivo* salt release and perception during the consumption of dairy gels. *Journal of Agricultural and Food Chemistry, 59*(6), pp. 2534–2542.

De Roos, K. B. (2006) Modeling aroma interactions in food matrices. in Voilley, A. and Etievant, P., (eds.) *Flavour in food*,Cambridge, England: Woodhead publishing limited. pp. 229–259.

Déléris, I., Guichard, E. and Salles, C. (2012) Libération des composés d'arôme et des composés sapides. in Guichard, E., Genot, C. and Voilley, A., (eds.) *Texture et flaveur des aliments : vers une conception maîtrisée*, Dijon (France): Éducagri éditions. pp. 203–231.

Druaux, C. and Voilley, A. (1997) Effect of food composition and microstructure on volatile flavour release. *Trends in Food Science and Technology, 8*(11), pp. 364–368.

Emorine, M., Mielle, P., Maratray, J., Septier, C., Thomas-Danguin, T. and Salles, C. (2012) The use of sensors to measure in-mouth salt release during food chewing. *Institute of Electrical and Electronics Engineers Sensors Journal, 12*, pp. 3124–3130.

Engel, E., Nicklaus, S., Septier, C., Salles, C. and Le Quéré, J.-L. (2001) Evolution of the taste of a bitter Camembert cheese during ripening: characterization of a matrix effect. *Journal of Agricultural and Food Chemistry, 49*(6), pp. 2930–2939.

Feron, G. and Salles, C. (2013) Rôle de la sphère orale dans la déstructuration de l'aliment, la libération des substances actives et de leurs fonctionnalités. in A. Fardet, I. Souchon, D. Dupond, (ed.) *Impact/rôle de la structure physique des produits alimentaire sur leur devenir digestif et les effets métaboliques et nutritionnels associés* Paris, France: Editions Quae. pp. 35–55.

Ferry, A. L. S., Mitchell, J. R., Hort, J., Hill, S. E., Taylor, A. J., Lagarrigue, S. and Valles-Pamies, B. (2006) In-mouth amylase activity can reduce perception of saltiness in starch-thickened foods. *Journal of Agricultural and Food Chemistry, 54*(23), pp. 8869–8873.

Floury, J., Jeanson, S., Aly, S. and Lortal, S. (2010) Determination of the diffusion coefficients of small solutes in cheese: A review. *Dairy Science and Technology, 90*(5), pp. 477–508.

Floury, J., Rouaud, O., Le Poullennec, M. and Famelart, M.-H. (2009) Reducing salt level in food: Part 2. Modelling salt diffusion in model cheese systems with regards to their composition. *LWT-Food Science and Technology, 42*(10), pp. 1621–1628.

Foucat, L., Donnat, J.P., and Renou, J.P. (2003) ^{23}Na and ^{35}Cl NMR studies of the interactions of sodium and chloride ions with meat products. in Belton, P. S., Gil, A. M., Webb, G. A. and Rutledge, D., (eds.) *Magnetic Resonance in Food Science: last developments*, Cambridge, UK: The Royal Society of Chemistry. pp. 180–185.

Guichard, E., Boisard, L. and Salles, C. (2013) Flavour release and sensory perception in cheeses. in open access book, (ed.) *Handbook for cheeses: Cheese: production, chemistry and sensory properties*, Hauppauge, NY, USA: Nova Science. pp. 313–340.

Harrison, S. M., Cleary, P. W., Eyres, G., Sinnott, M. D. and Lundin, L. (2014) Challenges in computational modelling of food breakdown and flavour release. *Food and Function, 5*(11), pp. 2792–2805.

Hiiemae, K. (2004) Mechanisms of food reduction, transport and deglutition: How the texture of food affects feeding behavior. *Journal of Texture Studies, 35*(2), pp. 171–200.

Hills, B. P. and Harrison, M. (1995) Two-film theory of flavour release from solids. *International Journal of Food Science and Technology, 30*(4), pp. 425–436.

Hui, G.-H., Mi, S.-S. and Deng, S.-P. (2012) Sweet and bitter tastant specific detection by the taste cell-based sensor. *Biosensors and Bioelectronics, 35*, pp. 429–438.

Jack, F. R., Piggott, J. R. and Paterson, A. (1995) Cheddar cheese texture related to salt release during chewing, measured by conductivity - Preliminary study. *Journal of Food Science, 60*(2), pp. 213–217.

Konitzer, K., Pflaum, T., Oliveira, P., Arendt, E., Koehler, P. and Hofmann, T. (2013) Kinetics of Sodium Release from Wheat Bread Crumb As Affected by Sodium Distribution. *Journal of Agricultural and Food Chemistry, 61*(45), pp. 10659–10669.

Lauverjat, C., de Loubens, C., Deleris, I., Trelea, I. C. and Souchon, I. (2009a) Rapid determination of partition and diffusion properties for salt and aroma compounds in complex food matrices. *Journal of Food Engineering, 93*(4), pp. 407–415.

Lauverjat, C., Deleris, I., Trelea, I. C., Salles, C. and Souchon, I. (2009b) Salt and Aroma Compound Release in Model Cheeses in Relation to Their Mobility. *Journal of Agricultural and Food Chemistry, 57*(21), pp. 9878–9887.

Lawrence, G., Buchin, S., Achilleos, C., Bérodier, F., Septier, C., Courcoux, P. and Salles, C. (2012a) *In vivo* sodium release and saltiness perception in solid lipoprotein matrices. 1. Effect of composition and texture. *Journal of Agricultural and Food Chemistry, 60*, pp. 5287–5298.

Lawrence, G., Septier, C., Achilleos, C., Courcoux, P. and Salles, C. (2012b) *In vivo* sodium release and saltiness perception in solid lipoproteic matrices. 2. Impact of oral parameters. *Journal of Agricultural and Food Chemistry, 60*, pp. 5299–5306.

Le Bleis, F., Chaunier, L., Della Valle, G., Panouille, M. and Reguerre, A. L. (2013) Physical assessment of bread destructuration during chewing. *Food Research International, 50*(1), pp. 308–317.

Le Reverend, B. J. D., Norton, I. T. and Bakalis, S. (2013) Modelling the human response to saltiness. *Food and Function, 4*(6), pp. 880–888.

Mosca, A. C., Andriot, I., Guichard, E. and Salles, C. (2015) Binding of Na^+ ions to proteins: Effect on taste perception. *Food Hydrocolloids, 51*, pp. 33–40.

Neyraud, E. (2014) Role of saliva in oral food perception. in Ligtenberg A.J.M, Veerman. E.C.I., (ed.) *Saliva: Secretion and Functions*, Basel Karger. pp. 61–70.

Neyraud, E., Prinz, J. and Dransfield, E. (2003) NaCl and sugar release, salivation and taste during mastication of salted chewing gum. *Physiology and Behavior, 79*(4-5), pp. 731–737.

Panouillé, M., Saint-Eve, A., Deleris, I., Le Bleis, F. and Souchon, I. (2014) Oral processing and bolus properties drive the dynamics of salty and texture perception of bread. *Food Research International, 62*, pp. 238–246.

Pedersen, A. M., Bardow, A., Jensen, S. B. and Nauntofte, B. (2002) Saliva and gastrointestinal functions of taste, mastication, swallowing and digestion. *Oral Diseases, 8*(3), pp. 117–129.

Pflaum, T., Konitzer, K., Hofmann, T. and Koehler, P. (2013) Influence of texture on the perception of saltiness in wheat bread. *Journal of Agricultural and Food Chemistry, 61*(45), pp. 10649–10658.

Phan, V. A., Yven, C., Lawrence, G., Chabanet, C., Reparet, J. M. and Salles, C. (2008) *In vivo* sodium release related to salty perception during eating model cheeses of different textures. *International Dairy Journal, 18*(9), pp. 956–963.

Pionnier, E., Chabanet, C., Mioche, L., Taylor, A. J., Le Queré, J. L. and Salles, C. (2004a) *In vivo* nonvolatile release during eating of a model cheese: relationships with oral parameters. *Journal of Agricultural and Food Chemistry, 52*(3), pp. 565–571.

Pionnier, E., Nicklaus, S., Chabanet, C., Mioche, L., Taylor, A. J., Le Quéré, J. L. and Salles, C. (2004b) Flavor perception of a model cheese: relationships with oral and physico-chemical parameters. *Food Quality and Preference, 15*(7-8), pp. 843–852.

Rama, R., Chiu, N., Da Silva, M. C., Hewson, L., Hort, J. and Fisk, I. D. (2013) Impact of salt crystal size on in-mouth delivery of sodium and saltiness perception from snack foods *Journal of Texture Studies, 44*(5), pp. 338–345.

Salles, C., Briand, L., Brachais, L. and Voilley, A. (2012) Molécules aromatisantes et sapides. in Guichard, E., Genot, C. and Voilley, A., (eds.) *Texture et flaveur des aliments : vers une conception maîtrisée*, Dijon (France): Éducagri éditions. pp. 31–58.

Salles, C., Chagnon, M. C., Feron, G., Guichard, E., Laboure, H., Morzel, M., Semon, E., Tarrega, A. and Yven, C. (2011) In-mouth mechanisms leading to flavor release and perception. *Critical Reviews in Food Science and Nutrition, 51*(1), pp. 67–90.

Tahara, Y. and Toko, K. (2013) Electronic tongues - A review. *Institute of Electrical and Electronics Engineers Sensors Journal, 13*, pp. 3001–3011.

Tian, X., Wang, J. and Zhang, X. (2013) Discrimination of preserved licorice apricot using electronic tongue. *Mathematical and computer modelling, 58*, pp. 743–751.

Tournier, C., Grass, M., Septier, C., Bertrand, D. and Salles, C. (2014) The impact of mastication, salivation and food bolus formation on salt release during bread consumption. *Food and Function, 5*, pp. 2969–2980.

Tournier, C., Grass, M., Zope, D., Salles, C. and Bertrand, D. (2012) Characterization of bread breakdown during mastication by image texture analysis. *Journal of Food Engineering, 113*, pp. 615–622.

World Health Organisation (2007) *Reducing salt intake in populations*. Report of a World Health Organization Forum and Technical Meeting. Geneva, Switzerland.

CHAPTER 12

Interactions between saliva and flavour compounds

Francis Canon & Eric Neyraud
Center for Taste and Feeding Behaviour, CNRS, INRA, University of Bourgogne Franche-Comté, Dijon, France

12.1 Introduction

Food flavour perception is a complex phenomenon which depends on the organoleptic properties of the food and on the physiology of the individual who experiences tasting. One of the critical steps leading to this perception occurs during "food oral processing" which consists of mixing food with saliva during mastication in order to form a bolus. During this step, flavour compounds are released from the food and can then interact with sensory receptors. The release of flavour compounds is governed by their interactions with both the food matrix (Chapter 9) and saliva. We can therefore say that over the course of food oral processing, we do not perceive the food itself but rather the result of the interactions between food and saliva. These interactions are of various kinds and depend on the chemical nature of the flavour compounds. Basically, these interactions can be separated into three different categories which are first, the role of saliva to dissolve and transport flavour compounds to the receptors; second, the capacity of salivary molecules to bind flavour compounds resulting in a decrease or a neutralization of the amount of active flavour molecules; and third the conversion of flavour compounds, especially by salivary enzymes, into molecules without or with a different sensory impact. In addition, it is important to keep in mind that the sensory consequences of these interactions depend on the variability of saliva composition. This variability is particularly high between individuals but saliva composition also varies intra-individually. Interestingly, flavour compounds themselves can stimulate salivation and modify saliva composition in such a way that saliva seems "adapted" to the nature of the stimulation. In this context, this chapter gives an overview of the different types of interactions between saliva and flavour compounds.

Flavour: From food to perception, First Edition.
Edited by Elisabeth Guichard, Christian Salles, Martine Morzel, and Anne-Marie Le Bon.

12.2 Influence of flavour compounds on salivation

Saliva secretions can be activated by different types of stimuli including flavour compounds. However, the impact of volatile olfactory compounds on salivary flow is not clearly established in the literature and seems to depend on the nature of the associated odour. For instance, odours associated to sour taste, like lemon, increase salivary flow (Lee and Linden 1992, Pangborn 1968, Pangborn et al. 1979, Spence 2011). Odours associated to palatable and familiar foods have been reported to induce an increase of salivary flow (Klajner et al. 1981, Rogers and Hill 1989, Sahakian et al. 1981), while other odours like almond or non-food related odours had no impact on salivary flow (Lashley 1916, Mattes 2000).

In contrast to odorants, the most admitted and efficient flavour compounds for increasing flow rate are tastants and particularly sour stimuli. Most of the tastants induce an increase of salivary flow, which increases with concentration (Froehlich et al. 1987, Hodson and Linden 2006, Neyraud et al. 2009, Watanabe and Dawes 1988). It has been also reported that beverages containing tannins such as wine increase the salivary flow rates (Fischer et al. 1994, Guinard et al. 1998). However, solution containing only tannins and perceived as astringent do not significantly increase salivary flow rates (Hyde and Pangborn 1978). Sourness, one of the most effective sialogogues, is probably at the origin of the sialogogic effect of wine.

Submandibular glands are the main contributors of whole saliva in absence of stimulation or during sleep (Aps and Martens 2005) and it is currently admitted that the increase of salivary flow following a mechanical or a taste stimulation is due to an increase of parotid glands secretions. Considering that the different salivary glands secrete different molecules, these differences of contribution explain in part differences observed in whole salivary composition following different types of stimulation. However, saliva composition is also affected by the nature of the stimulation. This has been observed through the modifications of the whole saliva proteome after stimulation by different tastants (Neyraud et al. 2006). The number of proteins differently expressed was found to be related to the degree of aversion of the taste following the decreasing order: sour, bitter, umami and sweet. Because some proteins over-expressed were involved in inflammatory response (e.g., Calgranulin A, Annexin A1), it was suggested that this specific secretion was a defence mechanism against potentially harmful food intake. Specific protective secretions of saliva following a potentially harmful stimulation were also suggested in other studies (Davies et al. 2009, Stokes and Davies 2007). The authors showed that elasticity of saliva was higher after drinking an acidic beverage compared to saliva secreted after drinking water or chewing. It was suggested that this increase of elasticity was due to over-secretion of salivary mucins with the result of protecting the oral mucosa. Other "omics" studies reported specific modifications of the salivary proteome following stimulation by different bitter tastants (Quintana et al. 2009) or by molecules having different tastes

(Lorenz et al. 2011, Silletti et al. 2012). Targeted studies have also followed the modification of specific proteins or properties of saliva based on the hypothesis that they could be involved in specific perceptions. Thus, Dinnella et al. (2009) have observed a wide variability in the modification of salivary protein profile in response to astringent stimulation, later shown to affect the sensitivity to astringency (Dinnella et al. 2010). The lipolytic activity and antioxidant capacity of saliva have been related to fatty acid (oleic acid) taste perception (Mounayar et al. 2013). The authors showed in this particular case that lipolytic activity decreases significantly and antioxidant capacity increases significantly after stimulation by oleic acid in saliva of subjects hypersensitive to the taste of this fatty acid. This work was completed by a combined proteomics and metabolomics approach which showed an over-abundance of organic acids, having likely a bacterial origin, after stimulation by oleic acid in saliva of hyper-sensitive subjects. This suggested a possible implication of the oral microbiota in fat perception (Mounayar et al. 2014a). Table 12.1 presents an overview of reported modifications of salivary proteins abundance following different oro-sensory stimulations.

12.3 Flavour compounds transport to sensory receptors

Because taste receptors located predominantly on the tongue are continuously bathed by saliva, it is necessary for taste compounds to be dissolved in saliva to reach these receptors. The simplest case concerns hydrophilic taste compounds, such as sodium ions or simple sugars, which only need to be dissolved in water. In this case, the volume of saliva appears crucial because it leads to dilution of the tastants and affects their concentration at the receptor's site. Some works showed that taste perception of sodium (Neyraud et al. 2003) or calcium (Neyraud and Dransfield 2004) is better correlated to tastant concentration in saliva than in the food matrix. In addition, a study performed on saltiness perception in solid lipoprotein matrices show that lower individual salivary flow rates are associated to higher perception of saltiness (Lawrence et al. 2012). This appears to be in accordance with the determinant role of dilution in the hydrophilic taste compound perception.

In contrast to hydrophilic compounds, hydrophobic compounds that cannot be dissolved in saliva need other systems of transport to reach taste receptors. For example, since the recent discovery of fatty acids taste receptors on the tongue (see Mattes 2009 for review), it is suggested that hydrophobic fatty acids could be considered as taste compounds. In this case, transporters are needed to ensure their transport to taste receptors through saliva. If mechanisms assuring the transport of fatty acids to taste receptors have not been shown yet, some proteins candidate can play this role of fatty acids transporters. This is the case of the lipocalin 1 (LCN1) protein also called VEGP for von Ebner's Gland Protein (Bläker

Table 12.1 Summary of the proteins whose abundance differs after a gustatory stimulation.

Protein	Type of stimulation	Effect [1]	Reference
Annexin A1	IMP (Umami)	+	Neyraud et al. (2006)
	$Ca(NO_3)_2$ (Bitter)	+	Neyraud et al. (2006)
	HNO_3 (Sour)	+	Neyraud et al. (2006)
	6-gingerol (Pungent)	–	Lorenz et al. (2011)
Annexin A2	HNO_3 (Sour)	+	Neyraud et al. (2006)
Lactoferrin	Fruit flavoured candy	–	Becerra et al. (2003)
Peroxidase	Fruit flavoured candy	–	Becerra et al. (2003)
Amylase	Fruit flavoured candy	–	Becerra et al. (2003)
	Tannic acid (Astringent)	–	Dinnella et al. (2010)
	HNO_3 (Sour)	+	Neyraud et al. (2006)
	Urea (Bitter)	+	Quintana et al. (2009)
	Sucrose (Sweet)	+	Harthoorn et al. (2009)
Carbonic anhydrase 2	Fruit flavoured candy	+	Becerra et al. (2003)
Carbonic anhydrase 6	6-gingerol (Pungent)	+	Lorenz et al. (2011)
	Urea (Bitter)	–	Quintana et al. (2009)
Prolin Rich Glycoprotein	Fruit flavoured candy	+	Becerra et al. (2003)
Glycosylated PRPs	Tannic acid (Astringent)	–	Dinnella et al. (2010)
PRPs	Tannic acid (Astringent)	–	Dinnella et al. (2010)
PRP3	Sucrose (Sweet)	–	Silletti et al. (2012)
PRH2 protein : acidic PRP	HNO_3 (Sour)	+	Neyraud et al. (2006)
bPRP	PROP (Bitter)	+[2]	Cabras et al. (2012)
Basic glycosylated Glycoprotein	Fruit flavoured candy	+	Becerra et al. (2003)
MUC5B	Fruit flavoured candy	+	Becerra et al. (2003)
Histatin	Tannic acid (Astringent)	–	Dinnella et al. (2010)
Cystatins	Tannic acid (Astringent)	–	Dinnella et al. (2010)
Cystatin S	$Ca(NO_3)_2$ (Bitter)	+	Neyraud et al. (2006)
Cystatin SN	Quinine (Bitter)	–	Quintana et al. (2009)
Calgranulin A	IMP (Umami)	+	Neyraud et al. (2006)
	$Ca(NO_3)_2$ (Bitter)	+	Neyraud et al. (2006)
	HNO_3 (Sour)	+	Neyraud et al. (2006)
	NaCl (Salty)	+	Silletti et al. (2012)
Calgranulin C	Sucrose (Sweet)	+	Silletti et al. (2012)
Beta-2-microglobulin	HNO_3 (Sour)	–	Neyraud et al. (2006)
Enolase 1	$Ca(NO_3)_2$ (Bitter)	–	Neyraud et al. (2006)
Glutathione S-transferase P	6-gingerol (Pungent)	–	Lorenz et al. (2011)
Heat shock protein beta 1	6-gingerol (Pungent)	–	Lorenz et al. (2011)
Heat shock 70kDa protein 1	6-gingerol (Pungent)	–	Lorenz et al. (2011)

Table 12.1 (*Continued*)

Protein	Type of stimulation	Effect [1]	Reference
Cytoplasmic beta-actin	6-gingerol (Pungent)	–	Lorenz et al. (2011)
Prolactin inductible proteins	6-gingerol (Pungent)	+	Lorenz et al. (2011)
	Urea (Bitter)	–	Quintana et al. (2009)
	Urea (Bitter)	+	Quintana et al. (2009)
SPLUNC2	6-gingerol (Pungent)	+	Lorenz et al. (2011)
Zinc-alpha 2 glycoproteins	6-gingerol (Pungent)	+	Lorenz et al. (2011)
Glyceraldehyde-3-phosphate dehydrogenase	Urea (Bitter)	–	Quintana et al. (2009)
	Quinine (Bitter)	–	Quintana et al. (2009)
Serum albumin	Urea (Bitter)	+	Quintana et al. (2009)
	Quinine (Bitter)	+	Quintana et al. (2009)
Transthyretin, chain A	Urea (Bitter)	+	Quintana et al. (2009)
Transferrin	Urea (Bitter)	+	Quintana et al. (2009)
	Quinine (Bitter)	–	Quintana et al. (2009)
Parotid secretory protein	Quinine (Bitter)	+	Quintana et al. (2009)
Thioredoxin	Quinine (Bitter)	–	Quintana et al. (2009)
Statherin	Sucrose (Sweet)	+	Silletti et al. (2012)
Alpha defensin 1 precursor	NaCl (Salty)	+	Silletti et al. (2012)
Polymeric immunoglobulin receptor	Oleic acid	–[3]	Mounayar et al. (2014b)
Rab GDP dissociation inhibitor beta	Oleic acid	–[3]	Mounayar et al. (2014b)

[1]: + indicates an increase after stimulation, - indicates a decrease after stimulation
[2]: in PROP supertaster subjects
[3]: in subjects hypersensitive to oleic acid taste

et al. 1993). LCN1 belongs to the lipocalin protein family. Proteins belonging to this family are typically small secreted proteins which exhibit several molecular recognition properties: the ability to bind small hydrophobic molecules, the binding to specific cell-surface receptors and the formation of macromolecular complexes (Flower 1996). Figure 12.1 presents the structure of LCN1. As the majority of the proteins belonging to the lipocalin family, LCN1 has a typical eight-stranded antiparallel β-barrel architecture, which defines a central cavity called calyx. A C-terminal α-helix, which is followed by an almost extended peptide, packs against one side of the β-barrel (Breustedt et al. 2005). At the open end of the pocket, four partially disordered loops connect each pair of the β-strands and form the entrance to the central cavity. Structural analysis have shown a

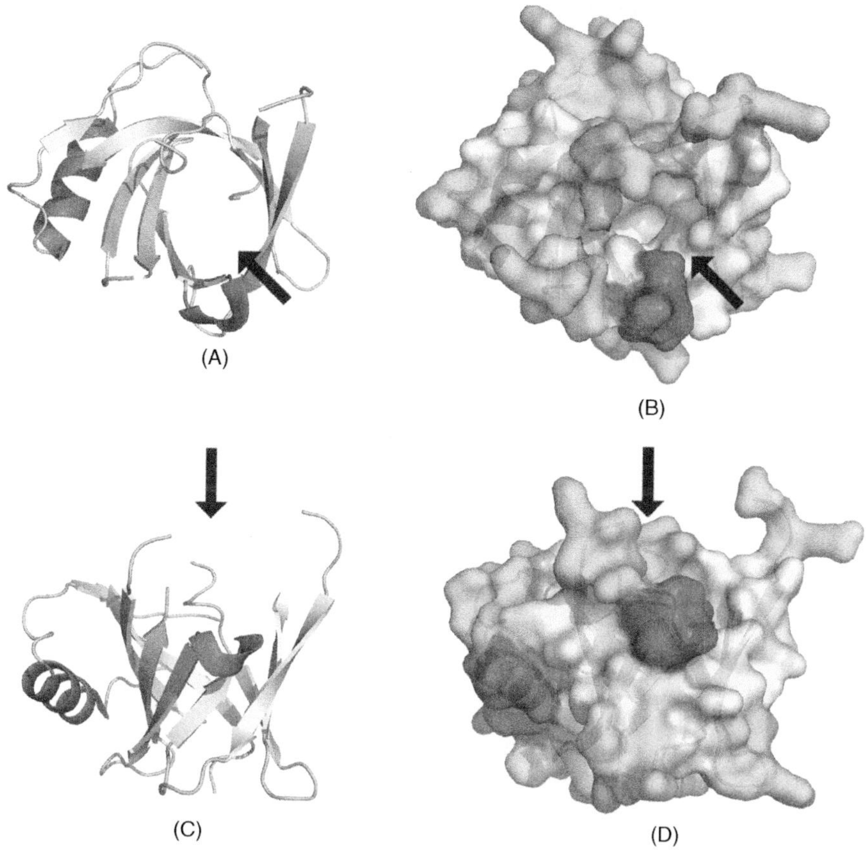

Figure 12.1 Structure of the lipocalin-1 (LCN1). **A**. ribbon diagram of topview into the calyx of LCN1; **B**. surface representation of topview into the calyx of LCN1; **C**. ribbon diagram of side view of LCN1; **D**. surface representation of side view of LCN1.

high conformational plasticity of the loop region and of the deeper parts of the ligand pocket (Breustedt et al. 2009). Such plasticity gives to LCN1 the ability to bind different ligands that differ vastly in size and shape (Glasgow et al. 1995). This feature is unusual in comparison with the other members of the lipocalin family. Among the different ligands bound by LCN1, LCN1 has demonstrated its ability to interact with fatty acids. Its affinity for fatty acids increases with the length of the aliphatic chain up to 18 carbons, corresponding to the maximal capacity of the hydrophobic cavity (Abduragimov et al. 2000). Interestingly, this protein is secreted by the von Ebner's glands located in the trenches of taste papillae close to the taste buds. Therefore, the concentration of LCN1 would be higher at proximity of taste receptors, where LCN1 could regulate the concentration of free fatty acids or create optimal conditions for fatty acids transport to taste receptors (Feron and Poette 2013, Gilbertson 1998).

12.4 Binding of flavour compounds by saliva

12.4.1 Binding of tannins and astringency perception

Beside the interactions of LCN1 with lipids, numerous interactions involving salivary proteins and flavour compounds have been reported. Among them, the interactions occurring between salivary proline-rich proteins and tannins are probably the most described. Tannins, which are phenolic compounds, are ubiquitous in plants (Cheynier 2005). These molecules play an important part in plant defence mechanisms due to their ability to bind proteins. Through their interactions, they inhibit digestive enzymes (Oh and Hoff 1986, Oh and Hoff 1988) and reduce protein digestibility (Horigome et al. 1988, Ozawa et al. 1987). In mammals, tannin-rich diets are generally avoided and less often consumed (Shimada and Saitoh 2003). Indeed, tannins are responsible for the sensation of astringency, which is thought to be at the origin of this behaviour. Astringency is defined as "the complex of sensations due to shrinking, drawing, or puckering of the epithelium as a result of exposure to substances such as alums or tannins" by the American Society for Testing of Material (ASTM 1989). It represents an important organoleptic property of plant-based food. Tea and red wine are the most commonly consumed products that are particularly rich in tannins. Despite the importance of astringency regarding consumer acceptability, the exact mechanism at the origin of this sensation is still not fully understood. Astringency is generally considered to be a tactile sensation (Breslin et al. 1993), detected through the activation of mechanoreceptors (Guinard and Mazzucchelli 1996), rather than a taste (Breslin et al. 1993). Indeed, contrarily to taste, astringency perception increases upon repeated expositions to astringent molecules (Lee and Lawless 1991). Moreover, it has been observed that astringency can be perceived on oral mucosa deprived of gustatory anatomical structures, and needs movement of oral tissues to be perceived (Rossetti et al. 2008). More recently, Schobel et al. (2014) have shown that anesthesia of the lingual trigeminal innervation, after the inferior alveolar nerve block, impairs astringency perception while loss of the taste nerve function does not. These observations reinforce the classification of astringency as a tactile trigeminal sensation, which could involve mechanoreceptors and also the activation of a trigeminal G protein-coupled receptor for galloylated molecules (Schobel et al. 2014). Regarding the activation of mechanoreceptors, it is probably caused by physical changes in saliva and in the pellicle of the oral mucosa, reducing the lubrication of the oral surfaces (Clifford 1997, Green 1993, Lyman and Green 1990).

Several salivary proteins have demonstrated their ability to interact with tannins (Perez-Gregorio et al. 2014). Among them, it has been shown that proline-rich proteins (PRPs) have a particular affinity for tannins (Hagerman and Butler 1981), which is estimated to be 5–80 and 1 000 times higher than that of bovine serum albumin and lysozyme, respectively (Asquith et al. 1987, Austin et al. 1989, Mole et al. 1990). PRPs belong to the intrinsically disordered proteins

(IDPs) family (Tompa 2002), which do not have a globular conformation but instead populate dynamic conformational ensembles (Dunker et al. 2002, Wright and Dyson 1999). Despite their lack of a well-defined three-dimensional conformation, IDPs fulfill a large range of vital cellular functions, which are linked to their disordered conformation (Dunker et al. 2002, Dunker et al. 2008, Tompa 2002). Salivary proline-rich proteins (PRPs) exhibit unusual extended conformations, which could be explained by the presence of polyproline II helices within the polypeptidic backbones (Boze et al. 2010). PRPs contain repeated sequences with high proportions of Pro, Gly, and Glu or Gln residues. PRPs are classically divided into three classes: acidic PRPs (*a*PRP), basic PRPs (*b*PRP) and glycosylated PRPs (*g*PRP) (Bennick 1982). *a*PRPs are involved in calcium homeostasis (Hay et al. 1987), *g*PRPs play a role in the lubrication of the oral cavity (Hatton et al. 1985) and could also prevent microbial colonization (Gillece-Castro et al. 1991), while the main function described for *b*PRPs is to bind tannins, even if they have been also reported to have anti-viral activity (Gu et al. 1995). In human, PRPs may represent up to 70% of the proteins secreted by the parotid glands (Kauffman and Keller 1979). In rodents the secretion of PRPs can be induced by a tannin-rich diet (Mehansho et al. 1985). The ability of PRPs to interact with tannins explains their involvement in adaptation to tannin-rich diet either as astringency mediator or/and as scavenger (Bennick 2002, McArthur et al. 1995). Indeed, *b*PRPs is the most important group of salivary proteins with histatins to form soluble complexes with tannins, depending of the tannin structure (Perez-Gregorio et al. 2014), while *a*PRPs have been reported to be the major protein group to be precipitated by tannins (Soares et al. 2012). Another study has reported that *b*PRPs have a higher capacity to bind tannins than *a*PRPs (Lu and Bennick 1998). These results show that it is not possible to evaluate correctly the ability of a protein to bind and scavenge tannins only based on its level of precipitation in presence of tannins. Indeed, the study of Yan and Bennick (1995) suggests that soluble protein•tannin complexes can efficiently scavenge tannins and protect other salivary proteins. It has been also reported that the ionic strength plays an important role in the level of protein-tannin aggregation and precipitation, while it does not affect their interactions and therefore the affinity of protein for tannins (Canon et al. 2013b).

Regarding the high affinity of PRPs for tannins, it is explained by their open and flexible conformation, which allows them to bind several tannins simultaneously (Canon et al. 2009) but also tannins with different structures (Canon et al. 2010). It has been proposed that the PRP-tannin interaction is driven both by hydrophobic effect and hydrogen bonding (Haslam 1996). However, recent investigations by mass spectrometry (Canon et al. 2010) or by nuclear magnetic resonance and molecular modeling (Cala et al. 2012, Simon et al. 2003) suggest mainly the involvement of hydrogen bonding. More recently, an innovating technique based on the coupling of mass spectrometry and synchrotron radiation has demonstrated that the procyanidin B2 3′OG has a preferential binding site on IB5, a *b*PRP (Canon et al. 2013a). Further investigations have revealed that the procyanidin B2 and the monomer epigallocatechin gallate (EgCG) bind also onto the same binding site (Canon et al. 2014). The binding site is present in the following sequence "KPQGPPPPPQGG", which contains the longest cluster of proline present in the protein sequence. This cluster very likely adopts a PPI or a PPII helix conformation in solution (Boze et al. 2010, Pascal et al. 2009, Simon et al. 2003) forming a rigid segment, which provides an initial contact point for binding (Charlton et al. 1996, Gu and Helms 2005, Pascal et al. 2009). Moreover, the carbonyl function in tertiary amides of proline residues is a more effective hydrogen bond acceptor than that in primary and secondary amides (Haslam 1996). This proline cluster is surrounded by glycine and alanine residues, which form flexible regions allowing the establishment of additional hydrogen bonds (Chatterjee et al. 2005). It is currently proposed that the clusters of proline are rigid regions, which are initial anchoring points favorable for effective tannin binding. The flexible hinges surrounding the proline cluster allow the reeling of the peptidic chain and the establishment of additional contact points. This model is in agreement with observations reporting that full PRPs have higher affinity for tannins as compared to shorter PRP peptides (Cala et al. 2012, Charlton et al. 1996). Interestingly, ion mobility experiments on PRP•tannin complexes have revealed that the binding of tannins onto a PRP lead to structural rearrangement and folding of the PRP as a function of the stoichiometry (Canon et al. 2011). Indeed, PRPs have several clusters of proline in their sequence explaining the observation of PRP•*n*Tannin complexes with several stoichiometries (Canon et al. 2013b, Canon et al. 2009). Increases in tannin concentration lead to the appearance of higher stoichiometries and, then to the formation of aggregates from the threshold of three tannins per protein (Canon et al. 2013b, Charlton et al. 2002). Aggregates coexist with free proteins that are recruited with the increase of tannin concentration until precipitation of the dense aggregates (Canon et al. 2013b, Pascal et al. 2007). It should be noticed that the presence of glycosylation prevents the precipitation of PRP-tannin complexes (Lu and Bennick 1998, Sarni-Manchado et al. 2008), as observed in the case of mucins (Davies et al. 2014). From these observations, the current model of PRP-tannin interaction proposes a three-stage mechanism. In the first stage, tannins bind to PRP forming soluble complexes, then in the second stage, when the tannin/protein molar ratio exceeds a threshold, aggregation of the PRP•tannin complexes takes place, and finally precipitation occurs, when a higher molar ratio is reached (Pascal et al. 2007). Interestingly, the aggregation threshold value of a small basic PRP in presence of EgCG is close to the concentration reported for the astringency perception of this tannin (Canon et al. 2013b). However, it is unknown if aggregates participate directly to the generation of the astringency sensation, or if from this threshold, PRPs are no longer able to protect the proteins of the pellicle against tannins. PRPs have already demonstrated their ability to protect other proteins against tannins (Yan and Bennick 1995). Indeed, alpha-amylase has been also reported to interact with tannins, leading to the inhibition of its enzymatic activity (Kandra et al. 2004, Zajacz et al. 2007), while the presence of PRP protects amylase from inhibition by tannins (Yan and Bennick 1995).

Histatins (Hsts), which have been called also histidine-rich polypeptides or proteins, are another group of salivary proteins that have been reported to highly interact with tannins (Shimada 2006, Yan and Bennick 1995). Histatins have only be reported in the saliva of humans and some primates (Sabatini and Azen 1989). This family of proteins is composed of 12 members, named Hst 1 to Hst 12 (Troxler et al. 1990). They are small cationic proteins with a molecular weight below 5000 Da. Their sequence is typically rich in basic residues, and especially in histidine. Histatins also contain numerous lysine residues in their sequence. In Histatin 5, histidine, arginine and lysine represent about 60% of the total amino-acids. Histatins are multifunctional proteins, whose main function appears to be their antifungal activity which is directed for example towards *Candida albicans* causing oral candidiasis (Puri and Edgerton 2014, Xu et al. 1991). Histatins have been also reported to be present in the acquired enamel pellicle (Jensen et al. 1992, Siqueira et al. 2010), where they can bind to hydroxyapatite and protect enamel against acid demineralization more efficiently when they are phosphorylated (Siqueira et al. 2010). Histatins have been also reported to form soluble complexes with tannins (Perez-Gregorio et al. 2014) and to efficiently precipitate tannins (Yan and Bennick 1995). Moreover, as PRPs, histatin 5 has demonstrated its ability to protect α-amylase from tannin inhibition (Yan and Bennick 1995). Therefore, histatins can be considered as tannin-binding salivary proteins like PRPs (Shimada 2006). However, considering the amount of histatins in saliva compared to the one of PRPs, it is expected that PRPs play a more important role in tannins scavenging than histatins, which is in agreement with the total amount of soluble tannin•protein complexes observed in saliva classified by salivary protein family (Perez-Gregorio et al. 2014).

A study on the astringency of tea has shown that removal of saliva by rinsing the mouth leads to increase of the perceived astringency (Nayak and Carpenter 2008). It suggests that the presence of the free salivary proteins can decrease astringency while the proteins of the mucosal pellicle can be more directly involved in this sensation. In this scenario, the "free" salivary proteins classified as tannin-binding salivary proteins play both a protective role of the pellicle by scavenging tannins, preventing their interaction with the proteins of the mucosal pellicle, and a role in oral lubrication (Dinnella et al. 2010). Indeed, higher sensitivity to astringency could be related to modification in PRPs profile (Dinnella et al. 2009).

Oral mucosae are exposed to intense friction during deglutition, mastication or speech. To limit such friction strains, lubrication is ensured by salivary compounds, either in flowing form or structured as biological films. Thus, a first layer of salivary proteins, called the mucosal pellicle is tightly anchored onto the epithelium. Lubrication is further ensured by a mobile salivary film (or residual fluid), defined as free saliva retained on oral mucosa after deglutition. The thickness of the mobile salivary film varies on different surfaces in the mouth (Pramanik et al. 2010) but on average, it ranges between 70 and 100 μm (Collins and Dawes 1987). As to the salivary mucosal pellicle, it is composed mainly of salivary proteins selectively adsorbed on the surface of epithelial cells (Bradway et al. 1989). MUC5B, amylase, cystatin and *a*PRPs (Bradway et al. 1992) but also secretory component and IgA (Gibbins and Carpenter 2013) have been identified in mucosal pellicles. The exact mechanism by which these proteins are adsorbed onto the epithelial cells and form a structured film is still unclear. It has been proposed that pellicle formation may be aided by the presence of transglutaminase (Bradway et al. 1989), which is able for example to catalyze a crosslink between the acidic PRP-1 (24 kDa) and statherin (8 kDa) *in vitro* (Yao et al. 2000). The involvement of transmembrane proteins such as cell surface mucins is also suggested. Indeed, the cell surface mucin MUC1 has been identified in the oral cavity (Pramanik et al. 2010) and has been suggested to be an anchoring point for salivary proteins constitutive of the mucosal pellicle (Gibbins and Carpenter 2013). Our own results have confirmed that the membrane-bound mucin MUC1 is a factor enhancing the formation of the mucosal pellicle by increasing the binding of another salivary mucin, MUC5B, to oral epithelial cells.

Mucins are highly hydrated, amphiphilic glycoproteins that coat hydrophobic surfaces, rendering them hydrophilic and electrostatically and sterically repulsive. These properties at interfaces are favourable for both boundary lubrication and aqueous fluid film lubrication (Coles et al. 2010). Mucins exhibit also a tendency to aggregate and to adsorb onto surfaces of different natures (Bansil and Turner 2006). These large proteins are highly glycosylated and have molecular weights ranging from 0.5 to 20 MDa. They consist of a peptidic chain, with serine- and threonine-rich mucin domain, onto which are attached oligosaccharide chains by *O*-glycosidic bonds to the hydroxyl side chains of serine and threonines and arranged in a "bottle brush" configuration (Bansil and Turner 2006, Hattrup and Gendler 2008). There are two types of mucin: membrane-bound mucin, such as MUC1, and secreted mucin, such as MUC5B. MUC1 is composed of two subunits, generated by autoproteolyis of the *MUC1* gene product, which non-covalently associate: the N-terminal subunit is entirely extracellular while the C-terminal subunit comprises a short extracellular domain, a transmembrane domain and a cytoplasmic domain. Concerning the secreted mucins, two are found in saliva: MUC7 and MUC5B. MUC7 is involved mainly in antibacterial protection of the oral cavity, while MUC5B, a large protein with a molecular weight above 1 000 kDa, plays a major role in the lubricating properties of saliva and of salivary pellicles (Cárdenas et al. 2007, Nieuw Amerongen and Veerman

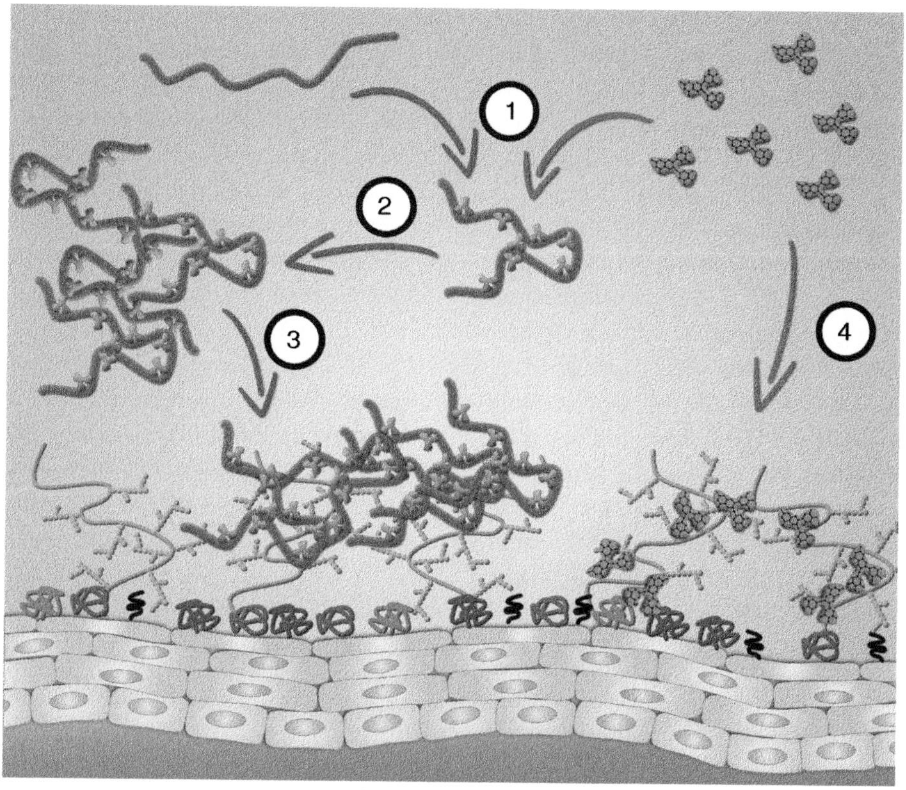

Figure 12.2 Theoretical mechanisms of astringency. 1-formation of soluble PRP•tannin noncovalent complexes; 2- formation of soluble PRP•tannin aggregates; 3- precipitation of PRP•tannin aggregates; 4- alteration of the properties of the salivary mucosal pellicle due to tannin binding.

2002). MUC5B has been recently reported to aggregate in presence of the galloylated tannin EgCG (Davies et al. 2014). Therefore, astringency could be, at least in part, due to changes in the pellicle, decreasing its lubricating properties as illustrated in Figure 12.2. Tribological and rheological investigations have been performed on a salivary protein layer preadsorbed onto an elastic hydrophobic substrate (Macakova et al. 2010, Macakova et al. 2011, Rossetti et al. 2009). Rossetti et al. (2009) have reported a significant increase in friction coefficient upon exposure of salivary protein layer with EgCG solutions due to a depletion of the lubricating proteins. However, these studies should be considered with caution as the properties and structure of salivary films absorbed on synthetic surfaces could be different than those of *in vivo* salivary films.

12.4.2 Binding of bitter tastants

Salivary proteins have been also reported to bind bitter compounds and therefore to be involved in sensitivity to bitterness. More particularly, the group of

Barbarossa investigated the link between the sensitivity to 6-n-propylthiouracil (PROP) bitterness and saliva composition. Their researches revealed that the amount of two *b*PRPs, II-2 and Ps-1, was significantly higher in PROP super-tasters than in non-tasters (Cabras et al. 2012). Oral supplementation with Ps-1 in individuals lacking this protein in saliva increased the intensity of their perception of PROP bitterness (Melis et al. 2013). Regarding the molecular mechanisms involved, ^{1}H-NMR experiments on the interaction between PROP and the free amino-acids composing the sequence of Ps-1 and II-1 revealed that only free L-Lys and L-Arg interact with PROP. This experiment has also shown that L-Arg interacts more strongly with PROP than L-Lys. DFT calculations showed that the $-NH_2$ group of L-Arg H^+ side chains interact with either the C=O or the C=S groups of PROP by forming two hydrogen bonds (Melis et al. 2015). In the same time, psychophysical investigation showed that L-Arg enhances the bitterness intensity of PROP in a dose-dependent manner and more than L-Lys (Melis et al. 2013). It has been also reported that PRP, but also Histatin, can interact with quinine, another bitter molecule (Arakida et al. 2008). All these observations suggest that salivary proteins can bind bitter compounds with a positive (*b*PRP and PROP) or negative (PRPs or histatins and quinine) effect on the availability of bitter molecules at receptor site.

12.4.3 Binding of aroma compounds

Aroma perception is another sensory modality which can be impacted by interaction between flavour compounds and salivary proteins. Several studies have reported a difference between retronasal and orthonasal perception of the same odorant molecule (see chapter 13), indicating an influence of oral physiology on aroma perception (Burdach et al. 1984, Kuo et al. 1993, Linforth et al. 2002, Voirol and Daget 1986). Among the different factors influencing the release of volatile compounds, the impact of saliva (Genovese et al. 2009, Munoz-Gonzalez et al. 2014) and of oral volume (Mishellany-Dutour et al. 2012) were evidenced. In the same time, numerous studies have demonstrated that proteins can interact with and bind aroma compounds (Guichard 2006). Therefore, some authors have studied the effects of salivary proteins on aroma release using artificial saliva (Friel and Taylor 2001, van Ruth et al. 2001, van Ruth and Roozen 2000). These investigations revealed that mucins modify the release of aroma compounds. Regarding the nature of the involved interactions, van Ruth et al. (2001) reported a higher retention for the more hydrophobic aroma compounds. However Friel and Taylor (2001) did not observe a correlation between the relative concentration of aroma compounds in air and their corresponding $\log P$ value in presence of mucins. Recent investigations on the effect of mucins and amylase on aroma release have revealed a linear correlation between the log P value of aroma and their retention by these two proteins for linear ketones and esters (Pagès-Hélary et al. 2014). Indeed, the structure of mucins presents hydrophobic domains (Bansil and Turner 2006), which could constitute binding sites of small molecules such as aroma compounds (Pagès-Hélary et al. 2014). Saliva has been also reported to strongly absorb free thiols, even if the mechanism involved (physicochemical interactions, chemical transformations, or covalent linkage to glycoproteins) is not clear (Starkenmann et al. 2008). Therefore, saliva can impact in mouth aroma release due to interactions between salivary proteins and aroma compounds. Moreover, recent results on the effect of free salivary proteins alone or in mixture show that no cumulative effect is observed with mucins and alpha-amylase on aroma release and suggest that the complexity of saliva composition has to be considered to understand the real effect of saliva (Pagès-Hélary et al. 2014).

Oral mucosa has been also suggested to explain aroma persistence in breath after food or beverage consumption and, therefore, could play an important role in aftertaste (Buffo et al. 2005, Hussein et al. 1983, Linforth and Taylor 2000, Wright et al. 2003). An *in vitro* investigation has reported no effect of epithelium on aroma release (Rabe et al. 2004). However, the epithelium model used in this study did not include the salivary pellicle anchored onto the oral epithelium. The salivary proteins anchored to the epithelium and forming the mucosal pellicle could be at the origin of aroma persistence, which result from an adsorption/desorption process involving noncovalent interactions. Among the salivary proteins forming the salivary pellicle, mucins could play a major role, as they represent the main family of salivary proteins forming the pellicle and as they have been previously reported to modify aroma release (Pagès-Hélary et al. 2014, van Ruth et al. 2001, van Ruth et al. 1995). Several studies have reported an effect of the chemical structure of aroma compounds on their persistence (Buettner 2004, Buettner and Welle 2004, Buffo et al. 2005, Linforth and Taylor 2000). Indeed, the physicochemical properties of aroma compounds depend on their structure and impact their ability to establish non-covalent interactions. Linforth and Taylor (2000) have shown that polarity, vapour pressure and hydrophobicity are important physicochemical properties of aroma compounds explaining differences in their persistence. It should be noticed that food matrix components can compete with the mucosal adsorption process particularly when fat is present (Buettner and Mestres 2005).

12.5 Modification of flavour compounds by saliva

Some saliva constituents have the capacity to modify taste compounds. Such modifications can have two consequences: a neutralization of taste compounds leading to a decrease of the sensation or an enzymatic conversion of molecules which can activate taste receptors. Neutralization of taste compounds concerns mainly the perception of sourness. This perception is generally associated to a low pH. However, it is also dependent on titrable acidity (Norris et al. 1984) since at equal pH, weak acids elicit longer sensory response than strong acids (Ganzevles and Kroeze 1987). Sourness perception in mouth is linked to the buffering capacity of saliva which is due to proteins and ions such as bicarbonate ions. Interestingly, bicarbonate concentration in saliva is increased at higher flow rate (Neyraud et al. 2009). Because acids stimulate saliva flow rate (Chauncey et al. 1963, Feller et al. 1965), perception of sourness will be strongly decreased by an increasing buffering capacity of saliva during acid consumption.

Saliva can also be at the origin of the formation of taste compounds due to enzymatic conversions. Different enzymatic activities, amylolysis (de Wijk et al. 2004, Engelen et al. 2007), proteolysis (Helmerhorst and Oppenheim 2007) and lipolysis (Drago et al. 2011, Neyraud et al. 2012, Poette et al. 2014, Stewart et al. 2010) have already been reported in saliva, with a possible impact on lysis of the main macronutrients. These enzyme activities are rather low and their contributions probably negligible within the entire digestive system with the possible exception of amylase. However, these enzymes could allow the release or generation of signal molecules detectable by the taste system in order to inform the digestive system that macronutrients are ingested. This has been shown by Mandel and Breslin (2012) who compared glycemic homeostasis after starch consumption of two groups of subjects having a high (HA) or a low (LA) salivary amylase activity. They reported that the LA group has a significantly lower post-prandial blood glucose concentration than the LA group. They also observed a preabsorptive insulin release in the first 9min after starch consumption in the HA group, that did not occur in the LA group. They therefore suggested that salivary amylase activity in the HA group allows a release of glucose or short chain oligosaccharides detected in the mouth and participating to an anticipatory response to eating. In other studies, some relationships have been found between enzyme activities and sensitivity to taste compounds possibly coming from the enzymatic degradation of macronutrients. In the case of fat perception, a correlation has been found between lipolytic activity and perception of a triglyceride solution (Neyraud et al. 2012). A possible explanation to such a correlation was that triglycerides are hydrolyzed in the mouth into free fatty acids and glycerol under the action of a lipase. This has been confirmed by Pepino et al. (2012), who showed that oral detection threshold for triolein was higher in the presence of orlistat, a lipase inhibitor. The concentration of the free fatty acids released and consequently their perception seems to be directly linked to the lipolytic activity. Direct relationships between amylase and protease activities with the sensory perception of their reaction products have not been shown yet. In the case of amylase it may be due to the fact that maltose, which is the final product of starch degradation by alpha-amylase, does not have a strong sweet taste. However, alpha-amylase can be implicated indirectly in flavour perception by facilitating the release of flavour compounds present in a starch matrix, as demonstrated for salt (Ferry et al. 2006). If no direct relationships were found between protease activity and flavour perception, it has been shown that a greater sensitivity to bitterness of caffeine was associated with a lower level of cystatin SN, a protease inhibitor (Dsamou et al. 2012). It was suggested by the authors that this reflects a higher proteolytic activity in saliva of the most sensitive subjects probably at the origin of peri-receptor events leading to a higher perception.

Saliva contains antioxidant enzymes that can also be involved in flavour perception and in particular in the metallic sensation. This sensation involves the taste of metal ions on the tongue and the retronasal odour perception of carbonyls produced as a result of lipid oxidation catalyzed by metal in the mouth (Omur-Ozbek et al. 2012). Modulation of this sensation by antioxidant enzymes such as peroxidases is theoretically possible (Salles et al. 2011). However, a study published later on did not report a significant correlation between individual saliva antioxidant capacity and the malondialdehyde formation, a marker of oxidation, in a panel of 18 subjects (Omur-Ozbek et al. 2012).

Regarding aroma perception, Buettner (2002a) has investigated the effect of saliva enzymatic activity on several aroma compounds. No enzymatic degradation of 3-alkyl-2-methoxypyrazines, methoxyphenols, sotolone and alcohols was observed, while aldehydes were degraded as a function of incubation time in presence of saliva. Esters and thiols can also undergo degradation in presence of saliva (Buettner 2002b). Degradation of thiols seems to result from enzymatic reactions occurring in saliva since it does not occur after heating of saliva (Buettner 2002b). Regarding esters, many esterolytic enzymes are present in saliva (Lindqvist and Augustinsson 1975, Lindqvist and Augustinsson 1980, Lindqvist et al. 1977), most of them originating from submandibular saliva (Lindqvist and Augustinsson 1975). For linear series of esters, a linear correlation has been reported between ester degradation and the length of the aliphatic chain (Pagès-Hélary et al. 2014), the most hydrophobic esters being the most degraded (Buettner 2002b, Pagès-Hélary et al. 2014).

Furthermore, Starkenmann et al. (2003, 2008) have shown that aroma compounds can be formed from non-odorant precursors (Starkenmann 2003, Starkenmann et al. 2008). These authors reported the transformation of cysteine-S-conjugates into free thiols due to enzymatic degradation by oral bacteria (Starkenmann et al. 2008). Free thiols are immediately perceived after consumption, whereas perception starts after a few seconds when their corresponding cysteine-S-conjugates are consumed. The times to reach the maximal intensity and to stop detecting the free thiol odour are also increased in the case of the consumption of the cysteine-S-conjugates. In the same study, the authors demonstrated that free thiols interact with salivary proteins and that the level of interaction is decreased by the denaturation of the salivary proteins. Therefore, it appears that the flavour of cysteine-S-conjugates is affected both by interactions with salivary proteins and enzymatic degradation (Starkenmann 2003, Starkenmann et al. 2008).

12.6 Conclusion

There is a huge diversity in the chemical structure of flavour compounds and therefore in their physicochemical properties. In the same line, saliva contains thousands of different salivary proteins, which have a large variety of functions and properties. As a result, numerous kinds of interactions can occur between saliva and flavour compounds with different effects such as transport of flavour stimuli up to their receptors, scavenging or degradation of flavour molecules, and formation of flavour compounds from non-flavour precursors. In addition, saliva secreted after chemosensory stimulation by flavour compounds can be modified as a function of the nature of the flavour compound. There is a growing interest for a better understanding of these interactions in particular in a context of the dynamic physiological changes occurring during food oral processing. This area of research has been underestimated for a long time, probably due to the fact that it requires a pluri-disciplinary approach (e.g., chemistry, physics, biochemistry, physiology) to be thoroughly understood, but it is now a booming field in food sciences. These researches are at the beginning and only the emerged part of the iceberg has been discovered so far. The recent development of tools to study *in vivo* release of flavour compounds represents an exciting challenge that still requires the emergence of innovating approaches. The development of *in vitro* models mimicking in mouth conditions or biological structures such as the mucosal pellicle could also favour promising research. However, to date, the sensory impact of these interactions is difficult to quantify and this remains a major challenge for future works as it affects the final acceptance of food products.

References

Abduragimov, A. R., Gasymov, O. K., Yusifov, T. N. and Glasgow, B. J. (2000) Functional cavity dimensions of tear lipocalin. *Current Eye Research, 21*(4), pp. 824–832.

Aps, J. K. M. and Martens, L. C. (2005) Review: The physiology of saliva and transfer of drugs into saliva. *Forensic Science International, 150*(2–3), pp. 119–131.

Arakida, Y., Shimazaki, N., Wada, H., Ando, I., Hayata, Y., Yamamori, T., Seino, K. and Marui, T. (2008) Analysis of bitter binding protein on quinine. *Chemical Senses, 33*(2), pp. J5–J5.

Asquith, T. N., Uhlig, J., Mehansho, H., Putnam, L., Carlson, D. M. and Butler, L. (1987) Binding of condensed tannins to salivary proline-rich glycoproteins: the role of carbohydrates. *Journal of Agricultural and Food Chemistry, 35*, pp. 331–334.

ASTM (1989) Standard definitions of terms relating to sensory evaluation of materials and products. in *Annual Book of ASTM Standards*, Philadelphia: American Society of Testing and Materials. p. 2.

Austin, P. J., Suchar, L. A., Robbins, C. T. and Hagerman, A. E. (1989) Tannin-binding proteins in saliva of deer and their absence in saliva of sheep and cattle. *Journal of Chemical Ecology, 15*(4), pp. 1335–1347.

Bansil, R. and Turner, B. S. (2006) Mucin structure, aggregation, physiological functions and biomedical applications. *Current Opinion in Colloid & Interface Science, 11*(2–3), pp. 164–170.

Becerra, L., Soares, R. V., Bruno, L. S., Siqueira, C. C., Oppenheim, F. G., Offner, G. D. and Troxler, R.F. (2003) Patterns of secretion of mucins and non-mucin glycoproteins in human submandibular/sublingual secretion. *Archives of Oral Biology, 48*(2), pp. 147–154.

Bennick, A. (1982) Salivary proline-rich proteins. *Molecular and Cellular Biochemistry, 45*, pp. 83–99.

Bennick, A. (2002) Interaction of plant polyphenol with salivary proteins. *Critical Reviews in Oral Biology & Medicine, 13*(2), pp. 184–196.

Bläker, M., Kock, K., Ahlers, C., Buck, F. and Schmale, H. (1993) Molecular cloning of human von Ebner's gland protein, a member of the lipocalin superfamily highly expressed in lingual salivary glands. *Biochimica et Biophysica Acta (BBA) - Gene Structure and Expression, 1172*(1-2), pp. 131–137.

Boze, H., Marlin, T., Durand, D., Pérez, J., Vernhet, A., Canon, F., Sarni-Manchado, P., Cheynier, V. and Cabane, B. (2010) Proline-Rich Salivary Proteins Have Extended Conformations. *Biophysical Journal, 99*, pp. 656–665.

Bradway, S. D., Bergey, E. J., Jones, P. C. and Levine, M. J. (1989) Oral mucosal pellicle - adsoprtion and transpeptidation of salivary components to buccal epithelial-cells. *Biochemical Journal, 261*(3), pp. 887–896.

Bradway, S. D., Bergey, E. J., Scannapieco, F. A., Ramasubbu, N., Zawacki, S. and Levine, M. J. (1992) Formation of salivary-mucosal pellicle: the role of transglutaminase. *Biochemical Journal, 284*, pp. 557–564.

Breslin, P. A., Gilmore, M., Beauchamp, G. K. and Green, B. G. (1993) Psychophysical evidence that oral astringency is a tactile sensation. *Chemical Senses, 18*, pp. 405–417.

Breustedt, D. A., Chatwell, L. and Skerra, A. (2009) A new crystal form of human tear lipocalin reveals high flexibility in the loop region and induced fit in the ligand cavity. *Acta Crystallogr D Biol Crystallogr, 65*(Pt 10), pp. 1118–1125.

Breustedt, D. A., Korndörfer, I. P., Redl, B. and Skerra, A. (2005) The 1.8-Å Crystal Structure of Human Tear Lipocalin Reveals an Extended Branched Cavity with Capacity for Multiple Ligands. *Journal of Biological Chemistry, 280*(1), pp. 484–493.

Buettner, A. (2002a) Influence of Human Saliva on Odorant Concentrations. 2. Aldehydes, Alcohols, 3-Alkyl-2-methoxypyrazines, Methoxyphenols, and 3-Hydroxy-4,5-dimethyl-2(5H)-furanone. *Journal of Agricultural and Food Chemistry, 50*(24), pp. 7105–7110.

Buettner, A. (2002b) Influence of Human Salivary Enzymes on Odorant Concentration Changes Occurring in Vivo. 1. Esters and Thiols. *Journal of Agricultural and Food Chemistry, 50*(11), pp. 3283–3289.

Buettner, A. (2004) Investigation of potent odorants and afterodor development in two Chardonnay wines using the buccal odor screening system (BOSS). *Journal of Agricultural and Food Chemistry, 52*, pp. 2339–2346.

Buettner, A. and Mestres, M. (2005) Investigation of the retronasal perception of strawberry aroma aftersmell depending on matrix composition. *Journal of Agricultural and Food Chemistry, 53*(5), pp. 1661–1669.

Buettner, A. and Welle, F. (2004) Intra-oral detection of potent odorants using a modified stir-bar sorptive extraction system in combination with HRGC-0 known as the buccal odor screening system (BOSS). *Flavour and Fragrance Journal, 19*(6), pp. 505–514.

Buffo, R. A., Rapp, J. A., Krick, T. and Reineccius, G. A. (2005) Persistence of aroma compounds in human breath after consuming an aqueous model aroma mixture. *Food Chemistry, 89*(1), pp. 103–108.

Burdach, K. J., Kroeze, J. H. A. and Koster, E. P. (1984) Nasal, retronasal, and gustatory perception - an experimental comparison. *Perception & Psychophysics, 36*(3), pp. 205–208.

Cabras, T., Melis, M., Castagnola, M., Padiglia, A., Tepper, B. J., Messana, I. and Tomassini Barbarossa, I. (2012) Responsiveness to 6-n-Propylthiouracil (PROP) Is Associated with Salivary Levels of Two Specific Basic Proline-Rich Proteins in Humans. *PLoS ONE, 7*(2), p. e30962.

Cala, O., Dufourc, E., Fouquet, E., Manigand, C., Laguerre, M. and Pianet, I. (2012) The colloidal state of tannins impacts the nature of their interaction with proteins: the case of salivary proline-rich protein/procyanidins binding. *LANGMUIR, 28*(50), pp. 17410–17418.

Canon, F., Ballivian, R., Chirot, F., Antoine, R., Sarni-Manchado, P., Lemoine, J. r. m. and Dugourd, P. (2011) Folding of a Salivary Intrinsically Disordered Protein upon Binding to Tannins. *Journal of the American Chemical Society, 133*(20), pp. 7847–7852.

Canon, F., Giuliani, A., Paté, F. and Sarni--Manchado, P. (2010) Ability of a salivary intrinsically unstructured protein to bind different tannin targets revealed by mass spectrometry. *Analytical and Bioanalytical Chemistry, 398,* pp. 815–822.

Canon, F., Milosavljević, A. R., van der Rest, G., Réfrégiers, M., Nahon, L., Sarni-Manchado, P., Cheynier, V. and Giuliani, A. (2013a) Photodissociation and dissociative photoionization mass spectrometry of proteins and noncovalent protein-ligand complexes. *Angewandte Chemie (International ed. in English), 52*(32), pp. 8377–8381.

Canon, F., Paté, F., Cheynier, V., Sarni-Manchado, P., Giuliani, A., Pérez, J., Durand, D., Li, J. and Cabane, B. (2013b) Aggregation of the Salivary Proline-Rich Protein IB5 in the Presence of the Tannin EgCG. *LANGMUIR, 29*(6), pp. 1926–1937.

Canon, F., Paté, F., Meudec, E., Marlin, T., Cheynier, V., Giuliani, A. and Sarni-Manchado, P. (2009) Characterization, stoichiometry and stability of salivary protein-tannin complexes by ESI-MS and ESI-MS/MS. *Analytical and Bioanalytical Chemistry, 395*(8), pp. 2535–2545.

Canon, F., Ployon, S., Mazauric, J.-P., Sarni-Manchado, P., Réfrégiers, M., Giuliani, A. and Cheynier, V. (2014) Binding site of different tannins on a human salivary proline-rich protein evidenced by dissociative photoionization tandem mass spectrometry. *Tetrahedron, 71*(2), 3039–3040.

Cárdenas, M., Elofsson, U. and Lindh, L. (2007) Salivary Mucin MUC5B Could Be an Important Component of in Vitro Pellicles of Human Saliva: An in Situ Ellipsometry and Atomic Force Microscopy Study. *Biomacromolecules, 8,* pp. 1149–1156.

Charlton, A. J., Baxter, N. J., Lilley, T. H., Haslam, E., McDonald, C. J. and Williamson, M. P. (1996) Tannin interactions with a full-length human salivary proline -rich protein display a stronger affinity than with single proline-rich repeats. *FEBS Letters, 382,* pp. 289–292.

Charlton, A. J., Baxter, N. J., Lokman Khan, M., Moir, A. J. G., Haslam, E., Davies, A. P. and Williamson, M. P. (2002) Polyphenol/peptide binding and precipitation. *Journal of Agricultural and Food Chemistry, 50,* pp. 1593–1601.

Chatterjee, A., Kumar, A., Chugh, J., Srivastava, S., Bhavesh, N. S. and Hosur, R. V. (2005) NMR of unfolded proteins. *Journal of Chemical Sciences, 117*(1), pp. 3–21.

Chauncey, H. H., Feller, R. P. and Shannon, I. L. (1963) Effect of acid solutions on human gustatory chemoreceptors as determined by parotid gland secretion rate. *Proceedings of the Society for Experimental Biology and Medicine, 112,* pp. 917–923.

Cheynier, V. (2005) Polyphenols in foods are more complex than often thought. *American Journal of Clinical Nutrition, supp 80*:1S, pp. 223S–229S.

Clifford, M. N. (1997) Astringency. in Tomas-Barberan, F. and Robins, R., (eds.) *Phytochemistry of fruits and vegetables,*Oxford: Clarendon Press. pp. 87–108.

Coles, J. M., Chang, D. P. and Zauscher, S. (2010) Molecular mechanisms of aqueous boundary lubrication by mucinous glycoproteins. *Current Opinion in Colloid & Interface Science, 15*(6), pp. 406–416.

Collins, L. M. C. and Dawes, C. (1987) The Surface Area of the Adult Human Mouth and Thickness of the Salivary Film Covering the Teeth and Oral Mucosa. *Journal of Dental Research, 66*(8), pp. 1300–1302.

Davies, G. A., Wantling, E. and Stokes, J. R. (2009) The influence of beverages on the stimulation and viscoelasticity of saliva: Relationship to mouthfeel? *Food Hydrocolloids, 23*(8), pp. 2261–2269.

Davies, H. S., Pudney, P. D., Georgiades, P., Waigh, T. A., Hodson, N. W., Ridley, C. E., Blanch, E. W. and Thornton, D. J. (2014) Reorganisation of the salivary mucin network by dietary components: insights from green tea polyphenols. *PLoS ONE, 9*(9), p. e108372.

de Wijk, R. A., Prinz, J. F., Engelen, L. and Weenen, H. (2004) The role of alpha-amylase in the perception of oral texture and flavour in custards. *Physiology & Behavior, 83*(1), pp. 81–91.

Dinnella, C., Recchia, A., Fia, G., Bertuccioli, M. and Monteleone, E. (2009) Saliva Characteristics and Individual Sensitivity to Phenolic Astringent Stimuli. *Chemical Senses, 34*(4), pp. 295–304.

Dinnella, C., Recchia, A., Vincenzi, S., Tuorila, H. and Monteleone, E. (2010) Temporary modification of salivary protein profile and individual responses to repeated phenolic astringent stimuli. *Chemical Senses, 35*(1), pp. 75–85.

Drago, S. R., Panouille, M., Saint-Eve, A., Neyraud, E., Feron, G. and Souchon, I. (2011) Relationships between saliva and food bolus properties from model dairy products. *Food Hydrocolloids, 25*(4), pp. 659–667.

Dsamou, M., Palicki, O., Septier, C., Chabanet, C., Lucchi, G., Ducoroy, P., Chagnon, M.-C. and Morzel, M. (2012) Salivary Protein Profiles and Sensitivity to the Bitter Taste of Caffeine. *Chemical Senses, 37*(1), pp. 87–95.

Dunker, A. K., Brown, C. J., Lawson, J. D., Iakoucheva, L. M. and Obradovic, Z. (2002) Intrinsic disorder and protein function. *Biochemistry, 41*(21), pp. 6573–6582.

Dunker, A. K., Silman, I., Uversky, V. N. and Sussman, J. L. (2008) Function and structure of inherently disordered proteins. *Current Opinion in Structural Biology, 18*(6), pp. 756–764.

Engelen, L., van den Keybus, P. A. M., de Wijk, R. A., Veerman, E. C. I., Amerongen, A. V. N., Bosman, F., Prinz, J. F. and van der Bilt, A. (2007) The effect of saliva composition on texture perception of semi-solids. *Archives of Oral Biology, 52*(6), pp. 518–525.

Feller, R. P., Sharon, I. M., Chauncey, H. H. and Shannon, I. L. (1965) Gustatory perception of sour, sweet, and salt mixtures using parotid gland flow rate. *Journal of Applied Physiology, 20*, pp. 1341–1344.

Feron, G. and Poette, J. (2013) In-mouth mechanism leading to the perception of fat in humans: from detection to preferences. The particular role of saliva. *OCL - Oleagineux, Corps Gras, Lipides, 20*(2), pp. 102–107.

Ferry, A.-L. S., Mitchell, J. R., Hort, J., Hill, S. E., Taylor, A. J., Lagarrigue, S. and Vallès-Pàmies, B. (2006) In-Mouth Amylase Activity Can Reduce Perception of Saltiness in Starch-Thickened Foods. *Journal of Agricultural and Food Chemistry, 54*(23), pp. 8869–8873.

Fischer, U., Boulton, R. B. and Noble, A. C. (1994) Physiological factors contributing to the variability of sensory assessments: Relationship between salivary flow rate and temporal perception of gustatory stimuli. *Food Quality and Preference, 5*(1–2), pp. 55–64.

Flower, D. R. (1996) The lipocalin protein family: Structure and function. *Biochemical Journal, 318*, pp. 1–14.

Friel, E. N. and Taylor, A. J. (2001) Effect of Salivary Components on Volatile Partitioning from Solutions. *Journal of Agricultural and Food Chemistry, 49*(8), pp. 3898–3905.

Froehlich, D. A., Pangborn, R. M. and Whitaker, J. R. (1987) The effect of oral-stimulation on human-parotid salivary flow-rate and alpha-amylase secretion. *Physiology & Behavior, 41*(3), pp. 209–217.

Ganzevles, P. G. J. and Kroeze, J. H. A. (1987) The sour taste of acids - the hydrogen-ion and the undissociated acid as sour agents. *Chemical Senses, 12*(4), pp. 563–576.

Genovese, A., Piombino, P., Gambuti, A. and Moio, L. (2009) Simulation of retronasal aroma of white and red wine in a model mouth system. Investigating the influence of saliva on volatile compound concentrations. *Food Chemistry, 114*(1), pp. 100–107.

Gibbins, H. L. and Carpenter, G. H. (2013) Alternative Mechanisms of Astringency - What is the Role of Saliva? *Journal of Texture Studies, 44*(5), pp. 364–375.

Gilbertson, T. A. (1998) Gustatory mechanisms for the detection of fat. *Current Opinion in Neurobiology, 8*(4), pp. 447–452.

Gillece-Castro, B. L., Prakobphol, A., Burlingame, A. L., Leffler, H. and Fisher, S. J. (1991) Structure and bacterial receptor activity of a human salivary proline-rich glycoprotein. *Journal of Biological Chemistry, 266*(26), pp. 17358–17368.

Glasgow, B. J., Abduragimov, A. R., Farahbakhsh, Z. T., Faull, K. F. and Hubbell, W. L. (1995) Tear lipocalins bind a broad array of lipid ligands. *Current Eye Research, 14*(5), pp. 363–372.

Green, B. G. (1993) Oral astringency: a tactile component of flavor. *Acta Psychologica, 84*, pp. 119–125.

Gu, M., Haraszthy, G. G., Collins, A. R. and Bergey, E. J. (1995) Identification of salivary proteins inhibiting herpes simplex virus 1 replication. *Oral Microbiology and Immunology, 10*(1), pp. 54–59.

Gu, W. and Helms, V. (2005) Dynamical binding of proline-rich peptides to their recognition domains. *Biochimica et Biophysica Acta (BBA) - Proteins & Proteomics, 1754*(1–2), pp. 232–238.

Guichard, E. (2006) Flavour retention and release from protein solutions. *Biotechnology Advances, 24*(2), pp. 226–229.

Guinard, J., Zoumas-Morse, C. and Walchak, C. (1998) Relation Between Parotid Saliva Flow and Composition and the Perception of Gustatory and Trigeminal Stimuli in Foods. *Physiology & Behavior, 63*, pp. 109–118.

Guinard, J. X. and Mazzucchelli, R. (1996) The sensory perception of texture and mouthfeel. *Trends in Food Science & Technology, 7*, pp. 213–219.

Hagerman, A. E. and Butler, L. G. (1981) The specificity of proanthocyanidin-protein interactions. *The Journal of Biological Chemistry, 256*(9), pp. 4494–4497.

Harthoorn, L. F., Brattinga, C., Van Kekem, K., Neyraud, E. and Dransfield, E. (2009) Effects of sucrose on salivary flow and composition: differences between real and sham intake. *International Journal of Food Science and Nutrition, 60*(8), pp. 637–646.

Haslam, E. (1996) Natural polyphenols (vegetable tannins) as drugs: possible modes of action. *Journal of Natural Products, 59*, pp. 205–215.

Hatton, M. N., Loomis, R. E., Levine, M. J. and Tabak, L. A. (1985) Masticatory lubrification. The role of carbohydrate in the lubricating property of a salivary glycoprotein-albumin complex. *Biochemical Journal, 230*(3), pp. 817–820.

Hattrup, C. L. and Gendler, S. J. (2008) Structure and function of the cell surface (tethered) mucins. *Annual Review of Physiology, 70*, pp. 431–57.

Hay, D., Carlson, E., Schluckebier, S., Moreno, E. and Schlesinger, D. (1987) Inhibition of calcium phosphate precipitation by human salivary acidic proline-rich proteins: Structure-activity relationships. *Calcified Tissue International, 40*(3), pp. 126–132.

Helmerhorst, E. J. and Oppenheim, F. G. (2007) Saliva: a dynamic proteome. *Journal of Dental Research, 86*(8), pp. 680–693.

Hodson, N. A. and Linden, R. W. A. (2006) The effect of monosodium glutamate on parotid salivary flow in comparison to the response to representatives of the other four basic tastes. *Physiology & Behavior, 89*(5), pp. 711–717.

Horigome, T., Kumar, R. and Okamoto, K. (1988) Effects of condensed tannins prepared from leaves of fodder plants on digestive enzymes *in vitro* and in the intestine of rats. *British J. Nutr., 60*, pp. 275–285.

Hussein, M. M., Kachikian, R. and Pidel, A. R. (1983) Analysis for Flavor Residuals in the Mouth by Gas Chromatography. *Journal of Food Science, 48*(6), pp. 1884–1885.

Hyde, R. J. and Pangborn, R. M. (1978) Parotid salivation in response to tasting wine. *American Journal of Enology and Viticulture, 29*(2), pp. 87–91.

Jensen, J. L., Lamkin, M. S. and Oppenheim, F. G. (1992) Adsorption of Human Salivary Proteins to Hydroxyapatite: A Comparison Between Whole Saliva and Glandular Salivary Secretions. *Journal of Dental Research, 71*(9), pp. 1569–1576.

Kandra, L., Gyemant, G., Zajacz, A., et al. and Batta, G. (2004) Inhibitory effects of tannin on human salivary alpha-amylase. *Biochemical and Biophysical Research Communications, 319*(4), pp. 1265–1271.

Kauffman, D. L. and Keller, P. J. (1979) The basic proline-rich proteins in human parotid saliva from a single subject. *Archives of Oral Biology, 24*, pp. 249–256.

Klajner, F., Herman, C. P., Polivy, J. and Chhabra, R. (1981) Human obesity, dieting, and anticipatory salivation to food. *Physiology & Behavior, 27*(2), pp. 195–198.

Kuo, Y.-L., Pangborn, R. M. and Noble, A. C. (1993) Temporal patterns of nasal, oral, and retronasal perception of citral and vanillin and interaction of these odourants with selected tastants. *International Journal of Food Science & Technology, 28*(2), pp. 127–137.

Lashley, K. S. (1916) Reflex secretion of the human parotid gland. *Journal of Experimental Psychology, 1*(6), pp. 461–493.

Lawrence, G., Septier, C., Achilleos, C., Courcoux, P. and Salles, C. (2012) In Vivo Sodium Release and Saltiness Perception in Solid Lipoprotein Matrices. 2. Impact of Oral Parameters. *Journal of Agricultural and Food Chemistry, 60*(21), pp. 5299–5306.

Lee, C. B. and Lawless, H. T. (1991) Time-course of astringent sensations. *Chemical Senses, 16*(3), pp. 225–238.

Lee, V. M. and Linden, R. W. A. (1992) The effect of odors on stimulated parotid salivary flow in humans. *Physiology & Behavior, 52*(6), pp. 1121–1125.

Lindqvist, L. and Augustinsson, K. B. (1975) Esterases in human saliva. *Enzyme, 20*(5), pp. 277–291.

Lindqvist, L. and Augustinsson, K. B. (1980) Esterase patterns in human-saliva established by chromatography on con-a-sepharose and by isoelectric-focusing. *Archives of Oral Biology, 25*(6), pp. 363–368.

Lindqvist, L., Nord, C. E. and Soder, P. O. (1977) Origin of esterases in human whole saliva. *Enzyme, 22*(3), pp. 166–175.

Linforth, R., Martin, F., Carey, M., Davidson, J. and Taylor, A. J. (2002) Retronasal Transport of Aroma Compounds. *Journal of Agricultural and Food Chemistry, 50*(5), pp. 1111–1117.

Linforth, R. and Taylor, A. J. (2000) Persistence of volatile compounds in the breath after their consumption in aqueous solutions. *Journal of Agricultural and Food Chemistry, 48*(11), pp. 5419–5423.

Lorenz, K., Bader, M., Klaus, A., Weiss, W., Gorg, A. and Hofmann, T. (2011) Orosensory Stimulation Effects on Human Saliva Proteome. *Journal of Agricultural and Food Chemistry, 59*(18), pp. 10219–10231.

Lu, Y. and Bennick, A. (1998) Interaction of tannin with human salivary proline-rich proteins. *Archives of Oral Biology, 43*(9), pp. 717–728.

Lyman, B. J. and Green, B. G. (1990) Oral astringency: effects of repeated exposure and interactions with sweeteners. *Chemical Senses, 15*, pp. 151–164.

Macakova, L., Yakubov, G. E., Plunkett, M. A. and Stokes, J. R. (2010) Influence of ionic strength changes on the structure of pre-adsorbed salivary films. A response of a natural multi-component layer. *Colloids and Surfaces B: Biointerfaces, 77*(1), pp. 31–39.

Macakova, L., Yakubov, G. E., Plunkett, M. A. and Stokes, J. R. (2011) Influence of ionic strength on the tribological properties of pre-adsorbed salivary films. *Tribology International, 44*(9), pp. 956–962.

Mandel, A. L. and Breslin, P. A. S. (2012) High Endogenous Salivary Amylase Activity Is Associated with Improved Glycemic Homeostasis following Starch Ingestion in Adults. *Journal of Nutrition, 142*(5), pp. 853–858.

Mattes, R. D. (2000) Nutritional implications of the cephalic-phase salivary response. *Appetite, 34*(2), pp. 177–183.

Mattes, R. D. (2009) Is there a fatty acid taste? *Annual Review of Nutrition, 29*, pp. 305–327.

McArthur, C., Sanson, G. D. and Beal, A. M. (1995) Salivary proline-rich proteins in mammals: role in oral homeostasis and counteracting dietary tannin. *Journal of Chemical Ecology, 21*, pp. 663–691.

Mehansho, H., Clements, S., Sheares, B. T., Smith, S. and Carlson, D. M. (1985) Induction of proline-rich glycoprotein synthesis in mouse salivary glands by isoproterenol and by tannins. *The Journal of Biological Chemistry, 260*, pp. 4418–4423.

Melis, M., Aragoni, M. C., Arca, M., Cabras, T., Caltagirone, C., Castagnola, M., Crnjar, R., Messana, I., Tepper, B. J. and Barbarossa, I. T. (2013) Marked increase in PROP taste responsiveness following oral supplementation with selected salivary proteins or their related free amino acids. *PLoS ONE, 8*(3), p. e59810.

Melis, M., Arca, M., Aragoni, M. C., Cabras, T., Caltagirone, C., Castagnola, M., Crnjar, R., Messana, I., Tepper, B. J. and Tomassini Barbarossa, I. (2015) Dose-Dependent Effects of L-Arginine on PROP Bitterness Intensity and Latency and Characteristics of the Chemical Interaction between PROP and L-Arginine. *PLoS ONE, 10*(6), p. e0131104.

Mishellany-Dutour, A., Woda, A., Laboure, H., Bourdiol, P., Lachaze, P., Guichard, E. and Feron, G. (2012) Retro-nasal aroma release is correlated with variations in the in-mouth air cavity volume after empty deglutition. *PLoS ONE*, 7(7), p. e41276.

Mole, S., Butler, L. G. and Iason, G. (1990) Defense against dietary tannins in herbivores: a survey for proline-rich salivary proteins in mammals. *Biochemical Systematics and Ecology, 18*, pp. 287–293.

Mounayar, R., Morzel, M., Brignot, H., Tremblay-Franco, M., Canlet, C., Lucchi, G., Ducoroy, P., Feron, G. and Neyraud, E. (2014a) Salivary markers of taste sensitivity to oleic acid: a combined proteomics and metabolomics approach. *Metabolomics, 10*(4), pp. 688–696.

Mounayar, R., Morzel, M., Brignot, H., Tremblay-Franco, M., Canlet, C., Lucchi, G., Ducoroy, P., Feron, G. and Neyraud, E. (2014b) Nutri-metabolomics Applied to Taste Perception Phenotype: Human Subjects with High and Low Sensitivity to Taste of Fat Differ in Salivary Response to Oleic Acid. *Omics-a Journal of Integrative Biology, 18*(11), pp. 666–672.

Mounayar, R., Septier, C., Chabanet, C., Feron, G. and Neyraud, E. (2013) Oral Fat Sensitivity in Humans: Links to Saliva Composition Before and After Stimulation by Oleic Acid. *Chemosensory Perception, 6*(3), pp. 118–126.

Munoz-Gonzalez, C., Feron, G., Guichard, E., Rodriguez-Bencomo, J. J., Martin-Alvarez, P. J., Moreno-Arribas, M. V. and Pozo-Bayon, M. A. (2014) Understanding the Role of Saliva in Aroma Release from Wine by Using Static and Dynamic Headspace Conditions. *Journal of Agricultural and Food Chemistry. 62*(33), pp. 8274–8288.

Nayak, A. and Carpenter, G. H. (2008) A physiological model of tea-induced astringency. *Physiology & Behavior, 95*(3), pp. 290–294.

Neyraud, E., Bult, J. H. and Dransfield, E. (2009) Continuous analysis of parotid saliva during resting and short-duration simulated chewing. *Archives of Oral Biology, 54*(5), pp. 449–456.

Neyraud, E. and Dransfield, E. (2004) Relating ionisation of calcium chloride in saliva to bitterness perception. *Physiology & Behavior, 81*(3), pp. 505–510.

Neyraud, E., Heinzerling, C. I., Bult, J. H. F., Mesmin, C. and Dransfield, E. (2009) Effects of Different Tastants on Parotid Saliva Flow and Composition. *Chemosensory Perception, 2*(2), pp. 108–116.

Neyraud, E., Palicki, O., Schwartz, C., Nicklaus, S. and Feron, G. (2012) Variability of human saliva composition: Possible relationships with fat perception and liking. *Archives of Oral Biology, 57*(5), pp. 555–566.

Neyraud, E., Prinz, J. and Dransfield, E. (2003) NaCl and sugar release, salivation and taste during mastication of salted chewing gum. *Physiological Behaviors, 79*(4-5), pp. 731–737.

Neyraud, E., Sayd, T., Morzel, M. and Dransfield, E. (2006) Proteomic analysis of human whole and parotid salivas following stimulation by different tastes. *Journal of Proteome Research, 5*(9), pp. 2474–2480.

Nieuw Amerongen, A. V. and Veerman, E. C. I. (2002) Saliva -the defender of the oral cavity. *Oral Diseases, 8*(1), pp. 12–22.

Norris, M. B., Noble, A. C. and Pangborn, R. M. (1984) Human-saliva and taste responses to acids varying in anions, titrable acidity, and pH. *Physiology & Behavior, 32*(2), pp. 237–244.

Oh, H. I. and Hoff, J. E. (1986) Effect of condensed grape tannins on the in vitro activity of digestive proteases and activation of their zymogens. *Journal of Food Science, 51*(3), pp. 577–580.

Oh, H. I. and Hoff, J. E. (1988) Interaction of condensed grape tannins with pepsin and trypsin in simulated human digestive system. *Nutr. Rep. Internat., 38*(3), pp. 445–453.

Omur-Ozbek, P., Dietrich, A. M., Duncan, S. E. and Lee, Y. (2012) Role of Lipid Oxidation, Chelating Agents, and Antioxidants in Metallic Flavor Development in the Oral Cavity. *Journal of Agricultural and Food Chemistry, 60*(9), pp. 2274–2280.

Ozawa, T., Lilley, T. H. and Haslam, E. (1987) Polyphenol interactions : astringency and the loss of astringency in ripening fruit. *Phytochemistry, 26*(11), pp. 2937–2942.

Pagès-Hélary, S., Andriot, I., Guichard, E. and Canon, F. (2014) Retention effect of human saliva on aroma release and respective contribution of salivary mucin and α-amylase. *Food Research International, 64*, pp. 424–431.

Pangborn, R. M. (1968) Parotid flow stimulated by the sight, feel and odor of lemon. *Perceptual and Motor Skills, 27*, pp. 1340–1342.

Pangborn, R. M., Witherly, S. A. and Jones, F. (1979) Parotid and whole-mouth secretion in response to viewing, handling, and sniffing food. *Perception, 8*, pp. 339–346.

Pascal, C., Paté, F., Cheynier, V. and Delsuc, M.-A. (2009) Study of the interactions between a proline-rich protein and a flavan-3-ol by NMR: Residual structures in the natively unfolded protein provides anchorage points for the ligands. *Biopolymers, 91*(9), pp. 745–756.

Pascal, C., Poncet-Legrand, C., Imberty, A., Gautier, C., Sarni-Manchado, P., Cheynier, V. and Vernhet, A. (2007) Interactions between a non glycosylated human proline-rich protein and flavan-3-ols are affected by protein concentration and polyphenol/protein ratio. *Journal of Agricultural and Food Chemistry, 55*(12), pp. 4895–4901.

Pepino, M. Y., Love-Gregory, L., Klein, S. and Abumrad, N. A. (2012) The fatty acid translocase gene CD36 and lingual lipase influence oral sensitivity to fat in obese subjects. *Journal of Lipid Research, 53*(3), pp. 561–566.

Perez-Gregorio, M. R., Mateus, N. and de Freitas, V. (2014) Rapid screening and identification of new soluble tannin-salivary protein aggregates in saliva by mass spectrometry (MALDI-TOF-TOF and FIA-ESI-MS). *LANGMUIR, 30*(28), pp. 8528–8537.

Poette, J., Mekoue, J., Neyraud, E., Berdeaux, O., Renault, A., Guichard, E., Genot, C. and Feron, G. (2014) Fat sensitivity in humans: oleic acid detection threshold is linked to saliva composition and oral volume. *Flavour and Fragrance Journal, 29*(1), pp. 39–49.

Pramanik, R., Osailan, S. M., Challacombe, S. J., Urquhart, D. and Proctor, G. B. (2010) Protein and mucin retention on oral mucosal surfaces in dry mouth patients. *European Journal of Oral Sciences, 118*, pp. 245–253.

Puri, S. and Edgerton, M. (2014) How does it kill?: understanding the candidacidal mechanism of salivary histatin 5. *Eukaryot Cell, 13*(8), pp. 958–964.

Quintana, M., Palicki, O., Lucchi, G., Ducoroy, P., Chambon, C., Salles, C. and Morzel, M. (2009) Short-Term Modification of Human Salivary Proteome Induced by Two Bitter Tastants, Urea and Quinine. *Chemosensory Perception, 2*(3), pp. 133–142.

Rabe, S., Krings, U. and Berger, R. G. (2004) In vitro Study of the Influence of Physiological Parameters on Dynamic In-mouth Flavour Release from Liquids. *Chemical Senses, 29*(2), pp. 153–162.

Rogers, P. J. and Hill, A. J. (1989) Breakdown of dietary restraint following mere exposure to food stimuli - interrelationships between restraint, hunger, salivation, and food-intake. *Addictive Behaviors, 14*(4), pp. 387–397.

Rossetti, D., Bongaerts, J. H. H., Wantling, E., Stokes, J. R. and Williamson, A. M. (2009) Astringency of tea catechins: More than an oral lubrication tactile percept. *Food Hydrocolloids, 23*(7), pp. 1984–1992.

Rossetti, D., Yakubova, G. E., Stokesa, J., Williamson, A. and Fullerb, G. (2008) Interaction of human whole saliva and astringent dietary compounds investigated by interfacial shear rheology. *Food Hydrocolloids, 22*, pp. 1068–1078.

Sabatini, L. M. and Azen, E. A. (1989) Histatins, a family of salivary histidine-rich proteins, are encoded by at least two loci (HIS1 and HIS2). *Biochemical and Biophysical Research Communications, 160*(2), pp. 495–502.

Sahakian, B. J., Lean, M. E. J., Robbins, T. W. and James, W. P. T. (1981) Salivation and insulin-secretion in response to food in nonobese men and women. *Appetite, 2*(3), pp. 209–216.

Salles, C., Chagnon, M. C., Feron, G., Guichard, E., Laboure, H., Morzel, M., Semon, E., Tarrega, A. and Yven, C. (2011) In-Mouth Mechanisms Leading to Flavor Release and Perception. *Critical Reviews in Food Science and Nutrition, 51*(1), pp. 67–90.

Sarni-Manchado, P., Canals-Bosch, J., Mazerolles, G. and Cheynier, V. (2008) Influence of the glycosylation of human salivary proline-rich proteins on their interactions with condensed tannins. *Journal of Agricultural and Food Chemistry, 56*(20), pp. 9563–9569.

Schobel, N., Radtke, D., Kyereme, J., Wollmann, N., Cichy, A., Obst, K., Kallweit, K., Kletke, O., Minovi, A., Dazert, S., Wetzel, C. H., Vogt-Eisele, A., Gisselmann, G., Ley, J. P., Bartoshuk, L. M., Spehr, J., Hofmann, T. and Hatt, H. (2014) Astringency Is a Trigeminal Sensation That Involves the Activation of G Protein-Coupled Signaling by Phenolic Compounds. *Chemical Senses, 39*(6), pp. 471–487.

Shimada, T. (2006) Salivary proteins as a defense against dietary tannins. *Journal of Chemical Ecology, 32*(6), pp. 1149–1163.

Shimada, T. and Saitoh, T. (2003) Negative effects of acorns on the wood mouse *Apodemus speciosus. Population Ecology, 45*(1), pp. 7–17.

Silletti, E., Bult, J. H. F. and Stieger, M. (2012) Effect of NaCl and sucrose tastants on protein composition of oral fluid analysed by SELDI-TOF-MS. *Archives of Oral Biology, 57*(9), pp. 1200–1210.

Simon, C., Barathieu, K., Laguerre, M., Schmitter, J. M., Fouquet, E., Pianet, I. and Dufourc, E. J. (2003) Three-dimensional structure and dynamics of wine tannin-saliva protein complexes. A multitechnique approach. *Biochemistry, 42*(35), pp. 10385–10395.

Siqueira, W. L., Margolis, H. C., Helmerhorst, E. J., Mendes, F. M. and Oppenheim, F. G. (2010) Evidence of Intact Histatins in the in vivo Acquired Enamel Pellicle. *Journal of Dental Research, 89*(6), pp. 626–630.

Soares, S., Sousa, A., Mateus, N. and de Freitas, V. (2012) Effect of condensed tannins addition on the astringency of red wines. *Chemical Senses, 37*(2), pp. 191–198.

Spence, C. (2011) Mouth-watering: The influence of environmental and cognitive factors on salivation and gustatory/flavor perception. *Journal of Texture Studies, 42*(2), pp. 157–171.

Starkenmann, C. (2003) Analysis of a Model Reaction System Containing Cysteine and (E)-2-Methyl-2-butenal, (E)-2-Hexenal, or Mesityl Oxide. *Journal of Agricultural and Food Chemistry, 51*(24), pp. 7146–7155.

Starkenmann, C., Le Calvé, B., Niclass, Y., Cayeux, I., Beccucci, S. and Troccaz, M. (2008) Olfactory Perception of Cysteine–S-Conjugates from Fruits and Vegetables. *Journal of Agricultural and Food Chemistry, 56*(20), pp. 9575–9580.

Stewart, J. E., Feinle-Bisset, C., Golding, M., Delahunty, C., Clifton, P. M. and Keast, R. S. J. (2010) Oral sensitivity to fatty acids, food consumption and BMI in human subjects. *British Journal of Nutrition, 104*(1), pp. 145–152.

Stokes, J. R. and Davies, G. A. (2007) Viscoelasticity of human whole saliva collected after acid and mechanical stimulation. *Biorheology, 44*(3), pp. 141–160.

Tompa, P. (2002) Intrinsically unstructured proteins. *Trends in Biochemical Sciences, 27*(10), pp. 527–533.

Troxler, R. F., Offner, G. D., Xu, T., Vanderspek, J. C. and Oppenheim, F. G. (1990) Structural Relationship Between Human Salivary Histatins. *Journal of Dental Research, 69*(1), pp. 2–6.

van Ruth, S. M., Grossmann, I., Geary, M. and Delahunty, C. M. (2001) Interactions between Artificial Saliva and 20 Aroma Compounds in Water and Oil Model Systems. *Journal of Agricultural and Food Chemistry, 49*(5), pp. 2409–2413.

van Ruth, S. M. and Roozen, J. P. (2000) Influence of mastication and saliva on aroma release in a model mouth system. *Food Chemistry, 71*(3), pp. 339–345.

van Ruth, S. M., Roozen, J. P. and Cozijnsen, J. L. (1995) Changes in flavour release from rehydrated diced bell peppers (*Capsicum annuum*) by artificial saliva components in three mouth model systems. *Journal of the Science of Food and Agriculture, 67*(2), pp. 189–196.

Voirol, E. and Daget, N. (1986) Comparative study of nasal and retronasal olfactory perception. *Food Science and Technology - Lebensmittel - Wissenschaft & Technologie, 19*, pp. 316–319.

Watanabe, S. and Dawes, C. (1988) The effects of different foods and concentrations of citric-acid on the flow-rate of whole saliva in man. *Archives of Oral Biology, 33*(1), pp. 1–5.

Wright, K. M., Hills, B. P., Hollowood, T. A., Linforth, R. S. T. and Taylor, A. J. (2003) Persistence effects in flavour release from liquids in the mouth. *International Journal of Food Science and Technology, 38*(3), pp. 343–350.

Wright, P. E. and Dyson, H. J. (1999) Intrinsically unstructured proteins : re-assessing the protein structure-function paradigm. *Journal of Molecular Biology, 293*(2), pp. 321–331.

Xu, T., Levitz, S. M., Diamond, R. D. and Oppenheim, F. G. (1991) Anticandidal activity of major human salivary histatins. *Infection and Immunity, 59*(8), pp. 2549–2554.

Yan, Q. and Bennick, A. (1995) Identification of histatins as tannin-binding proteins in human saliva. *Biochemical Journal, 311*, pp. 341–347.

Yao, Y., Lamkin, M. S. and Oppenheim, F. G. (2000) Pellicle precursor protein crosslinking: Characterization of an adduct between acidic proline-rich protein (PRP-1) and statherin generated by transglutaminase. *Journal of Dental Research, 79*(4), pp. 930–938.

Zajacz, A., Gyemant, G., Vittori, N. and Kandra, L. (2007) Aleppo tannin: structural analysis and salivary amylase inhibition. *Carbohydrate Research, 342*(5), pp. 717–23.

CHAPTER 13

Orthonasal and retronasal perception

Thomas Hummel[1] & Han-Seok Seo[2]

[1] *Smell & Taste Clinic, Department of Otorhinolaryngology, TU Dresden, Dresden, Germany*

[2] *Department of Food Science, University of Arkansas, Fayetteville, Arkansas, U.S.A*

13.1 Significance of ortho- and retronasal perception

Odorous volatile compounds reach the olfactory epithelium by two pathways: *via* the nostrils, during sniffing, and *via* the mouth, during eating or drinking (Figure 13.1). These two pathways are referred to as orthonasal and retronasal olfaction, respectively (Rozin 1982, Shepherd 2007). Because the retronasal sensations are typically perceived through the mouth (Lim and Johnson 2012), retronasal sensations are commonly referred to as 'taste'. In this way, people appear to frequently confuse retronasal sensation with taste, which is called as "smell–taste confusion"(Murphy and Cain 1980, Murphy et al. 1977). In fact, when people lose their sense of smell they would often describe this loss as a '*loss of taste function*' (Deems et al. 1991). While gustatory function and retronasal function are fundamentally different, there are also major differences between orthonasal and retronasal perception of odours, as reflected in the different cortical processing of ortho- or retronasal activation.

13.2 Methods to investigate ortho- and retronasal olfaction

Retronasal olfactory function can be assessed by psychophysical, electrophysiological, and imaging techniques. These methods allow researchers and clinicians to examine the clinical aspects of retronasal olfaction, as well as dynamics of retronasal olfaction at the perceptual and cognitive levels.

Several types of psychophysical tests have been addressed to explore retronasal olfactory function. "Flavour identification test" using food powders ("*Schmeckpulver*": Figure 13.2) is a test kit which is composed of 20 commercially-available food powders: bread, celery, cinnamon, cloves, cocoa,

Flavour: From food to perception, First Edition.
Edited by Elisabeth Guichard, Christian Salles, Martine Morzel, and Anne-Marie Le Bon.

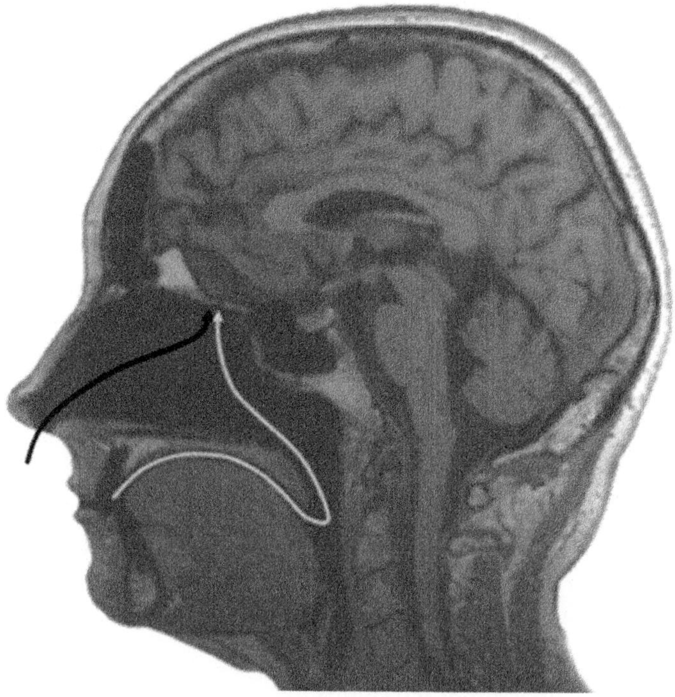

Figure 13.1 There are two pathways for odours to reach the olfactory epithelium. The orthonasal route (dark arrow) is used during sniffing, for example to identify odours like rose, smoke, or other odorants in the environment. Through the retronasal route (light arrow), the flavour of foods can be perceived (Source: Adapted from Bojanowski and Hummel 2012).

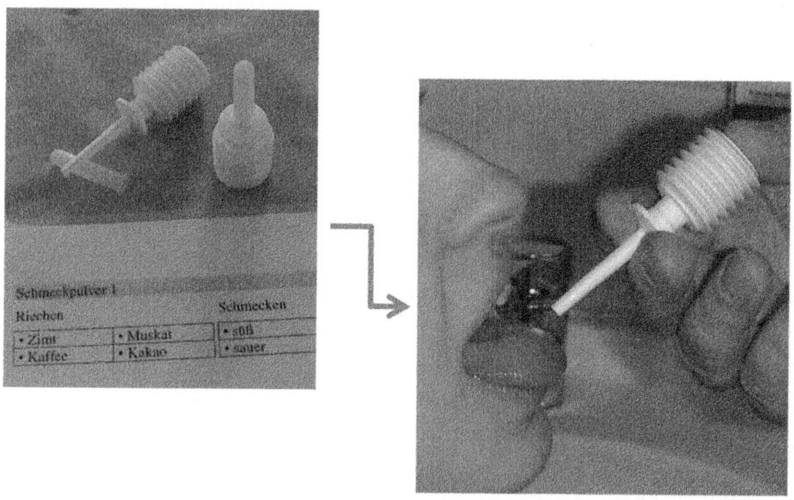

Figure 13.2 Flavour identification test ("Schmeckpulver"). Photograph of the psychophysical investigation of retronasal olfactory perception. Powders like cacao, paprika or garlic are applied to the tongue. Subjects have to select one of four items that best describes the flavour (Source: Adapted from Heilmann et al. 2002).

coffee, curry, garlic, ginger, grapefruit, milk, muscat, mushrooms, onion, orange, paprika, raspberry, smoked ham, strawberry, and vanilla (Croy et al. 2014, Heilmann et al. 2002). As shown in the right side of Figure 13.2, each powder sample is administered to the middle of the tongue by using a squeezable plastic vial. Subjects are then asked to identify the "flavour" of each powder from a list of four items (Heilmann et al. 2002). Another type of test is referred to as "candy smell test" (Haxel et al. 2011). As the name says, subjects are asked to identify the "flavour" of candies that are placed in their mouth. Both tests use food-related flavours since the retronasal olfaction occurs normally during eating. In addition, both tests are meritorious in terms of ease to handle and high reliability of test results. Thus, both methods appear to be suitable for the clinical routine work (Bojanowski and Hummel 2012).

When researchers investigate differences between orthonasal and retronasal perception of odours, it is necessary to consider how to present the odorous stimuli to subjects without accompanying additional gustatory or mechanical stimuli (Halpern 2004). If the olfactory stimuli produce gustatory or mechanical sensations in the mouth during their presentation, the researchers may have a difficulty in drawing the sole effect of olfactory stimuli on retronasal olfaction from the data obtained under the experimental design. In other words, the olfactory stimuli-mediated oral sensations may alter responses to the odours in a direct or indirect manner (Land 1994, Welge-Lüssen et al. 2005). For example, Sun and Halpern retronasally presented odours *via* a container placed on the tongue (Sun and Halpern 2005). As another example, subjects were asked to smell the headspace odours of an odorous liquid or inhale the same headspace odours through the mouth followed by nasal exhalation (Burdach et al. 1984, Masaoka et al. 2007, Voirol and Daget 1986). However, these methods have limitations since odour concentration in the oral cavity is non-predictable and the mechanical stimulation of intraoral surfaces is not fully suppressed (Bojanowski and Hummel 2012, Land 1994).

A more defined retronasal stimulation was available using a computer-controlled air-dilution olfactometer (Kobal 1981), which allows researchers to control the amount and flow rate of odour stimuli. Under endoscopic control, Heilmann and Hummel placed the opening of one nasal tube beyond the nasal valve and also placed the opening of another tube in the epipharynx (Heilmann and Hummel 2004). In this way, the odour stimuli could be presented to the olfactory epithelium *via* either orthonasal or retronasal pathway through one of the two tubes, with a similar concentration and time course (Small et al. 2005). It is worth noting that this technique allows researchers to solely compare ortho- and retronasal olfactory perceptions.

13.3 Differences between orthonasal and retronasal olfaction

Food odours like cheese odours may produce different sensations when presented in front of the nose or in the mouth. A number of possible reasons for these phenomena exist. Differences in airflow patterns, apart from cognitive factors, play an important role in perceptual differences between ortho- and retronasal presentations of odours (Mozell 1964, Rebello et al. 2015, Scott et al. 2007, Scott et al. 2014). The direction of odour stream (e.g., forward or backward) across the olfactory epithelium probably leads to differences in the processing of odorous information.

Differences between ortho- and retronasal olfactory functions also affect the salivary response. When a certain food odour is presented repeatedly, the salivary response to the odour decreases until a new food odour is presented. Presentation of the novel food odour induces an increase in salivary flow. The same pattern is found if the route of odour administration is changed from orthonasal to retronasal or vice versa. These results demonstrate that presenting an odour *via* different pathways represents a distinct sensory signal, whether the odours are similar or different (Bender et al. 2009).

Changes in the mechanical structure of olfactory route also show differences in ortho- and retronasal olfactory perception. For instance, there are patients with an isolated smell deficit only for one route of olfaction. Impaired smell with preserved flavour perception is found in patients with nasal polyposis (Landis et al. 2003). This is thought to be related to mechanical obstruction in the anterior portion of the olfactory cleft (Pfaar et al. 2006), whereas odours could still reach the olfactory epithelium through the retronasal route during eating and drinking. Orthonasal olfactory loss with little changes in retronasal olfactory function has also been reported in the absence of nasal polyposis (Cerf-Ducastel and Murphy 2001, Ogle 1870). In contrast, other studies also demonstrated examples of retronasal smell loss in the absence of orthonasal deficits (Landis et al. 2005, Landis et al. 2003). Thus, in these cases it is likely that—even subtle—changes in airflow are responsible for differences in ortho- and retronasal olfactory function.

Orthonasal and retronasal olfaction differ in terms of odour threshold, odour intensity, ability to localize an odour, and the neuronal processing. Thresholds to orthonasal stimuli are typically lower compared to retronasal thresholds; in other words, people are more sensitive to orthonasal stimuli than retronasal stimuli (Duffy et al. 1999, Voirol and Daget 1986). It has also been reported that the odour identification is less efficient when stimuli are presented retronasally (Burdach et al. 1984). Numerous studies indicate that pure olfactory stimuli cannot be lateralized when the stimulus is applied to one of the two nostrils (Kobal et al. 1989, von Skramlik 1924). Interestingly, subjects are able to localize ortho- and retronasal presentation of an odorant (Hummel 2008). At the level of the olfactory bulb, rats show differences in the processing of ortho- and retronasal information (Gautam and Verhagen 2012, Rebello et al. 2015). Cortical responses reflect differences between ortho- and retronasal stimuli. Electroencephalogram-derived event-related potentials (ERPs) indicated that, when a non-food related odour (lavender) is presented in a contextually unusual site, that is, retronasally, the response (peak amplitude of ERPs) is larger compared to the presentation of the same odourant through the orthonasal route (Hummel and Heilmann 2008). This is the other way around for chocolate, being a food-related odour. In addition, brain imaging studies demonstrate that orthonasal and retronasal stimuli produce different patterns of brain activation. Functional magnetic resonance imaging shows that brain activation by retronasal odour stimulation was found at the base of the central sulcus, corresponding to the primary representation of the oral cavity-related stimulation (Yamashita et al. 1999), possibly reflecting that retronasal odours are referred to the mouth (Green et al. 2012). Small et al. reported differential brain activations in response to taste and orthonasally / retronasally presented odour stimuli (Small et al. 2005, Small et al. 1997); they observed supra-additive responses at the orbitofrontal cortex, insula, and anterior cingulate cortex for the congruent combination of taste and retronasal stimuli.

13.4 Retronasal olfaction and gustatory or mechanosensory influences

Texture or taste of foods influence the perception of odours (Bult et al. 2007, Hollowood et al. 2002, Roudnitzky et al. 2011, Stevenson et al. 1999). For example, perceived intensity of retronasally presented marzipan odour is increased by the sweet taste of sucrose. This is not the case when the odour is contextually incongruent, like water or glutamate (Dalton et al. 2000, Green et al. 2012, Veldhuizen et al. 2010). Faster neural processing of a retronasal olfactory stimulus results from a congruent simultaneous taste as shown with olfactory ERPs (Welge-Lüssen et al. 2009b). With regard to the trigeminal activation induced by most odourants (Doty et al. 1978, Frasnelli et al. 2007), there is evidence that the sensitivity for chemosensory stimuli is higher in the anterior than the posterior portion of the nose, while this is the other way around for mechanical stimuli (e.g., air puffs) (Frasnelli et al. 2004, Melzner et al. 2011).

Visschers et al. (2006) showed the effect of rheological characteristics of food on olfactory perception. In their study, the intensity of aroma decreased with increasing consistency of the consumed food (Visschers et al. 2006). In addition, retronasal cream odour stimuli enhanced the intensities of thickness and creaminess of oral stimuli (Bult et al. 2007).

13.5 Clinical aspects of retronasal olfaction

Satiation may be modulated by the extent of retronasal aroma release. People differ in oral processing, bite size, and meal duration. Small bite sizes and longer duration of oral processing tend to increase aroma release in the mouth (Ruijschop et al. 2009), which may result in higher or faster satiation.

It is also interesting to note that adenoid/tonsillar hypertrophy seems to have a significant impact on retronasal olfactory function in children. In fact, retronasal olfactory function seems to improve significantly after surgery which may also imply that from then on children would experience a change in eating behavior, and accordingly, a higher quality of life because they then can enjoy differences in food which was much more difficult before surgery (Altundag et al. 2014, Cayonu et al. 2014, Konstantinidis et al. 2005).

Another clinical aspect is the effect of retronasal odours on swallowing. Compared to orthonasal stimulation, swallowing occurs significantly faster and more frequently when odours are presented retronasally (Welge-Lüssen et al. 2009a). In this sense, retronasal olfactory stimulation may be used as a clinical option to facilitate swallowing, aiming to prevent aspiration in the patients with swallowing disorders (Ryan and Hummel 2012).

13.6 Conclusion

Odours delivered *via* the nostrils are differently processed and perceived from those delivered *via* the mouth. These differences are mediated by differential activations at the level of the olfactory epithelium and contextual differences between odours and flavours. Most importantly, the two modes of action have a differential impact on eating and drinking.

References

Altundag, A., Salihoglum, M., Cayonum, M. and Tekeli, H. (2014) Clinical assessment of olfactory functions in children who underwent adenotonsillectomy during pre- and post-operative period. *International Journal of Pediatric Otorhinolaryngology, 78*(7), pp. 1138–1142.

Bender, G., Hummel, T., Negoias, S. and Small, D. M. (2009) Separate signals for orthonasal vs. retronasal perception of food but not nonfood odors. *Behavioral Neuroscience, 123*(3), pp. 481–489.

Bojanowski, V. and Hummel, T. (2012). Retronasal perception of odors. *Physiology & Behavior, 107*(4), pp. 484–487.

Bult, J. H. F., de Wijk, R. A. and Hummel, T. (2007) Investigations on multimodal sensory integration: texture, taste, and ortho- and retronasal olfactory stimuli in concert. *Neuroscience Letters, 411*(1), pp. 6–10.

Burdach, K. J., Kroeze, J. H. and Koster, E. P. (1984) Nasal, retronasal, and gustatory perception: an experimental comparison. *Perception and Psychophysics, 36*(3), pp. 205–208.

Cayonu, M., Salihoglu, M., Altundag, A., Tekeli, H. and Kayabasoglu, G. (2014) Grade 4 tonsillar hypertrophy associated with decreased retronasal olfactory function: a pilot study. *European Archives of Otorhinolaryngology, 271*(8), pp. 2311–2316.

Cerf-Ducastel, B. and Murphy, C. (2001) FMRI activation in response to odorants orally delivered in aqueous solutions. *Chemical Senses, 26*(6), pp. 625–637.

Croy, I., Hoffmann, H., Philpott, C., Rombaux, P., Welge-Luessen, A., Vodicka, J., Konstantinidis, I., Morera, E. and Hummel, T. (2014) Retronasal testing of olfactory function: an investigation and comparison in seven countries. *European Archives of Otorhinolaryngology, 271*(5), pp. 1087–1095.

Dalton, P., Doolittle, N., Nagata, H. and Breslin, P. A. (2000) The merging of the senses: integration of subthreshold taste and smell. *Nature Neuroscience, 3*(5), pp. 431–432.

Deems, D. A., Doty, R. L., Settle, R. G., Moore-Gillon, V., Shaman, P., Mester, A. F., Kimmelman, C. P., Brightman, V. J. and Snow, J. B. j. (1991) Smell and taste disorders: a study of 750 patients from the University of Pennsylvania Smell and Taste Center. *European Archives of Otorhinolaryngology Head and Neck Surgery, 117*(5), pp. 519–528.

Doty, R. L., Brugger, W. P. E., Jurs, P. C., Orndorff, M. A., Snyder, P. J. and Lowry, L. D. (1978) Intranasal trigeminal stimulation from odorous volatiles: psychometric responses from anosmic and normal humans. *Physiology and Behavior, 20*(2), pp. 175–185.

Duffy, V. B., Cain, W. S. and Ferris, A. M. (1999) Measurement of sensitivity to olfactory flavor: application in a study of aging and dentures. *Chemical Senses, 24*(6), pp. 671–677.

Frasnelli, J., Heilmann, S. and Hummel, T. (2004) Responsiveness of the human nasal mucosa to trigeminal stimuli depends on the site of stimulation. *Neuroscience Letters, 362*(1), pp. 65–69.

Frasnelli, J., Schuster, B. and Hummel, T. (2007) Interactions between olfaction and the trigeminal system: what can be learned from olfactory loss. *Cerebral Cortex, 17*(10), pp. 2268–2275.

Gautam, S. H. and Verhagen, J. V. (2012) Retronasal odor representations in the dorsal olfactory bulb of rats. *Journal of Neuroscience, 32*(23), pp. 7949–7959.

Green, B. G., Nachtigal, D., Hammond, S. and Lim, J. (2012) Enhancement of retronasal odors by taste. *Chemical Senses, 37*(1), pp. 77–86.

Halpern, B. P. (2004) Retronasal and orthonasal smelling. *ChemoSense, 6*, pp. 1–7.

Haxel, B. R., Bertz-Duffy, S., Faldum, A., Trellakis, S., Stein, B., Renner, B., Kobal, G., Letzel, S., Mann, W. J. and Muttray, A. (2011) The Candy Smell Test in clinical routine. *American Journal of Rhinology and Allergy, 25*(4), pp. e145–e148.

Heilmann, S. and Hummel, T. (2004) A new method for comparing orthonasal and retronasal olfaction. *Behavioral Neuroscience, 118*(2), pp. 412–419.

Heilmann, S., Strehle, G., Rosenheim, K., Damm, M. and Hummel, T. (2002) Clinical assessment of retronasal olfactory function. *Archives of Otorhinolaryngology Head and Neck Surgery, 128*(4), pp. 414–418.

Hollowood, T. A., Linforth, R. S. T. and Taylor, A. J. (2002) The effect of viscosity on the perception of flavour. *Chemical Senses, 27*(7), pp. 583–591.

Hummel, T. (2008) Retronasal perception of odors. *Chemistry and Biodiversity, 5*(6), pp. 853–861.

Hummel, T. and Heilmann, S. (2008) Olfactory event-related potentials in response to ortho- and retronasal stimulation with odors related or unrelated to foods. *International Dairy Journal, 18*(8), pp. 874–878.

Kobal, G. (1981) *Elektrophysiologische Untersuchungen des menschlichen Geruchssinns*, Stuttgart: Thieme Verlag.

Kobal, G., Van Toller, S. and Hummel, T. (1989) Is there directional smelling? *Experientia, 45*(2), pp. 130–132.

Konstantinidis, I., Triaridis, S., Triaridis, A., Petropoulos, I., Karagiannidis, K. and Kontzoglou, G. (2005) How do children with adenoid hypertrophy smell and taste? Clinical assessment of olfactory function pre- and post-adenoidectomy. *International Journal of Pediatric Otorhinolaryngology, 69*(10), pp. 1343–1349.

Land, D. G. (1994) Perspectives on the effects of interactions on flavor perception: an overview. in McGorrin, R. J. and Leland, J., (eds.) *Flavor-food interactions,* Washington (USA): ACS Symp. Ser. *633*. pp. 2–11.

Landis, B. N., Frasnelli, J., Reden, J., Lacroix, J. S. and Hummel, T. (2005) Differences between orthonasal and retronasal olfactory functions in patients with loss of the sense of smell. *Archives of Otolaryngology Head and Neck Surgery, 131*(11), pp. 977–981.

Landis, B. N., Giger, R., Ricchetti, A., Leuchter, I., Hugentobler, M., Hummel, T. and Lacroix, J. S. (2003) Retronasal olfactory function in nasal polyposis. *Laryngoscope, 113*(11), pp. 1993–1997.

Lim, J. and Johnson, M. B. (2012) The role of congruency in retronasal odor referral to the mouth. *Chemical Senses, 37*(6), pp. 515–522.

Masaoka, Y., Yoshimura, N., Inoue, M., Kawamura, M. and Homma, I. (2007) Impairment of odor recognition in Parkinson's disease caused by weak activations of the orbitofrontal cortex. *Neuroscience Letters, 412*(1), pp. 45–50.

Melzner, J., Bitter, T., Guntinas-Lichius, O., Gottschall, R., Walther, M. and Gudziol, H. (2011) Comparison of the orthonasal and retronasal detection thresholds for carbon dioxide in humans. *Chemical Senses, 36*(5), pp. 435–441.

Mozell, M. M. (1964) Evidence for sorption as a mechanism of the olfactory analysis of vapours. *Nature, 203*, pp. 1181–1182.

Murphy, C. and Cain, W. S. (1980) Taste and olfaction: independence vs interaction. *Physiology and Behavior, 24*(3), pp. 601–605.

Murphy, C., Cain, W. S. and Bartoshuk, L. M. (1977) Mutual action of taste and olfaction. *Sensory Processes, 1*(3), pp. 204–211.

Ogle, W. (1870) Anosmia or cases illustrating the physiology and pathology of the sense of smell. *Medico-Chirurgical Transactions, 53*, pp. 263.

Pfaar, O., Landis, B. N., Frasnelli, J., Huttenbrink, K. B. and Hummel, T. (2006) Mechanical obstruction of the olfactory cleft reveals differences between orthonasal and retronasal olfactory functions. *Chemical Senses, 31*(1), pp. 27–31.

Rebello, M. R., Kandukuru, P. and Verhagen, J. V. (2015) Direct behavioral and neurophysiological evidence for retronasal olfaction in mice. *PLoS One, 10*(2), p. e0117218.

Roudnitzky, N., Bult, J. H., de Wijk, R. A., Reden, J., Schuster, B. and Hummel, T. (2011) Investigation of interactions between texture and ortho- and retronasal olfactory stimuli using psychophysical and electrophysiological approaches. *Behavioural Brain Research, 216*(1), pp. 109–115.

Rozin, P. (1982) "Taste-smell confusions" and the duality of the olfactory sense. *Perception and Psychophysics, 31*(4), pp. 397–401.

Ruijschop, R. M., Boelrijk, A. E., de Graaf, C. and Westerterp-Plantenga, M. S. (2009) Retronasal aroma release and satiation: a review. *Journal of Agricultural and Food Chemistry, 57*(21), pp. 9888–9894.

Ryan, C. and Hummel, T. (2012) Smell, Taste, and Deglutition. in Shaker, R., Belafsky, P. C., Postma, G. N. and Easterling, C., (eds.) *Principles of Deglutition: A Multidisciplinary Text for Swallowing and its Disorders,* Heidelberg (Germany): Springer. pp. 19–24.

Scott, J. W., Acevedo, H. P., Sherrill, L. and Phan, M. (2007) Responses of the rat olfactory epithelium to retronasal air flow. *Journal of Neurophysiology, 97*(3), pp. 1941–1950.

Scott, J. W., Sherrill, L., Jiang, J. and Zhao, K. (2014) Tuning to odor solubility and sorption pattern in olfactory epithelial responses. *Journal of Neuroscience, 34*(6), pp. 2025–2036.

Shepherd, G. M. (2007) Perspectives on olfactory processing, conscious perception, and orbitofrontal cortex. *Annals of the New York Academy of Sciences, 1121*, pp. 87–101.

Small, D. M., Gerber, J. C., Mak, Y. E. and Hummel, T. (2005) Differential neural responses evoked by orthonasal versus retronasal odorant perception in humans. *Neuron, 47*(4), pp. 593–605.

Small, D. M., Jones-Gotman, M., Zatorre, R. J., Petrides, M. and Evans, A. C. (1997) Flavor processing: more than the sum of its parts. *Neuroreport, 8*(18), pp. 3913–3917.

Stevenson, R. J., Prescott, J. and Boakes, R. A. (1999) Confusing tastes and smells: how odours can influence the perception of sweet and sour tastes. *Chemical Senses, 24*(6), pp. 627–635.

Sun, B. C. and Halpern, B. P. (2005) Identification of air-phase retronasal and orthonasal odorant pairs. *Chemical Senses, 30*(8), pp. 1–14.

Veldhuizen, M. G., Shepard, T. G., Wang, M. F. and Marks, L. E. (2010) Coactivation of gustatory and olfactory signals in flavor perception. *Chemical Senses, 35*(2), pp. 121–133.

Visschers, R. W., Jacobs, M. A., Frasnelli, J., Hummel, T., Burgering, M. and Boelrijk, A. E. (2006) Cross-modality of texture and aroma perception is independent of orthonasal or retronasal stimulation. *Journal of Agricultural and. Food Chemistry, 54*(15), pp. 5509–5515.

Voirol, E. and Daget, N. (1986) Comparative study of nasal and retronasal olfactory perception. *Food Science and Technology, 19*, pp. 316–319.

von Skramlik, E. (1924) Über die Lokalisation der Empfindungen bei den niederen Sinnen. *Zeitschrififiir Sinnesphysiologie, 56*, pp. 69.

Welge-Lüssen, A., Drago, J., Wolfensberger, M. and Hummel, T. (2005) Gustatory stimulation influences the processing of intranasal stimuli. *Brain Research, 1038*(1), pp. 69–75.

Welge-Lüssen, A., Ebnother, M., Wolfensberger, M. and Hummel, T. (2009a) Swallowing is differentially influenced by retronasal compared with orthonasal stimulation in combination with gustatory stimuli. *Chemical Senses, 34*(6), pp. 499–502.

Welge-Lüssen, A., Husner, A., Wolfensberger, M. and Hummel, T. (2009b) Influence of simultaneous gustatory stimuli on orthonasal and retronasal olfaction. *Neuroscience Letters, 454*(2), pp. 124–128.

Yamashita, H., Kumamoto, Y., Nakashima, T., Yamamoto, T., Inokuchi, A. and Komiyama, S. (1999) Magnetic sensory cortical responses evoked by tactile stimulations of the human face, oral cavity and flap reconstructions of the tongue. *European Archives of Otorhinolaryngology, 256*(Suppl. 1), pp. S42–S46.

CHAPTER 14

Perception of mixtures of odorants and tastants: sensory and analytical points of view

Thierry Thomas-Danguin, Carmen Barba, Christian Salles & Elisabeth Guichard
Center for Taste and Feeding Behaviour, CNRS, INRA, University of Bourgogne Franche-Comté, Dijon, France

14.1 Introduction

Food flavour is a cerebral construction that results from the integration of gustatory, trigeminal and olfactory information elaborated from food chemical composition (Thomas-Danguin 2009). This multimodal representation is under the influence of the information transmitted by all the other senses as detailed in Chapter 16. The olfactory and trigeminal components of food can be perceived before or once food is put into the mouth, whereas the gustatory component arises only when food is located into the oral cavity. In a given food product or in a beverage, several molecules contribute to the elaboration of odour, aroma and taste. These molecules activate simultaneously the olfactory and the gustatory systems so that different types of interactions can occur along the sensory integration pathways, from competition towards the receptors to cognitive modulations in higher brain areas. In this chapter, we review the perceptual aspects of mixtures of odorants and mixtures of tastants and their mutual interactions in the context of food flavour perception.

14.2 Perceptual aspects of odour mixtures

14.2.1 The olfactory component of food

Food odour and aroma have a specific status within the flavour construction process since they were found to carry a large amount of food identity and typicality (Hornung and Enns 1989, Murphy 1985). Odour and aroma percepts originate from the olfactory processing of complex mixtures of volatile compounds, released from the food and reaching the olfactory mucosa thanks to

---

Flavour: From food to perception, First Edition.
Edited by Elisabeth Guichard, Christian Salles, Martine Morzel, and Anne-Marie Le Bon.

the respiration cycle (Thomas-Danguin et al. 2014). The very efficient neurobiological processing cascade within the olfactory system allows an immediate categorization and further recognition of the odour source and more specifically its discrimination from other odours, including background and/or non-food odours as presented in more details in Chapter 15. These mechanisms enable the olfactory system to very rapidly and efficiently discriminate complex overlapping mixtures of odorants. To achieve this critical processing in a few seconds, olfaction relies on odour objects segregation through receptors activation patterns separation (Gottfried 2010, Stevenson and Wilson 2007, Szyszka et al. 2012, Wilson and Sullivan 2011).

14.2.2 Processing of food odour mixtures

A major consequence of this object-oriented processing is that not all odorants released from food are perceived. Indeed, several studies have shown that humans are not able to identify more than three or four odorants in a mixture that include up to eight odorants (Laing and Francis 1989, Laing and Livermore 1992, Livermore and Laing 1998). This limitation is likely the result of physiological or processing constraints within the olfactory processing pathway and leads to the loss of most odorant's major characteristic once in a complex mixture (Laing 1991). Nevertheless, being not identifiable does not imply that a given odorant has no perceptual role in the mixture. Indeed, neurobiological interactions taking place during the olfactory information treatment lead to perceptual interactions (e.g., inhibition, synergy or perceptual blending) which contribute to elaborate the stereotyped pattern, specific to the complex chemical stimulus. This pattern contains the information on the odorants present in the mixture but also on their association (Thomas-Danguin et al. 2014). The perceptual interactions, induced by the sniffing of a mixture of at least two odorants, explain that perception of the mixture is not a linear combination of the elemental perception of the single odorants.

14.2.3 Perceptual interactions in food odour mixtures

The consequences of the non-linear integration of the chemical information carried by a mixture of odorants can be observed on all the sensory dimensions of odours, and particularly their intensity, quality and pleasantness.

In terms of odour quality, two different situations can occur when smelling a food: the perception is homogeneous if a single odour is perceived or heterogeneous if several odours are discernible (Berglund et al. 1976). The odour of a mixture of odorants can be considered homogeneous when one component completely overshadows the quality of the other components (Kay et al. 2005). This situation illustrated in Figure 14.1 is also refered to as complete masking (Cain and Drexler 1974). A homogeneous percept also appears when the mixed odours blend into a new odour perceived as an entity (Thomas-Danguin

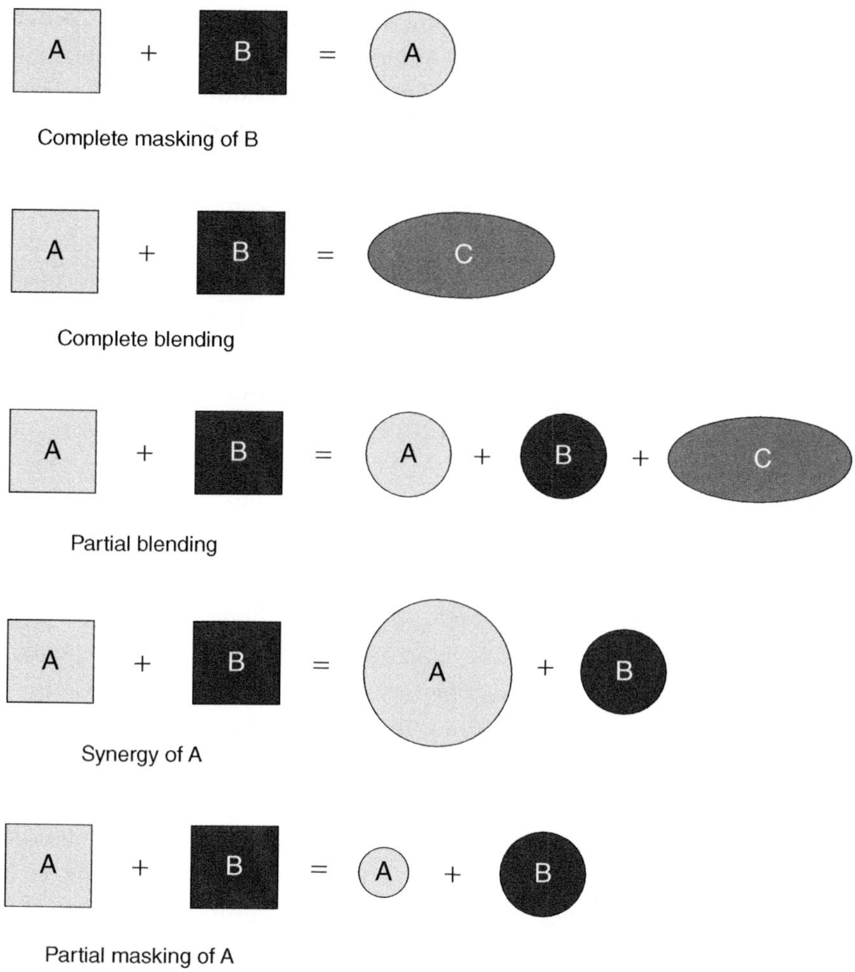

Figure 14.1 Different percepts that can arise from an odour mixture of two odours noted A and B. C is an odour dissimilar to A and B. The size of the circles reflects perceived intensity.

et al. 2007). This perceptual complete blending of odours refers to the configural processing of odour mixtures (Coureaud et al. 2014b, Kay et al. 2005, Thomas-Danguin et al. 2014). Configural processing is at the basis of what perfumers or flavourists call chords, which are mixtures of a few odours that give rise to a novel odour object. For example, a binary mixture of ethyl isobutyrate (strawberry odour) and ethyl maltol (caramel odour) has been shown to give rise to a pineapple odour that is not carried by the two components (Le Berre et al. 2008). The configural processing of this mixture has recently been confirmed by neurobiological studies (Coureaud et al. 2014a, Coureaud et al. 2014b, Schneider et al. 2016). Sometimes, a mixture can have blending properties that lead to the perception of a specific odour for the mixture, on top of the odours of the components. In such case, the blending is not complete but partial since the components' odour can still be perceived (Figure 14.1). When partial blending occurs, the perception is heterogeneous because several odours can be discernible when smelling the mixture, namely the odour specific to the mixture and the odour of the individual components. More generally, the perception is heterogeneous when at least some of the components' odour can be perceived within the mixture; in the case of a binary mixture the two components can be perceived. In that case, heterogeneous perceptions refer to the elemental processing of odour mixtures (Kay et al. 2005, Thomas-Danguin et al. 2014). Heterogenous perception has been particularly studied so that several perceptual effects have been listed such as perceptual dominance (Atanasova et al. 2005a, Brodin et al. 2009, Brossard et al. 2007, Ferreira 2012b), but also odour intensity related effects.

Regarding odour intensity, to characterize the perceptual effects of mixing odours, the mixture intensity is compared to the intensities of the single components (Berglund and Olsson 1993, Cain 1975, Patte and Laffort 1979, Thomas-Danguin and Chastrette 2002). For homogeneous percepts, hyper-addition, complete addition, or hypo-addition can be observed depending on whether the mixture intensity is higher, equal, or lower than the sum of the single components' odour intensities. In the case of heterogeneous percept, masking and synergy have been reported (Cain 1975, Ferreira 2012a, Laffort et al. 1989). Synergy refers to an increased intensity of one of the odour component after mixing (Figure 14.1). This phenomenon is usually observed in mixtures where odours are mixed at a low intensity level (Laing et al. 1984, Laska and Hudson 1992, Moskowitz 1976). Interestingly, such effect can also arise in mixtures where one of the components is at subthreshold level (Atanasova et al. 2005b), or even when all mixed odorants are at subthreshold level (Guadagni et al. 1963). Partial masking (Figure 14.1) refers to the opposite situation compared to synergy, namely to a decreased intensity of one of the odour component after mixing (Cain and Drexler 1974). As an illustration of the complexity of perceptual interactions in odour mixture, it has been observed that both masking and synergy can take place in the same binary mixture depending on the component concentrations and/or ratio. In the mixture of isoamyl acetate and whiskey lactone, synergy of the fruity odour of the ester was observed at low lactone concentration, whereas a masking of the fruity odour was observed at high lactone concentration (Chaput et al. 2012, Ishii et al. 2008). Only a few studies reported pleasantness of mixtures compared to their components. In binary mixtures, it has been found that the pleasantness of the mixture falls between the pleasantness of the components (Dravnieks and Jarke 1980, Moskowitz and Barbe 1977).

14.2.4 Predictability of perceptual interactions in food odour mixtures

In food flavour analysis, the challenge is to point out the molecules that actually contribute to the perception of the whole food aroma perceived by the consumer. The application of gas chromatography-olfactometry (GC-O) is efficient in providing a wide list of the odour-active compounds present in a given food sample (Delahunty et al. 2006). However, partly because of the above presented non-linear mixture effects, it is obvious that having the list of odour-active compounds of a food sample and knowing their single sensory features does not mean knowing the actual contributors to the whole food aroma. This raises the question of the predictability of a perceived odour knowing the components.

Several attempts have been made to predict the odour characteristics of binary or higher order mixtures. For odour intensity, several models were proposed but none of them can efficiently cover all the types of interactions reported (Berglund and Olsson 1993, Laffort and Dravnieks 1982, Thomas-Danguin and Chastrette 2002, Yan et al. 2014). Odour quality of binary heterogeneous mixtures could be predicted on the basis of the odour intensity of the components (Laing and Willcox 1983, Olsson 1998, Wise et al. 2000), but such an approach hardly takes into account perceptual dominance or partial overshadowing effects. For more complex mixtures, it has been observed that mixtures' quality is more frequently different from the components' one, likely due to higher blending probability, making the prediction even harder (Ferreira 2012b, Lindqvist et al. 2012, Livermore and Laing 1998). If pleasantness has been quite successfully predicted from the components' odour intensity in binary mixtures (Lapid et al. 2008), for complex odour mixtures pleasantness seems to be hardly predictable (Lindqvist et al. 2012).

To overcome the issue of perceptual interaction unpredictability, the flow chart of flavour analysis recommends performing omission or recombination experiments as a final step to verify the actual key odorants of an aroma mixture (Grosch 2001). These strategies, even if time-consuming, have proved relevant in providing further evidence of the role of perceptual interactions in mixture processing (Escudero et al. 2004, Lorrain et al. 2006). Therefore it is clear that the synergy, masking, dominance and blending are the kingpins of the perception of complex mixtures of odorants. Deciphering the biological mechanisms that underlie these perceptual effects would probably be the only way to reveal the rules that govern odour mixture perception. This appears today as the missing link in the development of a comprehensive theoretical model explaining the role of each odorant in the mixture and to predict the overall food aroma perception in its qualitative, quantitative and hedonic dimensions.

14.3 Recombination strategies

14.3.1 Off-line recombination strategies for odorants

As described in Chapter 6, a great number of volatile compounds are present in a food product. These compounds can be identified by mass spectrometry after separation by gas chromatography. However not all of these compounds contribute to the overall odour of the food. The compounds responsible for the typical aroma of a specific food product can be detected by the human nose after GC-O separation, a technique developed by Fuller, Steltenkamp and Tisserand (Fuller et al. 1964). For more information on GC-O techniques, the reader is invited to consult the book edited by Leland et al. (1999). The panellist is asked to give an odour descriptor, each time an odour is detected. Different procedures have been developed to build what is called an aromagram, which is a representation of the odour perception as a function of elution time or retention index (RI). The procedure called CHARM analysis (combined hedonic aroma response measurement) consists of the GC injection of increasing dilutions of the aroma extract until no odour is perceived at the sniffing port (Acree et al. 1984). From individual panellist's results, a computerized data treatment builds the global aromagram and assigns a greater peak area to odorants that are perceived at highest dilution. An alternative is the dilution technique called aroma extract dilution analysis (AEDA) which plots the dilution factor as a function of RI (Ullrich and Grosch 1987). Since these two techniques are based only on detection thresholds and do not take into account the Stevens law, that is, the non-linearity of the stimulus-response function (Stevens 1960, Chastrette et al. 1998), other procedures allow a direct measurement of intensity. The OSME technique is based on the continuous recording of the odour intensity perceived at the sniffing port by moving a cursor of a variable resistor as a function of the intensity of perception (Miranda-Lopez et al. 1992). This technique was improved by Etievant et al. (1999) using the finger span cross modality (FSCM) and allows evaluation of odour intensities with a significant number of panellists using only one dilution. OSME and FSCM are more efficient and rapid than CHARM or AEDA which are often used with only 2 or 3 panellists due to the great number of dilutions needed (Guichard et al. 1995). Another procedure consists in calculating peak detection frequency (DF), nasal impact frequency (NIF) or surface of nasal impact frequency (SNIF) of odours by a panel of more than 10 panellists (Pollien et al. 1997). This technique is rapid and repeatable (Pollien et al. 1999). However, GC-O allows evaluating the odour of isolated compounds and not their impact in the real food where they are present in mixture. Different methodologies were thus proposed to study aroma recombination to reproduce food aroma from a mixture of the individual compounds.

The classical approach is based on the concept of odour activity values (OAV), which is the ratio of the concentration of the aroma compound in the food divided by its detection threshold (Rothe 1975, Ullrich and Grosch 1987). Some "key aroma compounds" are mixed to create an aroma model as close as possible to the original product. From this model subsequent experiments are performed involving respectively omitting or adding one by one a subset of compounds (Grosch 2001). The main limitation of this method concerns the selection of the "key odorant compounds". The most common methodologies are based on the idea that the higher the intensity, the higher the contribution to the whole aroma. Odour intensities can be evaluated by the results from GC-O, using CHARM, DF, NIF or SNIF data. The calculated OAV can then be used to select the most potent aroma and prepare an aroma recombination using the calculated amounts of the selected aroma compounds in mixture. This sensomic approach has proven to be a successful concept in characterizing the aroma compounds responsible for the aroma signature of foods (Dunkel et al. 2014, Schieberle and Hofmann 2012). It has been applied to reproduce aroma of different foods such as wheat beer (Langos et al. 2013), Swiss cheese (Preininger et al. 1996) or roasted Arabica coffee (Czerny et al. 1999).

In some cases this approach was not successful to reproduce the perceived aroma (for example for Cola drinks) and omission tests were used successfully to rebalance the proportions of the aroma compounds in the mixture in order to better reproduce the global aroma (Lorjaroenphon and Cadwallader 2015). Moreover the approach based on OAV values does not take into account the impact of odorants present at sub and peri-threshold concentrations, which can also impact odour perception when present in mixture (Guadagni et al. 1963, Ryan et al. 2008). In wine for example, Escudero et al. (2004) demonstrated a significant impact of the addition of two mercaptans with low OAV to the global aroma. Based on the same assumptions, Escudero et al. (2007) constructed a wine model from a common base supplemented by six target odour families. They showed that the impact of a single family is highly conditioned by the blend context. Following this idea, Paravisini et al. (2014) developed an original approach based on odour qualities of the volatile compounds to evaluate their impact on caramel odour typicality using omission and addition tests. They observed that caramel typicality resulted from a complex balance between fruity, vegetal, sharp, nutty and caramel notes.

14.3.2 Online recombination strategies for odorants

Recently, several methods based on dynamic reconstitution of odorant mixtures online during GC-O analysis have been proposed to study the perception of key odorants in a mixture context.

In a first category of methods, GC-O is combined with systems producing a background odour in which the molecules eluted from GC-O can be embedded. The GC-PO method was developed to study the perception of odorants eluted from a GC-O in a uniform background odour called "pedestal" (Williams et al. 2009). The constant pedestal composition was generated by a dynamic headspace released from an odorant solution introduced into the auxiliary humidified air flow device connected to the GC-O sniffing port. This system has been validated using a single odorant (octanal) to produce the pedestal odour while hexanal was delivered by the GC-O. Trained panellists were asked to identify odours and rate their intensity at different times so that odorants were evaluated in- or out-of-mixture.

In a similar way, the OASIS method (Original Aroma Simultaneously Input to the Sniffing port) has been developed to embed the odorants, delivered singly from the GC-O, into the original whole aroma (Hattori et al. 2003, Hattori et al. 2005). This method was applied to identify the key odorants of Japanese green tea by testing the impact of target odorants identified after a classical GC-O approach (AEDA), once mixed with the original tea odour (OASIS). The results revealed that specific odorants, even those with high odour threshold, can enhance odour notes of Japanese green tea.

The olfactoscan method (Burseg and de Jong 2009) couples two devices: a GC-O apparatus and a multi-channel dynamic dilution olfactometer. In this system, the humidified air stream at the outlet of the olfactometer is connected to the GC-O sniffing port. Therefore, when controlled odorant phases are provided by the olfactometer, they can be mixed with the odorants coming from the GC-O. The olfactoscan system enables the scan ("screening") of the olfactory active compounds delivered during a GC-O run, mixed with a background odour or with selected odorant mixtures generated with the olfactometer. The olfactoscan method has been successfully applied to screen for odorants that could mask the off-odour induced by methional in orange juice (Burseg and de Jong 2009). Molecules to be tested were delivered by the GC-O and mixed, or not, with various concentrations of methional delivered by the olfactometer. A panel of judges familiar with the GC-O technique had to identify the smell of methional in each condition. The results showed that one of the screened molecules (octanal) significantly masked the methional odour.

Another category of techniques relies on modified GC-O systems that can perform online omission or recombination during the GC run. The gas chromatography-global olfactometry omission detection (GC-GOOD) allows the evaluation of mixtures of odorants at the outlet of the GC (Hallier et al. 2004). The mixture to be tested after GC-GOOD can include all odorants of a food extract eluted from the GC, or a mixture in which one or several odorants have been omitted. To do so, the GC effluent, usually directed to the sniffing port, can be directed to a PTFE bag or be omitted. The control of the omission is ensured by the FID trace. A panel of trained panellists can rate the odour similarity between the bag containing all the odorants and the ones in which at least one odorant was omitted. The panellists can also perform intensity judgments or quantitative descriptive analysis of the odorant mixtures trapped into the bags. This technique has been used to identify key families of volatile compounds in cooked catfish (sulfury, moldy, grassy, green and plastic odours) and revealed synergistic and masking effects between these families.

In the gas chromatography recomposition-olfactometry (GC-R) technique (Johnson et al. 2012), the approach is quite similar but GC online recombination is achieved instead of omission. During the GC analysis of a flavour extract (e.g., after SPME), individual compounds are selectively in-instrument recombined into a mixture as they elute from the column. The mixture can then be submitted to sensory analysis. The selective recombination is accomplished by a Deans Switch directly in-line with the column and enables the creation of mixtures including all or a subset of the volatiles present in a given sample. Applied to a flowering lavender SPME extract, GC-R was used to test sub-mixtures of lavender volatiles with a sensory panel of trained judges. The results showed evidence of additive, masking and synergistic effects in the tested mixtures but also of perceptual blending in specific mixtures that likely contribute to the lavender aroma character.

14.4 Mixture of tastants

The human gustatory system identifies mainly five basic taste qualities: sweet, salty, sour, bitter and umami (see Chapter 3). Binary taste-taste interactions have been reviewed by Keast and Breslin (2002), mainly in relation with psychophysical theories. However, taste-taste interactions are not well understood, especially in the context of real food perception. Generally, it is suggested that peripheral physiological interactions occur at the taste receptor cell level. However, the availability of the taste stimuli for the receptor may be modulated by chemical interactions occurring with the food matrix components. These taste-taste interactions were found to depend on both the components for a specific taste and the concentration (subthreshold or suprathreshold) of the considered taste stimuli (Breslin 1996, Keast and Breslin 2002). After some general aspects of taste-taste interactions, we hereafter focus on the impact of perceptual interactions between tastants in real conditions of food consumption, considering separately sweet and salty products.

14.4.1 Perceptual aspects of mixtures of tastants

Sweet, salty, sour, bitter and umami tastants are perceived through different biological mechanisms (see Chandrashekar et al. 2006). Sweet, bitter and umami tastants interact with G protein-coupled taste receptor in the apical part of taste receptor cells (Rawson and Li 2004). Salty and sour tastants are perceived after direct entry of Na^+ and H^+ in taste cells through specialized membrane channels on the apical surface of the cell. In spite of these different detection mechanisms at the peripheral level, between-tastes interactions have been observed at the perceptual level when tastants are present in mixture (Breslin 1996, Helleman 1992, Keast and Breslin 2002).

Most of the knowledge on taste-taste interactions has been acquired using simple binary mixtures (Breslin and Beauchamp 1997, Calvino et al. 1990, Calvino et al. 1993, Calvino and Garrido 1991, Curtis et al. 1984, Keast and Breslin 2002, Pangborn 1960, Schifferstein 1994, Schiffman et al. 1994). In such tastant mixtures, different types of interactions can occur such as masking, enhancing and synergistic effects which will affect the quality, intensity, temporality and spatiality of the overall perception (Breslin 1996, Keast and Breslin 2002).

The modifications in terms of quality highly depend on the tastants' concentration. Sweetness has been described as a dominant taste quality in binary, ternary and quaternary mixtures (Green et al. 2010). However, at low intensity and/or concentration of a sweet tastant (when the perceived intensity rises faster than the concentration), sweetness can either enhance or suppress other tastes in a binary mixture. At medium and high intensities and/or concentration (when intensity rises proportionally to the concentration), sweetness is suppressive of other basic tastants. At high concentration, where receptors are saturated, sweetness is symmetrically suppressive of bitter and sour perception.

In the same way, the interactions between sour and salty compounds are highly dependent on their intensities and/or concentration (Breslin 1996). Subthreshold concentrations of salt generally suppress the sourness perception due to acid compounds. Similarly, acids generally suppress the perceived saltiness but with two exceptions: subthreshold concentrations of many acids enhance the saltiness intensity of sodium chloride and tartaric acid enhances the perception of sodium chloride. Weak suprathreshold concentrations of sour and salty compounds generally show mutual enhancing effects while at strong suprathreshold concentrations, inversely, a mutual suppression is observed.

14.4.2 Examples of taste-taste interactions in real foods

Taste-taste interactions were observed or strongly suspected in several studies on food products. Several authors reported that taste–taste interactions in complex matrices can be predicted from aqueous solutions (Keast and Breslin 2002), but other authors argued that perception of tastants in aqueous solutions may relate more to the perception of impurities in water than to perception of tastants in foods and beverages, which are accompanied by trigeminal and retronasal olfactory stimulation (Green et al. 2010). The use of addition tests on real samples and omission tests from reconstituted fractions allowed highlighting taste-taste interactions (Engel et al. 2002). The degree of agreement between binary taste interactions in model systems and taste interactions in a real food during consumption has been less reported, although it is one way of better understanding the overall taste perception.

14.4.2.1 Perceptual aspects of taste-taste interactions in sweet food products

The quality of binary mixtures including a sweet component has been measured at low intensity/concentration in pear nectar, lima bean puree, wine, tomato juice and Champagne wine, showing that the suppressive effect of sweetness was stronger than the suppressive effect of sourness (Martin 2002, Pangborn and Chrisp 1964, Pangborn and Hansen 1963, Pangborn et al. 1964b, Pangborn and Trabue 1964). Opposite results were obtained in Chardonnay wine in which the suppressive effect of sourness on sweetness was stronger (Zamora et al. 2006). At higher intensity/concentration, the suppressive effect of sweetness on other basic tastes observed in model solutions has been confirmed in some real foods. For example, the sweet taste of sucrose suppressed the sourness of citric acid in lemon juice (McBride and Johnson 1987). This effect is reciprocal but it has been reported that higher level of acid is required to suppress sweetness in flavoured drinks (Bonnans and Noble 1993). Obviously, evaluation of the effect of quality of binary mixtures in real sample is carried out with the tastant and intensity/concentration appropriate for the target food product. Then, it is possible that even for the same taste category, depending on the tastant used, slight differences can be perceived on its quality (Schiffman 2000) therefore on its interaction in a mixture.

Generally speaking, sweet–bitter interactions evaluated from different drinks such as coffee, cabbage juice and citrus juice (Beck et al. 2014, Calvino et al. 1990, Dea et al. 2013, Guadagni et al. 1973) are in agreement with the results obtained from binary mixtures model solutions. In general, results are in line with the psychophysical function, the higher the intensity of sweetness or bitterness, the higher the suppressive effect. Beck et al. (2014) reported that, at low concentration of bitter tastant, sucrose masked bitterness of goitrin, sinigrin and quinine in aqueous sample and cabbage juice, except for sinigrin in cabbage juice. Calvino et al. (1990) determined that the suppression of bitterness and coffee flavour qualities increased when sucrose levels increased, and the suppression of sweetness increased when caffeine or coffee flavour rose. Guadagni et al. (1973) reported that the threshold of the bitter compound limonin was about six times higher in orange juice than in water, probably because of matrix effect which has some chemical components that are effective in masking or blending bitterness. Dea et al. (2013) observed a synergetic effect of limonin and nomilin that decreases each other's threshold when in mixture and found that these thresholds were lower in orange juice compared to sugars (sucrose, glucose and fructose) and acids (citric and malic) solutions. One explanation could be due to the sugar and acid concentration used, because increasing sugar levels was efficient in decreasing bitterness perception of limonin and nomilin in orange juices but the bitterness suppression was not effective when citric acid concentration increased. Another explanation could be that some chemical compounds presented in juices such as flavonoids may also contribute to bitterness in real sample. These results clearly demonstrate that it is always interesting to evaluate the impact of tastants mixtures in the most complete food matrix and/or in real food media.

14.4.2.2 Perceptual aspects of taste-taste interactions in salty food products

Various taste-taste interactions have been reported in several salty food systems. For instance in tomato juice, using omission tests, original results were found concerning the links between saltiness and sourness (Salles et al. 2003). Starting from a model tomato juice containing sugars, acids, mineral salts and amino acids at a pH adjusted with potassium hydroxide, it was shown that omission of the acids led to a total disappearance of sourness and saltiness, showing their involvement in these taste characteristics. The omission of acids implied that less potassium hydroxide was used for pH adjustment. Variation in potassium ion concentration may also affect taste characteristics, especially saltiness. The main part of saltiness seemed to arise from the presence of citric and malic acids, probably their anionic form, with potassium as main counter ion. Thus, the association of organic acids and potassium ions seemed to be determinant for the saltiness perception of tomato juice. In addition, it can be noted that the omission of all the sugars led to an almost total disappearance of sweetness and a significant increase in sourness and saltiness which had been partly masked by the sugars.

Studying the impact of different compounds on the taste quality of seafood, Fuke and Ueda (1996) reported the results of omission tests from model solutions representing the taste extract of the products. They showed in particular that the omission of sodium chloride was responsible for a decrease in sourness perception for clam and skipjack. In dairy products also, several studies report salt-acid interactions. Stampanoni and Noble (1991) studying in parallel the sourness perception of eight different model cheeses as a function of the concentration in citric acid and sodium chloride showed that their sourness intensity was explained by both the pH level and the concentration of sodium chloride of the model cheeses. In another example, Vangtal and Hammond (1986) studying 15 Swiss cheeses, tried to correlate physico-chemical and sensory data using multiple linear regression. Among other results the saltiness was found to be more correlated to the concentration of aromatic and iso-acids than to the sodium chloride concentration, thus showing the high impact of sourness perception on saltiness perception. Inversely, saltiness can also impact sourness perception. Breslin (1996) reported that weak suprathreshold concentrations of sodium chloride have the tendency to enhance sourness perception and higher concentrations in sodium chloride can mask sourness perception. A more complete and reliable study was carried out on a water-soluble extract of goat cheese. A model was constituted based on total physicochemical characterization of a water-soluble extract of goat cheese and the taste similarity between the model and the extract was controlled by sensory analyses (Engel et al. 2000a). The omission tests carried out on this model gave information on saltiness and sourness perception and on their interactions (Engel et al. 2000b). Among the main taste characteristics of a goat cheese water-soluble extract (salty, sour and bitter), saltiness was explained by an additive contribution of sodium, potassium, calcium and magnesium cations, whereas sourness appeared to be due to an enhancing effect of sodium chloride on compounds acting on the pH, phosphates and lactic acid. This conclusion was confirmed by the simultaneous omission of phosphates and sodium chloride, which led to a positive mean sourness deviation weaker than when phosphate was omitted alone. This results support the idea that the significant increase in sourness due to phosphate omission was partially compensated by the omission of sodium chloride, which tends to decrease the sour note because of its enhancing contribution. A camembert cheese extract was studied with the same approach as for goat cheese (Engel et al. 2001) and the results showed that sodium chloride dramatically increased the level of both sourness and saltiness. The sourness of the water-soluble extract of camembert cheese was mainly explained by the enhancing effect of sodium chloride on the slight sourness due to H_3O^+ concentration. This effect has already been shown for wheat bread (Helleman, 1992) or cheeses (Stampanoni and Noble, 1991).

Many other taste-taste interactions can be found in food products. As an example, salty taste was reported to suppress bitterness in salt and sour mixtures (Keast and Breslin 2002). This interaction was also reported from model water-soluble extract of goat cheese (Engel et al. 2000b) for which the omission of sodium salts significantly increased bitterness intensity. More generally, it is unquestionable that various taste–taste interactions occur in complex real foods according to their composition and significantly contribute to its overall flavour. In addition, odour perception can be combined to taste perception. These odour-taste interactions, also potentially influencing the overall flavour of food, are briefly presented below.

14.5 Odour-taste interactions in real foods

Odour-taste interaction basically depends on the capacity of the two stimuli to form an appropriate combination in a food product context (*congruency*) (Schifferstein and Verlegh 1996). Odour-taste integrated perception thus highly depends on learned associations, the context in which the food is consumed and the consumer's previous experience (Stevenson et al. 1998). In experimental situations, odour-taste interactions were found to be affected by the so-called dumping and halo effects (Clark and Lawless 1994), which resulted from the tendency for sensory panellists to dump their sensations on several available scales (taste and/or odour). Nevertheless, experiments performed with appropriate scales have shown how odour is able to modulate taste (Nguyen,

2001). Moreover, studies performed without scaling procedures have clearly demonstrated odour-taste integration (Dalton et al., 2000).

Odour-taste interactions have been extensively reported in model solution but less in real foods. The impact of aroma on sweet perception has been more studied in comparison to other taste attributes, maybe because of the common observation that certain odours smell "sweet". For example, Frank and Byram (1988) showed that, for the same sugar concentrations, subjects perceived whipped cream with strawberry aroma as being sweeter than whipped cream alone. Knoop et al. (2009) reported that the addition of a sweet smelling volatile compound (ethyl-hexanoate) significantly enhanced the sweetness of apple juice. Labbe et al. (2006) studied the influence of olfaction on taste perception in bitter coconut beverage. They found that vanilla flavour increased sweetness perception but it was highly dependent on the food matrix because the addition of vanilla to bitter milk did not enhance sweetness. Moreover, the authors found that taste-olfaction integration was dependent on previous food experience even for trained sensory panellists. Tournier et al. (2009) observed that aroma influenced taste perception and vice versa in custard desserts varying in viscosity, sucrose level and aroma nature.

Odour-taste perceptual interactions were also reported for saltiness and aroma, mainly in studies aiming to lower the salt content in foods without affecting the saltiness intensity (Lawrence et al. 2009). Interestingly, on the basis of previous results reporting that citric acid had an enhancing effect on saltiness at a low level of NaCl (Breslin 1996) and that salt–related odorants had an enhancing effect on saltiness perception (Lawrence et al. 2009), Nasri et al. (2013) examined if saltiness–sourness–aroma (Sardine) ternary cross–modal interactions could enhance saltiness intensity in model solutions. They reported that adding citric acid led to a significant increase in saltiness for salt and sardine salt-related aroma mixture only, thus showing the efficiency of sour–salty–aroma interaction in a strategy to enhance saltiness perception in foods with reduced salt content. In addition, in more complex cheese flavour mixtures, NaCl, lactic acid, and aroma were found to be able to enhance cheese flavour intensity and to compensate each other towards cheese flavour intensity (Niimi et al. 2014), thus suggesting complex taste-taste and taste-aroma interactions involved in the overall cheese flavour. Such interactions were also reported in real solid foods. The saltiness enhancement by aromas in model cheeses was found to be dependent on the food matrix composition and was no longer observed for high NaCl content and firm textures (Lawrence et al. 2011). Fat content seems to be also an important factor. Indeed, in model cheeses varying in fat, salt and pH levels, sardine odour was found to induce a significant saltiness enhancement (Syarifuddin et al. 2016), but the amplitude of the enhancement depended on the matrix composition and texture. For instance, sardine odour was found to induce higher saltiness enhancement in model cheeses with the low fat level while for the high fat level, saltiness enhancement was less pronounced.

Very few studies reported examples of perceptual interactions between aroma and fat perceptions. The main reason could be that fat perception is complex and probably a multimodal sensation in itself (Mattes, 2009). Bult et al. (2007) reported that the odour stimulus increased the intensities of oral texture perception such as thickness and creaminess when the odour was presented retronasally for dairy liquid. A clear demonstration of the perceptual impact of odorants on fat perception has been reported only by Syarifuddin et al. (2016). These authors found that butter aroma can enhance fat perception in model cheeses. Again, the enhancement amplitude depended on the product composition and texture and especially firmness supporting the idea of fat multimodal perception.

14.6 Conclusion

This chapter demonstrates that it is not easy to predict the global perception of a mixture of odorants and/or taste compounds from the sensory properties of the single elements because of perceptual and multimodal interactions. The sensory perception of mixtures of odorants and of tastants in real foods has to be controlled by recombination or omission studies, after quantification of the amount of odorants or tastants and/or by taking into account their perceptual qualities. The synergistic or masking effects observed in binary systems depend on the stimuli relative concentrations and are not always predictive of the effects in real food systems. Odour-taste interactions are mainly driven by learned associations but have been used as a strategy to enhance saltiness in low salt products or sweetness in low sugar products. Overall, it appears that a better understanding of the perceptual interactions between odorants and tastants is needed and can only be obtained through a deeper knowledge of the underlying neurobiological processes. Such research would be worth the effort since it would provide useful knowledge for reformulation of foods better satisfying nutritional recommendations.

References

Acree, T. E., Barnard, J. and Cunningham, D. G. (1984) A procedure for the sensory analysis of gas chromatographic effluents. *Food Chemistry, 14*, pp. 273–286.

Atanasova, B., Thomas-Danguin, T., Chabanet, C., Langlois, D., Nicklaus, S. and Etievant, P. (2005a) Perceptual interactions in odour mixtures: Odour quality in binary mixtures of woody and fruity wine odorants. *Chemical Senses, 30*(3), pp. 209–217.

Atanasova, B., Thomas-Danguin, T., Langlois, D., Nicklaus, S., Chabanet, C. and Etievant, P. (2005b) Perception of wine fruity and woody notes: influence of peri-threshold odorants. *Food Quality and Preference, 16*(6), pp. 504–510.

Beck, T. K., Jensen, S., Bjoern, G. K. and Kidmose, U. (2014) The masking effect of sucrose on perception of bitter compounds in brassica vegetables. *Journal of Sensory Studies, 29*(3), pp. 190–200.

Berglund, B., Berglund, U. and Lindvall, T. (1976) Psychological processing of odor mixtures. *Psychological Review, 83*(6), pp. 432–441.

Berglund, B. and Olsson, M. J. (1993) Odor-intensity interaction in binary and ternary mixtures. *Perception and Psychophysics, 53*(5), pp. 475–482.

Bonnans, S. and Noble, A. C. (1993) Effect of sweetener type and of sweetener and acid levels on temporal perception of sweetness, souness and fruitiness. *Chemical Senses, 18*(3), pp. 273–283.

Breslin, P. A. S. (1996) Interactions among salty, sour and bitter compounds. *Trends in Food Science and Technology, 7*, pp. 390–399.

Breslin, P. A. S. and Beauchamp, G. K. (1997) Salt enhances flavour by suppressing bitterness. *Nature, 387*(6633), pp. 563.

Brodin, M., Laska, M. and Olsson, M. J. (2009) Odor interaction between bourgeonal and its antagonist undecanal. *Chemical Senses, 34*(7), pp. 625–630.

Brossard, C., Rousseau, F. and Dumont, J. P. (2007) Perceptual interactions between characteristic notes smelled above aqueous solutions of odorant mixtures. *Chemical Senses, 32*(4), pp. 319–327.

Bult, J. H. F., de Wijk, R. A. and Hummel, T. (2007) Investigations on multimodal sensory integration: texture, taste and orthonasal olfactory stimuli in concert. *Neuroscience Letters, 411*, pp. 6–10.

Burseg, K. and de Jong, C. (2009) Application of the olfactoscan method to study the ability of saturated aldehydes in masking the odor of methional. *Journal of Agricultural and Food Chemistry, 57*(19), pp. 9086–9090.

Cain, W. S. (1975) Odor intensity - Mixtures and masking. *Chemical Senses & Flavour, 1*(3), pp. 339–352.

Cain, W. S. and Drexler, M. (1974) Scope and evaluation of odor counteraction and masking. *Annals New York Academy of Sciences, 237*, pp. 427–439.

Calvino, A. M., Garciamedina, M. R. and Comettomuniz, J. E. (1990) Interactions in caffeine-sucrose and coffee-sucrose mixtures - evidence of taste and flavor suppression. *Chemical Senses, 15*(5), pp. 505–519.

Calvino, A. M., Garciamedina, M. R., Comettomuniz, J. E. and Rodriguez, M. B. (1993) Perception of sweetness and bitterness in different vehicles. *Perception and Psychophysics, 54*(6), pp. 751–758.

Calvino, A. M. and Garrido, D. (1991) Spacial and temporal suppressive behaviour of bitter-sweet mixtures. *Perceptual and Motor Skills, 73*(3), pp. 1216–1216.

Chandrashekar, J., Hoon, M. A., Ryba, N. J. P. and Zuker, C. S. (2006) The receptors and cells for mammalian taste. *Nature, 444*, pp. 288–294.

Chaput, M. A., El Mountassir, F., Atanasova, B., Thomas-Danguin, T., Le Bon, A. M., Perrut, A., Ferry, B. and Duchamp-Viret, P. (2012) Interactions of odorants with olfactory receptors and receptor neurons match the perceptual dynamics observed for woody and fruity odorant mixtures. *European Journal of Neuroscience, 35*(4), pp. 584–597.

Clark, C. C. and Lawless, H. T. (1994) Limiting response alternatives in time-intensity scaling - an examination of the halo-dumping effect. *Chemical Senses, 19*(6), pp. 583–594.

Chastrette, M., Thomas-Danguin, T. and Rallet, E. (1998) Modelling the human olfactory stimulus-response function. *Chemical Senses. 23* (2), 181–196.

Coureaud, G., Thomas-Danguin, T., Datiche, F., Wilson, D. A. and Ferreira, G. (2014a) Differential memory persistence of odor mixture and components in newborn rabbits: competition between the whole and its parts. *Frontiers in Behavioral Neuroscience, 8*, p. 211.

Coureaud, G., Thomas-Danguin, T., Wilson, D. A. and Ferreira, G. (2014b) Neonatal representation of odour objects: distinct memories of the whole and its parts. *Proceedings of the Royal Society B-Biological Sciences, 281*(1789).

Curtis, D. W., Stevens, D. A. and Lawless, H. T. (1984) Perceived intensity of the taste of sugar mixtures and acid mixtures. *Chemical Senses, 9*(2), pp. 107–120.

Czerny, M., Mayer, F. and Grosch, W. (1999) Sensory study on the character impact odorants of roasted Arabica coffee. *Journal of Agricultural and Food Chemistry, 47*(2), pp. 695–699.

Dalton, P., Doolittle, N., Nagata, H. and Breslin, P.A.S. (2000) The merging of the senses: integration of subthreshold taste and smell. *Nature Neuroscience. 3*(5), pp. 431–432.

Dea, S., Plotto, A., Manthey, J. A., Raithore, S., Irey, M. and Baldwin, E. (2013) Interactions and thresholds of limonin and nomilin in bitterness perception in orange juice and other matrices. *Journal of Sensory Studies, 28*, pp. 311–323.

Delahunty, C. M., Eyres, G. and Dufour, J. P. (2006) Gas chromatography-olfactometry. *Journal of Separation Science, 29*(14), pp. 2107–2125.

Dravnieks, A. and Jarke, F. H. (1980) Hedonic values of odor mixtures as fuctions of hedonic values of their components. in Van der Starre, H., (ed.) *Olfaction and Taste VII*,London: IRL Press. pp. 393–396.

Dunkel, A., Steinhaus, M., Kotthoff, M., Nowak, B., Krautwurst, D., Schieberle, P. and Hofmann, T. (2014) Nature's chemical signatures in human olfaction: a foodborne perspective for future biotechnology. *Angewandte Chemie-International Edition, 53*(28), pp. 7124–7143.

Engel, E., Nicklaus, S., Garem, A., Septier, C., Salles, C. and Le Quéré, J.-L. (2000a) Taste active compounds in a goat cheese water-soluble extract. 1. Development and sensory validation of a model water-soluble extract. *Journal of Agricultural and Food Chemistry, 48*(9), pp. 4252–4259.

Engel, E., Nicklaus, S., Salles, C. and Le Quéré, J.-L. (2002) Relevance of omission tests to determine flavour-active compounds in food: application to cheese taste. *Food Quality and Preference, 13*(7-8), pp. 505–513.

Engel, E., Nicklaus, S., Septier, C., Salles, C. and Le Quéré, J.-L. (2000b) Taste active compounds in a goat cheese water-soluble extract. 2. Determination of the relative impact of water-soluble extract components on its taste using omission tests. *Journal of Agricultural and Food Chemistry, 48*(9), pp. 4260–4267.

Engel, E., Septier, C., Leconte, N., Salles, C. and Le Quéré, J.-L. (2001) Determination of taste-active compounds of a bitter Camembert cheese by omission tests. *Journal of Dairy Research, 68*(4), pp. 675–688.

Escudero, A., Campo, E., Farina, L., Cacho, J. and Ferreira, V. (2007) Analytical characterization of the aroma of five premium red wines. Insights into the role of odor families and the concept of fruitiness of wines. *Journal of Agricultural and Food Chemistry, 55*(11), pp. 4501–4510.

Escudero, A., Gogorza, B., Melus, M. A., Ortin, N., Cacho, J. and Ferreira, V. (2004) Characterization of the aroma of a wine from Maccabeo. Key role played by compounds with low odor activity values. *Journal of Agricultural and Food Chemistry, 52*(11), pp. 3516–3524.

Etievant, P. X., Callement, G., Langlois, D., Issanchou, S. and Coquibus, N. (1999) Odor intensity evaluation in gas chromatography-olfactometry by finger span method. *Journal of Agricultural and Food Chemistry, 47*(4), pp. 1673–1680.

Ferreira, V. (2012a) Revisiting psychophysical work on the quantitative and qualitative odour properties of simple odour mixtures: a flavour chemistry view. Part 1: intensity and detectability. A review. *Flavour and Fragrance Journal, 27*(2), pp. 124–140.

Ferreira, V. (2012b) Revisiting psychophysical work on the quantitative and qualitative odour properties of simple odour mixtures: a flavour chemistry view. Part 2: qualitative aspects. A review. *Flavour and Fragrance Journal, 27*(3), pp. 201–215.

Frank, R. A. and Byram, J. (1988) Taste smell interactions are tastant and odorant dependent. *Chemical Senses, 13*(3), pp. 445–455.

Fuke, S. and Ueda, Y. (1996) Interactions between umami and other flavor characteristics. *Trends in Food Science and Technology, 7,* pp. 407–411.

Fuller, G. H., Steltenkamp, R. and Tisserand, G. A. (1964) The gas chromatograph with human sensor: perfumer model. *Annals of the New-York Academy of Sciences, 116,* pp. 711–724.

Gottfried, J. A. (2010) Central mechanisms of odour object perception. *Nature Reviews Neuroscience, 11*(9), pp. 628–641.

Green, B. G., Lim, J., Osterhoff, F., Blacher, K. and Nachtigal, D. (2010) Taste mixture interactions: Suppression, additivity, and the predominance of sweetness. *Physiology and Behavior, 101*(5), pp. 731–737.

Grosch, W. (2001) Evaluation of the key odorants of foods by dilution experiments, aroma models and omission. *Chemical Senses, 26*(5), pp. 533–545.

Guadagni, D. G., Buttery, R. G., Okano, S. and Burr, H. K. (1963) Additive effect of sub-threshold concentrations of some organic compounds associated with food aromas. *Nature, 200*(4913), pp. 1288–1289.

Guadagni, D. G., Maier, V. P. and Turnbaugh, J. G. (1973) Effect of some citrus juice constituents on taste thresholds for limonin and naringin bitterness. *Journal of the Science of Food and Agriculture, 24,* pp. 1277–1288.

Guichard, H., Guichard, E., Langlois, D., Issanchou, S. and Abbott, N. (1995) GC sniffing analysis: olfactive intensity measurement by two methods. *Zeitschrift für Lebensmittel-Untersuchung und-Forschung, 201,* pp. 344–350.

Hallier, A., Courcoux, P., Serot, T. and Prost, C. (2004) New gas chromatography-olfactometric investigative method, and its application to cooked Silurus glanis (European catfish) odor characterization. *Journal of Chromatography A, 1056*(1-2), pp. 201–208.

Hattori, S., Takagaki, H. and Fujimori, T. (2003) Evaluation of Japanese green tea extract using GC/O with original aroma simultaneously input to the sniffing port method (OASIS). *Food Science and Technology Research, 9*(4), pp. 350–352.

Hattori, S., Takagaki, H. and Fujimori, T. (2005) Identification of volatile compounds which enhance odor notes in Japanese green tea using the OASIS (Original aroma simultaneously input to the sniffing port) method. *Food Science and Technology Research, 11*(2), pp. 171–174.

Helleman, U. (1992) Perceived taste of NaCl and acid mixtures in water and in bread. *Inernational Journal of Food Science & Technology, 27,* pp. 201–211.

Hornung, D. E. and Enns, M. P. (1989) Separating the contributions of smells and tastes in flavor perception. in Laing, D. G., Cain, W. S., McBride, R. L. and Ache, B. W., (eds.) *Perception of complex smells and tastes,*Sidney: Academic Press. pp. 285–296.

Ishii, A., Roudnitzky, N., Beno, N., Bensafi, M., Hummel, T., Rouby, C. and Thomas-Danguin, T. (2008) Synergy and masking in odor mixtures: An electrophysiological study of orthonasal vs. retronasal perception. *Chemical Senses, 33*(6), pp. 553–561.

Johnson, A. J., Hirson, G. D. and Ebeler, S. E. (2012) Perceptual characterization and analysis of aroma mixtures using gas chromatography recomposition-olfactometry. *PLoS ONE, 7*(8).

Kay, L. M., Crk, T. and Thorngate, J. (2005) A redefinition of odor mixture quality. *Behavioral Neuroscience, 119*(3), pp. 726–733.

Keast, R. S. J. and Breslin, P. A. S. (2002) An overview of binary taste-taste interactions. *Food Quality and Preference, 14*(2), pp. 111–124.

Knoop, J. E., Bult, J. H. F. and Smit, G. (2009) Masking undesirable effects of ethylhexanoate on apple flavour. *Chemical Senses, 34*(3), pp. E22.

Labbe, D., Damevin, L., Vaccher, C., Morgenegg, C. and Martin, N. (2006) Modulation of perceived taste by olfaction in familiar and unfamiliar beverages. *Food Quality and Preference, 17*(7-8), pp. 582–589.

Laffort, P. and Dravnieks, A. (1982) Several models of suprathreshold quantitative olfactory interaction in humans applied to binary, ternary and quaternary mixtures. *Chemical Senses, 7*(2), pp. 153–174.

Laffort, P., Etcheto, M., Patte, F. and Marfaing, P. (1989) Implications of power law exponent in synergy and inhibition of olfactory mixtures. *Chemical Senses, 14*(1), pp. 11–23.

Laing, D. G. (1991) Perception of complex smells. *Encyclopedia of Human Biology, 5*, pp. 759–767.

Laing, D. G. and Francis, G. W. (1989) The capacity of humans to identify odors in mixtures. *Physiology and Behavior, 46*(5), pp. 809–814.

Laing, D. G. and Livermore, B. A. (1992) Perceptual analysis of complex chemical signals by humans. in Doty, R. L. and Müller-Schwarze, D., (eds.) *Chemical signals in vertebrates VI*,New York (USA): Plenum Press. pp. 587–593.

Laing, D. G., Panhuber, H., Willcox, M. E. and Pittman, E. A. (1984) Quality and intensity of binary odor mixtures. *Physiology and Behavior, 33*, pp. 309–319.

Laing, D. G. and Willcox, M. E. (1983) Perception of components in binary odour mixtures. *Chemical Senses*, 7(3-4), pp. 249–264.

Langos, D., Granvogl, M. and Schieberle, P. (2013) Characterization of the key aroma compounds in two bavarian wheat beers by means of the sensomics approach. *Journal of Agricultural and Food Chemistry, 61*(47), pp. 11303–11311.

Lapid, H., Harel, D. and Sobel, N. (2008) Prediction models for the pleasantness of binary mixtures in olfaction. *Chemical Senses, 33*(7), pp. 599–609.

Laska, M. and Hudson, R. (1992) Ability to discriminate between related odor mixtures. *Chemical Senses, 17*(4), pp. 403–415.

Lawrence, G., Salles, C., Palicki, O., Septier, C., Busch, J. and Thomas-Danguin, T. (2011) Using cross-modal interactions to counterbalance salt reduction in solid foods. *International Dairy Journal, 21*(2), pp. 103–110.

Lawrence, G., Salles, C., Septier, C., Busch, J. and Thomas-Danguin, T. (2009) Odour-taste interactions: A way to enhance saltiness in low-salt content solutions. *Food Quality and Preference, 20*(3), pp. 241–248.

Le Berre, E., Thomas-Danguin, T., Beno, N., Coureaud, G., Etievant, P. and Prescott, J. (2008) Perceptual processing strategy and exposure influence the perception of odor mixtures. *Chemical Senses, 33*(2), pp. 193–199.

Leland, J. V., Schieberle, P., Buettner, A. and Acree, T. E. (1999) Gas chromatography-olfactometry. The state of the art. in *219th American Chemical Society National Meeting*, New Orleans, USA: American Chemical Society, Washington, DC.

Lindqvist, A., Hoglund, A. and Berglund, B. (2012) The role of odour quality in the perception of binary and higher-order mixtures. *Perception, 41*(11), pp. 1373–1391.

Livermore, A. and Laing, D. G. (1998) The influence of chemical complexity on the perception of multicomponent odor mixtures. *Perception and Psychophysics, 60*(4), pp. 650–661.

Lorjaroenphon, Y. and Cadwallader, K. R. (2015) Identification of character-impact odorants in a cola-flavored carbonated beverage by quantitative analysis and omission studies of aroma reconstitution models. *Journal of Agricultural and Food Chemistry, 63*(3), pp. 776–786.

Lorrain, B., Ballester, J., Thomas-Danguin, T., Blanquet, J., Meunier, J. M. and Le Fur, Y. (2006) Selection of potential impact odorants and sensory validation of their importance in typical Chardonnay wines. *Journal of Agricultural and Food Chemistry, 54*(11), pp. 3973–3981.

Martin, N. (2002) Sweet/sour balance in champagne wine and dependence on taste/odour interactions. *Food Quality and Preference, 13*(5), pp. 295–305.

Mattes, R.D. (2009) 'Is there a fatty acid taste?', in *Annual Review of Nutrition*. [Online]. Palo Alto: Annual Reviews. pp. 305–327.

McBride, R. L. and Johnson, R. L. (1987) Perception of sugar mixtures in lemon juice drink. *International Journal of Food Science and Technology, 22*(4), pp. 399–408.

Miranda-Lopez, R., Libbey, L. M., Watson, B. T. and McDaniel, M. R. (1992) Odor analysis of pinot noir wines from grapes of different maturities by a gas chromatography-olfactometry technique (osme). *Journal of Food Science, 57*(4), pp. 985–993 and 1019.

Moskowitz, H. R. (1976) Multidimensional scaling of odorants and mixtures. *Lebensmittel-Wissenschaft und –Technologie, 9*, pp. 232–238.

Moskowitz, H. R. and Barbe, C. D. (1977) Profiling of odor components and their mixtures. *Sensory Processes, 1*(3), pp. 212–226.

Murphy, C. (1985) Cognitive and chemosensory influences on age-related-changes in the ability to identify blended foods. *Journals of Gerontology, 40*(1), pp. 47–52.

Nasri, N., Septier, C., Béno, N., Salles, C. and Thomas-Danguin, T. (2013) Enhancing salty taste through odour-taste-taste interactions: Influence of odour intensity and salty tastants' nature. *Food Quality and Preference, 28*(1), pp. 134–140.

Nguyen, D. (2001) Contribution à l'etude des interactions entre entrées sensorielles: l'effet de la presence d'un arome sur la perception d'une saveur. Doctoral dissertation, Université de Dijon, France.

Niimi, J., Eddy, A. I., Overington, A. R., Silcock, P., Bremer, P. J. and Delahunty, C. M. (2014) Cross-modal interaction between cheese taste and aroma. *International Dairy Journal, 39*(2), pp. 222–228.

Olsson, M. J. (1998) An integrated model of intensity and quality of odor mixtures. *Olfaction and Taste Xii, 855*, pp. 837–840.

Pangborn, R. M. (1960) Taste interrelationships. *Food Research, 25*, pp. 245–256.

Pangborn, R. M. and Chrisp, R. B. (1964) Taste interrelationships VI: sucrose, sodium chloride and citric acid in canned tomato juice. *Journal of Food Science, 29*, pp. 490–498.

Pangborn, R. M. and Hansen, B. (1963) The influence of color on discrimination of sweetness and sourness in pear-necta. *American Journal of Psychology, 76*, pp. 315–317.

Pangborn, R. M., Ough, C. S. and Chrisp, R. B. (1964b) Taste interrelationships of sucrose, tartaric acid and caffeine in white table wine. *Journal of the American Society of Enology and Viticulture, 15*, pp. 154–161.

Pangborn, R. M. and Trabue, I. M. (1964) Taste interrelationships V: sucrose, sodium chloride and citric acid in lima bean puree. *Journal of Food Science, 29*, pp. 233–240.

Paravisini, L., Septier, C., Moretton, C., Nigay, H., Arvisenet, G., Guichard, E. and Dacremont, C. (2014) Caramel odor: Contribution of volatile compounds according to their odor qualities to caramel typicality. *Food Research International, 57*, pp. 79–88.

Patte, F. and Laffort, P. (1979) Alternative model of olfactory quantitative interaction in binary-mixtures. *Chemical Senses and Flavour, 4*(4), pp. 267–274.

Pollien, P., Fay, L. B., Baumgartner, M. and Chaintreau, A. (1999) First attempt of odorant quantitation using gas chromatography-olfactometry. *Analytical Chemistry, 71*(23), pp. 5391–5397.

Pollien, P., Ott, A., Montigon, F., Baumgartner, M., Muñoz-Box, R. and Chaintreau, A. (1997) Hyphenated headspace-gas chromatography-sniffing technique: screening of impact odorants and quantitative aromagram comparisons. *Journal of Agricultural and Food Chemistry, 45*(7), pp. 2630–2637.

Preininger, M., Warmke, R. and Grosch, W. (1996) Identification of the character impact flavour compounds of Swiss cheese by sensory studies of models. *Zeitschrift für Lebensmittel Untersuchung und Forschung, 202*, pp. 30–34.

Rawson, N. E. and Li, X. (2004) The cellular basis of flavour perception: taste and aroma. in Taylor, A. J. and Roberts, D. D., (eds.) *Flavor perception*,Oxford: Blackwell Publishing Ltd. pp. 57–85.

Rothe, M. (1975) Aroma values - a useful concept? In Maarse, H., Groenen, P.J. (eds.) *Aroma research*, Zeist, the Netherlands: Centre for Agricultural Publishing and Documentation. pp. 111–119.

Ryan, D., Prenzler, P. D., Saliba, A. J. and Scollary, G. R. (2008) The significance of low impact odorants in global odour perception. *Trends in Food Science & Technology, 19*(7), pp. 383–389.

Salles, C., Nicklaus, S. and Septier, C. (2003) Determination and gustatory properties of taste-active compounds in tomato juice. *Food Chemistry, 81*(3), pp. 395–402.

Schieberle, P. and Hofmann, T. (2012) Mapping the combinatorial code of food flavors by means of molecular sensory science approach. in Jelen, H., (ed.) *Food Flavors. Chemical, Sensory and Technological Properties; Chemical and Functional Properties of Food Components* Boca Raton (FL): CRC Press. pp. 413–438.

Schifferstein, H. N. J. (1994) Sweetness suppression in fructose citric-acid mixtures - a study of contextual effects. *Perception & Psychophysics, 56*(2), pp. 227–237.

Schifferstein, H. N. J. and Verlegh, P. W. J. (1996) The role of congruency and pleasantness in odor-induced taste enhancement. *Acta Psychologica, 94*(1), pp. 87–105.

Schiffman, S. S., Gatlin, L. A., Sattelymiller, E. A., Graham, B. G., Heiman, S. A., Stagner, W. C. and Erickson, R. P. (1994) The effect of sweeteners on bitter taste in young and elderly subjects. *Brain Research Bulletin, 35*(3), pp. 189–204.

Schiffman, S.S. (2000) Taste quality and neural coding. *Physiology & Behavior, 69*(1-2), pp. 147–159.

Schneider, N. Y., Datiche, F., Wilson, D. A., Gigot, V., Thomas-Danguin, T., Ferreira, G. and Coureaud, G. (2016) Brain processing of a configural vs elemental odor mixture in the newborn rabbit. *Brain Structure and Function, 221*(5), pp. 2527–2539.

Stampanoni, C. R. and Noble, A. C. (1991) The influence of fat, acid, and salt on the perception of selected taste and texture attributes of cheeses analogs: a scalar study. *Journal of Texture Studies, 22*, pp. 367–380.

Stevens, S. S. (1960) The psychophysics of sensory function. *American Scientist, 48*, pp. 226–253.

Stevenson, R. J., Boakes, R. A. and Prescott, J. (1998) Changes in odor sweetness resulting from implicit learning of a simultaneous odor-sweetness association: An example of learned synesthesia. *Learning and Motivation, 29*(2), pp. 113–132.

Stevenson, R. J. and Wilson, D. A. (2007) Odour perception: An object-recognition approach. *Perception, 36*(12), pp. 1821–1833.

Syarifuddin, A., Septier, C., Salles, C. and Thomas-Danguin, T. (2016) Reducing salt and fat while maintaining taste: An approach on a model food system. *Food Quality and Preference, 48*, pp. 59–69.

Szyszka, P., Stierle, J. S., Biergans, S. and Galizia, C. G. (2012) The speed of smell: Odor-object segregation within milliseconds. *PLoS ONE, 7*(4).

Thomas-Danguin, T. (2009) Flavor. in Binder, M. D., Hirokawa, N. and Windhorst, U., (eds.) *Encyclopedia of Neuroscience*,Berlin Heidelberg, Germany: Springer-Verlag GmbH.

Thomas-Danguin, T. and Chastrette, M. (2002) Odour intensity of binary mixtures of odorous compounds. *Comptes Rendus Biologies, 325*(7), pp. 767–772.

Thomas-Danguin, T., Le Berre, E., Barkat, S., Coureaud, G. and Sicard, G. (2007) Evidence for odor blending in odorant mixtures. In Paper Presented to the AChemS Annual Meeting 2007, Sarasota, FL, USA.

Thomas-Danguin, T., Sinding, C., Romagny, S., El Mountassir, F., Atanasova, B., Le Berre, E., Le Bon, A. M. and Coureaud, G. (2014) The perception of odor objects in everyday life: a review on the processing of odor mixtures. *Frontiers in Psychology, 5*, p. 504.

Tournier, C., Sulmont-Rosse, C., Semon, E., Vignon, A., Issanchou, S. and Guichard, E. (2009) A study on texture-taste-aroma interactions: Physico-chemical and cognitive mechanisms. *International Dairy Journal, 19*(8), pp. 450–458.

Ullrich, F. and Grosch, W. (1987) Identification of the most intense volatile flavour compounds formed during autoxidation of linoleic acid. *Zeitschrift für Lebensmittel Untersuchung und Forschung, 184*, pp. 277–282.

Vangtal, A. and Hammond, E. G. (1986) Correlation of the flavor characteristics of Swiss-type cheeses with chemical parameters. *Journal of Dairy Science, 69*, pp. 2892–2993.

Williams, R. C., Sartre, E., Parisot, F., Kurtz, A. J. and Acree, T. E. (2009) A gas chromatograph-pedestal olfactometer (GC-PO) for the study of odor mixtures. *Chemosensory Perception, 2*(4), pp. 173–179.

Wilson, D. A. and Sullivan, R. M. (2011) Cortical processing of odor objects. *Neuron, 72*(4), pp. 506–519.

Wise, P. M., Olsson, M. J. and Cain, W. S. (2000) Quantification of odor quality. *Chemical Senses, 25*(4), pp. 429–443.

Yan, L. C., Liu, J. M., Wang, G. H. and Wu, C. D. (2014) An odor interaction model of binary odorant mixtures by a partial differential equation method. *Sensors, 14*(7), pp. 12256–12270.

Zamora, M. C., Goldner, M. C. and Galmarini, M. V. (2006) Sourness-sweetness interactions in different media: White wine, ethanol and water. *Journal of Sensory Studies, 21*(6), pp. 601–611.

CHAPTER 15

Odour mixture coding from the neuronal point of view

Patricia Duchamp-Viret

Centre de Recherche en Neurosciences de Lyon (CRNL), UMR CNRS 5292, INSERM U1020, Université Lyon1, Lyon Cedex, France

15.1 Introduction

In neurophysiology, the odour is the percept which is initially born from the interaction between volatile molecules with olfactory receptor neurons (ORNs) laying in the nasal olfactory mucosa. In the everyday environment, most odours are multicomponent since raised by mixtures of chemical compounds which can be very different with regard to their volatilization properties. The immediate consequence is that the composition defining real olfactory objects can be quantitatively dominated by the most volatile compounds, while the least volatile ones are present in the vapour phase in a ratio being all the more low than their volatility is weak. However, despite the fact that concentration ratios are always quantitatively in favour of the most volatile compounds, less volatile compounds can also bear important qualitative notes of fragrance. The olfactory system, as a phylogenetically very ancient system, likely adapted to the chemical reality. Thus, faced to behavioural evidences of odorant detection at very low concentration, olfactory molecular receptors (OR) endowing ORNs must be hypothesized as having very high affinities for some stimuli, compensating for their very low vaporizing properties. In the real world of multicomponent odours, the most concentrated compounds are not always the ones which win and define the major note of an olfactory object (see Chapter 14).

15.2 Peripheral neural coding of mixtures

15.2.1 Pioneer studies

On the olfactory perception forefront are the ORs/ORNs. They process sensory information by integrating simultaneously, as soon as the first sniff, quantitative, qualitative and temporal features and send to higher levels a well-defined

Flavour: From food to perception, First Edition.
Edited by Elisabeth Guichard, Christian Salles, Martine Morzel, and Anne-Marie Le Bon.

description of stimulus attributes. They are also the first involved in detecting odour stimuli encountered on top of a background and in alerting for changes in the signal, over a wide range of background intensities. Altogether, the task of the olfactory peripheral level is critical and decisive, since it conditions all the subsequent processes.

In the beginning of the 80s, some of the olfaction researches in lobsters at ORN levels started to pay attention to the fact that the olfactory sensory system mainly processes complex stimuli (Derby and Ache 1984). Thus, by using mixtures of amino acids, the authors observed that ORNs' responses to multicomponent mixtures were not readily predictable from responses to individual components. These studies revealed that mixture interactions commonly occurred and pointed out that suppressive interactions prevailed on synergistic ones (Daniel and Derby 1991). Furthermore, by analyzing ORNs' responses to binary mixtures at different concentrations, the "direction" of a mixture interaction was proposed as depending on the concentration; synergism would rather occur at low mixture concentrations, while suppression would occur at high concentrations (Derby et al. 1991a, Derby et al. 1991b). Regarding intracellular mechanism of these suppressions, patch-clamp studies revealed that the inward current amplitude could decrease in response to mixtures, by comparison with the current induced by the most effective component (Michel and Ache 1994, Simon and Derby 1995): suppressions were here described for two excitatory or one excitatory and one inactive chemicals' binary combinations. In addition to such suppressions, different stimuli could activate either an inward (excitatory) or outward (inhibitory) conductance in the same ORN (McClintock and Ache 1989), and extracellular single-unit data support that this was a common mechanism in spiny lobsters (Derby et al. 1991a, Derby et al. 1991b, Michel and Ache 1994). In other invertebrates as in fishes, ORNs' extracellular inhibitory responses (Kaissling 1986) and intracellular hyperpolarization (Ivanova and Caprio 1993) were described in response to odorants, suggesting that activation of opposing ionic conductance may be a general mechanism of producing mixture suppression and therefore in shaping the responses to mixtures (Ache 1994). However, other mechanisms are necessary to explain suppression, described as occurring between two excitatory compounds (which both induced inward currents) or between one excitatory compound and an inactive one. This is why Simon and Derby (1995) proposed that such suppressions could engage either competitive or non-competitive mechanisms as well as inhibitory cross-talk between two transduction pathways.

In parallel to Derby's group studies, a dynamic approach of the odour mixture was proposed by Kurahashi et al. (1994) in newts. To do so, the currents elicited by two successive applications of two molecules (among which d-limonene, amyl acetate and acetophenone) were recorded in isolated ORNs. The paradigm consisted in delivering the second molecule once the response to the first one had started. In most cases, this resulted in transitorily interrupting the time-course of the first compound response. How can one odorant stimulus interact with ORNs during the time course of depolarization or inward current induced by another one? In agreement with Simon and Derby (1995), the authors hypothesized that the transduction-current timeless suppression could be due to a direct effect of odorants on ion channels or on second messengers of transductory pathways. In a similar protocol, Yamada and Nakatani (2001) applied one odorant against IBMX (3-isobutyl-1-methylxanthine), a specific and direct agonist of cyclic adenosine monophosphate (cAMP) pathway. They described that the odorant suppressed cAMP-induced depolarization or inward current, providing evidence that odorants can directly inhibit cyclic nucleotide-gated ion channels (CNGCs). They therefore concluded that most mixture suppressions are not based on the hyperpolarizing current genesis, that is, classic inhibition, but on the inward-current reduction or suppression.

15.2.2 Coding of binary mixtures in rats

In 2003, we published the first study on odour mixture coding in rats *in vivo* (Duchamp-Viret et al. 2003). Here, the choice of the stimuli set was founded on the hypothesis that some odorants may specifically elicit the cAMP transduction pathway, while others would elicit the PI3 transduction pathway (Breer and Boekhoff, 1991). Our binary mixtures were thus designed by associating one putative cAMP compound and one putative PI3 compound, aiming at observing interactions between the two transduction pathways. By doing that, we incidentally mixed a high saturated vapour pressure (SVP) compound with a low SVP compound (i.e a highly with a poorly volatile compound). Indeed, the three putative PI3 tested compounds, namely ethyl vanillin, lyral and lilial are compounds with low SVP (2.3 10^{-3} to 4.8 10^{-1} Pa at 25°C) while the others (limonene, anisol, isoamylacetate ...) are high SVP compounds (1.9 10^{2} to 7.6 10^{2} Pa).

In psychophysiology when the perceived intensities of single odours and mixtures are compared, three types of mixture interactions are defined: hypoaddition, suppression and synergy (Cain and Drexler 1974, Laing et al. 1984, Cometto-Muniz et al. 1999). In our studies, the "perceived intensity" parameter was replaced by the single ORN responses' spike-discharge frequency in order to make our observations fit these three categories. Thus, hypo-addition corresponded to the case where the mixture response frequency was equivalent to that induced by the most effective component. Suppression was defined when the mixture response frequency was lower than that induced by the most effective component or less than the response to either component; this corresponds to the "compromise" or "subtraction" levels of hypoaddition (Thomas-Danguin et al. 2014). Lastly, synergy was defined when the mixture response frequency was higher than the one induced by the most effective component; synergy comprising partial and complete additivity mechanisms termed by Cometto-Muniz et al. (1999). The interaction type could evolve over the stimulus concentration range. With our odour set, ORNs' responses frequently reflected hypoaddition for a given range of concentration and suppression or synergy for another range. On average, mixture responses reflected suppression for 20 to 30% of ORNs and synergy for 2 to 20%.

Since inhibitory responses to pure molecules are rare at the ORNs' level in rats (about 3% in Duchamp-Viret et al. 2003 and 5% in Duchamp-Viret et al. 1999), interactions in binary mixtures were mostly observed between two excitatory compounds or between one excitatory compound and an inactive one. Synergy implies at least that the two odour molecules do not counteract each other in binding processes and, at best that they work in positive cooperation. Nevertheless, synergy would be limited by the fact that the recorded neuron reached its maximum depolarization or by OR saturation. This may partly account for why synergy turned into hypoaddition at high concentrations. By contrast, suppression reflects negative cooperation between odorants, the less effective compound counteracting the more effective one.

In order to be closer to real stimuli for ORNs, Chaput et al. (2012) focused on two molecules known as bearing some of the fruity and woody notes of the red wine of Burgundy: isoamyl acetate (IAA) and whiskey lactone (WL), smelling like banana and coconut respectively. This pair in which the low SVP of WL (2.05 Pa) sharply contrasts with the high SVP of IAA (7.56 10^2 Pa) is described by human subjects, from a perceptual point of view, as evolving from the one compound dominance to that of the other one, depending on the ratio and concentration of each of the compounds. The experimentally induced-changes are assumed as reflecting natural aging of some Burgundy red wines, where aroma progressively shifts from rather fruity to woody dominating notes. In rats *in vivo*, our electrophysiological study of responses to IAA and WL (singly and in mixture) comprised recordings of single ORN and electro-olfactogram (EOG; mucosa field potential). In single recordings, ORNs were selected if they were excited by at least one of the two compounds. This aimed at observing interactions between one excitatory compound and one putative silent agonist (or antagonist) compound as described in several *in vitro* works (Oka et al. 2004, Sanz et al. 2005, Sanz et al. 2008; Ukhanov et al. 2010, Ukhanov et al. 2011). For WL and IAA, no inhibitory response was detected. Thus, given that IAA was far more effective than WL to induce ORNs' excitatory response, this selection criterion resulted in selecting all IAA and some IAA+WL responding ORNs. On the whole, IAA+WL mixtures' responses displayed about 40% of hypoaddition (mirroring IAA dominating action in mixture), 20% of synergy and 40% of suppression. Suppression exclusively reflected extensive suppressive action of WL on IAA responses and more specifically concerned ORNs showing no response to WL. The WL suppressive action was also observed in OR1G1-transfected HEK cells by calcium imaging studies (Chaput et al. 2012).

Because IAA has an about 370-fold higher SVP than WL, the use of different WL/IAA ratios de facto resulted in mixtures where IAA was the most concentrated compound, which is the ecological situation of all mixtures comprising these compounds, in wine in particular (Spillman 1997). In spite of this, IAA did not systematically dominate the mixture perception. Thus, similarly to what we observed for ethyl vanillin, lyral and lilial (Duchamp-Viret et al. 2003), the suppressive influence of WL cannot be merely ascribed to a quantitative superiority. As explained in the next paragraph, low SVP molecules can exert suppressive action and dominate an odour blend *via* syntopic interactions. Regardless of the underlying mechanism, the strong influence of WL on responses to IAA mostly matched the perceptive interferences between WL-related woody and IAA-related fruity notes in human subjects. Indeed, the fruity note perception is enhanced in dose-dependent manner by low concentrations of woody odorants and decreased by high concentrations of woody odorants while ORN responses to IAA are likewise reduced or enhanced depending on the OR type and/or the concentration of the WL in the mixture. Thus, our cell level study sheds light on the peripheral cellular mechanisms that may account for (i) the observed perceptual interactions between the fruity and woody notes in wines (Atanasova et al. 2005) and (ii) the subtle aromatic balance between these two notes observed for red wines (Thomas-Danguin et al. 2009, Moio et al. 1993).

Again *in vivo* in rats, aiming at observing interactions at ORNs' level of two compounds described as being able to bind the same OR, we chose the pair octanal/citral (SVP: 1.99 10^2 and 4.8 Pa). These odorants are described as acting as agonist or antagonist especially with the mouse ORI7 (Zhao et al. 1998, Araneda et al. 2000) and with human OR1G1 and OR52D1 (Sanz et al. 2005, Araneda et al. 2000). Noteworthy is that OR reactivity for octanol and octanal is rather similar (Sanz et al. 2005). We therefore postulated that octanol and octanal could be assimilated as similar excitatory agonists for ORs (Sanz et al. 2005). Thus, citral will be CIT and octanol/al will be OCT thereafter. In 80% of unit and EOG responses, OCT was found more potent to excite ORNs than CIT, as previously mentioned by Araneda et al. (2000) and Ukhanov et al. (2011, 2013) and congruently with OCT quantitative superiority in mixture. In our *in vivo* studies, CIT, although little effective, was observed as attenuating or suppressing (in about 60% of cases) both EOG (Figure 15.1) and single ORN's excitatory responses to OCT (Figure 15.2), and this in a ratio-dependent manner. However, CIT and OCT could also act in synergy (in 25% of recordings), at least over the middle concentration range. As we previously observed (Duchamp-Viret et al. 2003, Chaput et al. 2012), a salient and intriguing feature of our results is the fact that hypoaddition, suppression and synergy are observed in similar proportion at single ORNs' and at EOGs' levels for a given binary mixture. This means that the interaction types described at one single neuron level are representative of events occurring in broad ensembles of ORNs and thus widely concern the OR repertoire. Strikingly, EOG accurately reflects every facet of peripheral mixture processes. In addition, *in vitro* studies in human OR-transfected HEK cells mentioned that mixture interaction types can be more easily defined for a given odorant pair than for a given specific OR. For example, the pair OCT-methional was

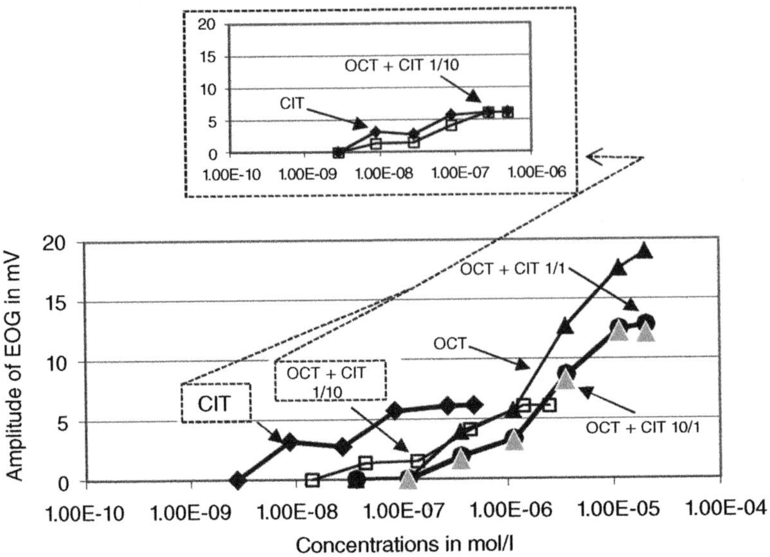

Figure 15.1 Curves of EOGs' amplitudes as a function of concentrations for OCT and CIT, singly and in mixture at different ratios. The curves show the CIT suppressive action on mixture response. This suppression is similar for 1/1 and 10/1 ratios whereas for the 1/10 ratio, EOGs reflect the dominating action of CIT. Indeed in the insert box, the two EOG curves, for CIT alone and OCT/CIT at 1/10 ratio, are similar when they are represented as a function of CIT concentrations alone. It is as if, at 1/10 mixture ratio, OCT effect was totally masked by the presence of CIT. As the mixture was designed in vapour phase, whatever the ratio and the concentrations to which compounds are delivered, CIT was always less concentrated than OCT. So the ratios 1/1 10/1 and 1/10 only proportionally modulated this difference but never resulted in equalizing or getting CIT more concentrated than OCT. These experimental conditions perfectly respected ecological conditions.

found to express synergy (according to our definition) regardless the OR, among OR1G1, OR2W1, OR52D1 or OR1A1 (El Mountassir and colleagues, personal communication). Altogether, this compels us to propose two key assumptions: first, a given binary mixture would be similarly processed by large ensembles of ORNs/ORs throughout mixture interaction widespread mechanisms, and second, mixture interaction could depend more on the binary mixture than on ORN/OR identity and properties.

15.2.3 Mechanisms hypothesized as supporting mixture suppressions in ORNs

How do the different molecules docking the same ORs/ORNs interfere to give rise to either suppression or synergy? Data modeling suggests that both competitive and non-competitive interactions occur at the OR level (Münch et al. 2013) and may account for the suppressive interactions (Figure 15.3). There is competitive interaction if two molecules bind to the same receptor site (= syntopy) and

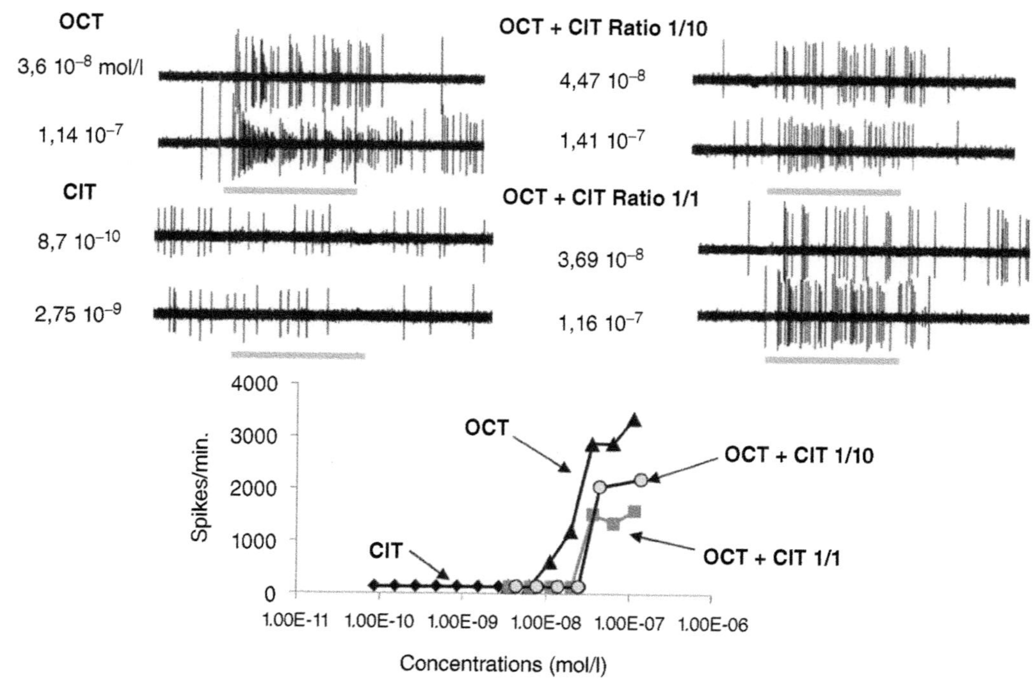

Figure 15.2 Single ORN recording *in vivo* in rats stimulated with OCT, CIT and OCT/CIT mixtures. The rough spike recordings as well as the concentration/response frequency curves show that, even though a silent actor when delivered alone, CIT suppressed in a ratio-dependent manner the OCT response in mixture. The grey horizontal bars under the single recordings indicate the 2-second stimulation delivery.

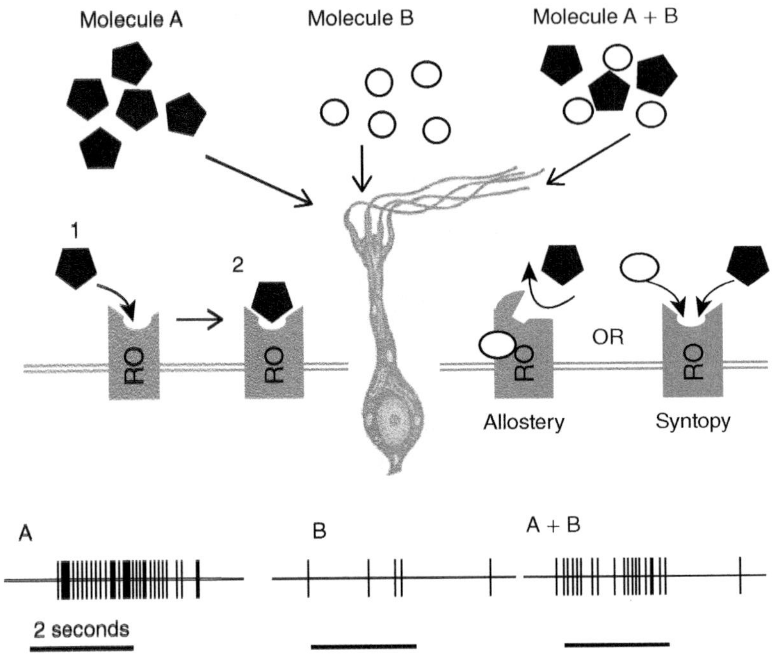

Figure 15.3 Hypothetical isosteric-syntopic (competitive) or allosteric interactions of two odorant molecules (A and B) at single ORN level. The two types of interactions can result in mixture suppression: the AB mixture response frequency is lower than that of the best compound (A). Here, when delivered alone, B is silent and thus unable to induce an observable ORN excitation. When delivered simultaneously to A, the silent actor B actively concurs in reducing the mixture ORN response.

if this interaction does not induce any conformation change (= isostery). The model of isosteric-syntopic interaction seemed to match about half of our mixture curves (Rospars et al. 2008). Competition could occur between two agonist odorants, that is, molecules that are able to activate the receptor, or one agonist and one silent compound (Spehr et al. 2002, Oka et al. 2004, Sanz et al. 2008, Jacquier et al. 2006). In contrast, for the other half of our mixture curves (Rospars et al. 2008), the shifts of curves along the concentration axis (response threshold changes) and the changes of maximal responses would indicate that noncompetitive interactions occur and modulate OR conformation (allosteric modulation). Allosterism is a common modulation in G-protein-coupled receptors (GPRC), including the rhodopsin family to which ORs belong (Gao and Jacobson 2006), and is consistent with the effects we observed (Duchamp-Viret et al. 2003; Rospars et al. 2008). In noncompetitive mechanisms linked to allosteric phenomena, one component of the mixture would be able to change the affinity (binding step) and/or the efficacy (activation step) of the OR for another component (Rospars et al. 2008). Whatever the step and, according to the direction of the changes following the noncompetitive modifications, allostery could account for both suppression and synergy. The probability of allosteric changes would increase logarithmically according to the number of components in mixture (0.51 for binary mixture, 0.66 for ternary, 0.76 for quaternary, etc.) and would be quite common with natural complex odours. Noncompetitive molecular interactions scale up the mixture combinatorial dimension, which may contribute to the emergence of synthetic attributes.

By which intracellular mechanisms one odorant would suppress the action of another? Spehr et al. (2002) showed that PI3K transduction pathway induced PIP3 release which negatively controls cyclic nucleotide-gated channels (CNGC) and could intervene in odorant suppression phenomena. CNGC would be the downstream target of PI3K-dependent suppression. How do odorants trigger cAMP rather than PI3K pathways (and conversely) and, how do the transduction pathways interplay? Ukhanov et al. (2010, 2011) tackled this question by studying single rats' ORNs responses by calcium imaging *in vitro* (Figure 15.4). The authors mainly used the odorant pair octanol/citral (OCT/CIT) and reported suppressive actions between these molecules similarly to our observations, and demonstrated that CIT, partially or completely, antagonized single ORN's OCT response. OCT was shown as engaging cAMP pathway (Figure 15.4A) whereas CIT engaged PI3K pathway. Indeed, when PI3K pathway was pharmacologically blocked, CIT lost the ability to reduce or suppress OCT excitatory action while OCT response was rescued (Figure 15.4B). Moreover, most unexpectedly, once the PI3K blocked, CIT became a cAMP excitatory agonist (Figure 15.4C). Because PI3K dependent inhibition and cAMP dependent excitation induced by CIT were both shown as being G-protein dependent, CIT necessarily acts through canonical OR. As a matter of fact, in accordance with Yamada and Nakatani (2001), Ukhanov et al. (2011) described how various pure molecules could reduce or inhibit cAMP-dependent responses (induced by IBMX/forskolin thus bypassing OR binding step) acting throughout the PI3K pathway. In 2013, Ukhanov et al expanded the reach of their results, by recording single ORNs excitatory responses to a large set of pure chemicals, in control condition and under PI3k blockers. They observed that blocking PI3K could result in an increase in amplitude of their responses.

Overall, observations at the periphery led to envision that suppressive action of one odorant molecule (silent or not) on one another (rather strongly excitatory) would be a common widespread component of peripheral coding, exerted regardless of OR/ORN entities. In a simple binary mixture, the two components could engage two signaling pathways acting in opposite directions on CNGC channels, thus on ORN polarization. By blocking the inhibitory action of PI3K pathway, the excitatory-response annihilated by the inhibitory compound could be restored; here, even if the ORN was stimulated with the binary mixture, it was as if only the excitatory odorant molecule was delivered. Far more unexpected was that, when the inhibitory compound wass alone delivered in ORNs where

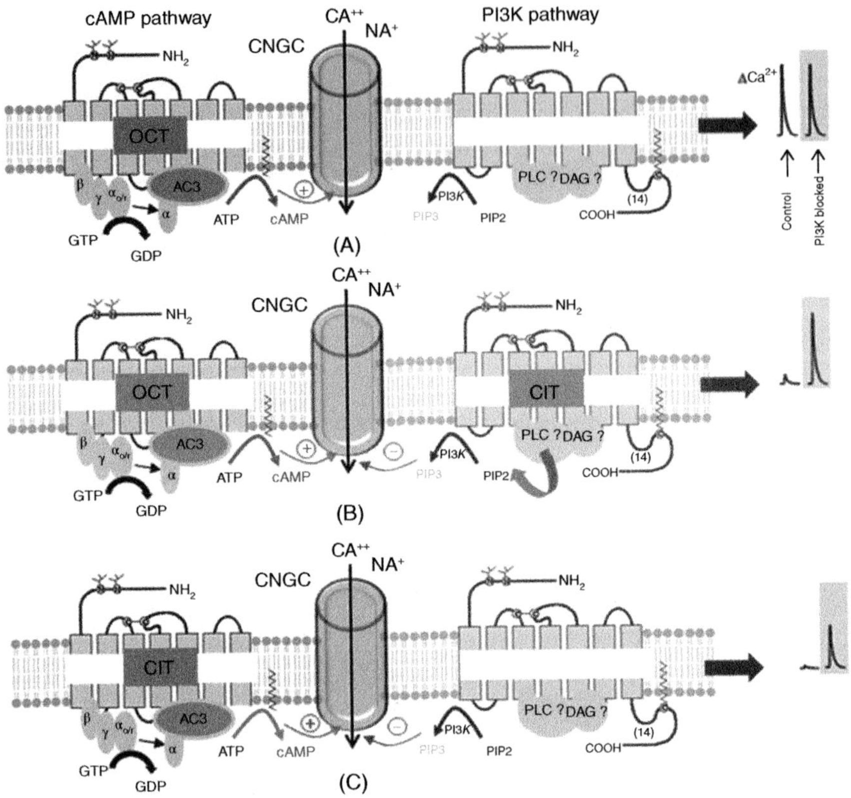

Figure 15.4 Hypothetical engagement of PI3K and cAMP signaling pathways in ORN response to odorants (Source: Adapted from Ukhanov et al. 2011). On the left, schematic representation of OR and signaling pathways. On the right, calcium excitatory responses of one isolated ORN: the left trace was obtained in control condition and the right one in presence of PI3K pharmacological blocker. **A**: OCT delivered alone is assumed to bind and excite the OR/ORN exclusively throughout the cAMP pathway. So the blocking of PI3 kinase (PI3K) pathway (symbolized by the rectangular grey mask) does not impact OCT response. **B**: OCT/ CIT mixture is assumed to engage both cAMP by OCT and PI3K by CIT, the latter wholly or partly suppressing the OCT response. In mixture condition, the pharmacological blocking of PI3K activity restores an excitatory response similar to this induced by OCT alone. **C**: CIT delivered alone is mainly observed to be a silent actor. However PI3K blocking revealed an excitatory response mediated through cAMP, demonstrating that CIT can become an excitatory actor by switching from PI3K pathway to cAMP if the former is unavailable.

PI3K pathway was blocked, its action shifted on cAMP pathway and became excitatory.

Altogether, the interplay mechanisms at docking binding OR level (Figure 15.3) and then at signaling pathway levels, offer a large variety of possibilities which may confer some synthetic aspect to the ORN processing and could explain why mixture interactions could be mainly qualified from the odorant pair identity, regardless of the OR. This result was rather unexpected compared to the literature in molecular biology which has advocated for years that the OR identity dictates all specific properties against ligands (Axel 1995).

One true question which remains is: how does the selection of PI3K (over cAMP) pathway occur regardless of the OR type? The striking fact that odorants can inhibit ORNs' responses by bypassing OR binding step supports a ligand-induced selective signaling (LiSS) concept (Millar and Newton 2010). LiSS was demonstrated for many G-protein-coupled receptors (GPCRs) to which ORs belong, and states that the nature of the ligand and the dynamically changing intracellular environment can alter the manner of signaling for many different GPCRs (Millar and Newton 2010). This implies that receptor would be not simply activated or inactivated, but would adopt many conformations, each of them leading to a highly specific ligand-receptor interaction and each conformation acting in turn with a specific intracellular signaling complex. Millar and Newton wrote: "The LiSS concept is creating a new level of sophistication, which challenges the dogma that ligand engagement of a GPCR consistently elicits a specific intracellular signal". However, the preponderance of a signaling pathway may stabilize one conformation rather than another, leading the receptor to increase its specificity for certain ligands. Such a receptor functioning appears of a particular interest for ORs which have to face limitless number of odorant molecular structures since it would offer many conformation possibilities. ORs could be therefore viewed as a signaling pathway gate (Ukhanov et al. 2011) where cAMP, the prominent signaling pathway, may be engaged or disengaged in favour of PI3K pathway, or conversely triggered according to "the nature of the ligand and the dynamically changing intracellular environment". Based on this functioning and our results, we propose that when CIT suppresses or decreases OCT responses, CIT might induce an OR specific conformation that would selectively result in engaging PI3K pathway which would block the excitation downstream of the OR. By contrast, when CIT acts with OCT in synergy, both of them may engage the cAMP pathway resulting in a response larger than that obtained for the best odorant alone. The LiSS concept is not only especially well adapted to ORs' functioning but further offer an additional margin of plasticity. According to this assumption, because both PI3K and cAMP pathways are functional in all ORNs, it may be further hypothesized that the qualification of an odorant as inhibitory, excitatory or dual, may change according to many circumstances, this offering a way for exerting metabolic influences, experience, learning, plasticity, and so on. on ORNs' reactivity. The LiSS concept may be also involved in epigenetic maturation of newborn ORN/ORs (Yu et al. 2004), and may support some plasticity phenomena impacting ORs/ORNs (Wysocki et al. 1977, Yee and Wysocki 2001, Wang et al. 1993, Cadiou et al. 2014). Such a LiSS concept may be also partly involved in the extreme adaptation reported in mice where OR expression was experimentally restricted to M71, which can behaviourally distinguish all stimuli except the favourite ligand of

M71 (Fleishmann et al. 2008). M71 would then trigger signaling pathways differently from it does in wild mice. Altogether these receptor and post receptor mechanisms linked to the GPRC functioning appear as extremely promising explanations for the sophistication of odorant processing as observed at ORN/OR level.

15.3 Olfactory bulb coding of mixtures

15.3.1 The glomerular map: relevant or irrelevant?

The olfactory bulb (OB), as the first relay, receives the primary sensory signal at level of discrete dense sub-spherical neuropil structures, called glomeruli. The projection rule of the ORNs to glomeruli is determined by a molecular addressing mechanism using ORs themselves and all ORNs expressing the same OR project to one or two glomeruli (Treloar et al. 2002). Many OB functional coding studies were based on the hypothesis that OB glomerular map relates the odour processing and, thus, attempted to interpret the spatial patterning as a chemotopic stimulus representation on which quality coding and discrimination would be based. This major interest in odour-induced glomerular patterns was logically reinforced by the fact that glomeruli seem to create apparent columnar functional organization of the OB output neurons connected thereto (Buonviso and Chaput 1990, Luo and Katz 2001).

A large set of studies showed that peripheral inputs, elicited by one pure molecule as well as by mixtures of odorant molecules, result in activating complex stereotyped activation patterns across the OB's input layer, the glomeruli (Carlsson et al. 2007, Cruz and Lowe 2013, Deisig et al. 2006, Fletcher 2011, Frederick et al. 2009, Grossman et al. 2008, Johnson et al. 2010, Krofczik et al. 2009, Kuebler et al. 2011, Meyer and Galicia 2012, Schubert et al. 2014, Tabor et al. 2004). Thus, in this body of research, the question was to know if OB mixture processing rather prioritizes the elemental odorants than engenders a global code, or the reverse, or ensures both. Preserving information about each elemental component would imply linear summation of single compound patterns. By contrast, if OB processes throughout nonlinear summation of single compound patterns, it gives birth to a completely novel quality from mixture. Of course, depending on compounds present in the mixture, all intermediate situations could be envisaged. Analysis of the bibliography reveals that data and conclusions do not lead to consensus.

A linear coding of stimulus complexity throughout glomerular patterning was reported in invertebrates (Carlsson et al. 2007, Deisig et al. 2006) as well as in vertebrates (Tabor et al. 2004, Grossman et al. 2008, Johnson et al. 2010, Fletcher 2011, Schubert et al. 2014). However, Grossman et al. (2008) noted that behavioural performances could not be easily correlated to glomerular spatial patterning. Important overshadowing maps for 2 monomolecular compounds could not be correlated to low discriminative abilities or, on the contrary, to a configural perception. The general consensus of the authors is that all glomeruli excited by single components were also excited by the mixture, and no new glomeruli were recruited. Differently, some other works in invertebrates (Lei and Vickers 2008, Capurro et al. 2012, Kuebler et al. 2011) as well as in vertebrates (Frederick et al. 2009, Cruz and Lowe 2013) rather support that glomeruli patterning rather reflects a nonlinear coding. High overlapping patterns induced by close monomolecular stimuli reflect hypo-additive or suppressive interactions with the binary mixture (Cruz and Lowe 2013). The latter work supports that mixture interactions would all the more occur for mixtures comprising chemically related compounds.

Peripheral suppressive and synergistic interactions engendered by mixture at ORNs' level (Duchamp-Viret et al. 2003; Chaput et al. 2012; Ukhanov et al. 2011) were expected to result in nonlinear combinations of the single compound glomerular patterns. Why would these changes be less obvious at the OB level? A possible reason may be ascribed to the high convergence ratio of ORNs onto glomeruli. Quantitatively, such a convergence allows increasing statistically mitral cells' (MC) sensitivities against those of ORNs (Duchamp-Viret et al. 1989). However, at the same time, by collecting numerous excitatory inputs, the convergence may result in an apparent loss of interactions (suppressive in particular) observed throughout single ORNs' responses. Thus, considering that ORN reshaping events cannot be denied, it is proposed that to observe them would require a finer scale of analysis. At such a scale, the glomerular-patterning activity which seems as organizing the OB processes throughout functional mitral cell columns might be not as relevant as it appears throughout anatomy. Such an hypothesis is supported by Soucy et al. (2009) who demonstrated that despite apparent glomerular spatial distribution of primary inputs, on a finer scale, no chemotopic order is apparent: "nearby glomeruli had strongly differing or nonoverlapping response properties", so nearby glomeruli are as diverse in their odor sensitivity as distant ones.

What might be the implications of having a non chemotopic order in glomeruli? First and foremost the role of OB lateral inhibitions must be deeply revisited, especially the notion of the center-surround contrast enhancement which would enhance the odorant quality (Schoppa, 2009). Fantana et al. (2008) also proposed that the lateral inhibition concept, dominating from the 1980s to the present day, should be abandoned or redefined as suggested by Pain et al. (2011). For these authors, mitral cells receive excitatory inputs in their glomerulus but only few inhibitory ones from neighbouring glomeruli, and OB sensory signal integration processes would not be based on glomerulus-mitral cell columns functional entities. This appears as fully consistent with a recent work showing that two different pure odours (hexanal and pentylacetate) induced spatially segregated activation in OB while activating the entire anterior piriform cortex, in all layers, with no distinguishable odour-specific areas

(Osmanski et al. 2014). On the whole, it is highly tempting to think that glomerular map per se is not relevant for deciphering OB odour coding and there is no reading central structure of this map. The absence of spatial obvious organization of functional meaningful entities shakes the notion encountered in many sensory brain areas and leads us to admit that OB network should be based on other principles. Finally, olfactory stimuli would not need spatial cues to be encoded, other than this inherent to anatomy constraints. Glomerular map would be only the consequence of the ORNs and OB neuron's wiring rules, which, being based on OR molecular addressing, ensures the wiring continuation in the lifelong learning of individuals, despite the continual renewing of ORNs. Based on this recognition, the question is how OB gets sharp and meaningful information from such sparse odour activations patterns in glomeruli.

15.3.2 What else at the single mitral cells' scale?

15.3.2.1 In invertebrates and anesthetized naïve mammals

First let us start with invertebrates. In the honeybee antennal lobe (AL), Krofczik et al. (2009) distinguished two types of projecting neurons (PNs, phylogenetically equivalent to mitral cells in vertebrates): lateral PNs which responses to mixture showed suppressive interaction between compounds, and median PNs which responses to mixture were dominated by the most effective compound. Similarly, in locust AL, Shen et al. (2013) also reported that most PNs' responses reflect hypoaddition, showing one component "preference". From Krofczik et al. (2009), lateral and median PNs would ensure elemental and synthetic coding, respectively. Another study in the honeybee AL (Meyer and Galizia 2012) reported that about half of neurons responded as "elemental neurons"; the other half responding as "configural neurons" but they were local neurons (LN) and not PN.

In the moth AL, Kuebler et al. (2011) reported that odour mixtures are represented by each cell as a unique combinatorial response, underlying the lack of general rule by which the network computes the mixture in comparison to single components. At the single PN or LN level, mixture responses exhibited several interaction types, mainly hypoaddition and suppression and rarely synergy, which establish an original neuronal representation. Again in moth, Capurro et al. (2012) mainly observed nonlinear interactions in responses to blend components.

Shen et al. (2013) improved knowledge at single AL neuron by examining numerous PN responses to mixtures of molecules classically used in olfaction, among which citral and octanol which were combined in various ratios. The authors examined the extent to which responses to binary mixtures could be predicted from component responses. Doing so, they showed that mixture response pattern smoothly morphed from the first component pattern to the second component one, reflecting more or less their respective ratio. However, mixtures PN responses deviated from the theoretical sums of the corresponding component responses, so that they mostly fit with response to their "favourite" component (i.e., inducing the strongest excitation), regardless of the respective ratio. In other words, component dominance or hypoaddition is the main interaction type observed in mixture PN responses. In a computational model, Shen et al. (2013) confirmed what Khan et al. (2008) had reported in vertebrates: when one odour is morphed into another, the population trajectory shifts gradually toward the other odour, passing through their 1:1 mixture trajectory. For more complex mixtures, Shen et al.'s computation further indicates that, over the qualitative space drawn from PNs' responses, odour representations are not randomly distributed: the more similar odours are, the more similar their corresponding representations. In rats (Giraudet et al. 2002) and fish (Tabor et al. 2004), MCs' responses to mixture are reported as being mostly hypoadditions. In most cases, the mixture pattern was closely similar to one of the component patterns (Giraudet et al. 2002). Wilson (2003) reported that MCs' responses to mixtures reflected mainly suppression and hypoaddition (72%) but also some synergy (28%).

Altogether, at OB single neuron level as well as at the periphery, the elemental mixture coding concept can be associated to hypo-additive mechanisms, reflecting some independence between compounds in ORNs' responses, whereas the synthetic coding concept would rather engage both suppression and synergy, real interactions impacting more obviously OB inputs.

15.3.2.2 In awake or experienced animals

In vertebrate OB, even at single MC level, meaningful information about mixture coding seems difficult to be extracted in anesthetized animals. In awake animals, odorant sampling is shorter (Gervais et al. 2007) and dynamic, since cadenced by the natural respiratory rhythm (Buonviso et al. 2006, Roux et al. 2006, Bathellier et al. 2008) and active sniffing (Kepecs et al. 2007, Courtiol et al. 2011). The shorter stimulus-sampling could result in much less stimulating MCs than in anaesthetized animals. This may explain, at least partly, why their responses appear briefer and less sustained (Rinberg et al. 2006).

In mice, Kato et al. (2012) reported that wakefulness lowers MC responses to stimuli which constitute more discriminative odour representations, throughout a diversification of their temporal dynamics, partly ascribable to a granule-cell activity enhancement. Also in awake animals, MC responses appear to be rather characterized by a re-organization of firing activity during respiratory cycles than by an absolute quantitative change in the firing rate (Gschwend et al. 2012). As proposed by several authors, breathing cycles, which temporally pre-format ORNs' inputs (Rospars et al. 2008), offer to MCs a basis for temporal realignment while OB intrinsic oscillations forming gamma and beta neuron assemblies could also be involved (Cenier et al. 2009). In this line, Gschwend et al. (2012) proposed that MC odour code would be based on fine temporal tuning of spiking activity within cell assemblies, which would transfer sensory information throughout burst sequences using the gamma oscillation as a clock. A dynamic timing of MC spike discharges regarding OB intrinsic gamma and beta oscillations was proposed as being involved in olfactory discrimination task, differently according to the animal experience (Ravel et al. 2003, Martin et al. 2004). In experienced mice, Kato et al. (2012) showed that MC tuning was specifically changed: responses to experienced odour were reduced compared to not experienced odours. Interestingly, these changes did not correspond to a definitive imprint: indeed, if odour experience-dependent plasticity is not maintained it could extinguish over months, the cycle being repeatable. This form of plasticity can be induced and observed only in awake animals.

Tackling again the experience impact, Doucette et al. (2011) recently reported that after a single training session, MCs appear as conveying both odour identity and its reward meaning, by coupling experience and precise sub-millisecond synchrony. Such a synchrony appears between a small fraction of spikes in MC pairs (on average about 0.09%). Noteworthy was that such pair synchrony could occur between MCs being at any distance apart in OB, this shaking a little more the notion of OB ordered-spatial map. Several authors favoured the view that precise synchrony would be mainly established by coincident excitatory centrifugal inputs coming from the olfactory anterior nucleus and cortex (Matsutani 2010, Restrepo et al. 2009). Thus, by operating concurrently and simultaneously, synchrony, rate- and sniff-locked coding would specify various stimulus features giving birth to a multiplexed code (Gire et al. 2013). Such a multiplexed code at different timescales appears especially relevant in the olfactory system. Indeed the latter has to face, due the high number or OR types (~1200 in mice), multidimensional stimuli which inputs are fragmented across OB (Cleland and Sethupathy 2006). Moreover, multidimensional stimuli must be linked to associational significance to arbitrary combinations of theses inputs; sniff-phase MC discharges may constitute dense identity coding and sparse synchrony may convey behavioural relevant information.

Sensory coding implies both discrimination and stable representation of stimuli. In anaesthetized rats, Khan et al. (2008) examined how MC responses changed according to different ratio of compounds in binary mixtures. To do so, the authors "morphed" from one odorant to another through a series of intermediate binary mixtures. This led them to formulate a model explaining multiple dimensions of odour identity, intensity, and combinations. The model mainly pleads in favour of additivity of different odours' contributions at the level of MCs, such additivity being attributed to the input stage rather than to intrinsic OB circuitry. During odour morphing, some MC responses to mixtures shifted with the respiratory phase, this phase-position providing neuron-level mechanisms for configural odour mixtures. By observing both simple additivity and novel time-locked representations of odour mixture, the authors concluded that MC responses to mixtures could morph from elemental to configural representations.

Could such morphing support the olfactory perceptual stability? Indeed, the olfactory system needs to discriminate but also to recognize and to categorize stimuli according to experience or familiarity, the latter process being an appropriate "generalization". Barnes et al. (2008) addressed this issue in the rat OB and cortex by using a ten-mixture laboratory blend. This blend was morphed by removing or replacing one of constituents. From a behavioural point of view, the 10-1 (1 removed constituent) morphed stimulus was considered as being similar to the conditioning stimulus whereas MC responses appeared as very sensitive to any change, and encoded it throughout spike patterning. Here, MCs' processing appears as favouring discrimination to the detriment of generalization. Differently, in insect antennal lobe (OB in vertebrates), PN neurons (MCs in vertebrates) can also favour generalization (Riffel et al. 2009). In *Manducasexta*, 3 and 6 key odorants were identified as enough to define 2 desert flowers (natural nectar sources for these insects) and elicited flight and feeding behaviours similar to those observed with natural essences. The authors therefore concluded that PN responses, throughout a combination of mean-rate coding and synchrony of firing, underlie generalization among effective mixtures but also maintain sufficient contrast for discrimination. Also in insects, the ordered representation of complex mixtures from PNs' responses described by Shen et al. (2013) could also reflect that some completion or generalization mechanisms could be already established in AL whereas some further sophistication could occur only in higher centers.

15.3.3 Conclusion on OB output neuron abilities to encode odour mixtures

Altogether, studies on OB mixture coding lead to conclude that OB does not host a spatial code per se whereas ORNs' inputs draw glomerular maps. Consensually, the meaningful information would be brought by temporal and dynamic MC activities elicited by sensory inputs. MC temporal activities would be modulated by the OB intrinsic network, synchronized on respiratory, gamma or beta oscillations' rhythm and highly under centrifugal control. Throughout their response patterns, single MC responses could encode both elementary information about mixtures' chemical composition or synthetic olfactory objects. Furthermore, if MC mixture responses throughout progressive patterns appear as being especially gifted to perform discrimination, they can also reflect some generalization. Because mammals dynamically sample odorants and can discriminate them in one breath, further understanding of OB mixture coding would require working in breathing conscious animals, especially since actual OB performance regarding odorant discrimination and plasticity phenomena seem to be expressed only in awake animals. In these conditions, MC responses could be more difficult to detect, and their relevant parameters less obvious than a simple rate-coding. However, a multiplexed code based simultaneously on MCs' time-locked firing and synchrony appears capable of ensuring simultaneously the odour quality coding and behavioural relevance.

15.4 What remains to be achieved by the olfactory cortex for giving birth to the perceptual object?

15.4.1 Response properties to monomolecular odorants

The mammalian piriform cortex (PCx), the largest sub-region of the olfactory cortex, receives direct synaptic input from OB and is likely the locus for the formation of the odour percept. MCs project broadly throughout the olfactory PCx without apparent topographic rule (Ghosh et al. 2011, Sosulski et al. 2011) in such a manner that single odours activate a distributed subpopulation of neurons across the cortex without spatial preference (Poo and Isaacson 2009, Rennaker et al. 2007, Stettler and Axel 2009; Zhan and Luo, 2010). The MCs' projection mode on PCx neurons is both spatially divergent (Ojima et al. 1984) and quantitatively convergent (Franks and Isaacson 2006). Because PCx samples all OB activities without obvious spatial bias (Sosulski et al. 2011), coding mechanisms should be based on different odour representations in OB and PCx (Carey and Wachowiak 2011, Shusterman et al. 2011, Miura et al. 2012). Stettler and Axel (2009) literally wrote: "PCx discards OB spatial segregation as well as chemotopy and returns to a highly dispersed organization in which different odorants activate unique ensembles of cortical neurons".

Compared to OB, relatively little is known about how monomolecular odour information is coded by neurons in the PCx and much less how odour mixtures are processed. Briefly, by using monomolecular stimuli, two categories of PCx neurons were reported: broadly tuned and narrowly tuned (Litaudon et al. 2003, Poo and Isaacson 2011). Most studies underlined that PCx odour representation is mostly governed by global inhibition and sparse activated neurons (Stettler and Axel 2009, Poo and Isaacson 2011). The global inhibition was attributed to broad and high convergence of MCs on PCx. Furthermore, Poo and Isaacson proposed that inhibitory synaptic activity would be largely exerted on PCx neurons and would contribute to their selectivity. Indeed the authors showed that blocking and agonizing GABAergic inhibition resulted in a broadening (2009) or a narrowing of PCx and neurons' response profiles (2011) respectively. So GABAergic inhibition clearly shapes the extent of excitation in PCx (Isaacson and Scanziani, 2011) throughout both feedback and feedforward circuits.

To understand how olfactory signals are processed and encoded in PCx, it appeared critical to study PCx neurons' responses in awake animals. In mice during behavioural experiments, Miura et al. (2012) observed responses in broadly distributed, moderately sparse neural population which ratios are fully consistent with other results in awake animals (Zhan and Luo 2010) and rather higher than these observed in anaesthetized animals (Poo and Isaacson 2009, Rennaker et al. 2007, Stettler and Axel 2009). Moreover, Miura et al. (2012) concluded that at individual PCx neurons' level, brief and early bursts, tightly locked to inhalation onset, support a rate-based code apparently clearer and simpler than in MCs: odour identity can be better decoded using spike frequency than latency or temporal patterns. Similar brief burst of PCx responses have been previously described (Duchamp-Viret et al. 1996, McCollum et al. 1991, Wilson 1998). Again in awake mice, Zhan and Luo (2010) could precisely categorize PCx neurons in 3 subtypes according to their response profiles. Type I (25%) were broadly excited (69%) and rarely inhibited (0.4%) and support an odour representation which contrasts to the concept of "sparse" representation pattern generally attributed to PCx. Type II (25%) exhibited only inhibition, supporting the odorant-evoked global inhibition noticed by Poo and Isaacson (2009). Type III (50%) responded extremely selectively. Thus, odour cortical representation appeared as accomplished by both broadly and narrow-tuned PCx neurons.

15.4.2 Response properties to mixtures

In 2003, in anaesthetized rats, Wilson recorded single anterior PCx neurons' responses to "laboratory bench" mixtures and reported that most neurons responding to single odorants responded to their mixtures. Similarly to ORNs and MCs, PCx neurons' mixture responses reflected either synergy or suppression. Suppression was observed more commonly (72%) than synergy (28%) exactly like, in OB. What Wilson termed as suppression included what we defined as being either hypoaddition or suppression in ORNs' mixture responses. In anaesthetized rats, Yoshida and Mori (2007) recorded single cortical neurons in a small area of the anterior PCx dorsal part in response to mixtures and to mixtures of mixtures. The authors used a panel of 8 different category mixtures, each gathering core odorants of foods (fruit and vegetable). Using these stimuli, two sets of neurons were distinguished: single and multiple category neurons. The former responded to only one mixture, the latter to a combination of mixtures. In terms of responsivity, 76% responded at least to one mixture category, 62% being excited and 14% inhibited. Among the excited neurons, 42% were single and 58% were multiple category neurons. For multiple category neurons, all combinations of the 8 stimuli were observed as defining their response profiles. In agreement with other studies, the authors observed that same PCx neuron excitatory responses to the combination of mixture could reflect synergy (Wilson 2003, Kadohisa and Wilson, 2006, Lei et al. 2006) or suppression (Wilson 2003, Kadohisa and Wilson 2006). The neuron excitatory response to a mixture could be suppressed by another one which evoked itself no response. Suppression dominated the mixture interaction type and the authors proposed that mixture synergy and mixture inhibition may enhance the category-profile selectivity of individual neurons.

Stettler and Axel (2009) analyzed PCx neuron responses to monomolecular odorants, laboratory bench mixtures or urine odour, by using two-photon calcium imaging in mice. Even if they could visualize several hundreds of neurons on the PCx, the authors failed to evidence any spatial segregation of responsive PCx neurons. 20-40% of PCx neurons could respond to a mixture whereas they did not respond to components. Shen et al. (2013) also reported, but did not show, such "mixture-specific" neurons. In their study, the neurons were Kenyon cells (KCs) in the mushroom body (phylogenetically equivalent to PCx). From Stettler and Axel (2009), the most prominent feature of PCx neurons' responses to mixture was also suppression; synergy being rare. Indeed, the authors observed numerous mixture-induced suppressions regardless the resemblance or dissemblance of paired odorants. For example, 90% of octanal responses were suppressed by alpha-pinene which itself elicited only 4% of excitation.

Which is of particular interest in the study of Kadohisa and Wilson (2006) is that the cortical mixture coding issue was immediately situated in relation to experience, memory and learning. To do so, they compared what happened when MC and PCx neurons were submitted to 10-50 sec mixture exposure (habituation paradigm) regarding their responsivity to single compounds. After such a habituation paradigm, MC responses reflected similarly self and cross habituation, regardless of the odorant molecule. Their responses were less robust to both components and mixture. Differently, PCx neurons showed a greater habituation when stimuli were associations of "alkane molecules". Moreover, after the habituation paradigm PCx neurons showed loss of reactivity for the mixture but remained excited to components and thus, became capable of discriminating between mixture and components. This ability increased with the habituation duration. The differences between MC and PCx neurons were interpreted in term of higher performances of PCx neurons to discriminate, thanks to their stimulating history which the authors assimilated to experience. Naive PCx neurons would not be better than MC ones and experienced PCx neurons would gain in competence. However it cannot be ruled out that some experience-dependent amelioration of performances could already take place in OB which functioning is also clearly impacted by learning and memory (Ravel et al. 2003, Buonviso and Chaput 2000). From Wilson (2003), in PCx neurons, 10 or 50 seconds of stimulation would be enough to establish synthetic representation of mixture and to discriminate it from individual components. This hypothesis involves very short-term plasticity mechanisms which could occur within cortical association fibers (Haberly 2001), mainly throughout cholinergic modulation (Hasselmo and Bower 1992, Hasselmo and Barkai 1995).

For Kadohisa and Wilson (2006), PCx is characterized by its rapid neural plasticity on which discrimination mechanisms' performances could mainly depend. So, implicit memory and learning would highly contribute to PCx processing in a way that odours are analyzed according to where, when and in which context their perception occurred. Wilson and Stevenson (2003) proposed that this functioning rather promotes a synthetic than an analytic coding, and that analytical feature discrimination is only an initial step. However, because the different molecules composing a scent simultaneously interact at ORNs' level and elicit mixture responses which are different to these induced by single compounds, it can be accepted that elemental information is compromised since the system's periphery. However, "compromised" does not mean lost. So, if physicochemical properties of molecules do not wholly dictate the odour percept, physicochemical properties constitute at least implicit information, if not explicit. As a matter of fact, most common tasks that mammals should achieve are mixture identifications and involve a synthetic treatment of olfactory information (gestalt recognition) rather than an elementary analysis of distinct odorants. Doing so, the olfactory mixture processing may partly involve the same neuronal resources associated with symbolic processing when single odours do not (Lorig et al. 2005). Nevertheless the mixture elemental information is kept along the entire olfactory pathways. Of course, each of us knows that to decompose an olfactory stimulus in elements is not easy and immediate, whereas to decompose a visual image in components is. However, after a specific training, the perfumers are perfectly able to dissect out or, on the contrary, to build a perfume complex composition using their memorized engrams of both mono-and multi-molecular odorant percepts.

The olfactory system discrimination abilities are high and could be exerted on very similar stimuli. However, if some tasks or behaviours require discriminating between very close odours, others require assimilating these similar odours as being the same general percept. The olfactory system abilities should balance the two main perceptual necessities which are stability (for recognition despite missing element) and discrimination (for relevant choice). Barnes et al. (2008) reported that PCx neurons reactivity changes are more important when one constituent was substituted in the mixture than when one was simply removed. The authors demonstrated that, despite the fact that single PCx neurons respond differentially to similar overlapping mixtures, the cortical neuron ensemble disregards this difference and functions consistently with pattern completion. As reported by Wilson (2009), the cortical pattern completion allows minor variation in the stimulus to be disregarded, a function that is critical for stability of olfactory percepts of complex stimuli. Thus, Barnes et al. (2008) showed that PCx neuron ensemble responses progressively de-correlated when the initial mixture was progressively depleted in components. However PCx neurons easily "refilled" one missing element, by processing some "pattern completion". By contrast, if one mixture element was replaced by a new one, the PCx perceptual balance became clearly in favour of discrimination: PCx neurons clearly detected the novel element by processing some "pattern separation". This suggests that one compound replacement in an odour mixture should be easily detected. The rats' (Barnes et al. 2008, Lovitz et al. 2012) and humans' (Ferreira et al. 2002) behavioural discrimination abilities confirm this suggestion.

From the literature emerge some common features. Thus, altogether PCx odour mixture coding is accomplished by both broadly- and narrowly-tuned neurons which could display weak or robust responses. PCx would ensure a relatively sparse odour coding, sparseness being often attributed to broad and unspecific inhibitions exerted on PCx neurons (Poo and Isaacson 2009). So, the wide and discontinuous features of receptive fields described by Stettler and Axel (2009) could be modulated. The sparse coding may be a fundamental strategy of both OB and PCx/OC (Davison and Katz 2007, Fantana et al. 2008, Laurent 2002, Rinberg et al. 2006), which would favour associative learning (Laurent 2002) and would allow to encode a virtually limitless number of stimuli. Moreover, sparse coding facilitates discrimination by reducing activated pattern overlaps (Laurent et al. 2001).

Regarding mixture responses, PCx neurons responses show mainly suppression (gathering hypoaddition and suppression according to data reported in ORNs) and rarely synergy. However, "mixture-specific" PCx neurons responses described by Stettler and Axel (2009) or Shen et al. (2013) likely involved synergy. The important features of PCx neurons may be linked to experience, learning and memory. PCx neurons' ensembles appear as capable of ensuring the perception stability throughout "completion" processing while it remains acute for discriminating the perception variability throughout "separation". Interestingly, the third stage of the olfactory system could also favour one component of complex stimuli and drove some neurons to respond to any mixture containing it (Shen et al. 2013). What is the involved process here, completion or discrimination? We propose that such neuron behaviour could be rather termed as a generalization which would be based on one component recognition.

Thus, overall, the olfactory cortex appears as an archeo-cortex which the pure associative processing (Wilson and Rennaker 2010) may be based on random convergent excitatory inputs from MCs (Stettler and Axel 2009). The associative processing would promote rather synthetic than analytical coding modes (Wilson and Rennaker 2010). The processing of complex stimulus patterns defines objects by associative circuits, allowing the robust stimulus recognition despite missing inputs and discrimination of overlapping patterns (Whitfield 1979).

15.5 Conclusion on mixture coding performances over the olfactory system

Unlike the visual and auditory systems, which process information from immutable physical objects, the olfactory system tracks, detects and discriminates an almost infinite number of stimuli representing the moving world of odours (Bargmann 1997). A very recent study, based on psychophysical test and mathematical extrapolations, has proposed that the human nose is able to discriminate at least one trillion of odour mixtures (Bushdid et al. 2014). This work further shows that mixtures are easier to discriminate when they increase in complexity. Such performances place the olfactory system far ahead of the vision and audition with several orders of magnitude more than distinguishable colours or tones.

Over the phylogenetic evolution, given the crucial involvement of olfaction in vital behaviours, the olfactory system has likely given priority to the functioning's sustainability and robustness. In addition, given the limitless number of potential stimuli, the olfactory system also gave priority to plasticity. Such priorities may account for some of the exceptional features which are expressed all along the olfactory pathways, from the periphery to cortices.

At the periphery, ORNs are continuously renewed along the lifetime of individuals and the renewal is likely paired with the OR expression plasticity. ORNs which are directly exposed to the ambient air could be damaged by toxic and harmful agents. To compensate, ORNs, which are of placodal origin, maintain in adult vertebrates a neurogenetic matrix that allows for their replacement without impairment of function, both in normal circumstances (turnover every 6-8 weeks) and after accidental or experimental injuries (Graziadei 1973). Moreover, during maturation, neo-ORNs would be "educated" by the surrounding adult ORNs' odour-activities which might direct their OR expression choice (Mombaerts 2004, Yu et al. 2004). Such a mechanism would ensure both continuation and plasticity. The next step is the wiring of ORNs to the OB. The latter appeared as based on OR addressing which ensures the OB input stability and can be further used to order the OB amygdala inputs or other limbic structures. One thing is certain: the biological reality generally chooses clever and frugal strategies. The dual role of ORs in odorant molecular binding and primary sensory axons wiring is consistent with such cleverness and frugality. Once in the OB, ORNs' axons form glomeruli where they synapse with MCs. The glomerular structuring is so robustly and intrinsically induced by ORNs' endings (Takeuchi and Sakano 2014) that, if OB is experimentally removed, they form ectopic glomeruli on the first brain targets they encounter (Graziadei et al. 1978, Graziadei and Samanen 1980). As at periphery, plasticity phenomena are also present in OB where, as in the hippocampus, a continuous renewal of interneurons which are critical actors in olfactory learning and memory occurs (Mandairon et al. 2011).

In OB, the ORNs' ordered wiring gives birth to an incident functional mapping. Because in humans, our dominating sense is vision, the obvious functional maps have long time focused our attention and, to date, it is still difficult to separate the wheat from the chaff. Functionally, the different ORs expressed are scattered all over the olfactory mucosa. The wiring rules throughout a quantitative convergence ensure that all ORNs project on the OB. Then, OB outputs project according to a widely quantitatively and qualitatively divergent mode on the olfactory cortex. Thus, to admit that OB hosts strict odotopic relevant glomerular maps would result in admitting that phylogenetical natural-selection mechanisms consecutively established the OB map and promptly dismantled it in the cortex. It is hardly conceivable and the recent literature comforts that from the periphery to the center, the olfactory meaningful information would be computed throughout dynamic and synchronized neuron specific assemblies, and not throughout a spatial map for reading. In this respect, in PCx, MCs project broadly over the olfactory piriform cortex thought convergence-divergence random projections. This absence of cortical topology added to the mosaic distribution of OR types over the mucosa reinforce the idea that the convergence ratio leading to a spatial glomerular map does not sign a spatial code *per se* but merely aimed at increasing statistically the sensitivity of MCs compared to that of ORNs (Duchamp-Viret et al. 1989) while simultaneously ensuring the OB wiring continuation.

The redundancy of odour representations over the OB network (Hudson and Distel 1987) and the PCx random processing both prevent injury consequences (which could greatly impact the system functioning) and emphasize the role of experience in odour processing. OB and OC are qualified as paleo- or archeo-cortex, depending on the authors, and may be seen as structures wherein the processing major roles are given to the experience, plasticity and memory. In the olfactory mucosa, the physico-chemical attributes of odorant compounds direct their access to ORNs and then their binding rules with ORs. ORs are not selective and display some affinity-dependent preferences for some odorants which more or less overlap between each other. This reality imposes a combinatory neural coding mode for pure odorant as for mixtures. For the latter, without subsequent processing, since each odorant can simultaneously interact with a large set of ORNs, some synthetic attributes of odorants could emerge notwithstanding the elemental information is conserved.

The limitless number of items defining the odorant world likely explains why ORs should be necessarily unselective, and why OB and PCx structures may match a pure associative processing. Let's admit for a while that the olfactory function would be based on absolutely not pre-formatted neuron networks (on a functional point of view) and let us follow the related line of reasoning. Since the organism birth, the olfactory hard-wiring, managed by OR-dependent immutable rules, ensures the anatomical stability of OB inputs. On this basis, the representation of the stimulus at the OB level would mainly depend on, and evolve with, the animal experience. In naive animals, the hard-wiring would ensure an olfactory perception since the first odorant molecule docks ORNs. No recognition, no discrimination, only detection and perception. Then, molecule by molecule, scent by scent, the olfactory experience would enrich by matching sensory information with encountered olfactory items and segregating odorous signals; relevant from irrelevant, toxic from innocuous, palatable from non-palatable, and so on. Progressively the olfactory software will develop, establishing in the OB and cortex, working networks highly adapted and adaptable, which would process the information according to the organism holistic point of view. If the plasticity of OB and cortex are immediately conceivable, it should be added that of ORNs' level which the OR expression could also be modulated by the environment (phenotypic plasticity: Wysocki et al. 1989, Wang et al. 1993, Cadiou et al. 2014). Of course, such a hypothesis although proposing OB and PCx processes entirely based on experience networks does not exclude some invariants which are imposed by wiring rules: reproducible glomerular maps can be observed in several animals. Altogether the ancient sense of olfaction is robust, suitably efficient, and especially highly adaptable to the changing extrinsic or intrinsic environment of living organisms, its remarkable properties being partly ensured by both phenotypical induced plasticity and neuron renewal.

References

Ache, B. W. (1994) Towards a common strategy for transducing olfactory information. *Seminars in Cellular Biology*, *5*(1), pp. 55–63.

Araneda, R. C., Kini, A. D. and Firestein, S. (2000) The molecular receptive range of an odorant receptor. *Nature Neuroscience*, *3*(12), pp. 1248–1255.

Atanasova, B., Thomas-Danguin, T., Chabanet, C., Langlois, D., Nicklaus, S. and Etievant, P. (2005) Perceptual interactions in odour mixtures: odour quality in binary mixtures of woody and fruity wine odorants. *Chemical Senses*, *30*(3), pp. 209–217.

Axel, R. (1995) The molecular logic of smell. *Scientific American*, *273*(4), pp. 154–159.

Bargmann, C. I. (1997) Olfactory receptors, vomeronasal receptors, and the organization of olfactory information. *Cell*, *90*(4), pp. 585–587.

Barnes, D. C., Hofacer, R. D., Zaman, A. R., Rennaker, R. L. and Wilson, D. A. (2008) Olfactory perceptual stability and discrimination. *Nature Neuroscience*, *11*(12), pp. 1378–1380.

Bathellier, B., Buhl, D. L., Accolla, R. and Carleton, A. (2008) Dynamic ensemble odor coding in the mammalian olfactory bulb: sensory information at different timescales. *Neuron*, *57*(4), pp. 586–598.

Breer, H. and Boekhoff, I. (1991) Odorants of the same odor class activate different 2nd messenger pathways. *Chemical Senses*, *16*(1), pp. 19–29.

Buonviso, N., Amat, C. and Litaudon, P. (2006) Respiratory modulation of olfactory neurons in the rodent brain. *Chemical Senses*, *31*(2), pp. 145–154.

Buonviso, N. and Chaput, M. (2000) Olfactory experience decreases responsiveness of the olfactory bulb in the adult rat. *Neuroscience*, *95*(2), pp. 325–332.

Buonviso, N. and Chaput, M. A. (1990) Response similarity to odors in olfactory bulb output cells presumed to be connected to the same glomerulus. Electrophysiological study using simultaneous single unit recordings. *Journal of Neurophysiology, 63*(3), pp. 447–454.

Bushdid, C., Magnasco, M. O., Vosshall, L. B. and Keller, A. (2014) Humans can discriminate more than 1 trillion olfactory stimuli. *Science, 343*(6177), pp. 1370–1372.

Cadiou, H., Aoude, I., Tazir, B., Molinas, A., Fenech, C., Meunier, N. and Grosmaitre, X. (2014) Postnatal odorant exposure induces peripheral olfactory plasticity at the cellular level. *Journal of Neuroscience, 34*(14), pp. 4857–4870.

Cain, W. S. and Drexler, M. (1974) Scope and evaluation of odor counteraction and masking. *Annals of the New-York Academy of Sciences, 237*, pp. 427–439.

Capurro, A., Baroni, F., Olsson, S. B., Kuebler, L. S., Karout, S., Hansson, B. S. and Pearce, T. C. (2012) Non-linear blend coding in the moth antennal lobe emerges from random glomerular networks. *Frontiers in Neuroengineering, 5*, p. 6.

Carey, R. M. and Wachowiak, M. (2011) Effect of sniffing on the temporal structure of mitral/tufted cell output from the olfactory bulb. *Journal of Neuroscience, 31*(29), pp. 10615–10626.

Carlsson, M. A., Chong, K. Y., Daniels, W., Hansson, B. S. and Pearce, T. C. (2007) Component information is preserved in glomerular responses to binary odor mixtures in the moth Spodoptera littoralis. *Chemical Senses, 32*(5), pp. 433–443.

Cenier, T., David, F., Litaudon, P., Garcia, S., Amat, C. and Buonviso, N. (2009) Respiration-gated formation of gamma and beta neural assemblies in the mammalian olfactory bulb. *European Journal of Neuroscience, 29*(5), pp. 921–930.

Chaput, M. A., El Mountassir, F., Atanasova, B., Thomas-Danguin, T., Le Bon, A. M., Perrut, A., Ferry, B. and Duchamp-Viret, P. (2012) Interactions of odorants with olfactory receptors and receptor neurons match the perceptual dynamics observed for woody and fruity odorant mixtures. *European Journal of Neuroscience, 35*(4), pp. 584–597.

Cleland, T. A. and Sethupathy, P. (2006) Non-topographical contrast enhancement in the olfactory bulb. *BMC Neurosciences, 7*, p. 7.

Cometto-Muniz, J. E., Cain, W. S., Abraham, M. H., and Gola J. M. (1999) Chemosensory detectability of 1-butanol and 2-heptanone singly and in binary mixtures. *Physiology and Behavior, 67*(2), pp. 269–276.

Courtiol, E., Hegoburu, C., Litaudon, P., Garcia, S., Fourcaud-Trocme, N. and Buonviso, N. (2011) Individual and synergistic effects of sniffing frequency and flow rate on olfactory bulb activity. *Journal of Neurophysiology, 106*(6), pp. 2813–2824.

Cruz, G. and Lowe, G. (2013) Neural coding of binary mixtures in a structurally related odorant pair. *Scientific Reports, 3*, p. 1220.

Daniel, P. C. and Derby, C. D. (1991) Chemosensory responses to mixtures: a model based on composition of receptor cell types. *Physiology and Behavior, 49*(3), pp. 581–589.

Davison, I.G. and Katz, L.C. (2007) Sparse and selective odor coding by mitral/tufted neurons in the main olfactory bulb. *Journal of Neuroscience, 27*(8), pp. 2091–2101.

Deisig, N., Giurfa, M., Lachnit, H. and Sandoz, J. C. (2006) Neural representation of olfactory mixtures in the honeybee antennal lobe. *European Journal of Neuroscience, 24*(4), pp. 1161–1174.

Derby, C. D. and Ache, B. W. (1984) Quality coding of a complex odorant in an invertebrate. *Journal of Neurophysiology, 51*(5), pp. 906–924.

Derby, C. D., Girardot, M. N., and Daniel, P. C. (1991a) Responses of olfactory receptor cells of spiny lobsters to binary mixtures. *I. Intensity mixture interactions. Journal of Neurophysiology, 66*(1), pp. 112–130.

Derby, C. D., Girardot, M. N., and Daniel, P. C. (1991b) Responses of olfactory receptor cells of spiny lobsters to binary mixtures. *II. Pattern mixture interactions. Journal of Neurophysiology, 66*(1), pp. 131–139.

Doucette, W., Gire, D. H., Whitesell, J., Carmean, V., Lucero, M. T. and Restrepo, D. (2011) Associative cortex features in the first olfactory brain relay station. *Neuron, 69*(6), pp. 1176–1187.

Duchamp-Viret, P., Chaput, M. A. and Duchamp, A. (1999) Odor response properties of rat olfactory receptor neurons. *Science, 284*(5423), pp. 2171–2174.

Duchamp-Viret, P., Duchamp, A. and Chaput, M. A. (2003) Single olfactory sensory neurons simultaneously integrate the components of an odour mixture. *European Journal of Neuroscience, 18*(10), pp. 2690–2696.

Duchamp-Viret, P., Duchamp, A. and Vigouroux, M. (1989) Amplifying role of convergence in olfactory system. A comparative study of receptor cell and second order neuron sensitivities. *Journal of Neurophysiology, 61*(5), pp. 1085–1094.

Duchamp-Viret, P., Palouzier Paulignan, B. and Duchamp, A. (1996) Odor coding properties of frog olfactory cortical neurons. *Neuroscience, 74*(3), pp. 885–895.

Fantana, A. L., Soucy, E. R. and Meister, M. (2008) Rat olfactory bulb mitral cells receive sparse glomerular inputs. *Neuron, 59*(5), pp. 802–814.

Ferreira, V., Ortin, N., Escudero, A., Lopez, R. and Cacho, J. (2002) Chemical characterization of the aroma of Grenache rose wines: aroma extract dilution analysis, quantitative determination, and sensory reconstitution studies. *Journal of Agricultural and Food Chemistry, 50*(14), pp. 4048–4054.

Fleischmann, A., Shykind, B. M., Sosulski, D. L., Franks, K. M., Glinka, M. E., Mei, D. F., Sun, Y., Kirkland, J., Mendelsohn, M., Albers, M. W. and Axel, R. (2008) Mice with a "monoclonal nose": perturbations in an olfactory map impair odor discrimination. *Neuron, 60*(6), pp. 1068–1081.

Fletcher, M. L. (2011) Analytical processing of binary mixture information by olfactory bulb glomeruli. *PLoS One, 6*(12), p. e29360.

Franks, K. M. and Isaacson, J. S. (2006) Strong single-fiber sensory inputs to olfactory cortex: implications for olfactory coding. *Neuron, 49*(3), pp. 357–363.

Frederick, D. E., Barlas, L., Ievins, A. and Kay, L. M. (2009) A critical test of the overlap hypothesis for odor mixture perception. *Behavioral Neuroscience, 123*(2) pp. 430–437.

Gao, Z. G. and Jacobson, K. A. (2006) Keynote review: allosterism in membrane receptors. *Drug Discovery Today, 11*(5-6), pp. 191–202.

Gervais, R., Buonviso, N., Martin, C. and Ravel, N. (2007) What do electrophysiological studies tell us about processing at the olfactory bulb level? *Journal of Physiology (Paris), 101*(1-3), pp. 40–45.

Ghosh, S., Larson, S. D., Hefzi, H., Marnoy, Z., Cutforth, T., Dokka, K. and Baldwin, K. K. (2011) Sensory maps in the olfactory cortex defined by long-range viral tracing of single neurons. *Nature, 472*(7342), pp. 217–220.

Giraudet, P., Berthommier, F. and Chaput, M. (2002) Mitral cell temporal response patterns evoked by odor mixtures in the rat olfactory bulb. *Journal of Neurophysiology, 88*(2), pp. 829–838.

Gire, D. H., Whitesell, J. D., Doucette, W. and Restrepo, D. (2013) Information for decision-making and stimulus identification is multiplexed in sensory cortex. *Nature Neuroscience, 16*(8), pp. 991–993.

Graziadei, P. P. and Samanen, D. W. (1980) Ectopic glomerular structures in the olfactory bulb of neonatal and adult mice. *Brain Research, 187*(2), pp. 467–472.

Graziadei, P. P. C. (1973) Cell dynamics in the olfactory mucosa. *Tissue and Cell, 5*, pp. 113–131.

Graziadei, P. P. C., Levine, R. R. and Monti-Graziadei, G. A. (1978) Regeneration of olfactory axons and synapse formation in the forebrain after bulbectomy in neonatal mice. *Proceedings of the National Academy of Science of USA, 75*, pp. 5230–5234.

Grossman, K. J., Mallik, A. K., Ross, J., Kay, L. M. and Issa, N. P. (2008) Glomerular activation patterns and the perception of odor mixtures. *European Journal of Neuroscience, 27*(10), pp. 2676–2685.

Gschwend, O., Beroud, J. and Carleton, A. (2012) Encoding odorant identity by spiking packets of rate-invariant neurons in awake mice. *PLoS One, 7*(1), p. e30155.

Haberly, L. B. (2001) Parallel-distributed processing in olfactory cortex: new insights from morphological and physiological analysis of neuronal circuitry. *Chemical Senses, 26*(5), pp. 551–576.

Hasselmo, M. E. and Barkai, E. (1995) Cholinergic modulation of activity-dependent synaptic plasticity in the piriform cortex and associative memory function in a network biophysical simulation. *Journal of Neuroscience, 15*(10), pp. 6592–6604.

Hasselmo, M. E. and Bower, J. M. (1992) Cholinergic suppression specific to intrinsic not afferent fiber synapses in rat piriform (olfactory) cortex. *Journal of Neurophysiology, 67*(5), pp. 1222–1229.

Hudson, R. and Distel, H. (1987) Regional autonomy in the peripheral processing of odor signals in newborn rabbits. *Brain Research, 421*(1-2), pp. 85–94.

Isaacson, J. S. and Scanziani, M. (2011) How inhibition shapes cortical activity. *Neuron, 72*(2), pp. 231–243.

Ivanova, T. T. and Caprio, J. (1993) Odorant receptors activated by amino acids in sensory neurons of the channel catfish *Ictalurus punctatus. Journal of General Physiology, 102*(6), pp. 1085–1105.

Jacquier, V., Pick, H. and Vogel, H. (2006) Characterization of an extended receptive ligand repertoire of the human olfactory receptor OR17-40 comprising structurally related compounds. *Journal of Neurochemistry, 97*(2), pp. 537–544.

Johnson, B. A., Ong, J. and Leon, M. (2010) Glomerular activity patterns evoked by natural odor objects in the rat olfactory bulb are related to patterns evoked to major components. *Journal of Comparative Neurology, 518*(9), pp. 1542–1555.

Kadohisa, M. and Wilson, D. A. (2006) Separate encoding of identity and similarity of complex familiar odors in piriform cortex. *Proceedings of the National Academy of Sciences of USA, 103*(41), pp. 15206–15211.

Kaissling, K. E. (1986) Chemo-electrical transduction in insect olfactory receptors. *Annual Review of Neuroscience, 9*, pp. 121–145.

Kato, H. K., Chu, M. W., Isaacson, J. S. and Komiyama, T. (2012) Dynamic sensory representations in the olfactory bulb: modulation by wakefulness and experience. *Neuron, 76*(5), pp. 962–975.

Kepecs, A., Uchida, N. and Mainen, Z. F. (2007) Rapid and precise control of sniffing during olfactory discrimination in rats. *Journal of Neurophysiology, 98*(1), pp. 205–213.

Khan, A.G., Thattai, M. and Bhalla, U. S. (2008) Odor representations in the rat olfactory bulb change smoothly with morphing stimuli. *Neuron, 57*(4), pp. 571–585.

Krofczik, S., Menzel, R. and Nawrot, M. P. (2009) Rapid odor processing in the honeybee antennal lobe network. *Frontiers in Computational Neurosciences, 2*, p. 9.

Kuebler, L. S., Olsson, S. B., Weniger, R. and Hansson, B. S. (2011) Neuronal processing of complex mixtures establishes a unique odor representation in the moth antennal lobe. *Frontiers in Neural Circuits, 5*, p. 7.

Kurahashi, T., Lowe, G. and Gold, G. H. (1994) Suppression of odorant responses by odorants in olfactory receptor cells. *Science, 265*(5168), pp. 118–120.

Laing, D. G., Panhuber, H., Willcox, M. E. and Pittman, E. A. (1984) Quality and intensity of binary odor mixtures. *Physiology and Behavior, 33*(2), pp. 309–319.

Laurent, G. (2002) Olfactory network dynamics and the coding of multidimensional signals. *Nature Reviews Neuroscience, 3*(11), pp. 884–895.

Laurent, G., Stopfer, M., Friedrich, R. W., Rabinovich, M. I., Volkovskii, A. and Abarbanel, H. D. (2001) Odor encoding as an active, dynamical process: experiments, computation, and theory. *Annual Review of Neuroscience, 24*, pp. 263–297.

Lei, H. and Vickers, N. (2008) Central processing of natural odor mixtures in insects. *Journal of Chemical Ecology, 34*(7), pp. 915–927.

Lei, H., Mooney, R. and Katz, L. C. (2006) Synaptic integration of olfactory information in mouse anterior olfactory nucleus. *Journal of Neuroscience, 26*(46), pp. 12023–12032.

Litaudon, P., Amat, C., Bertrand, B., Vigouroux, M. and Buonviso, N. (2003) Piriform cortex functional heterogeneity revealed by cellular responses to odours. *European Journal of Neuroscience, 17*(11), pp. 2457–2461.

Lorig, T. S., Malin, E. L. and Horwitz, J. E. (2005) Odor mixture alters neural resources during symbolic problem solving. *Biological Psychology, 69*(2), pp. 205–216.

Lovitz, A. M., Sloan, A. M., Rennaker, R. L. and Wilson, D. A. (2012) Complex mixture discrimination and the role of contaminants. *Chemical Senses, 37*(6), pp. 533–540.

Luo, M. and Katz, L. C. (2001) Response correlation maps of neurons in the mammalian olfactory bulb. *Neuron, 32*(6), pp. 1165–1179.

Mandairon, N., Sultan, S., Nouvian, M., Sacquet, J. and Didier, A. (2011) Involvement of newborn neurons in olfactory associative learning? The operant or non-operant component of the task makes all the difference. *Journal of Neuroscience, 31*(35), pp. 12455–12460.

Martin, C., Gervais, R., Chabaud, P., Messaoudi, B. and Ravel, N. (2004) Learning-induced modulation of oscillatory activities in the mammalian olfactory system: the role of the centrifugal fibres. *Journal of Physiology Paris, 98*(4-6), pp. 467–478.

Matsutani, S. (2010) Trajectory and terminal distribution of single centrifugal axons from olfactory cortical areas in the rat olfactory bulb. *Neuroscience, 169*(1), pp. 436–448.

McClintock, T. S. and Ache, B. W. (1989) Ionic currents and ion channels of lobster olfactory receptor neurons. *Journal of General Physiology, 94*(6), pp. 1085–1099.

McCollum, J., Larson, J., Otto, T., Schottler, F., Granger, R., and Lynch, G. (1991) Short-latency single unit processing in olfactory cortex. *Journal of Cognitive Neuroscience, 3*(3), pp. 293–299.

Meyer, A. and Galizia, C. G. (2012) Elemental and configural olfactory coding by antennal lobe neurons of the honeybee (*Apis mellifera*). *Journal of Comparative Physiology A, 198*(2), pp. 159–171.

Michel, W. C. and Ache, B. W. (1994) Odor-evoked inhibition in primary olfactory receptor neurons. *Chemical Senses, 19*(1), pp. 11–24.

Millar, R. P. and Newton, C. L. (2010) The year in G protein-coupled receptor research. *Molecular Endocrinology, 24*(1), pp. 261–274.

Miura, K., Mainen, Z. F. and Uchida, N. (2012) Odor representations in olfactory cortex: distributed rate coding and decorrelated population activity. *Neuron, 74*(6), pp. 1087–1098.

Moio, L., Schlich, P., Issanchou, S., Etiévant, P. and Feuillat, M. (1993) Description de la typicité aromatique de vins de Bourgogne issus du cépage Chardonnay. *Journal International des Sciences de la Vie et du Vin, 27*, pp. 179–189.

Mombaerts, P. (2004) Odorant receptor gene choice in olfactory sensory neurons: the one receptor-one neuron hypothesis revisited. *Current Opinion in Neurobiology, 14*(1), pp. 31–36.

Münch, D., Schmeichel, B., Silbering, A. F. and Galizia, C. G. (2013) Weaker ligands can dominate an odor blend due to syntopic interactions. *Chemical Senses, 38*(4), pp. 293–304.

Ojima, H., Mori, K. and Kishi, K. (1984) The trajectory of mitral cell axons in the rabbit olfactory cortex revealed by intracellular HRP injection. *Journal of Comparative Neurology, 230*(1), pp. 77–87.

Oka, Y., Nakamura, A., Watanabe, H. and Touhara, K. (2004) An odorant derivative as an antagonist for an olfactory receptor. *Chemical Senses, 29*(9), pp. 815–822.

Osmanski, B. F., Pezet, S., Ricobaraza, A., Lenkei, Z. and Tanter, M. (2014) Functional ultrasound imaging of intrinsic connectivity in the living rat brain with high spatiotemporal resolution. *Nature Communications, 5*, p. 5023.

Pain, F., L'heureux, B. and Gurden, H. (2011) Visualizing odor representation in the brain: a review of imaging techniques for the mapping of sensory activity in the olfactory glomeruli. *Cellular and Molecular Life Sciences, 68*(16), pp. 2689–2709.

Poo, C. and Isaacson J. S. (2009) Odor representations in olfactory cortex: "sparse" coding, global inhibition, and oscillations. *Neuron, 62*(6), pp. 850–861.

Poo, C. and Isaacson, J.S. (2011) A major role for intracortical circuits in the strength and tuning of odor-evoked excitation in olfactory cortex. *Neuron, 72*(1), pp. 41–48.

Ravel, N., Chabaud, P., Martin, C., Gaveau, V., Hugues, E., Tallon-Baudry, C., Bertrand, O. and Gervais, R. (2003) Olfactory learning modifies the expression of odour-induced oscillatory responses in the gamma (60-90 Hz) and beta (15-40 Hz) bands in the rat olfactory bulb. *European Journal of Neuroscience, 17*(2), pp. 350–358.

Rennaker, R. L., Chen, C. F., Ruyle, A. M., Sloan, A. M. and Wilson, D. A. (2007) Spatial and temporal distribution of odorant-evoked activity in the piriform cortex. *Journal of Neuroscience, 27*(7), pp. 1534–1542.

Restrepo, D., Whitesell, J. and Doucette, W. (2009) Need for related multipronged approaches to understand olfactory bulb signal processing. *Annals of the New-York Academy of Sciences, 1170*, pp. 298–305.

Riffell, J. A., Lei, H., Christensen, T. A. and Hildebrand, J. G. (2009) Characterization and coding of behaviorally significant odor mixtures. *Current Biology, 19*(4), pp. 335–340.

Rinberg, D., Koulakov, A., and Gelperin, A. (2006) Sparse odor coding in awake behaving mice. *Journal of Neuroscience, 26*(34), pp. 8857–8865.

Rospars, J. P., Lansky, P., Chaput, M. and Duchamp-Viret, P. (2008) Competitive and non-competitive odorant interactions in the early neural coding of odorant mixtures. *Journal of Neuroscience, 28*(10), pp. 2659–2666.

Roux, S. G., Garcia, S., Bertrand, B., Cenier, T., Vigouroux, M., Buonviso, N., and Litaudon, P. (2006) Respiratory cycle as time basis: an improved method for averaging olfactory neural events. *Journal of Neuroscience Methods, 152*(1-2), pp. 173–178.

Sanz, G., Schlegel, C., Pernollet, J. C. and Briand, L. (2005) Comparison of odorant specificity of two human olfactory receptors from different phylogenetic classes and evidence for antagonism. *Chemical Senses, 30*(1), pp. 69–80.

Sanz, G., Thomas-Danguin, T., Hamdani el, H., Le Poupon, C., Briand, L., Pernollet, J. C., Guichard, E. and Tromelin, A. (2008) Relationships between molecular structure and perceived odor quality of ligands for a human olfactory receptor. *Chemical Senses, 33*(7), pp. 639–653.

Schoppa, N. E. (2009) Making scents out of how olfactory neurons are ordered in space. *Nature Neuroscience, 12*(2), pp. 103–104.

Schubert, M., Hansson, B. S. and Sachse, S. (2014) The banana code-natural blend processing in the olfactory circuitry of *Drosophila melanogaster*. *Frontiers in Physiology, 5*, p. 59.

Shen, K., Tootoonian, S. and Laurent, G. (2013) Encoding of mixtures in a simple olfactory system. *Neuron, 80*(5), pp. 1246–1262.

Shusterman, R., Smear, M. C., Koulakov, A. A. and Rinberg, D. (2011) Precise olfactory responses tile the sniff cycle. *Nature Neuroscience, 14*(8), pp. 1039–1044.

Simon, T. W. and Derby, C. D. (1995) Mixture suppression without inhibition for binary mixtures from whole cell patch clamp studies of in situ olfactory receptor neurons of the spiny lobster. *Brain Research, 678*(1-2), pp. 213–224.

Sosulski, D. L., Bloom, M. L., Cutforth, T., Axel, R. and Datta, S. R. (2011) Distinct representations of olfactory information in different cortical centres. *Nature, 472*(7342), pp. 213–216.

Soucy, E.R., Albeanu, D.F., Fantana, A.L., Murthy, V.N., and Meister, M. (2009) Precision and diversity in an odor map on the olfactory bulb. *Nature Neuroscience, 12*(2), pp. 210–220.

Spehr, M., Wetzel, C. H., Hatt, H. and Ache, B. W. (2002) 3-phosphoinositides modulate cyclic nucleotide signaling in olfactory receptor neurons. *Neuron, 33*(5), pp. 731–739.

Spillman, P. J. (1997) Oak wood contribution to wine aroma. Thesis, Department of Horticulture, Viticulture and Oenology. University of Adelaïde, pp. 1–367.

Stettler, D. D. and Axel, R. (2009) Representations of odor in the piriform cortex. *Neuron, 63*(6), pp. 854–864.

Tabor, R., Yaksi, E., Weislogel, J. M. and Friedrich, R. W. (2004) Processing of odor mixtures in the zebrafish olfactory bulb. *Journal of Neuroscience, 24*(29), pp. 6611–6620.

Takeuchi, H. and Sakano, H. (2014) Neural map formation in the mouse olfactory system. *Cellular and Molecular Life Science, 71*(16), pp. 3049–3057.

Thomas-Danguin, T., Ishii, A., Etiévant, P. and Atanasova, B (2009) Le bouquet du vin. *Résultat d'une intégration sensorielle différente au nez et en bouche. Revue des Oenologues et des Techniques Vitivinicoles et Oenologiques, 132*, pp. 53–55.

Thomas-Danguin, T., Sinding, C., Romagny, S., El Mountassir, F., Atanasova, B., Le Berre, E., Le Bon, A. M. and Coureaud, G. (2014) The perception of odor objects in everyday life: a review on the processing of odor mixtures. *Frontiers in Psychology, 5*, p. 504.

Treloar, H. B., Feinstein, P., Mombaerts, P. and Greer, C. A. (2002) Specificity of glomerular targeting by olfactory sensory axons. *Journal of Neuroscience, 22*(7), pp. 2469–2477.

Ukhanov, K., Brunert, D., Corey, E. A. and Ache, B. W. (2011) Phosphoinositide 3-kinase-dependent antagonism in mammalian olfactory receptor neurons. *Journal of Neuroscience, 31*(1), pp. 273–280.

Ukhanov, K., Corey, E. A. and Ache, B.W. (2013) Phosphoinositide 3-kinase dependent inhibition as a broad basis for opponent coding in mammalian olfactory receptor neurons. *PLoS One, 8*(4), p. e61553.

Ukhanov, K., Corey, E. A., Brunert, D., Klasen, K. and Ache, B. W. (2010) Inhibitory odorant signaling in mammalian olfactory receptor neurons. *Journal of Neurophysiology, 103*(2), pp. 1114–1122.

Wang, H. W., Wysocki, C. J. and Gold, G. H. (1993) Induction of olfactory receptor sensitivity in mice. *Science, 260*(5110), pp. 998–1000.

Whitfield, I. C. (1979) The object of the sensory cortex. *Brain, Behavior and Evolution, 16*(2), pp. 129–154.

Wilson, D. A. (1998) Habituation of odor responses in the rat anterior piriform cortex. *Journal of Neurophysiology, 79*(3), pp. 1425–1440.

Wilson, D. A. (2009) Pattern separation and completion in olfaction. *Annals of the New York Academy of Sciences, 1170*, pp. 306–312.

Wilson, D. A. (2003) Rapid, experience-induced enhancement in odorant discrimination by anterior piriform cortex neurons. *Journal of Neurophysiology, 90*(1), pp. 65–72.

Wilson, D. A. and Rennaker, R. L. (2010) Cortical activity evoked by odors. in: Menini, A. , (ed.) *The Neurobiology of Olfaction*. Boca Raton (Florida, USA), pp. 353–366.

Wilson, D. A. and Stevenson, R. J. (2003) The fundamental role of memory in olfactory perception. *Trends in Neuroscience, 26*(5), pp. 243–247.

Wysocki, C. J., Dorries, K. M. and Beauchamp, G. K. (1989) Ability to perceive androstenone can be acquired by ostensibly anosmic people. *Proceedings of the National Academy of Sciences of USA, 86*(20), pp. 7976–7978.

Wysocki, C. J., Whitney, G. and Tucker, D. (1977) Specific anosmia in the laboratory mouse. *Behavior Genetics*, 7(2), pp. 171–188.

Yamada, H. and Nakatani, K. (2001) Odorant-induced hyperpolarization and suppression of cAMP-activated current in newt olfactory receptor neurons. *Chemical Senses, 26*(1), pp. 25–34.

Yee, K. K. and Wysocki, C. J. (2001) Odorant exposure increases olfactory sensitivity: olfactory epithelium is implicated. *Physiology and Behavior, 72*(5), pp. 705–711.

Yoshida, I. and Mori, K. (2007) Odorant category profile selectivity of olfactory cortex neurons. *Journal of Neuroscience, 27*(34), pp. 9105–9114.

Yu, C. R., Power, J., Barnea, G., O'Donnell, S., Brown, H. E., Osborne, J., Axel, R. and Gogos, J. A. (2004) Spontaneous neural activity is required for the establishment and maintenance of the olfactory sensory map. *Neuron, 42*(4), pp. 553–556.

Zhan, C. and Luo, M. (2010) Diverse patterns of odor representation by neurons in the anterior piriform cortex of awake mice. *Journal of Neuroscience, 30*(49), pp. 16662–16672.

Zhao, H. Q., Ivic, L., Otaki, J. M., Hashimoto, M., Mikoshiba, K. and Firestein, S. (1998) Functional expression of a mammalian odorant receptor. *Science, 279*(5348), pp. 237–242.

CHAPTER 16

Multisensory flavour perception

Charles Spence
Crossmodal Research Laboratory, Department of Experimental Psychology, University of Oxford, Oxford, UK

16.1 Introduction

Flavour perception is one of the most multisensory of our everyday experiences. Taste plays a central role, as does the smell (aroma) of food and drink. Tactile cues concerning the mouthfeel, texture, temperature, and trigeminal cues relating for example to the burning sensation associated with spicy foods are important too; On top of that, visual cues, primarily those related to food colour, sometimes exert a dramatic effect on the perceived identity and intensity of a food's flavour as well (see Spence, 2015, Spence et al 2010, for reviews). Finally, the sounds that we make when eating have also been shown to influence our perception of the crispiness, crackliness, and crunchiness of foods such as potato chips, crackers, and breakfast cereals (see Spence 2015a, for a recent review).

This article is designed to provide an overview of the role of multisensory integration in flavour perception in healthy adult humans. The interested reader is directed to Spence (2012) for a developmental perspective on multisensory flavour perception. Evidence concerning the integration of olfactory and gustatory cues is reviewed. Research showing that changing the colour of a wine can completely change the perceived aroma (an example of sensory dominance) will also be highlighted. Multisensorially-incongruency leading to foods that are rated as not tasting particularly good is an example of multisensory suppression (De Araujo et al. 2003, Skrandies and Reuther 2008, Spence and Piqueras-Fiszman 2014).

The contribution of multisensory integration to flavour perception in humans is an area that has seen a rapid growth of interest from the multisensory research community over the last decade or so (see Auvray and Spence 2008, Small 2012, Small and Prescott 2005, Spence and Piqueras-Fiszman 2014, Verhagen and Engelen 2006 for reviews). In fact, the techniques of cognitive neuroscience (such as neuroimaging, neurophysiology, computational modelling, and psychophysics) are increasingly coming to complement the more traditional techniques of sensory science when it comes to understanding why it is that food

Flavour: From food to perception, First Edition.
Edited by Elisabeth Guichard, Christian Salles, Martine Morzel, and Anne-Marie Le Bon.

tastes the way it does. The converging methodologies approach has led to a substantial advance in terms of our understanding of the neural substrates underlying multisensory flavour perception (see Small 2012, for a review).

According to the International Standards Organization (ISO 5492 1992, 2008), flavour can be defined as a "Complex combination of the olfactory, gustatory and trigeminal sensations perceived during tasting. The flavour may be influenced by tactile, thermal, painful and/or kinaesthetic effects." According to this definition, then, gustatory, olfactory, and trigeminal cues (which represent the activity of distinct interoceptive physiological systems), are the only senses contributing directly to the perception of flavour. Of course, this is not to say that oral-somatosensory, visual, and auditory cues can't modify a food's flavour, they most certainly can. It is just that they are not integral to it, at least not according to the ISO definition. However, this narrow definition is increasingly being questioned by researchers (Auvray and Spence 2008, Spence et al. 2015, Stevenson 2009). Part of the reason for this being the emergence of a growing body of empirical research, some of which will be reviewed here, demonstrating just how profoundly the visual attributes of food (such as its colour; Spence 2015), not to mention the sounds of consumption (Spence 2015a), can affect our perception of both the sensory-discriminative and hedonic attributes of food and drink.

When it comes to considering the role of olfaction in flavour perception, one critical point to note is that there are actually two relatively distinct sensory systems: The orthonasal, system associated with the inhalation of external odours and the retronasal system (involving the posterior nares), associated with the detection of the olfactory stimuli given off from the food we eat, as odours are periodically forced out of the nasal cavity when we chew or swallow food or drink. A number of differences between orthonasal and retronasal smell have been highlighted at both the subjective/perceptual level (Diaz 2004, Rozin 1982, Heilman and Hummel 2004), and in terms of the underlying neural substrates (Small et al. 2005, Small et al. 2008). Both orthonasal and retronasal odour perception are clearly important in terms of multisensory flavour perception. It is, though, striking to see how little attention researchers have traditionally given to thinking about whether the odours in their experiments ought to be presented orthonasally or retronasally (Bojanowski and Hummel 2012). More often than not, stimuli have been delivered orthonasally simply because it was much easier to do so. However, this approach may be more problematic than first realized, given the mounting evidence showing that substantially different patterns of multisensory integration can sometimes be observed as a function of whether an odour is presented *via* the orthonasal or retronasal route (Bult et al. 2007, Koza et al. 2005, Pfeiffer et al. 2005).

When thinking about the senses and their role in multisensory flavour perception, it can be helpful to distinguish between the exteroceptive senses of vision, audition, and orthonasal olfaction that are stimulated prior to the consumption of food, and the interoceptive senses of gustation, retronasal olfaction, trigeminal stimulation, oral-somatosensation, and the sounds that are experienced while eating and drinking (Stevenson 2009). Different neural mechanisms may be involved in these two cases (see also Small et al. 2008). One possibility here is that the multisensory integration of interoceptive flavour cues might be less open to cognitive penetration (i.e., it might be more automatic) than the combination of cues involved in interpreting exteroceptive cues (though see Stevenson 2012b). In the sections that follow, evidence concerning the role of each of the human senses in the sensory-discriminative aspects of multisensory flavour perception will be reviewed.

Given that gustation has been covered in some detail in a number of the other chapters in this volume, I will jump directly to looking at crossmodal, or multisensory, interactions between olfaction and gustation. Before I do though, it is perhaps just worth flagging up the fact that individual differences in taster status are potentially interesting in relation to the contents of this chapter in that they have been shown to exert some influence over certain aspects of multisensory flavour perception (Bajec and Pickering 2008, Bajec et al. 2012, Eldeghaidy et al. 2011, Essick et al. 2003, Hayes and Pickering 2012, Zampini et al. 2008). Several of these studies will be discussed in the relevant sections below.

16.2 Olfactory-gustatory interactions in multisensory flavour perception

While it is certainly true that taste (gustation) plays an important part in the multisensory perception of flavour, smell is also critical. In fact, olfaction is probably the more important sense, with some researchers claiming that olfaction contributes as much as 80-90% to what people normally report as flavour (Chartier 2012, Martin 2004). It should, though, be remembered that it is unclear whether there is actually any scientific evidence to back up this particular claim (see Spence 2015c, for a review). Furthermore, people often confuse the relative role of smell and taste to flavour perception (Davidson et al. 1999, Rozin 1982, Stevenson et al. 1999). Such confusions led some to claim that we may all be synesthetic for flavour (Stevenson and Boakes 2004, Stevenson and Tomiczek 2007, Verhagen and Engelen 2006; though see Deroy and Spence 2013, Spence et al. 2015).

Some of the most convincing evidence for the multisensory integration of orthonasal olfactory and gustatory cues in multisensory flavour perception comes from research conducted by Dalton et al. (2000). The participants in this particular study had to sniff two pairs of bottles containing clear solutions. One of the bottles contained benzaldehyde which has an almond-cherry like aroma, while the other three only contained odorless diluent. On each trial, the participants had to try and determine in which pair of bottles the bottle with benzaldehyde had been presented. The concentration of the olfactant was varied on a trial-by-trial basis in order to determine each participant's nasal

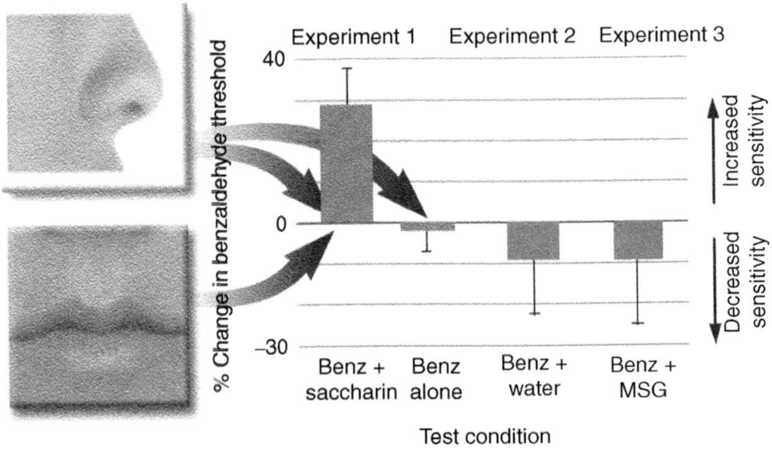

Figure 16.1 This figure highlights the results of a series of experiments conducted by Dalton and colleagues, showing the integration of orthonasal olfactory and gustatory cues (Source: Adapted from Dalton et al. 2000). When a sub-threshold saccharin solution was placed on the participant's tongue, a significant increase in olfactory sensitivity (to Benzaldehyde; Benz) was observed, despite the fact that the tastant had no odour, thus demonstrating the multisensory interaction of olfaction and taste. By contrast, holding a small amount of water in the mouth had no effect on olfactory thresholds, nor did holding a solution that happened to contain monosodium glutamate (MSG). These latter results highlight the stimulus dependency (both olfactory and gustatory) of multisensory integration in human flavour perception, a result that has now been demonstrated in many other studies.

detection threshold. In one experiment, the participants had to perform this olfactory discrimination task while holding a 10 ml sub-threshold concentration solution of saccharin in their mouths. The saccharin solution had no detectable taste or, importantly, odour. Under such conditions, the benzaldehyde was perceived as more intense, and the participants' sensitivity to the odorant went up (see Figure 16.1), when assessed relative to a baseline condition in which no tastant was present. The results of a follow-up study revealed that when the participants kept a little water in their mouth instead, no such change in olfactory detection thresholds was observed. Similarly, holding a sub-threshold solution of monosodium glutamate (MSG) in the mouth had no effect on orthonasal olfactory detection thresholds either. These results therefore demonstrate that orthonasal olfactory and sub-threshold gustatory stimuli are subject to multisensory integration. However, this effect appears to be specific to the particular combination of olfactants and tastants used.

This study provides convincing evidence that sub-threshold gustatory cues can enhance orthonasal odour perception. Similar results have now been reported in several other studies: For instance, Pfeiffer et al. (2005) reported a 50% lowering in the olfactory threshold (that is, complete additivity was observed) in three-quarters of the participants they tested, provided that the sub-threshold tastant and orthonasal olfactory stimuli (sucrose and benzaldehyde, just as in Dalton et al.'s research) were presented simultaneously. Pfeiffer et al. (2005) used a computer-controlled olfactometer and gustometer in order to deliver the stimuli. Similar effects were observed when the odour was delivered retronasally at the same time as the tastant. Meanwhile, Delwiche and Heffelfinger (2005) reported similar multisensory effects with solutions actually being tasted by the participants.

This kind of multisensory integration has been shown to depend on both the particular combination of stimuli used, and on the cultural background of the participants studied: For instance, Japanese participants have been shown to exhibit multisensory enhancement effects for the combination of MSG and benzaldehyde (Breslin et al. 2001, Spence 2008), but not for the combination of saccharine and benzaldehyde. Note that this is the opposite pattern of results to that shown by the North American participants in Dalton et al.'s (2000) seminal study. Why should this be so? Well, while the typical Western consumer is frequently exposed to the combination of a sweet taste and an almond-cherry odour in desserts such as Bakewell tart, combining a salty taste with the almond smell is far less common over there. In Japan, the salty taste - almond odour combination is far more common in foods such as pickled condiments.

According to Labbe et al. (2007), subliminally presented olfactory stimuli can also enhance sweetness perception. The fact that such effects occur for sub-threshold stimuli rules out a cognitive/expectancy-based account and the data are thus more consistent with an explanation in terms of multisensory integration (Dalton et al. 2000, Labbe et al. 2007). The results that have been reported to date with sub-threshold gustatory stimuli, highlighting the importance of stimulus congruency, map on to earlier results demonstrating similar multisensory interactions between congruent suprathreshold combinations of odour and taste. For instance, Frank and Byram (1988) reported that a strawberry odour enhanced the sweetness of sucrose delivered in a whipped cream substrate that the participants swallowed (in order to enhance the retronasal olfactory hit) while the odour of peanut butter did not have any such effect. That is, stimulus congruency (defined by Schifferstein and Verlegh 1996, as the extent to which two stimuli are appropriate for combination in a food product) appears to play an important role at both the sub- and supra-threshold levels. Interestingly, Labbe et al. (2006) have shown that odour congruency also plays a role in people's perception of bitterness in real food products, such as a cocoa beverage (see also Lawrence et al. 2009 on salt perception).

Taken together, then, the results reported in this section support the view that no matter where a person was born (or where they grew up), their brain uses the same rules of multisensory integration in order to combine the olfactory and gustatory (and presumably also oral-somatosensory) cues that contribute to the perception of flavour. However, the specific combination of tastants and olfactants (and, as we will see below, colour cues) that are integrated in a seemingly superadditive manner will depend on the particular ingredients that tend to be combined in the cuisine that the participant is familiar with. It turns out that certain of our preferences for specific flavours are established while we are still in the womb. Newborn babies have, for instance, been shown to prefer those odours that are associated with the foods that were consumed by their mothers while pregnant (see Schaal and Durand 2012, for a review).

16.3 Oral-somatosensory contributions to multisensory flavour perception

The oral-somatosensory system plays a crucial role in multisensory flavour perception, telling us about the temperature of a food, as well as its texture (Szczesniak 2002), not to mention the burning sensation associated with eating chili (Lawless et al. 1985). Tactile cues appear to play an important role in helping to localize flavour experiences to the oral cavity (Todrank and Bartoshuk 1991). Research over the last decade has shown that the oral texture (or mouthfeel) of what we eat and drink also influences multisensory flavour perception (Bult et al. 2007, Frost and Janhoj 2007, Weel et al. 2002). While the results of early studies often led to the suggestion that increased viscosity reduced flavour perception, it has always been difficult to determine whether such effects had a physicochemical or psychological/neurological origin (since increased viscosity is likely to reduce volatility at the food-air interface). However, recent technological advances have allowed researchers to demonstrate the genuinely psychological nature (of at least a part) of this crossmodal effect.

Bult et al. (2007), for instance, conducted an elegant study in which a creamy odour was presented either orthonasally or retronasally using a computer-controlled olfactometer. At the same time, milk-like foods with different viscosities were delivered to the participant's mouth. The participants had to rate the intensity of the flavour, as well as the thickness and creaminess of the resulting in-mouth experience. The key result to emerge from this research was that the perceived flavour intensity decreased as the viscosity of the solution increased. This was true regardless of whether the odour was presented orthonasally or retronasally. Given this study's independent control of texture and odour delivery, the results highlight the key role that texture (mouthfeel) plays in driving multisensory flavour perception. Bult et al.'s results also suggest that the presence of a retronasal odour could alter the perceived thickness of a foodstuff in the mouth (Stevenson and Mahmut 2011, Sundqvist et al. 2006, Tournier et al. 2009).

Everyone knows about the ventriloquist: the illusionist who projects his/her voice to the articulated lips of a nearby dummy. This illusion provides an example of the visual capture of auditory localization (Alais and Burr 2004). A similar effect may also occur in the mouth whenever a food or drink is consumed. That is, the perceived localization of a tastant has been shown to follow the location of a tactile stimulus as it is moved across the tongue (Green 2002, Lim and Green 2008, Murphy and Cain 1980, Todrank and Bartoshuk 1991). Given the pronounced differences in transduction latencies between the senses (Spence and Squire 2003), the tactile sensations associated with eating and drinking will normally arrive centrally (in the human brain) some time before either the associated gustatory or olfactory stimuli. Hence this 'prior entry' of the tactile signal may also play a role in localizing the resulting multisensory flavour experience to the mouth as well (see also Pfeiffer et al. 2005; Small et al. 2005). Recently, it has also been suggested that attention and stimulus congruency may also play an important role in the oral localization of our flavour experiences (Lim and Johnson 2011, 2012, Stevenson 2012b, Stevenson et al. 2011). Recently, researchers have started to investigate whether, knowing about the rules of multisensory integration, one can take flavour experiences out of the mouth (Michel et al. 2014).

Another kind of crossmodal interaction involves the interaction between temperature and taste. It turns out that roughly 33-50% of the population experience what is known as the 'thermal-taste' illusion (Cruz and Green 2000, Green and George 2004). Green and his colleagues have elegantly demonstrated that raising or lowering the temperature at various points on the tongue can elicit sensations of sweetness, sourness, saltiness, and bitterness—that is, all four basic tastes. What is more, Bajec and Pickering (2008) have shown that those individuals who experience the thermal-taste illusion tend to experience other tastes as being more intense as well.

Having reviewed the evidence concerning the multisensory integration of gustatory, orthonasal and retronasal olfactory, and oral-somatosensory cues contributing to the multisensory perception of flavour, I will now move on to look at the crossmodal influence of visual and auditory cues.

16.4 Eating with our eyes: on the visual modulation of multisensory flavour perception

Visual cues play an important role in multisensory flavour perception. In fact, more than 150 studies have now been published on this topic (see Spence 2015, Spence et al. 2010, for reviews). The research that has been published to date certainly supports the view that people's judgment of the identity of a food's taste, aroma, and flavour can all be influenced by changing the colour of the foodstuff that people happen to be evaluating. The latest evidence even suggests that colour cues may contribute to the expectation (and hence perception) of piquancy (Shermer and Levitan 2014). Interestingly, those studies that have investigated the effect of varying the intensity of the colour added to a food (typically a coloured beverage) have revealed rather mixed results (see Spence et al.

2010, for a review). As yet, far less research has attempted to investigate the influence of other visual appearance cues, such as, for example, the opacity of a drink (Spence 2015).

One of the classic studies demonstrating the influence of colour on taste detection thresholds was conducted by Maga (1974). He investigated the consequences on perceptual thresholds for the four basic tastes (sweet, sour, salty, and bitter) of colouring aqueous solutions red, yellow, or green. The concentration of the tastant often had to be increased in order for participants to be able to correctly detect its presence in a coloured, as compared to a clear, solution. The addition of yellow colouring to a sweet solution, decreased the taste detection threshold significantly while the addition of green colouring increased it (interestingly red had no effect; Frank et al. 1989). No matter whether yellow, green, or red colouring was added to sodium chloride solutions, it had no effect on the detection threshold. The addition of red colouring reduced the threshold to a bitter-tasting solution, while the addition of yellow and green colouring had no such effect. With respect to the sour taste detection threshold, the addition of either yellow or green colouring significantly decreased participants' threshold, with red colouring again having no effect. Interestingly, taste detection thresholds for the salt solutions were unaffected by colouring, perhaps because salty foods come in so many colours.

Morrot et al. (2001) conducted an oft-cited study in which they investigated how colour affected people's perception of wine aroma. These researchers 'fooled' more than 50 students enrolled on a university wine degree course in Bordeaux, France into believing that they were holding a glass of red wine, simply by colouring a white Bordeaux wine artificially red (using an odourless food dye)! The participants were first given a glass of white wine and instructed to describe its aroma having sniffed it. Next, they were given a glass of red wine and requested to do the same. Different terms were used by the students to describe the aromas of the two wines – terms like citrus, lychee, straw, and lemon for the white wine and chocolate, berry, tobacco, etc for the red wine. Finally, the students were given a third glass of wine, and were once again asked to describe the aroma. This glass looked like red wine but was, in fact, the same white wine that they had been given originally, but now coloured so as to make it look indistinguishable from a red. Surprisingly, the wine was described using red wine odour descriptors again. That is, they apparently no longer perceived the aromas in the coloured wine that they had previously reported when drinking the untainted white wine. This result powerfully demonstrates vision's dominance over orthonasal olfaction (see also Parr et al. 2003, for similar results from New Zealand). One interesting question here that has yet to receive a satisfactory answer is whether the experts would have been fooled by the colour if they had been informed in advance that the colour might have been misleading (i.e., rather than just being instructed to ignore the colour Zampini et al. 2007, 2008).

Moving away from the study of wine, many other researchers have demonstrated a similarly profound effect of visual cues on people's olfactory discrimination performance (Stevenson and Oaten 2008) and on people's identification of various beverages (Blackwell 1995, Davis 1981, Shankar et al. 2010c, Zellner et al. 1991, Zellner and Durlach 2003), as well as on their ratings of odour intensity (Zellner and Kautz 1990, Zellner and Whitten 1999).

The story regarding vision's influence over odour perception has, however, been complicated somewhat by the results of a study by Koza et al. (2005). These researchers found that colour had a qualitatively different effect on the perception of orthonasally- versus retronasally-presented odours associated with a commercially-available mixed fruit flavoured drink. In particular, colouring the solutions red led to odour enhancement in one group of participants when an odour was sniffed orthonasally, while leading to reduced odour intensity when presented retronasally (to another group of participants). Koza et al. tried to explain these results by suggesting that it may be more important to correctly evaluate foods once they have entered our mouths, since they pose a greater risk of poisoning. By contrast, the threat of poisoning from foodstuffs outside the mouth is less severe. Thus people may simply attend more to stimuli originating from within our bodies as compared to those situated elsewhere, and that this influenced the pattern of sensory dominance that was observed. Nevertheless, should Koza et al.'s findings be replicated (preferably using a within-participants experimental design), they would certainly add weight to the argument that qualitatively different patterns of multisensory integration/perception can be observed following the presentation of orthonasal as compared to retronasal odours.

Given the significant effect that colour has been shown to exert on both taste detection sensitivity (Maga, 1974) and on various aspects of odour perception (Blackwell 1995, Davis 1981, Koza et al. 2005, Shankar et al. 2010b, 2010c), it should come as little surprise that colour cues also exert a robust effect on people's flavour identification responses. For example, the participants in DuBose et al.'s (1980) classic study had to try and identify the flavours of a variety of differently-coloured fruit-flavoured drinks. However, while certain of the colour-flavour pairings were assumed to be 'appropriate' (e.g., a cherry-flavoured drink coloured red), others were deemed to be 'inappropriate' (e.g., as when a lime-flavored drink was coloured red; though see Shankar et al. 2010a, on the problematic notion of colour 'appropriateness'). DuBose et al. reported that participants misidentified the flavours of many of the drinks when coloured inappropriately. The results suggested that participants' incorrect answers were often driven by the colours of the drinks. That is, participants often made what might be classed as visual-flavour responses. So, for example, 26% of the participants reported that a cherry-flavoured drink tasted of lemon/lime when coloured green, as compared to no lime-flavour responses when the drink was coloured red or orange instead. Meanwhile, 19% of the participants detected an orange flavour when the drink was coloured orange, but none of the participants made such a response when the drink was coloured red, green, or was presented as a colourless drink.

Similar results have been reported in a number of other studies over the years (see Spence 2015, Spence et al. 2010, for reviews). An interesting, and as yet unanswered, question here though is why only a proportion of participants' responses are seemingly influenced by the changing of the colour of the drinks (as in DuBose et al.'s study). One suggestion is that people simply differ in the specific flavour expectations elicited by particular colours (Shankar et al. 2010a, 2010c, Zampini et al. 2007). Such individual differences in the flavour expectations that are elicited by particular colours may explain why certain colours appear to have a more pronounced effect on people's flavour identification responses (Wan et al. 2014a). In this regard, it is interesting to note that people are likely to have shared and strong expectations regarding the kinds of bouquet (flavour) characteristics that are signified by seeing a red wine, hence perhaps explaining why such robust results were reported by Morrot et al. (2001) in their study of the miscoloured white wine (see also Parr et al. 2003).

One particularly interesting result is that PROP (6-*n*-propylthiouracil) supertasters appear to show less visual dominance over their perception of flavour than do medium tasters, who, in turn, show less visual dominance than non-tasters (Zampini et al. 2008). The participants in the latter study had to try to identify the flavour of fruit-flavoured drinks presented amongst flavourless drinks. The drinks were coloured red, orange, yellow, grey, or else presented as colourless solutions. Colouring the orange-flavoured solutions orange led to a significant increase in the accuracy of participants' flavour identification responses (see Figure 16.2). Similarly, colouring the blackcurrant-flavoured solutions a greyish

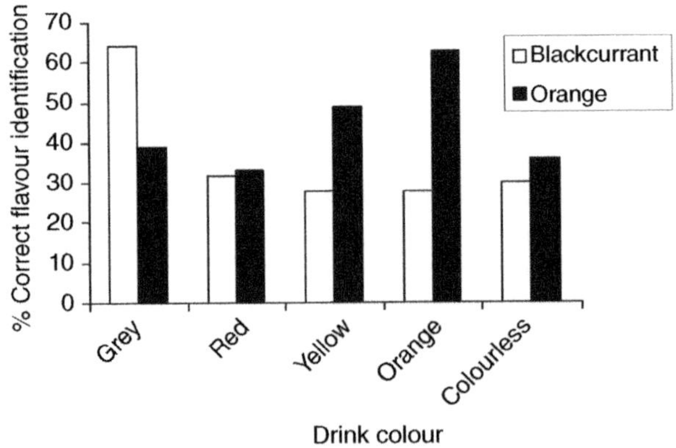

Figure 16.2 The results of Zampini et al.'s (2008) study highlighting the influence of colour on people's ability to correctly identify orange- and blackcurrant-flavoured solutions (Source: Adapted from Zampini et al., 2008). The results are separated by the taster status of the participant (non-taster, medium taster, or supertaster). Overall, non-tasters identified 19% of the solutions correctly, the medium tasters 31%, and the supertasters 67%.

purple led to a significant increase in the number of blackcurrant flavour responses. Note here that many blackcurrant-flavoured products, such as, for example, yoghurt are typically coloured a greyish-purple. What is more, the participants in Zampini et al.'s study also rated the congruently-coloured drinks as having a more intense flavour than when presented either with no colour or with an incongruent colour. By contrast, no such effect of congruent colouring was reported on participants' judgments of sourness or sweetness intensity. Interestingly, Zampini et al. also found that the addition of food colouring had the largest effect (in terms of participants' correctly identifying the flavour) on the performance of the non-tasters, less of an effect on the medium-tasters, and very little effect on the colour identification responses of the supertasters.

In summary, then, the addition of colour (be it appropriate or inappropriate) can exert a profound effect on people's flavour identification responses. Interestingly, the manipulation of the intensity of the colour added to a foodstuff does not, however, appear to have such a clear-cut effect on people's judgments of flavour (or, for that matter, taste) intensity (Petit et al. 2007; see Spence et al. 2010, for a review). One general point to note here is that if colour's effect on flavour perception is more cognitive in nature than the multisensory integration effects that appear to govern oral-somatosensory-gustatory-orthonasal-olfactory interactions (i.e., if it depends on an expectancy set-up by the presence of a particular colour in the substrate), then the exact testing protocol may influence the results of studies in ways that, as yet, have not been fully thought through (Frank et al. 1989, Spence et al. 2010).

One criticism of much of the research that has been conducted to date on visual influences over taste and flavour perception relates to the fact that fairly arbitrary/artificial colours have often been used in much of the research that has been published to date. It is, however, important to note that we may be exposed to particular colours in the marketplace and these colours may have a special meaning in terms of setting our expectations concerning taste and flavour perception (Garber et al. 2001, Wan et al. 2014b). One other limitation with much of the older research is that little thought has been given to the influence of the glassware and plateware (Spence and Piqueras-Fiszman 2014, Spence and Wan 2015). The latest research shows that the expectations elicited by beverage colour are sometimes modulated by the glass or cup in which that drink happens to be presented (Wan et al. 2014a, Wan et al. 2015). What is more, there is now a growing body of research to show just how much of an influence the colour of the plateware can have on taste perception (Piqueras-Fiszman et al. 2012, Stewart and Goss 2013).

16.5 Auditory contributions to multisensory flavour perception

Try eating a crisp (or potato chip) without making a noise. It's pretty much impossible! The question therefore arises as to what role such food-related eating sounds exert on the perception of food. One might wonder, for example, whether crispy or crunchy foods would taste different if eaten at a noisy party, say, or whilst listening to loud white noise (Spence 2014). While interest in the role of audition in multisensory flavour (and texture) perception goes back at least as far as the 1950's, the majority of published studies have correlated subjective ratings of the crispiness/crunchiness of various foods with objective measures of the sound elicited by biting into, or mechanically crushing, those foods (see Spence 2015a, for a review).

Zampini and Spence (2004) reported a study in which they demonstrated that the sounds that people make when they bite into dry food products such as crisps (potato chips), contribute as much as 15% to their perception of crispness and/or freshness. The participants in this study had to bite into 180 potato chips, and rated each one in terms of its crispness and freshness using labelled anchored scales. Real-time auditory feedback of the sound made by participants while biting into each crisp was played back over closed-ear headphones. Intriguingly, the participants perceived the sounds as coming from the potato chips (rather than from the headphones) due, presumably, to the audiotactile ventriloquism effect. The sound was altered in terms of its overall loudness and/or frequency composition. The sound that participants heard when biting into each potato chip could either be attenuated by 0, 20, or 40 dB across the entire frequency range, or else only those frequencies above 2 kHz could either be boosted or cut (by 12 dB) or left unaltered.

The potato chips were rated as tasting significantly crisper and fresher when the overall sound level was increased and/or when just the high frequency sounds were amplified (see Figure 16.3). By contrast, the crisps were rated as significantly staler and significantly softer when the overall intensity of the biting sounds was reduced and/or when the high frequency sounds were attenuated. Zampini and Spence's results therefore highlight the important role that food-eating sounds play in modulating people's perception and evaluation of foods (at least for those foods that make a 'noise' when we bite into them). Recently, Demattè et al. (2014) reported similar results when the sounds that people heard when biting into apples were modified. Meanwhile, Zampini and Spence (2005) have gone on to show that people's ratings of the carbonation of fizzy beverages can also be modified by changing the loudness of the popping sounds that they hear when evaluated a beverage presented in a plastic cup.

It should be noted that it is not only the overall loudness or frequency composition of food-eating sounds that modulate the perception of food texture. The temporal qualities of food sounds, such as, for example, how uneven or

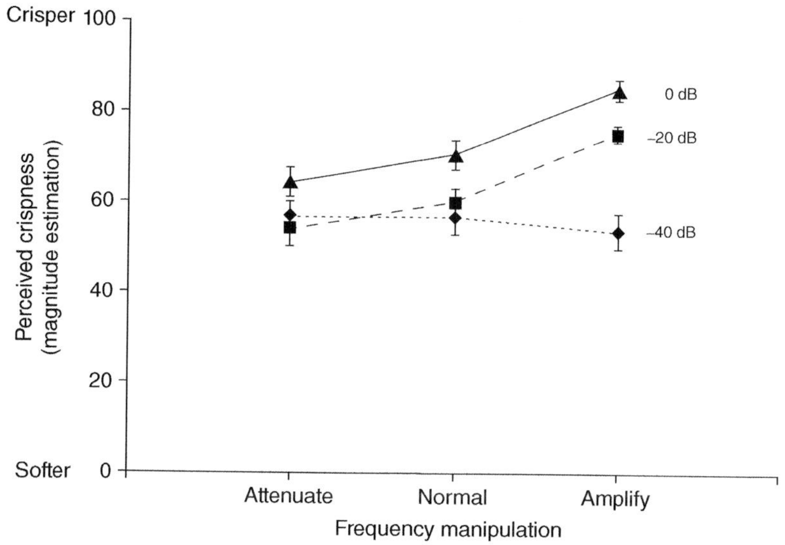

Figure 16.3 Results highlighting the effect of changing the food-eating sounds heard by participants on their multisensory perception of the crispness of potato chips (Source: Zampini and Spence, 2004. Reproduced with permission of John Wiley & Sons.). The self-produced biting sounds were fedback over headphones and attenuated by 0, −20 and −40 dB. Potato chips were judged as crisper (and fresher) when the overall sound level was boosted, and/or when just the high frequency components of the sound (> 2kHz) were amplified.

discontinuous they are, can also influence how "crispy" people perceive a food to be (Vickers and Wasserman 1979, Spence 2015a), or how carbonated a drink is judged to be (Zampini and Spence, 2005). Environmental sounds and background music have also been shown to exert a surprisingly large effect on our food-eating and drinking behaviours as well (see Spence 2014, for a review). Even the sounds of product packaging, such as the rattling of a noisy crisp packet, can influence people's perception of the crispness of potato chips (Spence et al. 2011). The available evidence now suggests that environmental sounds can influence our perception of both the sensory-discriminative and the hedonic attributes of food and drink items (see Spence 2014, for a review). Intriguingly, environmental noise might not affect the perception of all tastes equally (Spence et al. 2014).

16.6 The cognitive neuroscience of multisensory flavour perception

The last few years have seen a rapid growth in our understanding of the neural networks that underlie multisensory flavour perception. We now know that gustatory stimuli project to the primary taste cortex (the anterior insula and the frontal-or-parietal operculum, see Simon et al. 2006), while olfactory stimuli project to the primary olfactory (i.e., piriform) cortex. From there, the inputs from both modalities project to the orbitofrontal cortex (OFC), a small brain structure located behind the eyes. Gustatory stimuli project to caudolateral OFC while olfactory stimuli project to caudomedial OFC. The OFC plays a central role in mediating multisensory interactions in flavour perception (see Small 2012, for a review). The pleasantness (and reward value) of food and drink appears to be represented in the OFC (Rolls and Baylis 1994, Small 2004, Small et al. 1997; Small and Prescott 2005, Small et al. 2001).

The participants in one influential study had to rate the pleasantness, and congruency, of various different pairings of orthonasal olfactory and gustatory stimuli while lying in a functional magnetic resonance imaging (fMRI) brain scanner (de Araujo et al. 2003). The olfactory stimuli consisted of methianol (which smells of chicken-broth) and strawberry odour, while the tastants consisted of sucrose and MSG. Participants were presented with congruent (e.g., sucrose and strawberry odour) and incongruent (e.g., sucrose and chicken-broth odour) combinations of orthonasal olfactory and gustatory stimuli. Increased neural activity in the OFC was correlated with increased ratings of the pleasantness and congruency of the olfactory-gustatory stimulus pairing that the participants happened to be evaluating.

In another study, Small and her colleagues presented their participants with familiar/unfamiliar combinations of retronasal smell and taste stimuli (Small et al. 2004). Super-additive neural interactions were observed in the OFC in their event-related fMRI study for familiar (or congruent, sweet-vanilla), but not for unfamiliar (or incongruent) combinations of stimuli (salty-vanilla). Several other areas – including the dorsal insula, frontal operculum and anterior cingulate cortex – also lit-up in what appeared to be a putative 'flavour network'. Thus, it would seem that familiar combinations of olfactory (both orthonasal and retronasal) and gustatory flavour stimuli lead to enhanced neural responses in those parts of the brain that code for the hedonic (i.e., pleasantness) and reward value of food. Behaviourally, the presentation of familiar (and/or congruent) stimulus combinations results in enhanced flavour perception and increased pleasantness. By contrast, unfamiliar (and/or incongruent) combinations of taste and smell (delivered either orthonasally or retronasally) will likely lead to reduced liking and the suppression of neural response in those parts of the brain coding for the intensity and pleasantness of food (De Araujo et al. 2003, Skrandies and Reuther, 2008, Small et al. 2005). In an early neuroimaging study, Small et al. (1997) reported a significant decrease in regional cerebral blood flow in primary and secondary gustatory cortex and in secondary olfactory cortex during simultaneous presentation of orthonasal olfactory and gustatory stimuli as compared to when they were presented individually.

Oral-somatosensory information regarding food in the mouth is transferred to the brain by means of the trigeminal nerve, which projects directly to the primary somatic sensory cortex (Simon et al. 2006). This projection carries information concerning touch, texture (mouthfeel) (Christensen 1984), temperature, proprioception, nociception and chemical irritation from the receptors in the mouth. Given what we have seen so far, it should come as little surprise that oral texture (including the perception of fattiness in food) is also represented in the OFC (Eldeghaidy et al. 2011). Congruent combinations of colour and orthonasally-presented odour give rise to enhanced activation in the OFC (Österbauer et al. 2005). As yet, though, neuroimaging techniques have not been applied to the case of somatosensory-olfactory interactions (Stevenson 2012a).

16.7 Conclusions

Flavour perception in one of the most multisensory of our everyday experiences, involving as it does direct contributions from taste (gustation), smell (orthonasal olfaction), and trigeminal inputs (Small and Prescott 2005). Oral-somatosensory, visual, and auditory cues also influence multisensory flavour perception, though they have typically been excluded from standard definitions of flavour (e.g., ISO 5492 1992, 2008). Such definitions now seem overly narrow (at least to some researchers), and should perhaps be expanded to include proximal auditory cues elicited by foods in the mouth (and food-eating sounds). The influence of such auditory cues on multisensory flavour perception would appear to obey the temporal rule that marks out many effects as reflecting the consequences of multisensory integration rather than some form of expectancy effect (Spence 2015a), which are rarely sensitive to the precise synchronization of the constituent sensory inputs (Piqueras-Fiszman and Spence 2015).

It is currently an open question as to whether one should also include exteroceptive cues, such as those provided by the distal sounds associated with foods (e.g., the sizzling steak) or food-preparation, visual appearance cues, and orthonasal olfaction in one's definition of flavour. The answer to this latter question may turn on what exactly is meant by the term 'flavour' (see also Auvray and Spence 2008, Spence et al. 2015, Stevenson 2009). Nevertheless, however one chooses to define flavour, the point remains that the human brain appears to effortlessly bind all of the available sensory cues (regardless of their modality of occurrence) into a multisensory flavour percept/object. Whether flavours should be considered as multisensory Gestalts is a theoretical question that some researchers are now starting to address (Spence 2015b).

The evidence that has been reviewed here demonstrates just how important multisensory integration is to multisensory perception of flavour. That said, it is currently unclear whether the mechanisms underlying cue combination (multisensory integration) in the case of multisensory flavour perception are necessarily the same as those outlined elsewhere for the combination of auditory, visual, and tactile cues in non-food perception (Ernst and Bülthoff 2004, Small 2004,

Stein and Meredith 1993, Stein and Stanford 2008, Trommershäuser et al. 2011). One possibility here is that the interoceptive flavour cues may be combined along similar lines to those observed by Stein and his colleagues at the single-cell level (see also Small et al. 2004), while exteroceptive cues may influence performance by means of expectancy effects instead (Piqueras-Fiszman and Spence 2015). According to this account, visual appearance cues (especially colour), orthonasal olfactory cues, and distal food (and food preparation) sounds set-up an expectation regarding the food that is about to be eaten. When the food or drink is then evaluated (in terms of its flavour, aroma, or taste), assimilation may occur if the discrepancy between what was expected and what was perceived is small, whereas contrast may occur if the discrepancy between one's expectation and actual interoceptive information is too great. Indeed, the assimilation/contrast model has already been used previously to account for the results of multisensory integration on people's hedonic responses to (and perception of) food and drink (Yeomans et al. 2008). There seems no reason why it cannot be extended to the case of people's sensory-discriminative food judgments as well (Piqueras-Fiszman and Spence 2015, for a recent review).

The majority of the research that has been reviewed here has focused on the sensory-discriminative aspects of multisensory flavour perception. Equally important, if currently somewhat less thoroughly studied are, of course, are the multisensory contributions to people's hedonic responses to food and drink (i.e., how much do they like it). All that there is time to say here is that the available research shows that multisensory interactions appear to play just as important a role in people's hedonic responses to food and drink as they do to our sensory-discriminative responses (Piqueras-Fiszman and Spence 2015, Yeomans et al. 2008). That said, the results of the available cognitive neuroscience research now converge on the conclusion that the OFC plays a central role in the multisensory integration of gustatory, olfactory, tactile, and visual (and presumably also auditory) cues (e.g., De Araujo et al. 2003, Österbauer et al. 2005, Rolls and Baylis 1994, Small et al. 1997, 2004). The OFC appears to code the reward value of food. However, the OFC should be seen as just one node in a complex "flavour network" that spans multiple loci in the human brain, including the anterior insula, frontal operculum, anteromedial right temporal lobe, anterior cingulated cortex, posterior parietal cortex, and possibly also the ventral lateral prefrontal cortex (Small et al. 2004; Small 2012).

While it has been known for many years that taster status influences the experience of certain bitter tastes (Bartoshuk 2000), there is now a growing body of evidence to suggest that it also affects people's responses to other tastants (Bajec and Pickering 2008), as well as their responses to the oral-somatosensory attributes of foods (see Eldeghaidy et al. 2011, Essick et al. 2003, Prescott et al. 2004). The latest evidence has also provided some intriguing preliminary evidence that taster status might influence how much weight people put on what they see when trying to identify a food or drink's flavour as well (Zampini et al.

2008). There is also growing interest in the individual/cultural differences that pervade our experience of multisensory flavour (Shankar et al. 2010a, 2010c).

References

Alais, D. and Burr, D. (2004) The ventriloquist effect results from near-optimal bimodal integration. *Current Biology, 14,* pp. 257–262.

Auvray, M. and Spence, C. (2008) The multisensory perception of flavor. *Consciousness and Cognition, 17,* pp. 1016–1031.

Bajec, M. R. and Pickering, G. J. (2008) Thermal taste, PROP responsiveness, and perception of oral sensations. *Physiology & Behavior, 95,* pp. 581–590.

Bajec, M. R., Pickering, G. J. and DeCourville, N. (2012) Influence of stimulus temperature on orosensory perception and variation with taste phenotype. *Chemosensory Perception, 5,* pp. 243–265.

Bartoshuk, L. M. (2000) Comparing sensory experiences across individuals: Recent psychophysical advances illuminate genetic variation in taste perception. *Chemical Senses, 25,* pp. 447–460.

Blackwell, L. (1995) Visual clues and their effects on odour assessment. *Nutrition and Food Science, 5,* pp. 24–28.

Bojanowski, V. and Hummel, T. (2012) Retronasal perception of odors. *Physiology & Behavior, 107,* pp. 484–487.

Breslin, P. A., Doolittle, N. and Dalton, P. (2001) Subthreshold integration of taste and smell: The role of experience in flavour integration. *Chemical Senses, 26,* p. 1035.

Bult, J. H. F., de Wijk, R. A. and Hummel, T. (2007) Investigations on multimodal sensory integration: Texture, taste, and ortho- and retronasal olfactory stimuli in concert. *Neuroscience Letters, 411,* pp. 6–10.

Chartier, F. (2012) *Taste buds and molecules: The art and science of food, wine, and flavor (translated by Levi Reiss).* Hoboken, NJ: John Wiley and Sons.

Christensen, C. M. (1984) Food texture perception. In E. Mark (Ed.), *Advances in food research.* New York, NY: Academic Press. pp. 159–199.

Cruz, A. and Green, B. G. (2000) Thermal stimulation of taste. *Nature, 403,* pp. 889–892.

Dalton, P., Doolittle, N., Nagata, H. and Breslin, P. A. S. (2000) The merging of the senses: Integration of subthreshold taste and smell. *Nature Neuroscience, 3,* pp. 431–432.

Davidson, J. M., Linforth, R. S. T., Hollowood, T. A. and Taylor, A. J. (1999) Effect of sucrose on the perceived flavor intensity of chewing gum. *Journal of Agricultural & Food. Chemistry, 47,* pp. 4336–4340.

Davis, R. G. (1981) The role of nonolfactory context cues in odor identification. *Perception & Psychophysics, 30,* pp. 83–89.

De Araujo, I. E. T., Rolls, E. T., Kringelbach, M. L., McGlone, F. and Phillips, N. (2003) Taste-olfactory convergence, and the representation of the pleasantness of flavour, in the human brain. *European Journal of Neuroscience, 18,* pp. 2059–2068.

Delwiche, J. and Heffelfinger, A. L. (2005) Cross-modal additivity of taste and smell. *Journal of Sensory Studies, 20,* pp. 137–146.

Demattè, M. L., Pojer, N., Endrizzi, I., Corollaro, M. L., Betta, E. Aprea, E., Charles, M., Biasioli, F., Zampini, M. and Gasperi,. F. (2014) Effects of the sound of the bite on apple perceived crispness and hardness. *Food Quality and Preference, 38,* pp. 58–64.

Deroy, O. and Spence, C. (2013) Weakening the case for 'weak synaesthesia': Why crossmodal correspondences are not synaesthetic. *Psychonomic Bulletin & Review, 20,* pp. 643–664.

Diaz, M. E. (2004) Comparison between orthonasal and retronasal flavour perception at different concentrations. *Flavour and Fragrance Journal, 19,* pp. 499–504.

DuBose, C. N., Cardello, A. V. & Maller O. (1980) Effects of colorants and flavorants on identification, perceived flavor intensity, and hedonic quality of fruit-flavored beverages and cake. *Journal of Food Science, 45*, pp. 1393–1399, 1415.

Eldeghaidy, S., Marciani, L., McGlone, F., Hollowood, T., Hort, J., Head, K., et al. (2011) The cortical response to the oral perception of fat emulsions and the effect of taster status. *Journal of Neurophysiology, 105*, pp. 2572–2581.

Ernst, M. O. and Bülthoff, H. H. (2004) Merging the senses into a robust percept. *Trends in Cognitive Sciences, 8*, pp. 162–169.

Essick, G. K., Chopra, A., Guest, S. and McGlone, F. (2003) Lingual tactile acuity, taste perception, and the density and diameter of fungiform papillae in female subjects. *Physiology & Behavior, 80*, pp. 289–302.

Frank, R. A. and Byram, J. (1988) Taste-smell interactions are tastant and odorant dependent. *Chemical Senses, 13*, pp. 445–455.

Frank, R. A., Ducheny, K. and Mize, S. J. S. (1989) Strawberry odor, but not red color, enhances the sweetness of sucrose solutions. *Chemical Senses, 14*, pp. 371–377.

Frost, M. B. and Janhoj, T. (2007) Understanding creaminess. *International Dairy Journal, 17*, pp. 1298–1311.

Garber, L. L., Jr.,, Hyatt, E. M. and Starr, R. G., Jr., (2001) Placing food color experimentation into a valid consumer context. *Journal of Food Products Marketing, 7(3)*, pp. 3–24.

Green, B. G. (2002) Studying taste as a cutaneous sense. *Food Quality and Preference, 14*, pp. 99–109.

Green, B. G. and George, P. (2004) 'Thermal taste' predicts higher responsiveness to chemical taste and flavor. *Chemical Senses, 29*, pp. 617–628.

Hayes, J. E. and Pickering, G. J. (2012) Wine expertise predicts taste phenotype. *American Journal of Enology & Viticulture, 63*, pp. 80–84.

Heilman, S. and Hummel, T. (2004) A new method for comparing orthonasal and retronasal olfaction. *Behavioral Neuroscience, 118*, pp. 412–419.

ISO (1992) *Standard 5492: Terms relating to sensory analysis*. International Organization for Standardization. Vienna: Austrian Standards Institute.

ISO (2008) *Standard 5492: Terms relating to sensory analysis*. International Organization for Standardization. Vienna: Austrian Standards Institute.

Koza, B. J., Cilmi, A., Dolese, M. and Zellner, D. A. (2005) Color enhances orthonasal olfactory intensity and reduces retronasal olfactory intensity. *Chemical Senses, 30*, pp. 643–649.

Labbe, D., Damevin, L., Vaccher, C., Morgenegg, C. and Martin, N. (2006) Modulation of perceived taste by olfaction in familiar and unfamiliar beverages. *Food Quality and Preference, 17*, pp. 582–589.

Labbe, D., Rytz, A., Morgenegg, C., Ali, C. and Martin, N. (2007) Subthreshold olfactory stimulation can enhance sweetness. *Chemical Senses, 32*, pp. 205–214.

Lawless, H., Rozin, P. and Shenker, J. (1985) Effects of oral capsaicin on gustatory, olfactory and irritant sensations and flavor identification in humans who regularly or rarely consume chili pepper. *Chemical Senses, 10*, pp. 579–589.

Lawrence, G., Salles, C., Septier, C., Busch, J. and Thomas-Danguin, T. (2009) Odour–taste interactions: A way to enhance saltiness in low-salt content solutions. *Food Quality and Preference, 20*, pp. 241–248.

Lim, J. and Green, B. G. (2008) Tactile interaction with taste localization: Influence of gustatory quality and intensity. *Chemical Senses, 33*, pp. 137–143.

Lim, J. and Johnson, M. B. (2011) Potential mechanisms of retronasal odor referral to the mouth. *Chemical Senses, 36*, pp. 283–289.

Lim, J. and Johnson, M. (2012) The role of congruency in retronasal odor referral to the mouth. *Chemical Senses, 37*, pp. 515–521.

Maga, J. A. (1974) Influence of color on taste thresholds. *Chemical Senses and Flavor, 1*, pp. 115–119.

Martin, G. N. (2004) A neuroanatomy of flavour. *Petits Propos Culinaires, 76*, pp. 58–82.

Michel, C., Velasco, C., Salgado, A. and Spence, C. (2014) The butcher's tongue illusion. *Perception, 43*, pp. 818–824.

Morrot, G., Brochet, F. and Dubourdieu, D. (2001) The color of odors. *Brain & Language, 79*, pp. 309–320.

Murphy, C. and Cain, W. S. (1980) Taste and olfaction: Independence vs. interaction. *Physiology and Behavior, 24*, pp. 601–605.

Österbauer, R. A., Matthews, P. M., Jenkinson, M., Beckmann, C. F., Hansen, P. C. and Calvert, G. A. (2005) Color of scents: Chromatic stimuli modulate odor responses in the human brain. *Journal of Neurophysiology, 93*, pp. 3434–3441.

Parr, W. V., White, K. G. and Heatherbell, D. (2003) The nose knows: Influence of colour on perception of wine aroma. *Journal of Wine Research, 14*, pp. 79–101.

Petit, C. E. F., Hollowood, T. A., Wulfert, F. and Hort, J. (2007) Colour-coolant-aroma interactions and the impact of congruency and exposure on flavour perception. *Food Quality and Preference, 18*, pp. 880–889.

Pfeiffer, J. C., Hollowood, T. A., Hort, J. and Taylor, A. J. (2005) Temporal synchrony and integration of sub-threshold taste and smell signals. *Chemical Senses, 30*, pp. 539–545.

Piqueras-Fiszman, B., Alcaide, J., Roura, E. and Spence, C. (2012) Is it the plate or is it the food? Assessing the influence of the color (black or white) and shape of the plate on the perception of the food placed on it. *Food Quality & Preference, 24*, pp. 205–208.

Piqueras-Fizman, B. and Spence, C. (2015) Sensory expectations based on product-extrinsic food cues: An interdisciplinary review of the empirical evidence and theoretical accounts. *Food Quality & Preference, 40*, pp. 165–179.

Prescott, J., Bartoshuk, L. M. and Prutkin, J. (2004) 6-n-Propylthiouracil tasting and the perception of nontaste oral sensations. In J. Prescott & B. Tepper (Eds.), *Genetic variations in taste sensitivity*. New York, NY: Marcel Dekker, pp. 89–104.

Rolls, E. T. and Baylis, L. L. (1994) Gustatory, olfactory, and visual convergence within the primate orbitofrontal cortex. *Journal of Neuroscience, 14*, pp. 5437–5452.

Rozin, P. (1982) "Taste-smell confusions" and the duality of the olfactory sense. *Perception & Psychophysics, 31*, pp. 397–401.

Schaal, B. and Durand, K. (2012) The role of olfaction in human multisensory development. In A. J. Bremner, D. Lewkowicz and C. Spence (Eds.), *Multisensory development*. Oxford: Oxford University Press, pp. 29–62.

Schifferstein, H. N. J. and Verlegh, P. W. J. (1996) The role of congruency and pleasantness in odor-induced taste enhancement. *Acta Psychologica, 94*, pp. 87–105.

Shankar, M. U., Levitan, C. and Spence, C. (2010a) Grape expectations: The role of cognitive influences in color-flavor interactions. *Consciousness & Cognition, 19*, pp. 380–390.

Shankar, M., Simons, C., Levitan, C., Shiv, B., McClure, S. and Spence, C. (2010b) An expectations-based approach to explaining the crossmodal influence of color on odor identification: The influence of temporal and spatial factors. *Journal of Sensory Studies, 25*, pp. 791–803.

Shankar, M., Simons, C., Shiv, B., Levitan, C., McClure, S. and Spence, C. (2010c) An expectations-based approach to explaining the influence of color on odor identification: The influence of degree of discrepancy. *Attention, Perception and Psychophysics, 72*, pp. 1981–1993.

Shermer, D. Z. and Levitan, C. A. (2014) Red hot: The crossmodal effect of color intensity on piquancy. *Multisensory Research, 27*, pp. 207–223.

Simon, S. A., de Araujo, I. E., Gutierrez, R. and Nicolelis, M. A. L. (2006) The neural mechanisms of gustation: A distributed processing code. *Nature Reviews Neuroscience, 7*, pp. 890–901.

Skrandies, W. and Reuther, N. (2008) Match and mismatch of taste, odor, and color is reflected by electrical activity in the human brain. *Journal of Psychophysiology, 22,* pp. 175–184.

Small, D. M. (2004) Crossmodal integration - Insights from the chemical senses. *Trends in Neurosciences, 27,* pp. 120–123.

Small, D. M. (2012) Flavor is in the brain. *Physiology and Behavior, 107,* pp. 540–552.

Small, D. M. and Prescott, J. (2005) Odor/taste integration and the perception of flavour. *Experimental Brain Research, 166,* pp. 345–357.

Small, D. M., Gerber, J. C., Mak, Y. E. and Hummel, T. (2005) Differential neural responses evoked by orthonasal versus retronasal odorant perception in humans. *Neuron, 47,* pp. 593–605.

Small, D. M., Jones-Gotman, M., Zatorre, R. J., Petrides, M. and Evans, A. C. (1997) Flavor processing: More than the sum of its parts. *Neuroreport, 8,* pp. 3913–3917.

Small, D. M., Veldhuizen, M. G., Felsted, J., Mak, Y. E. and McGlone, F. (2008) Separable substrates for anticipatory and consummatory food chemosensation. *Neuron, 57,* pp. 786–797.

Small, D. M., Voss, J., Mak, Y. E., Simmons, K. B., Parrish, T. and Gitelman, D. (2004) Experience-dependent neural integration of taste and smell in the human brain. *Journal of Neurophysiology, 92,* pp. 1892–1903.

Small, D. M., Zatorre, R. J., Dagher, A., Evans, A. C. and Jones-Gotman, M. (2001) Changes in brain activity related to eating chocolate: From pleasure to aversion. *Brain, 124,* pp. 1720–1733.

Spence, C. (2008) Multisensory perception. In H. Blumenthal, *The big Fat Duck cook book*. London: Bloomsbury, pp. 484–485.

Spence, C. (2012) The development and decline of multisensory flavour perception. In A. J. Bremner, D. Lewkowicz and C. Spence (Eds.), *Multisensory development*. Oxford: Oxford University Press, pp. 63–87.

Spence, C. (2014) Noise and its impact on the perception of food and drink. *Flavour, 3*:9.

Spence, C. (2015) Visual contributions to taste and flavour perception. In M. Scotter (Ed.), *Colour additives for food and beverages*. Cambridge: Woodhead Publishing, pp. 189–210.

Spence, C. (2015a) Eating with our ears: Assessing the importance of the sounds of consumption to our perception and enjoyment of multisensory flavour experiences. *Flavour, 4,* p. 3.

Spence, C. (2015b) Cross-modal perceptual organization. In J. Wagemans (Ed.), *The Oxford handbook of perceptual organization*. Oxford: Oxford University Press, pp. 649–664.

Spence, C. (2015c) Just how much of what we taste derives from the sense of smell? *Flavour, 4*: 30.

Spence, C., Levitan, C., Shankar, M. U. and Zampini, M. (2010) Does food color influence taste and flavor perception in humans? *Chemosensory Perception, 3,* pp. 68–84.

Spence, C., Michel, C. and Smith, B (2014) Airplane noise and the taste of umami. *Flavour, 3,* p. 2.

Spence, C. and Piqueras-Fiszman, B. (2014) *The perfect meal: The multisensory science of food and dining*. Oxford: Wiley-Blackwell.

Spence, C., Shankar, M. U. and Blumenthal, H. (2011) 'Sound bites': Auditory contributions to the perception and consumption of food and drink. In F. Bacci & D. Melcher (Eds.), *Art and the senses*. Oxford: Oxford University Press, pp. 207–238.

Spence, C., Smith, B. and Auvray, M. (2015) Confusing tastes and flavours. In D. Stokes, M. Matthen & S. Biggs (Eds.), *Perception and its modalities*. Oxford: Oxford University Press, pp. 247–274.

Spence, C. and Squire, S. B. (2003) Multisensory integration: Maintaining the perception of synchrony. *Current Biology, 13,* pp. R519–R521.

Spence, C. and Wan, I. (2015) Beverage perception & consumption: The influence of the container on the perception of the contents. *Food Quality & Preference, 39,* pp. 131–140.

Stein, B. E. and Meredith, M. A. (1993) *The merging of the senses*. Cambridge, MA: MIT Press.

Stein, B. E. and Stanford, T. R. (2008) Multisensory integration: Current issues from the perspective of the single neuron. *Nature Reviews Neuroscience, 9,* pp. 255–267.

Stevenson, R. J. (2009) *The psychology of flavour*. Oxford: Oxford University Press.

Stevenson, R. J. (2012a) Multisensory interactions in flavor perception. In B. E. Stein (Ed.), *The new handbook of multisensory processes*. Cambridge, MA: MIT Press, pp. 283–299.

Stevenson, R. J. (2012b) The role of attention in flavour perception. *Flavour, 1,* p. 2.

Stevenson, R. J. and Boakes, R. A. (2004) Sweet and sour smells: Learned synaesthesia between the senses of taste and smell. In G. A. Calvert, C. Spence and B. E. Stein (Eds.), *The handbook of multisensory processing*. Cambridge, MA: MIT Press, pp. 69–83.

Stevenson, R. J. and Mahmut, M. K. (2011) Experience dependent changes in odour-viscosity perception. *Acta Psychologica, 136,* pp. 60–66.

Stevenson, R. J. and Oaten, M. (2008) The effect of appropriate and inappropriate stimulus color on odor discrimination. *Perception & Psychophysics, 70,* pp. 640–646.

Stevenson, R. J. and Oaten, M. J. and Mahmut, M. K. (2011) The role of attention in the localization of odors to the mouth. *Attention, Perception and Psychophysics, 73,* pp. 247–258.

Stevenson, R. J., Prescott, J. and Boakes, R. A. (1999) Confusing tastes and smells: How odours can influence the perception of sweet and sour tastes. *Chemical Senses, 24,* pp. 627–635.

Stevenson, R. J. and Tomiczek, C. (2007) Olfactory-induced synesthesias: A review and model. *Psychological Bulletin, 133,* pp. 294–309.

Stewart, P. C. and Goss, E. (2013) Plate shape and colour interact to influence taste and quality judgments. *Flavour, 2,* p. 27.

Sundqvist, N. C., Stevenson, R. J. and Bishop, I. R. J. (2006) Can odours acquire fat-like properties? *Appetite, 47,* pp. 91–99.

Szczesniak, A. S. (2002). Texture is a sensory property. *Food Quality and Preference, 13,* pp. 215–225.

Todrank, J. and Bartoshuk, L. M. (1991) A taste illusion: Taste sensation localized by touch. *Physiology & Behavior, 50,* pp. 1027–1031.

Tournier, C., Sulmont-Rossé, C., Sémon, E., Vignon, A., Issanchou, S. and Guichard, E. (2009) A study on texture-taste-aroma interactions: Physico-chemical and cognitive mechanisms. *International Dairy Journal, 19,* pp. 450–458.

Trommershäuser, J., Landy, M. S. and Körding, K. P. (Eds.). (2011) *Sensory cue integration*. New York, NY: Oxford University Press.

Verhagen, J. V. and Engelen, L. (2006) The neurocognitive bases of human multimodal food perception: Sensory integration. *Neuroscience and Biobehavioral Reviews, 30,* pp. 613–650.

Vickers, Z. M. and Wasserman, S. S. (1979) Sensory qualities of food sounds based on individual perceptions. *Journal of Texture Studies, 10,* pp. 319–332.

Wan, X., Velasco, C., Michel, C., Mu, B., Woods, A. T. and Spence, C. (2014a) Does the shape of the glass influence the crossmodal association between colour and flavour? A cross-cultural comparison. *Flavour, 3*: 3.

Wan, X., Woods, A. T., Seoul, K.-H., Butcher, N. and Spence, C. (2015) When the shape of the glass influences the flavour associated with a coloured beverage: Evidence from consumers in three countries. *Food Quality & Preference, 39,* pp. 109–116.

Wan, X., Zhou, X., Mu, B., Du, D., Velasco, C., Michel, C. and Spence, C. (2014b) Crossmodal expectations of tea colour based on its flavour. *Journal of Sensory Studies, 29,* pp. 285–293.

Weel, K. G. C., Boelrijk, A. C., Alting, P. J. J. M., van Mil, J. J., Burger, H., Gruppen, H., Voragen, A. G. J. and Smit, G. (2002) Flavor release and perception of flavored whey protein gels: Perception is determined by texture rather than by release. *Journal of Agricultural and Food Chemistry, 50,* pp. 5149–5155.

Yeomans, M., Chambers, L., Blumenthal, H. and Blake, A. (2008) The role of expectancy in sensory and hedonic evaluation: The case of smoked salmon ice-cream. *Food Quality and Preference, 19,* pp. 565–573.

Zampini, M., Sanabria, D., Phillips, N. and Spence, C. (2007) The multisensory perception of flavor: Assessing the influence of color cues on flavor discrimination responses. *Food Quality & Preference, 18,* pp. 975–984.

Zampini, M. and Spence, C. (2004) Multisensory contribution to food perception: The role of auditory cues in modulating crispness and staleness in crisps. *Journal of Sensory Science, 19,* pp. 347–363.

Zampini, M. and Spence, C. (2005) Modifying the multisensory perception of a carbonated beverage using auditory cues. *Food Quality and Preference, 16,* pp. 632–641.

Zampini, M., Wantling, E., Phillips, N. and Spence, C. (2008) Multisensory flavor perception: Assessing the influence of fruit acids and color cues on the perception of fruit-flavored beverages. *Food Quality & Preference, 19,* pp. 335–343.

Zellner, D. A., Bartoli, A. M. and Eckard, R. (1991) Influence of color on odor identification and liking ratings. *American Journal of Psychology, 104,* pp. 547–561.

Zellner, D. A. and Durlach, P. (2003) Effect of color on expected and experienced refreshment, intensity, and liking of beverages. *American Journal of Psychology, 116,* pp. 633–647.

Zellner, D. A. and Kautz, M. A. (1990) Color affects perceived odor intensity. *Journal of Experimental Psychology: Human Perception and Performance, 16,* pp. 391–397.

Zellner, D. A. and Whitten, L. A. (1999) The effect of color intensity and appropriateness on color-induced odor enhancement. *American Journal of Psychology, 112,* pp. 585–604.

Food Science and Technology from Wiley Blackwell

Decontamination of Fresh and Minimally Processed Produce	Gomez-Lopez	9780813823843
Progress in Food Preservation	Bhat	9780470655856
Food Safety for the 21st Century: Managing HACCP and Food Safety throughout the Global Supply Chain	Wallace	9781405189118
The Microbiology of Safe Food, 2nd edition	Forsythe	9781405140058

SENSORY SCIENCE, CONSUMER RESEARCH & NEW PRODUCT DEVELOPMENT

Olive Oil Sensory Science	Monteleone	9781118332528
Quantitative Sensory Analysis: Psychophysics, Models and Intelligent Design	Lawless	9780470673461
Product Innovation Toolbox: A Field Guide to Consumer Understanding and Research	Beckley	9780813823973
Sensory and Consumer Research in Food Product Design and Dev, 2nd Ed	Moskowitz	9780813813660
Sensory Evaluation: A Practical Handbook	Kemp	9781405162104
Statistical Methods for Food Science	Bower	9781405167642
Concept Research in Food Product Design and Development	Moskowitz	9780813824246
Sensory and Consumer Research in Food Product Design and Development	Moskowitz	9780813816326

FOOD INDUSTRY SUSTAINABILITY & WASTE MANAGEMENT

Food and Agricultural Wastewater Utilization and Treatment, 2nd Edition	Liu	9781118353974
Sustainable Food Processing	Tiwari	9780470672235
Food and Industrial Bioproducts and Bioprocessing	Dunford	9780813821054
Handbook of Sustainability for the Food Sciences	Morawicki	9780813817354
Sustainability in the Food Industry	Baldwin	9780813808468
Lean Manufacturing in the Food Industry	Dudbridge	9780813810072

FOOD LAWS & REGULATIONS

Guide to US Food Laws and Regulations, 2nd Edition	Curtis	9781118227787
Food and Drink - Good Manufacturing Practice: A Guide to its Responsible Management (GMP6), 6th Edition	Manning	9781118318201
The BRC Global Standard for Food Safety: A Guide to a Successful Audit, 2nd Edition	Kill	9780470670651
Food Labeling Compliance Review, 4th edition	Summers	9780813821818

DAIRY FOODS

Lactic Acid Bacteria: Biodiversity and Taxonomy	Holzapfel	9781444333831
From Milk By-Products to Milk Ingredients: Upgrading the Cycle	de Boer	9780470672228
Milk and Dairy Products as Functional Foods	Kanekanian	9781444336832
Milk and Dairy Products in Human Nutrition: Production, Composition and Health	Park	9780470674185
Manufacturing Yogurt and Fermented Milks, 2nd Edition	Chandan	9781119967088
Sustainable Dairy Production	de Jong	9780470655849
Advances in Dairy Ingredients	Smithers	9780813823959
Membrane Processing: Dairy and Beverage Applications	Tamime	9781444333374
Analytical Methods for Food and Dairy Powders	Schuck	9780470655986
Dairy Ingredients for Food Processing	Chandan	9780813817460
Processed Cheeses and Analogues	Tamime	9781405186421
Technology of Cheesemaking, 2nd edition	Law	9781405182980

SEAFOOD, MEAT AND POULTRY

Seafood Processing: Technology, Quality and Safety	Boziaris	9781118346211
Should We Eat Meat? Evolution and Consequences of Modern Carnivory	Smil	9781118278727
Handbook of Meat, Poultry and Seafood Quality, second edition	Nollet	9780470958322
The Seafood Industry: Species, Products, Processing, and Safety , 2nd Edition	Granata	9780813802589
Organic Meat Production and Processing	Ricke	9780813821269
Handbook of Seafood Quality, Safety and Health Effects	Alasalvar	9781405180702

BAKERY & CEREALS

Oats Nutrition and Technology	Chu	9781118354117
Cereals and Pulses: Nutraceutical Properties and Health Benefits	Yu	9780813818399
Whole Grains and Health	Marquart	9780813807775
Gluten-Free Food Science and Technology	Gallagher	9781405159159
Baked Products - Science,Technology and Practice	Cauvain	9781405127028

BEVERAGES & FERMENTED FOODS/BEVERAGES

Encyclopedia of Brewing	Boulton	9781405167444
Sweet, Reinforced and Fortified Wines: Grape Biochemistry, Technology and Vinification	Mencarelli	9780470672242
Technology of Bottled Water, 3rd edition	Dege	9781405199322
Wine Flavour Chemistry, 2nd edition	Bakker	9781444330427
Wine Quality: Tasting and Selection	Grainger	9781405113663

PACKAGING

Handbook of Paper and Paperboard Packaging Technology, 2nd Edition	Kirwan	9780470670668
Food and Beverage Packaging Technology, 2nd edition	Coles	9780813812748
Food and Package Engineering	Morris	9780813814797
Modified Atmosphere Packaging for Fresh-Cut Fruits and Vegetables	Brody	9780813812748

For further details and ordering information, please visit www.wiley.com/go/food

Index

Flavour: From food to perception, First Edition.
Edited by Elisabeth Guichard, Christian Salles, Martine Morzel, and Anne-Marie Le Bon.

S

T

Printed and bound by CPI Group (UK) Ltd, Croydon, CR0 4YY
17/09/2026
14964304-0003